The Impact of Long-Term Monitoring on Variable Star Research

NATO ASI Series

Advanced Science Institutes Series

A Series presenting the results of activities sponsored by the NATO Science Committee, which aims at the dissemination of advanced scientific and technological knowledge, with a view to strengthening links between scientific communities.

The Series is published by an international board of publishers in conjunction with the NATO Scientific Affairs Division

A Life Sciences **B Physics**	Plenum Publishing Corporation London and New York
C Mathematical and Physical Sciences **D Behavioural and Social Sciences** **E Applied Sciences**	Kluwer Academic Publishers Dordrecht, Boston and London
F Computer and Systems Sciences **G Ecological Sciences** **H Cell Biology** **I Global Environmental Change**	Springer-Verlag Berlin, Heidelberg, New York, London, Paris and Tokyo

NATO-PCO-DATA BASE

The electronic index to the NATO ASI Series provides full bibliographical references (with keywords and/or abstracts) to more than 30000 contributions from international scientists published in all sections of the NATO ASI Series.
Access to the NATO-PCO-DATA BASE is possible in two ways:

– via online FILE 128 (NATO-PCO-DATA BASE) hosted by ESRIN, Via Galileo Galilei, I-00044 Frascati, Italy.

– via CD-ROM "NATO-PCO-DATA BASE" with user-friendly retrieval software in English, French and German (© WTV GmbH and DATAWARE Technologies Inc. 1989).

The CD-ROM can be ordered through any member of the Board of Publishers or through NATO-PCO, Overijse, Belgium.

Series C: Mathematical and Physical Sciences - Vol. 436

The Impact of Long-Term Monitoring on Variable Star Research

Astrophysics, Instrumentation, Data Handling, Archiving

edited by

Christiaan Sterken

Belgian Fund for Scientific Research (NFWO)
University of Brussels,
Brussels, Belgium

and

Mart de Groot

Armagh Observatory,
College Hill,
Armagh, Northern Ireland

Kluwer Academic Publishers

Dordrecht / Boston / London

Published in cooperation with NATO Scientific Affairs Division

Proceedings of the NATO Advanced Research Workshop on
The Impact of Long-Term Monitoring on Variable Star Research:
Astrophysics, Instrumentation, Data Handling, Archiving
Ghent, Belgium
November 15–18, 1993

A C.I.P. Catalogue record for this book is available from the Library of Congress.

ISBN 0-7923-2913-9

Published by Kluwer Academic Publishers,
P.O. Box 17, 3300 AA Dordrecht, The Netherlands.

Kluwer Academic Publishers incorporates the publishing programmes of
D. Reidel, Martinus Nijhoff, Dr W. Junk and MTP Press.

Sold and distributed in the U.S.A. and Canada
by Kluwer Academic Publishers,
101 Philip Drive, Norwell, MA 02061, U.S.A.

In all other countries, sold and distributed
by Kluwer Academic Publishers Group,
P.O. Box 322, 3300 AH Dordrecht, The Netherlands.

Printed on acid-free paper

Printed in the Netherlands

TABLE OF CONTENTS

SESSION III: MISCELLANEOUS ASPECTS OF SPECTROGRAPHIC, SPECTROMETRIC AND PHOTOGRAPHIC MONITORING – Chair: L. Szabados

PREFACE

The theme of this workshop was about the kind of variable-star research that can only be succesfull if carried out in long-term monitoring (*LTM*) mode—that is, with observations covering a time span of a decade or longer. The scientific topics discussed included photometric and spectrographic observations of importance for the understanding of the physics of emission-line stars (Luminous Blue Variables, Herbig Ae/Be stars, Be and WR stars), binaries, and pulsating stars (β Cephei stars, δ Scuti stars, Ap stars, mid-B variables, Cepheids, RR Lyrae variables).

Besides reports based on the analysis of data collected in the wavelength range of visible light (photographic, photoelectric and spectrometric data, as well as visual estimates), conclusions based on infrared and on radio measurements were presented.

The design and history of well-known photometric systems, viz. the Geneva and the Strömgren photometric systems, is described in a couple of papers. Special emphasis was placed on novel designs of stable photometers and spectrometers with high spectral fidelity, and their application to *LTM*.

Several talks focused attention on secular changes of periods, and on period determination techniques for data collected on a very long-term basis.

Detailed discussions on pittfalls in observational work with relation to accuracy, homogeneity and precision of the data, with emphasis on the prevention of systematic error, took place.

Though the reporting of novel results was an important ingredient of the meeting, the workshop was not organised to communicate new findings only, but also to discuss discoveries on the background of *LTM* as a method to obtain those new results. The contents of 8 Discussion Sessions—which are integrally described in these Proceedings—not only refer to the talks and lectures, but also cover more general areas, such as extinction monitoring, the problem of archiving of astronomical data, service observing, the role long-term monitoring plays in graduate teaching and thesis supervision, the interplay between the large observational effort and theory, the contribution of *LTM* to new knowledge of fundamental data (i.e. masses, radii), and the increasing decommissioning of telescopes of modest aperture used for long-term monitoring.

The meeting valued the great wealth of information collected by many working groups involved in *LTM* campaigns covering a wide variety of astrophysical aspects, and realized the potential of long-term observations achievable with small instruments for the identification of scientific problems through complementary approaches with the new generation of very-large telescopes.

The Workshop's conclusion was the recognition of the fundamental significance of long-term observations for solving many important problems in astrophysics, and the unequaled value of extended observing runs with small telescopes for the education of young astronomers in order to teach them how to secure—over many years—observational data of high quality. The participants of the Workshop have unanimously adopted a resolution in support of the continuation of the operation of those observing facilities that are involved with long-term monitoring programmes.

The Workshop was supported by a generous grant from NATO, and by additional grants from the Belgian Fund for Scientific Research (NFWO) and the University of Brussels (VUB).

Acknowledgement

The Editors express their gratitude to the NATO Science Committee for allocating a NATO-ARW grant, and to the other members of the Scientific Organising Committee for their most valuable support during the preparation of the scientific programme of the meeting. One of us (MdG) acknowledges financial support from the Department of Education in Northern Ireland through an annual grant-in-aid for the operation of the Armagh Observatory. Miss Aileen Brannigan provided invaluable assistance in the editing of the many, not always quite camera-ready, manuscripts.

SCIENTIFIC ORGANISING COMMITTEE

A. BRUCH, Westf. Wilhelms-Universität, Münster, Germany

N. CRAMER, Observatoire de Genève, Genève, Switzerland

M. de GROOT, Armagh Observatory, Armagh, United Kingdom

E.H. OLSEN, Copenhagen University Observatory, Copenhagen, Denmark

O. STAHL, Landessternwarte Heidelberg, Heidelberg, Germany

C. STERKEN (Chair), University of Brussels (VUB), Brussels, Belgium

P.S. THÉ, University of Amsterdam, Amsterdam, The Netherlands

B. WOLF, Landessternwarte Heidelberg, Heidelberg, Germany

A.T. YOUNG, San Diego State University, San Diego, USA

Conference Photograph

1	Otmar Stahl	16	Drahomir Chochol	31	Karl-Heinz Mantel
2	Noel Cramer	17	Don Kurtz	32	Alex Schwarzenberg-Czerny
3	Jean-Michel Le Contel	18	Hans Van Winckel	33	Chris Lloyd
4	Nikolaus Vogt	19	Laszlo Szabados	34	Herman Hensberge
5	Karel van der Hucht	20	Holger Mandel	35	Peter Kroll
6	Ennio Poretti	21	Bernhard Wolf	36	Erik Heyn Olsen
7	Theodore Walraven	22	Michael Breger	37	Mart de Groot
8	Albert Bruch	23	Antonio Delgado	38	Hilmar Duerbeck
9	Arnout van Genderen	24	Petr Harmanec	39	Alain Jorissen
10	Nadine Rons	25	Koen Vyverman	40	Gilbert Burki
11	Chris Sterken	26	Andreas Kaufer	41	Douglas S. Hall
12	Amadeu Fernandes	27	Thomas Schimpke	42	Jacco van Loon
13	Panos Niarchos	28	Peredur Williams	43	Anthony Moffat
14	Ladislav Hric	29	Jan Cuypers	44	Dainis Dravins
15	Osman Demircan	30	Andy T. Young	45	Holger Pedersen

List of participants

Michael Breger	Universitäts-Sternwarte, Türkenschanzstr. 17 1180 Wien, AUSTRIA
Albert Bruch	Universität Münster, Astronomisches Institut, Wilhelm-Klemmstrasse 10 4400 Münster, GERMANY
Gilbert Burki	Observatoire de Genève, 51 Chemin des Maillettes 1290 Sauverny, SWITZERLAND
Drahomir Chochol	Astronomical Institute, Slovak Academy of Sciences 05960 Tatranska Lomnica, SLOVAC REPUBLIC
Noel Cramer	Observatoire de Genève, 51 Chemin des Maillettes 1290 Sauverny, SWITZERLAND
Jan Cuypers	Koninklijke Sterrenwacht, Ringlaan 3 1180 Ukkel-Brussel, BELGIUM
Mart de Groot	Armagh Observatory, College Hill BT61 9DG Armagh, UNITED KINGDOM
Antonio Delgado	Instituto de Astrofisica, de Andalucia, Apartado de Correos 18080 Granada, SPAIN
Osman Demircan	Ankara University Observatory, Faculty of Science, Dep. of Astronomy 06100 Tandogan ANKARA, TURKEY
Dainis Dravins	Lund Observatory, Box 43, 22100 Lund, SWEDEN
Hilmar Duerbeck	Universität Münster, Astronomisches Institut, Wilhelm-Klemmstrasse 10 4400 Münster, GERMANY
Amadeu Fernandes	Centro de Astrofisica, Rua do Campo Alegre Porto, PORTUGAL
Eric Gosset	Institut d'Astrophysique, Université de Liège, Avenue de Cointe 5 4000 Liège, BELGIUM
Elisabeth Griffin	Institute of Astronomy, Univ. Cambridge, Madingley Road Cambridge, UNITED KINGDOM
Douglas Hall	Arthur J. Dyer Observatory, Vanderbilt University, Box 1803 Nashville Tennessee, U.S.A.
Petr Harmanec	Astronomical Institute, Czech Academy Sciences Ondřejov 25165, CZECH REPUBLIC
Herman Hensberge	Koninklijke Sterrenwacht, Ringlaan 3 1180 Brussel, BELGIUM
Ladislav Hric	Astronomical Institute, Slovak Academy of Sciences 05960 Tatranska Lomnica, SLOVAC REPUBLIC
Alain Jorissen	Institut d'Astronomie ULB, C.P. 165, Av. F. Roosevelt 50 1050 Bruxelles, BELGIUM
Andreas Kaufer	Landessternwarte Heidelberg, Königstuhl 6900 Heidelberg, GERMANY
Peter Kroll	Thuringer Landessternwarte, Tautenburg, Sternwartestrasse 32 6400 Sonneberg, GERMANY
Don Kurtz	Department of Astronomy, University of Cape Town Rondebosch 7700, SOUTH AFRICA
Jean-Michel Le Contel	Observatoire de la Côte d'Azur, Departement Fresnel, B.P. 229 06304 Nice Cedex, FRANCE
Chris Lloyd	Rutherford Appleton Laboratory, Chilton, Didcot Oxon OX11 OQ1, UNITED KINGDOM
Holger Mandel	Landessternwarte Heidelberg, Königstuhl 6900 Heidelberg, GERMANY

Jean Manfroid	Institut d'Astrophysique, Université de Liège, Avenue de Cointe 5 4000 Liège, BELGIUM
Karl-Heinz Mantel	Inst. für Astronomie & Astrophysik, Universitäts-Sternwarte Scheinerstr. 8000 Munchen 80, GERMANY
Anthony Moffat	Département de Physique, Université Montréal, CP6128-SUC A QC H3C 357 Montreal, CANADA
Panos Niarchos	Department of Physics, University of Athens, Panepistimiopolis 15 Athens, GREECE
Erik Heyn Olsen	University of Copenhagen, Brorfeldevej 23 DK-4340 Tølløse, DENMARK
Holger Pedersen	Copenhagen University Observatory, Oster Voldgade 3 1350 Copenhagen K, DENMARK
Ennio Poretti	Osservatorio Astronomico, Via Bianchi 46 22055 Merate, ITALY
Gregor Rauw	Institut d'Astrophysique, Université de Liège, Avenue de Cointe 5 4000 Liège, BELGIUM
Nadine Rons	Astrofysisch Instituut, Vrije Universiteit Brussel, Pleinlaan 2 1050 Brussels, BELGIUM
Thomas Schimpke	Astronomisches Institut, Universität Münster, Wilhelm-Klemmstrasse 10 4400 Münster, GERMANY
Alex Schwarzenberg-Czerny	Warsaw University Observatory, AL. Ujadowskie 4 Warsaw, POLAND
Otmar Stahl	Landessternwarte Heidelberg, Königstuhl 6900 Heidelberg, GERMANY
Chris Sterken	Vrije Universiteit Brussel, Pleinlaan 2 1050 Brussels, BELGIUM
Laszlo Szabados	Konkoly Observatory, P.O. Box 67 1525 Budapest XII, HUNGARY
Pik Sin Thé	Sterrenkundig Instituut, Universiteit Amsterdam, Kruislaan 403 1098 SJ Amsterdam, THE NETHERLANDS
Karel van der Hucht	SRON Ruimteonderzoek Utrecht, Sorbonnelaan 2 3584 CA Utrecht, THE NETHERLANDS
Arnout van Genderen	Huygens Laboratory, Rijksuniversiteit Leiden, P.O. Box 9513 2300 RA Leiden, THE NETHERLANDS
Jacco van Loon	Sterrenkundig Instituut, Universiteit Amsterdam, Kruislaan 403 1098 SJ Amsterdam, THE NETHERLANDS
Hans Van Winckel	Astronomisch Instituut, K.U. Leuven, Celestijnenlaan 200B 3001 Heverlee, BELGIUM
Nikolaus Vogt	Pontif. Universidad Católica, Instituto de Astronomia, Casilla 6014 Santiago, CHILE
Jean-Marie Vreux	Institut d'Astrophysique, Universite de Liège, Avenue de Cointe 5 4000 Liège, BELGIUM
Koen Vyverman	Astrofysisch Instituut, Vrije Universiteit Brussel Pleinlaan 2 1050 Brussels, BELGIUM
Christoffel Waelkens	Astronomisch Instituut, Katholieke Universiteit, Celestijnenlaan 200B 3001 Heverlee, BELGIUM
Theodore Walraven	920 Keyter Street, Daspoort 0082 SOUTH AFRICA
Peredur Williams	Royal Observatory, Blackford Hill EH9 3HJ Edinburgh, UNITED KINGDOM
Bernhard Wolf	Landessternwarte Heidelberg, Königstuhl 6900 Heidelberg, GERMANY
Andy T. Young	San Diego State University, Dep. of Astronomy, 4906, 63rd Street San Diego CA 92115, U.S.A.

LONG-TERM PHOTOMETRY OF VARIABLES: THE STORY OF HD 160529

C. STERKEN*
University of Brussels
Pleinlaan 2
1050 Brussels
Belgium

ABSTRACT. This paper gives an account of the origin and evolution of the *Long-Term Photometry of Variables* (LTPV) project that, for more than a decade, has been running at the European Southern Observatory in Chile. The LTPV project is the scientific pacemaker for this meeting.

Keywords: HD 160529 – LTPV – Long-Term Monitoring

1. Introduction

This introductory paper gives an account of the origin and evolution of the *Long-Term Photometry of Variables* (LTPV) programme that, since October 1982, has been running at the European Southern Observatory (ESO) in Chile. The LTPV project is the scientific pacemaker for this meeting.

Before going into the particulars of the story of LTPV, I wish to accentuate that LTPV is only one of several existing projects of long-term monitoring of variable stars, an activity which belongs to the range of pursuits of long-term monitoring in astronomy. Long-term monitoring—which we coin *LTM*—in astronomy, in turn, is part of *LTM* in science and—in its broadest sense—in every-day life. *LTM* is not a fashion, *LTM* is not a vogue. But *LTM* is a *discipline*, a *method*, a *process*.

When I look back over my own previous work of observing variable stars, I recognise that I am still interested in the same variables I already observed 22 years ago. So, in a sense, I have been pursuing *LTM* activities all the time, and it is, therefore, not so easy to trace the roots of a project like LTPV. The situation certainly is similar to the following:

> *The reason why a young student in astronomy gets involved in precisely that kind of research in which he finally accomplishes a Ph.D. thesis, is often not clear to the student himself. One suspects that in many cases a senior colleague, who has a better view of the speciality, calls the attention of the younger researcher to a specific unsolved problem that offers ample opportunities...* (de Groot 1969)

* Belgian Fund for Scientific Research (NFWO)

C. Sterken and M. de Groot (eds.), The Impact of Long-Term Monitoring on Variable Star Research, 1–18.

The seed of action, I think, was exactly Mart de Groot offering me in 1971 a copy of his Ph.D. thesis on the very interesting variable supergiant P Cygni, a star whose variability had already been recorded almost four centuries ago.

One year later I was involved in monitoring the β Cephei star σ Scorpii from ESO at La Silla (Chile), when Bernhard Wolf asked me to determine the Strömgren colour indices of HD 160529 ($A2Ia^+, V = 6.67$), a highly-reddened extreme supergiant located in the direction of the Galactic Centre. Though I had put a lot of care in determining the colour-transformation equations and, thus, had obtained a fairly good internal precision, I had no idea what the real accuracy of my measurement was. So I decided not to measure HD 160529 just once, but to measure it once a night, for several weeks. I must add that my decision to spend more time than necessary for measuring a constant star was eased by the fact that, for a considerable part of my three-month observing run, the dome of the Danish 50-cm telescope could not rotate, so that I was technically restricted to observe only stars that pass near the zenith (such as σ Sco and HD 160529).

To my surprise, the standard deviations in $b - y$, m_1 and c_1 for HD 160529 amounted to several hundredths of a magnitude, and in the y, b, v and u bands they were even close to a tenth of a magnitude. The truth was that the star was varying rather irregularly and, without realizing it, I had obtained one of the first light curves of a Luminous Blue Variable—see Fig. 1. In December 1972, Wolf and myself submitted for publication a short paper entitled *Irregular variations of the extreme A supergiant HD 160529*, and we hit immediately one of the major publishing obstacles in *LTM* when we received the following reply from the Editor:

> *I am sure you can realize that it is very difficult for us to publish a research note every time you have observed some variation in the light of a star ...*

The following year (using two well-selected comparison stars), I monitored the star during three months, and another year later I did so again for six weeks. It soon became apparent that the light and colour of the star varied with a characteristic time of 2 to 3 months. Consequently, for almost a decade I pursued measuring the star from half a dozen different sites around the world, an enterprise that led me only to more frustration: though I collected state-of-the-art measurements, at good sites, with good photometers, and applied a well-accepted reduction procedure, there seemed no way to arrive at a homogeneously-covered phase diagram of the photometric variations of this star (see Fig 2). Consequently, I ascribed this failure completely to the impossibility to obtain photometric observing time on a night-to-night basis over a period of time spanning several years. At the ESO Workshop *The Most Massive Stars* (1981), other colleagues, facing similar problems, joined forces and a scheme of service observing was proposed for the coming decade (Sterken 1982, 1983). Thus, since October 1982, a considerable amount of photometric observing time at the European Southern Observatory has been allotted to the LTPV programme.

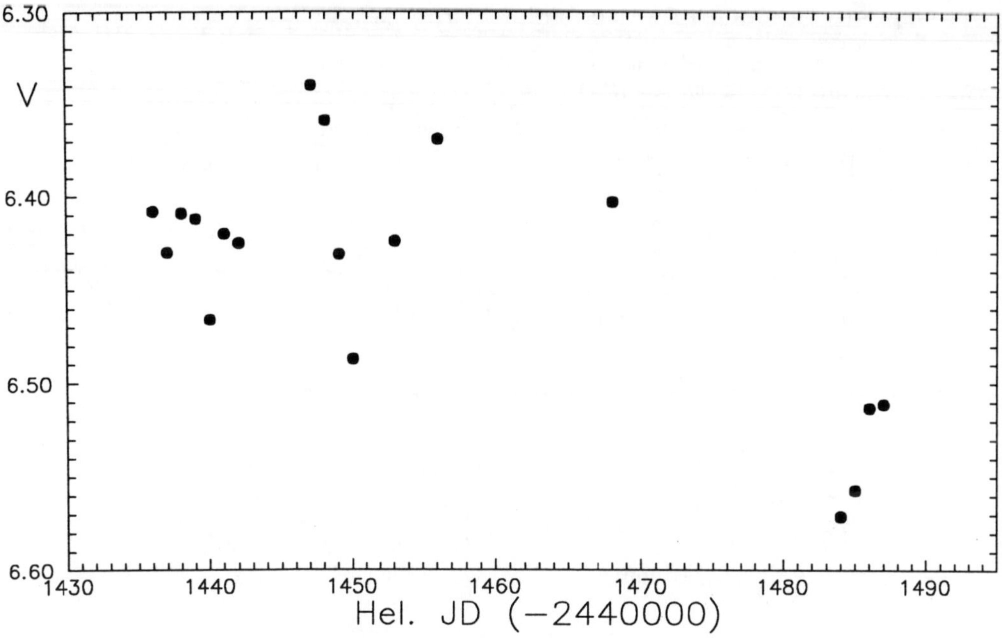

Fig. 1. Discovery light curve of HD 160529

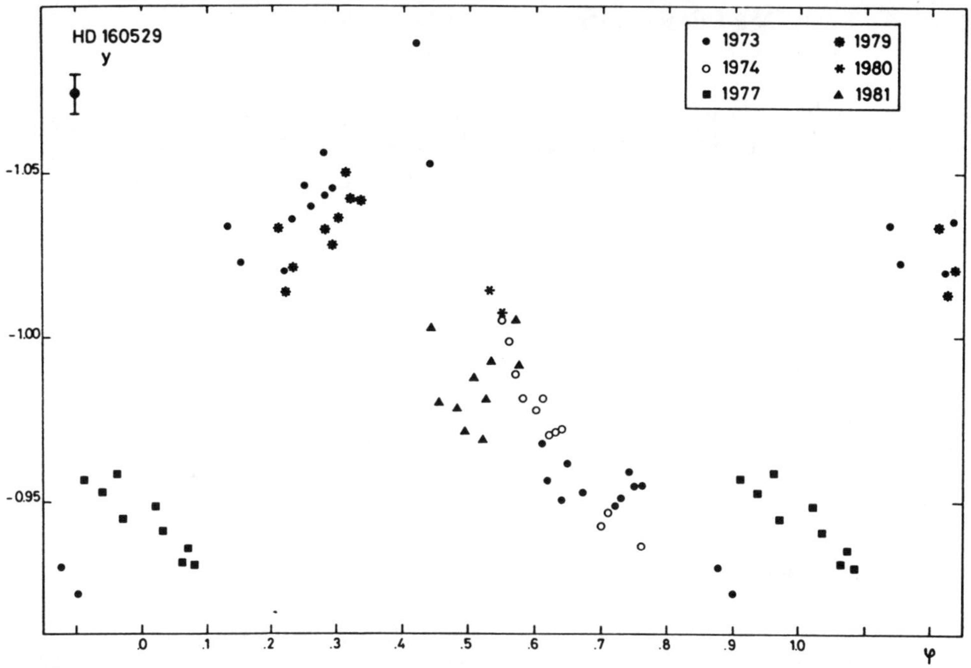

Fig. 2. Phase diagram of HD 160529 after almost a decade of private monitoring (Sterken 1982). Note the large gap at phases 0.1–0.2

2. Organisation of LTPV: the rules of the game

The project is a combination of several individual observing interests where the most exciting stars were grouped into several separate research topics, viz. Pre-main sequence stars, Ap and CP Stars, Eclipsing binaries, Be stars, Supergiants, X-ray sources, Peculiar late-type stars, and Wolf-Rayet stars. For each research topic, a Principal Investigator and a co-investigator were appointed (see Table 1), and these persons were responsible for the coordination of the work within their own field of interest (Sterken 1986). The Principal Investigators attract further participants, so that at any time the LTPV team is made up of about 30 to 40 members, and over the years more than 100 people have belonged to this organised group.

Table 1. Sections and today's Principal Investigators of LTPV

	Section	Principal Investigator(s)
1	Pre-main sequence stars	P.S. Thé
2	CP stars	H. Hensberge, J. Manfroid
3	Eclipsing binaries	A. Bruch, H. Duerbeck
4	Be stars	N. Vogt
5	Supergiants, LBVs	B. Wolf, O. Stahl
6	X-ray binaries	M. Burger
7	Objects of opportunity	
8	Peculiar late-type stars	A. Jorissen
9	Wolf-Rayet stars	J.-M. Vreux, C. Sterken

Knowing that, in private firms, the most efficient companies allow a great deal of de-centralised decision-making[1], we choose an approach that involves repeated discussion with the core of the team—that is, the Principal Investigators. Such consultations happen at our annual one-day meetings (12 such meetings, so far, were held) and through group-email contact. The observers were, sometimes, the Principal Investigators themselves but, more often, volunteering M.Sc. and Ph.D. students and young postdocs. From the very beginning, the observers have not been considered as night asssistants, but as fully responsible and autonomous *service observers*.

All participants have repeatedly expressed their opinion about the working rules of the group. The most important agreements which concern the observer are:

- The observer shares co-authorship of any **data** paper which is made with data to the collection of which he or she contributed
- The observer will *not* share co-authorship of the **analysis** papers which were made on the basis of his or her observations, *unless (s)he has contributed to the analysis.* Observers, though, are encouraged to get involved with the scientific analysis of some objects in which they may be interested; the interested observer, therefore, contacts the relevant Principal Investigator

[1] this fact, unfortunately, is less and less understood by those who continuously restructure and reorganize scientific organisations and universities

- The authors of analysis papers do not publish the observations in tabular form, since this is done in the data papers. Authors of analysis papers are never co-authors of a data paper, unless they contributed as an observer
- An analysis paper must explicitly mention the names of the observers who collected the observations, together with the dates and the designations of the telescopes
- If observers wish to incorporate extra objects in which they are interested themselves, they may include these objects in the programme

Though these rules have been respected by most of the participants, there have—perhaps inevitably—been a few problems and infringements. As can be recognised, our agreements about co-authoring are markedly different from those in other co-ordinated projects (see, for example, Hall's paper at this meeting).

So far, we have sent 73 observers to the mountain (each for a period from two to five weeks). In addition, we had several "inter-mission" observing runs that served to strengthen the solutions of the colour equations[2]. Fig. 3 illustrates the number of observing runs (observers) per participating Institute. The pattern of the histogram clearly indicates that in some specific Institutes (Heidelberg, Münster, Leiden, Amsterdam) active participation as observers is very strongly encouraged, whereas in some other Institutes that take part in LTPV, participation in the form of observing is rather on an individual basis. Not surprisingly, those Institutes that encourage their students to carry out observations are those places where students receive active schooling in observational practice.

Fig. 4 illustrates the distribution of the "expert level" of observers: short vertical bars represent observers who had no prior observing practice the first time they took care of an observing run, long bars stand for observers with previous observing experience[3]. The picture shows that the *majority* of our observers were, the first time, novices, and that for some of these young people the long observing run for LTPV was their first (and, sometimes, only) mission. As to the element of quality of performance, we have found that there is, indeed, a factor that strongly depends on training.

3. Scientific achievements

So far, the project has produced a number of interesting scientific discoveries, and dozens of scientific papers. Scientific papers are of four types, *viz.:*

1. scientific-analysis papers, which are papers produced by the Principle Investigators and LTPV participants and associated investigators
2. technical papers, which are mainly by-products—that is, reports on results that escape to be noticed by individual participants, but that emerge when inspecting the complete databank, such as
    - the monitoring of atmospheric extinction (Sterken & Manfroid 1992)
    - the assessment of instrumental stability (Manfroid & Sterken 1992)
    - the "discovery" of constant stars (Sterken et al. 1995)

[2] these were mainly carried out by C. Sterken and J. Manfroid

[3] it should be stressed, though, that some of these *novices* perform as well, or even better, as some of the *veterans*

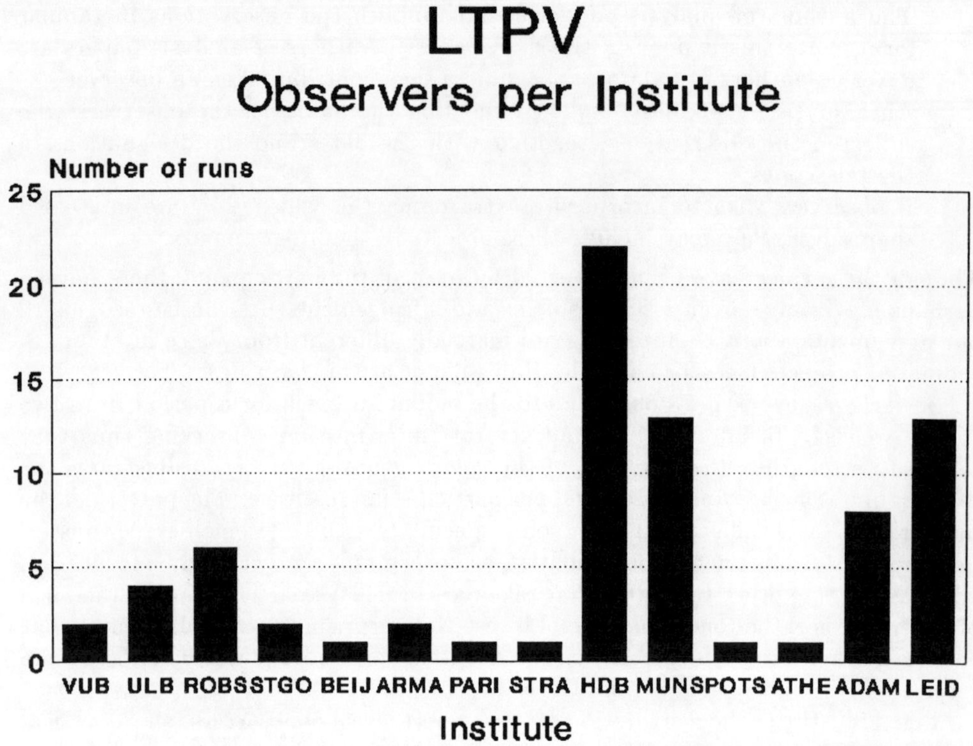

Fig. 3. Number of observing runs (observers) per participating Institute. Names of Institutes are alphabetically arranged per country along the horizontal axis: VUB, ULB and ROB, respectively, stand for the Universities of Brussels, and for the Royal Observatory of Belgium, STGO is Universidad Católica de Chile, BEIJ is Beijing Observatory, ARMA is Armagh Observatory, PARI and STRA represent Paris and Strasbourg Observatories, HDB, MUNS and POTS are Landessternwarte Heidelberg, University of Münster, and Potsdam Observatory, ATHE is Athens, and ADAM and LEID are the Astronomical Institutes of Amstersdam and Leiden University

3. data catalogues: two data catalogues have been released, a third is in the making.
4. indirect spin-offs: these are scientific papers by non-participants, but based on published and unpublished LTPV data (e.g., de Koter 1993)

To date, LTPV contributed to 118 scientific papers, which are signed by at least 425 authors[4]. Fig. 5 gives the number of scientific papers as a function of time, and Fig. 6 shows the distribution of the number of papers as a function of the number of authors per paper. Excluding the catalogue papers (see the rightmost entries in Fig. 6), there remain 114 papers with 329 authors, of which 120 are external to LTPV (co-authors that are in no way associated with the formal LTPV team, but who have been associated in some way or other with the scientists working on our

[4] non-distinct authors

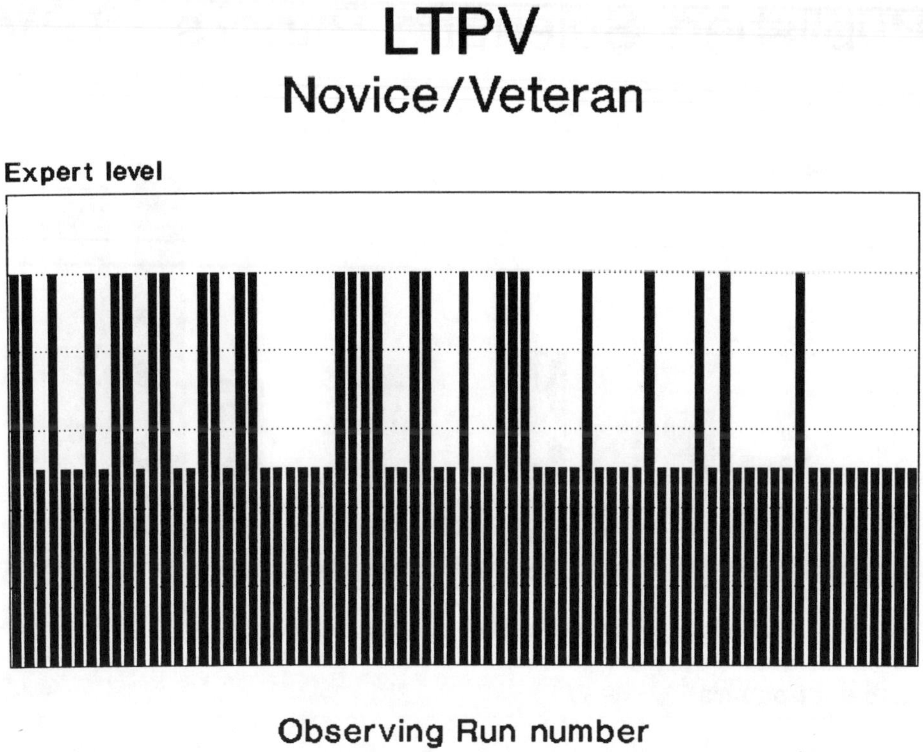

Fig. 4. Distribution of "experience" of LTPV observers in function of observing-run number (increases with time)

data). From Fig. 6 it is seen that the majority of papers are single-authored, or are written by duos or trios. The scientific production, measured by the number of papers published, is impressive, though one should realise that about 60 to 70 % of our data are still dormant—that is, have never been analysed properly.

As to HD 160529, the star that started it all, we reproduce in Fig. 7 the composite light-curve from Spoon et al. (1995): whereas the pre-LTPV data would suggest the presence of a wavy pattern with medium amplitude and a pseudo-period of about 100 days (Sterken 1977), the complete picture now reveals the existence of larger-amplitude variations of more irregular nature, where a characteristic time of 100 days is much more difficult to recognise.

4. Our scientific blunders

Principal Investigators and associates are supposed to be productive scientists, so there must be plenty of errors, and we recognise them. Scientific blunders are taken care of in the literature, so there is no need to go into detail here, and I refer to the papers of Kurtz, Young and Breger at this meeting for a description of the fields where many mistakes occur. It should, however, be remembered that it is easy to

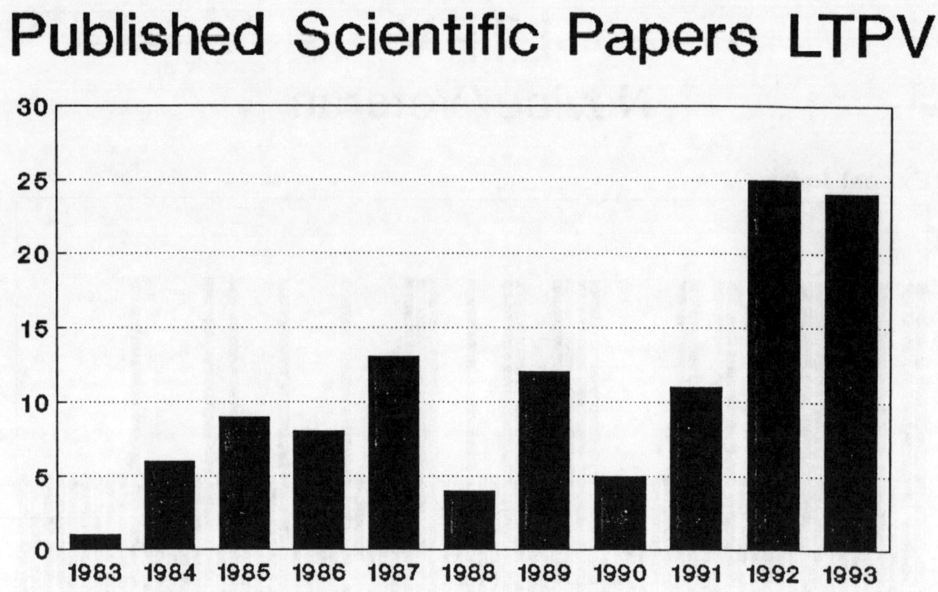

Fig. 5. Number of published scientific papers as a function of time

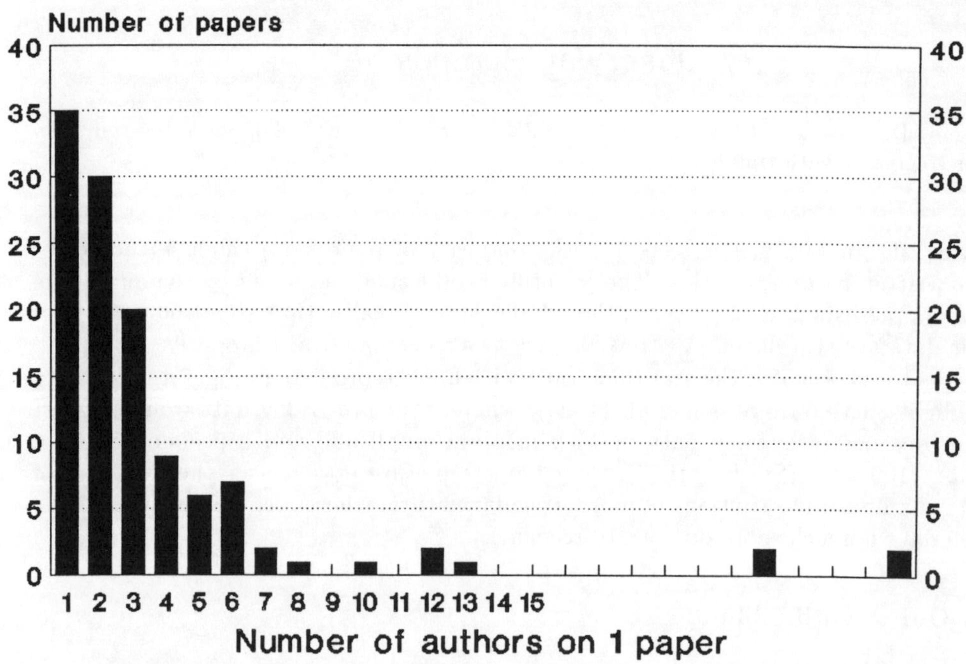

Fig. 6. Distribution of the number of papers as a function of the number of authors per paper; the rightmost entries are the two data-catalogues

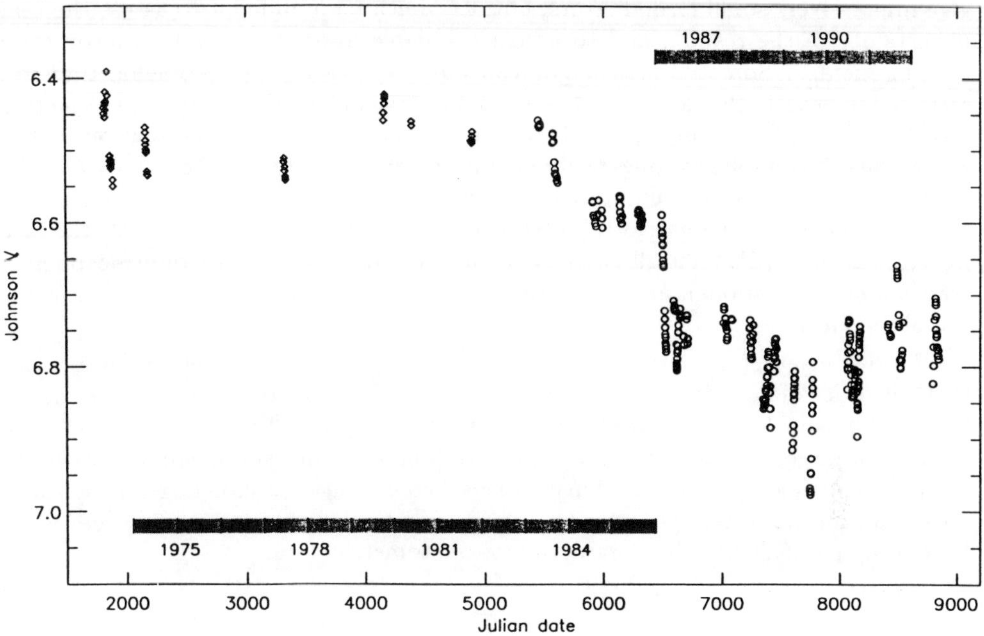

Fig. 7. Light curve in V for HD 160529: pre-LTPV (prior to 1982) and LTPV data depict a different portrait of the star's characteristics of variability

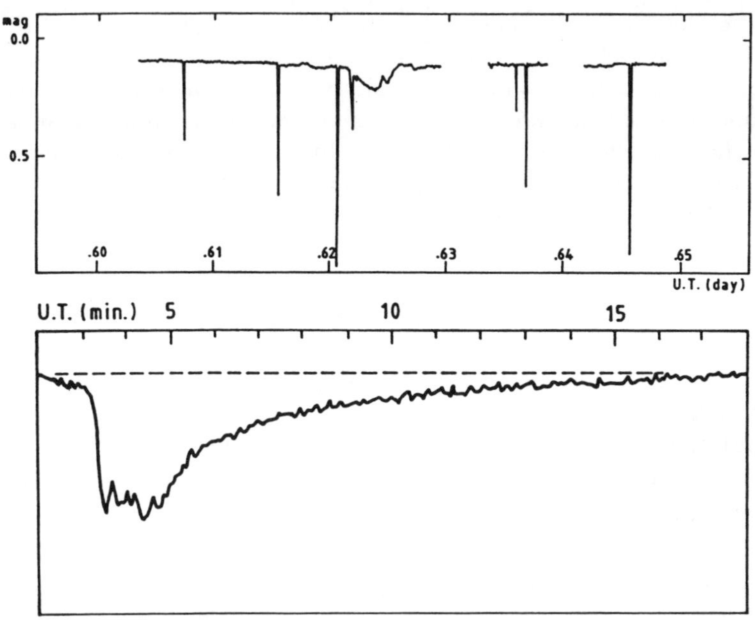

Fig. 8. Top: 15-min monitoring of HD 160529, with checks of the comparison star. Spikes are due to the interruptions for checking the centring. Bottom: text-book example of a similar monitoring of a flare star (note that, for reasons of argument, the direction of the magnitude axis has been reversed)

give impressively small formal errors, but it is much more difficult to assess the real uncertainty of the results, and to assign a reliable absolute error to photometric measurements.[5] Sometimes the significance of two observed phenomena, that are recorded in exactly the same way, is accepted or rejected on almost purely subjective grounds. Take, for example, Fig. 8, where the top panel is a reproduction of the raw counts of HD 160529 (single-channel photometry, B filter) and comparison-star readings during an almost uninterrupted 15-min monitoring, and where the bottom panel is a recording of a flare of YZ CMi (same procedure, same time-scale): the fast light variations in HD 160529 (which are, indeed, on the time-scale of atmospheric-transparency variations), are met with much more scepticism than is the case with the observed flares.

Besides meeting our preset goal—that is, adequate phase coverage on a time-scale of a decade or longer—we acquired experience from the mistakes we made. The first lesson learned is that we had assumed *a priori* that phase coverage is by far and large a *sufficient* condition to run long-term measurements successfully, without realizing that aspects of instrumental conformity, methodological rigorousness and procedural stringency constitute an equally important *necessary* condition to accomplish long time-base gathering of photometric data.

5. Our problems, and what we learned from them

5.1. COMMUNICATION

Communication has drastically improved over the last decade. Where, at the onset of LTPV, distribution of reduced data was done on the basis of computer listings (regular mail), we now use mainly electronic mail for data transfer. In parallel, the interaction with the observer and with the other participants has advanced a lot between the time when messages were sent by telex (no possibility for transfer of finding charts) and today, when the principal medium of communication is fax and e-mail. These improved communication facilities, however, entail their own problems:

- students suffer from *e-mailitis*, and spend an amazing part of their free day-time moments penning
- at the same time, their superiors—some of which are rather lordly in the control of their juniors—keep command through e-mail, not realising the pressure they put on someone observing for such a long period
- on the other side, the improved communication channels are in no way a guarantee that instructions, if they are read at all, are carried out properly
- electronic mail is a vulnerable and delicate procedure, leading to many misunderstandings and problems

[5] for a precise delineation of the difference between precision and accuracy, see the papers of Young and of Dravins at this meeting

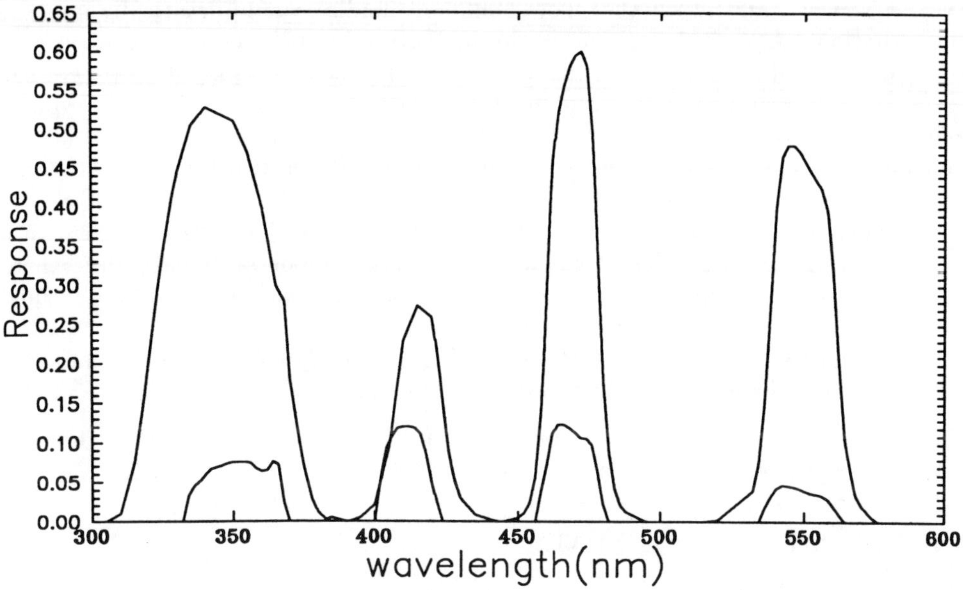

Fig. 9. Non-normalised transmission curves for two *uvby* instrumental configurations

5.2. STANDARDISATION AND CALIBRATION

We have had several difficulties with the standard instrumental system, and with calibration. Fig. 9 shows the transmission curves of the Strömgren *uvby* system for two different filter sets used by LTPV in 1982–1983. The problem is not only a fundamental change in the instrumental photometric system (which, by the way, makes any transformation to the standard system impossible), but also that the changes are poorly documented, or even not documented at all. We have, for instance, cases where the observer noted completely incompatible and non-existent photomultiplier-cathode types. For a critical discussion of fundamental aspects of calibration and standardisation, we refer to Young's paper in this meeting.

5.3. SOFTWARE

One should realize that, nowadays, photometric data reductions are carried out with complex computer programs, and that the verification of their reliability is practically impossible. Despite long-term testing, large programs, like the LTPV reduction package, *must* contain residual bugs; besides giving a false sense of security, such errors and bugs introduce non-random flaws in the final data, and such errors may completely bias any scientific conclusions. The only remedy is *protective redundancy* by developing two independent sets of software at different sites, but such refinement is far beyond the financial scope of projects such as LTPV. Moreover, it is even doubtful whether more abundant funding would drastically improve the situation. A typical example is the MIDAS software, where—despite a centralised and well-funded development—a major problem remains the existence of several sites with implementation of incompatible clone-versions of the mother package. The situation is not better where observers use "standard", non-interactive, and poorly-documented photometric-reduction algorithms that are now available from about everywhere.

Our experience only illustrates the correctness of A.P. Fairall's statement in a letter to the Editors of "The Observatory". Commenting on systematic differences between manual measurements of lines in spectra of galaxies and results obtained from cross-correlation methods, he writes:

> *...few of us who use such software packages actually wrote them, or have examined the coding in minute detail...* (Fairall 1992)

Not without reason, he asks whether this is an isolated case, or whether this is a widespread problem.

5.4. RAW DATA/REDUCED DATA

It is a very general practice in science not to publish raw data. Most often this is done because publication is not feasible, sometimes the reason is to render re-analysis impossible. One reason, among many, for making available the raw data, is because the analysis of variance of *transformed* data (i.e. data in a standard photometric system) where the transformation—for some reason—is inappropriate, will yield results that are totally different from the results that came from untransformed (or properly transformed) data. We have looked into this problem of publishing the raw LTPV data, but have not found a solution yet, though we plan to publish raw data in some form eventually.

The publication of reduced data from a *LTM* project evokes a particular difficulty: since—by definition—one should expect significant results only after a relatively long period of observing, one should not bring the data into the open too early. Divulgation of data, on the other hand, generates additional papers, and telescope-time allocation committees will not grant observing time if a project is not scientifically productive. There is another dilemma: it is very difficult to keep a private archive[6] secure and tight, and external competing teams entice or squeeze the supervisor of the data bank, so that leakage takes place. Such possibilities open the door for hard internal competition inside the team, a situation that quickly leads to deterioration of the working atmosphere, and even to quick decay of the project itself.

In the LTPV approach we publish *in extenso* data releases every couple of years: we published the *First Catalogue of Stars Measured in the Long-Term Photometry of Variables Project (1982–1986)* (Manfroid et al. 1991), and the *Second Catalogue of Stars Measured in the Long-Term Photometry of Variables Project (1986–1990)* (Sterken et al. 1993). The printed versions of these data catalogues appeared as ESO Scientific Reports; they list the individual V magnitudes and $b-y$, m_1, and c_1

[6] in the case of LTPV the archive is owned by a group

color indices of all programme and comparison stars, and have been distributed to the libraries of the principal astronomical institutes around the world. A machine-readible copy of these catalogues can be obtained from the Strasbourg Data Centre (CDS).

Let me stress that we publish **all** observations that have been collected under normal circumstances, in good faith, and according to the preset instructions. That means that the catalogues do contain a number of measurements that are—for some reason unknown to us—far off the expected values. Such single-instant *outliers* show up from time to time, and occur during the observation of programme stars as well as during measurement of comparison stars. We decided not to clean out such deviant points, because we wished to avoid unintentional rejection of real events (such as eclipse phenomena, or flare occurrences) by suppression. That means that *we have an unbiased approach to the data*, an element that I consider as a cardinal point of utmost relevance and which is, unfortunately, emphasized only every now and then.

Publishing the data (raw and processed) has—next to re-analysis and besides stimulation of research—other advantages. Trimming and cooking[7], if done, is inescapably revealed one day. There is also no chance for forging data, nor for keeping off the market specific data that may weaken the tenability of a certain theory, nor for suppressing data that are not supported by theoretical positions. The only bias I see is a kind of selective sampling during the observing, which is caused by some strong ideological commitment of the observer (we have, indeed, noticed that observers feel more warmly toward observing the stars of their—or their thesis supervisor's—own field of interest, or sometimes think negatively of a particular part of the programme, or of a given Principal Investigator).[8]

5.5. PRINCIPAL INVESTIGATORS

Principal Investigators, even if they are experiment–oriented, from time to time display a high degree of negligence for what concerns preparation of finding charts, coordinate lists, and positions for measuring the sky. Sometimes, Principal Investigators are less active than expected: we have one case that produced no papers in more than a decade. Another phenomenon of inactivity is the case of the guest Principal Investigator who, in July 1984, asked us to observe "his" star, but *never* asked for the data, and *never* analysed the catalogued data.

There is another, very natural, aspect related to Principal Investigators. A *LTM* project, such as LTPV, that has a life time of more than a decade—that is, a fraction of ~25% of a complete career—will see more than one Key Investigator leave (retirement or withdrawal due to assignment of other duties), thus invoking significant changes or even a break in the continuity of the research. Indeed, those Sections where a change of Principal Investigator occurred, turned out to be less productive than the other Sections.

[7] trimmed and cooked data are made by altering slightly, and by selective sampling

[8] LTPV does have some build-in bias by some categorizing and pigeonholing in Sections

5.6. FUNDING

LTPV has had rather limited finance, characterised by the absence of specific recurrent funding for operating the project. Though we managed, so far, to operate without any financial contribution from the participants, the tight financial environment has had some repercussions on the performance, especially for the aspect of training of the observers. Though our operating budget is only *peanuts* compared to the "megabuck" science we see around us, a second decade of operation of LTPV will have to cope with really unsurmountable difficulties, since it has been decided that the ESO share in operating the Danish 50-cm telescope will—from the end of 1994 on—be cut off.

5.7. TRAINING

Theoreticians are not the only ones who consider data as gospel truth, *untrained* observers, also, tend to regard data as unproblematical. We are convinced that dealing properly with uncertainty in photometry can only come after long years of practice, and after developing an almost second nature. What really matters is the training, and the philosophy, of observers. A well-trained observer is less likely to be in error, and the erring behaviour is more stable.

Training in LTPV is done in one of the following ways:

- at the observer's home Institute by real observing, or by "dry-swimming";
- through written instructions prepared by C. Sterken;
- through private tutoring at the telescope (mostly by C. Sterken at Jungfraujoch Observatory, Switzerland, or at ESO, La Silla);
- by osmosis, and by mutual contacts between the observers.

Positively-motivated students are helped through the observing runs by some kind of parenting, and this surveillance effect contributes to good performance, perhaps because they know that their work is scrutinised. Also, observers belonging to those institutes that regularly send out their students have an additional motivation to perform well, because they have to keep up the prestige of their group.

The principal observer-related problems we encountered were:

- integration on bright stars is often of too short duration;
- there are too few measurements of bright red standards;
- there are too few sky measurements and, frequently, the position where the sky is measured is too near to the star, or its position relative to the observed target star is not consistent;
- "bad-weather stress"—that is, the strain caused by long (or even short) periods of poor weather that lead to impatience and a strong desire to attempt observation and make a contribution— produce data of inferior quality.

Such defects, which vary in number and in amount from one observer to the other, lead to a personal equation and an "observer weight" (note that a robot, too, has its personal equation: see, for example, Sterken & Manfroid 1992).

Though—except for two totally unproductive cases—motivation of our observers was very positive and though there was always a high integrity of performance, our experience shows the importance of mastering the fundamentals of a technique before meaningful work can be done. But it is clear that less training will result from three nights at a large telescope, than from 30 nights *LTM* at a very small telescope.

In this respect it should be stressed that the tendency to close small observing facilities to provide for the development of large telescopes is a dangerous philosophy. Quoting Mel Schwartz, who shared the Nobel Prize Physics in 1988, commenting on the fact that facilities, where young people can work, exercise their imaginations and look for major surprises today, are being sacrificed to the tremendous cost of the superconducting supercollider:

> *There is no reason surprises can't be found by examining phenomena* **with great precision** *at existing machines ...*

This is not a statement from a time in the past, it is taken from a rather recent issue of *Newsweek.*[9] Leaders of very-large-telescope projects, who see the birth of their dinosaurs at the expense of sacrificing existing machines, should realize that—unlike what is true for small and middle-sized telescopes—the contributions to quantitative astrophysics of these heavily-armoured constructions will not even live up to their physical size, let alone to the second or third power of it. This statement only complements Lloyd Evans's (1993):

> *If the community of space astronomers and galaxy observers obtain their expensive new instruments at the cost of closing existing small telescopes, they will find themselves at the apex of a pyramid which lacks the base essential to calibrate their observations*

Observing (*with great accuracy*) at small instruments is not only a sufficient condition to carry out crucial science, it is also a necessary condition—as much from the point of view of helping young observers grow into mature scientists as from the point of view of furnishing fundamental astrophysical data. Many of us know that a strong element that propels scientific large-scale projects is a political and social prestige-driven force. Few of us realise that the importance of that kind of "large scale" science is often measured with inappropriate parameters. But considering that

- the cost to put objects in space by the Space Shuttle is now more than ten times as expensive as was anticipated in 1972,
- the construction of the Supercollider, seven years after its start, had a cost overrun of a factor of more than 3 (12 billion $ in 1993)[10], and
- the Hubble Space Telescope has the largest cost overrun ever seen,

and recognising that such deficits are—to no effect—muffled by "streamlining" more modest scientific undertakings and by excising productive and very cost-effective projects, one should really start thinking about what course we are navigating.

I do realise the veracity of the very-large-telescopes protagonists' argument, "You cannot make an omelette without breaking eggs", but I wonder whether the

[9] "In Physics, Dinosaur Days", Newsweek/February 1, 1993, p.49
[10] the project was stopped with $ 2 billion spent and 20% complete (Time, Nov 1, 1993 p. 65)

folk wisdom "You must not put all your eggs in one basket" is not more appropriate at a time of national and international belt tightening.

6. Conclusions

From our experience of running the LTPV project, we see a large number of hidden bonuses of *LTM*, viz.:

- *LTM* generates vast resources of data to be used in Ph.D. studies;
- *LTM* projects may serve as observer training-ground for large telescopes;
- young observers receive very appropriate education in astronomy due to many contacts with a wide spectrum of scientists belonging to a variety of disciplines;
- students identify with scientific values of their instructors and—if the *LTM* project is properly set up—it contributes to the enrichment of scientific ethics and morality of the participants;
- participation in *LTM* provides unique training in *desinterestedness* of the observer—that is, the ability to carry out *service observing* in a non-selective way and without direct personal interest—in other words, a shift from the classic "outcome-conscious" observing that is typical for short-term observing to "procedure-conscious" observing that is a requirement, not only for *LTM* but also for precise and accurate work at large telescopes.

There is one very excellent school example of a hidden bonus of *LTM*. In 1893, exactly 100 years ago, Miss H. S. Leavitt initiated systematic photography of the Large and Small Magellanic Clouds. She used a 60-cm telescope in Peru, and continued the monitoring for 12 years. The information from that data archive led to the discovery of 1777 new variables; the brightest of these stars had the longest periods (Leavitt 1908) and the resulting period-luminosity relation was calibrated by Hertzsprung (1913); these discoveries led to the understanding that pulsations are the cause of the variation (Plummer 1913), to the establishment of the true size of our galaxy (Shapley 1914), to the subsequent discovery that two classes of Cepheids exist, and to many more secondary research projects and spin-offs.

Though we may not expect current *LTM* programmes to have an impact on astrophysics as large as the Leavitt project, this Workshop will illustrate that the impact of *LTM* on variable-star research is vast, with fundamental and lasting results.

Acknowledgements

The author is indebted to the many observers of the LTPV program for their openness and for their great help. Thanks also to ESO and to the ESO Observing Programmes Committee—in particular, to Dr. Jacques Breysacher—for their full and continuous support, which has made this project possible, and to the University of Copenhagen for providing the telescope and photometer. Gratitude is expressed to the technical staff at La Silla for their support, and for the necessary night-time interventions at the telescope. Financial support from the Belgian Government, the Vrije Universiteit Brussel, and the Belgian Fund for Collective Basic Research (grant 2.0060.91) is acknowledged.

References

de Groot, M., 1969, in *On the Spectrum and Nature of P Cygni*, Ph. D. Thesis, University of Utrecht, The Netherlands
de Koter, A., 1993, in *Studies of the Variability of Luminous Blue Variable Stars*, Ph. D. Thesis, University of Utrecht, The Netherlands
Fairall, A.P. 1992, *The Observatory*, 112, 286
Hertszprung, E., *Astronomische Nachrichten*, 196, 201
Leavitt, H.S., 1908, *Ann. Harvard Coll. Obs.*, 60, 87
Lloyd Evans, T., 1993, in *Precision Photometry*, Eds. D. Kilkenny, E. Lastovica, J.W. Menzies, South African Astronomical Observatory, 71
Manfroid, J., Sterken, C., Bruch, A., Burger, M., de Groot, M., Duerbeck, H.W., Duemmler, R., Figer, A., Hageman, T., Hensberge, H., Jorissen, A., Madejsky, R., Mandel, H., Ott, H.A., Reitermann, A., Schulte-Ladbeck, R.E., Stahl, O., Steenman, H., vander Linden, D., Zickgraf, F.-J. 1991, *Astronomy and Astrophysics, Supplement Series*, 87, 481.
Manfroid, J., Sterken, C., 1992, *Astronomy and Astrophysics*, 258, 600
Plummer, H.C., 1913, *Monthly Notices of the RAS*, 73, 661
Shapley, H., 1914, *Astrophysical Journal*, 40, 448
Spoon, H.W.W., de Koter, A., Lamers, H.J.G.L.M., Sterken, C., Stahl, O., 1994, *Astronomy and Astrophysics, Supplement Series*, in press
Sterken, C., 1977, 1977, *Astronomy and Astrophysics*, 57, 361
Sterken, C., 1982, ESO Workshop *The Most Massive Stars*, Eds. S. D'Odorico, D. Baade and K. Kjär, 147
Sterken, C., 1983, *The Messenger*, 33, 10
Sterken, C., 1986, *Monitoring of variable stars on a long time-baseline*, IAU Symp. 118, eds. J.B. Hearnshaw, P.L. Cottrell, Kluwer Academic Publishers, 255
Sterken, C., Manfroid, J., 1992, *Astronomy and Astrophysics*, 266, 619
Sterken, C., Manfroid, J., 1992, in *Automatic Telescopes for Photometry and Imaging*, eds. S.J. Adelman, R.J. Dukes Jr., C.J. Adelman. A.S.P. Conf. Series 28, 57
Sterken, C., Manfroid, J., Anton, K., Barzewski, A., Bibo, E., Bruch, A., Burger, M., Duerbeck, H.W., Duemmler, R., Heck, A., Hensberge, H., Hiesgen, M., Inklaar, F., Jorissen, A., Juettner, A., Kinkel, U., Liu Zongli, Mekkaden, M.V., Ng, Y.K., Niarchos, P., Püttmann, M., Szeifert, T., Spiller, F., van Dijk, R., Vogt, N., Wanders, I., 1993, *Astronomy and Astrophysics, Supplement Series*, 102, 79
Sterken, C., Hensberge, H., Manfroid, J., 1995, in preparation

DISCUSSION

Moffat: In view of the relatively large time gaps (a nightmare for many projects!) in the ESO-LTPV program, I am surprised that more effort has not been invested in the APT (Automatic Photometric Telescope) approach. The cost of setting up an APT is mainly only "one shot", and surely well worth it on the long term.

Sterken: The problem, in my case, is that our funding is done in the form of an annual contribution to ESO (by our governments), and it is not easy to find the extras for supporting an additional facility, like a private APT.

Hall: Let me just point out that an APT is not the **perfect** solution to the problem of long-term photometric monitoring. Although they do produce a strong data flow and can be "dedicated" to a specific project and work year-around, any one APT will suffer from the limitation of "just one telescope at a single geographic site".

Periods of cloudy weather, the longitude effect, etc. can result in ugly gaps in otherwise continuous coverage.

Harmanec: Regarding the Phoenix APT, I would like to get your support to encourage the people operating it to reduce the data for nightly extinction and zero-point shifts using the measurements of comparison and check stars.

Sterken: Sure, I support your suggestion, but you should not forget that setting up an operational reduction code is not something that everyone is willing to invest time in. However, there is Andy Young's PEPSYS program in MIDAS that will enable anyone to do what you ask for. Alternatively, if raw data are published *in extenso*, anyone wishing to do so may proceed as you suggest.

Waelkens: A question about HD 160529. You mentioned the detection of a 100-day periodicity in the pre-LTPV data. Is that period still there?

Sterken: It depends on whose analysis you are looking at. The 100-day period is not present in the analysis of Sterken et al. (1991, Astron. Astrophys. 247, 383), but it is seen in an analysis made by H. Spoon, which is based on a longer time interval, and on data (*V* only) that were carefully homogenised (see Spoon et al. 1994, Astron. Astrophys. Suppl. Ser., in press).

Le Contel: It should be possible to get funding from national authorities to put in operation a telescope at ESO for a well-defined scientific project.

Sterken: I doubt it but, perhaps, an international project could be set up.

Le Contel: Why are we discussing only a site at ESO? To me, the best site in the world for *LTM* is the South Pole.

Sterken: I agree, let's go to the South Pole and obtain uninterrupted sequences of data!

van der Hucht, Delgado, Moffat, Kurtz

NEW ASTROPHYSICAL INSIGHTS BASED ON HISTORICAL AND MODERN DATA OF ETA CARINAE 1935-1992

A.M. van GENDEREN
Leiden Observatory
Postbus 9513
NL-2300 RA Leiden
The Netherlands

M. de GROOT
Armagh Observatory
College Hill
Armagh
BT61 9DG
Northern Ireland

and

P.S. THÉ
Astronomical Institute "Anton Pannekoek"
Kruislaan 403
1098 SJ Amsterdam
The Netherlands

ABSTRACT. We discuss the procedure to arrive at a more or less homogeneous visual light-curve of η Carinae, based on different kinds of photometric observations made between 1935 and 1992. The decrease of the circumstellar extinction by the expansion of the Homunculus can explain in a satisfactory manner the secular brightness increase during this time interval. Based on an analysis of different types of homogeneous long-term photometric programmes we come to a number of astrophysical explanations for the various types of light and colour variations between $\lambda = 0.32$ - 4.8 μm. They are summarized in the last section.

Keywords: photometric techniques – individual stars – variable stars – supergiants – circumstellar matter

1. Introduction

The behaviour of η Car is so remarkable that it is not parallelled by any other known galactic object. It also is the most luminous object in our neighbourhood: L$\sim$ 6$\pm$2 10^6 $L_\odot$ (distance 2.8 kpc), thus implying that it is the most massive star: $M_i \sim 170$ $M_\odot$ according to the extrapolation of the models of Maeder and Meynet

C. Sterken and M. de Groot (eds.), The Impact of Long-Term Monitoring on Variable Star Research, 19–30.

(1987). It is also the brightest infrared source in the sky. Even if the nucleus is not a single star, we need two stars of 120 $M_\odot$, or three of 85 $M_\odot$ to explain the high luminosity.

Having varied for at least a few centuries, η Car reached a climax in brightness in 1843 during a violent eruption, with $M_{bol} \sim -14.8$, which is in excess of its present bolometric magnitude by $\sim 2^{m}.5$ (van Genderen & Thé 1984, hereafter called Paper I). On that occasion η Car expelled a nebula which is now a pure reflection nebula, called the Homunculus, with a diameter of 16″or 23000 AU. Since the eruption the brightness faded by $\sim 9^m$, for a large part through dust condensation. Since the 1930s, the object has been brightening continuously; we call this the 'secular variation'. It is our goal to find an explanation for it, as well as for the many superimposed erratic light variations with time-scales ranging from days to years.

2. The available photometric data

For our analysis we had at our disposal the following data:

a. Photographic and photoelectric observations from various sources from 1935 until 1970.

b. Homogeneous Walraven *VBLUW* photometry, regularly made between 1974 and 1991 (see Fig. 1) (van Genderen et al. 1994, hereafter called Paper III).

c. Homogeneous Strömgren *uvby* data, made between 1983 and 1992 and acquired by observers of the LTPV (Long-Term Photometry of Variables) project of Sterken (1983) (Manfroid et al. 1991; Sterken et al. 1993).

d. Homogeneous near-infrared *JHKL* observations, regularly made between 1974 and 1992 (Whitelock et al. 1983; Carter et al. 1992).

e. Scattered near-infrared *JHKLM* photometry made by various observers between 1971 and 1987.

A complete analysis is presented in Paper III (van Genderen et al. 1994) and will not be repeated here. We shall confine ourselves to giving a short description of the procedure to arrive at a homogeneous visual light curve of the secular variation, to offer an explanation, and to list the main conclusions and astrophysical explanations for the various types of light and colour variations of which a more detailed description can be found in Paper III.

The results and analysis of the daily monitoring campaign, from October 1992 until July 1993 with the 50-cm Danish telescope equipped with *uvby* filters (and sometimes also for Hβ) at the ESO, will be published elsewhere. This campaign, partly an LTPV project, but largely supported voluntarily by many observers, resulted in nearly completely covered homogeneous u, v, b and y light curves, with a total duration of not less than 9 months (van Genderen et al. in preparation)!

3. The homogeneous visual light curve of η Car for the secular variation 1935-1992 and its interpretation (based on the data mentioned in 2a., b. and c.)

To transform the observations of various sources into as much as possible homogeneous visual magnitudes and to produce a long-term light-curve free from short-term fluctuations, we followed a procedure similar to that used by Lamers and de Groot (1992) in the case of P Cygni.

From the light-curve in Paper I (Fig. 20), it is clear that the brightness before about 1940 was likely influenced by the after-effects of the outburst in the 19th century and that it is not well-documented. Therefore, only observations made after that time were included in the study of the long-term light variations. The only exceptions to this rule are the photographic observations of O'Connell (1956), which cover the years 1935-1952. His values were transformed to visual magnitudes by subtracting η Car's value of C_p (= 0.42) from the annual mean m_{pg}. The m_{vis} value for 1952 thus obtained differs by about $0\overset{m}{.}03$ from the value given by de Vaucouleurs and Eggen (1952), thus providing a homogeneous data set.

Then, seven data sets for the interval 1952-1992 were collected from the literature. The measurements made by different observers at different times, with different instruments and diaphragms and in different photometric systems, must be brought to a uniform scale. Since the Leiden observations (refs. 4 and 5 in Table 1 of Paper III) cover the longest interval, its V-values have a well-defined transformation to the Johnson V-value (V_J), and include almost-simultaneous measurements through diaphragms of different apertures, we were able to reduce all other observations to this standard and to a diaphragm size of $16\overset{''}{.}5$. This was achieved partly by prescribing that the average V-value of measurements done in different systems but on very nearly identical dates, should give the same value of V_J. In this way differences between different sets of observations were eliminated, but the variations within each data set were retained.

The *uvby* observations of the LTPV project (ref. 6 in Table 1 of Paper III) overlap in the years 1983-1991. This allows us to bring the $y = V$, values of the LTVP observations onto the same scale as the Leiden V values by subtracting $0\overset{m}{.}112$ from the former. Part of this difference arises because of the difference in diaphragm sizes used and the brightness of the Homunculus beyond its nominal diameter of 16″. A straight interpolation between the values given in Table IV of Paper I gives $0\overset{m}{.}034$ for this difference between the two photometric systems. After all, η Car is a strong emission-line source and, therefore, the effective wavelengths of the y and V filters will be influenced in different ways. Thus, if reduced with extinction coefficients obtained with normal standard stars, systematic zero point errors are introduced in the magnitudes.

From the mean error in the value of $0\overset{m}{.}112$ above, we conclude that the adaptation of the various measurements can be done to an accuracy in V of $0\overset{m}{.}04$.

As in the case of P Cygni (Lamers and de Groot 1992) we believe that the secular brightness variation of η Car becomes sufficiently homogeneous to warrant conclusions about the star's behaviour in visual light. The method thus produces a light-curve relatively free from the rapid irregular variations and the small differ-

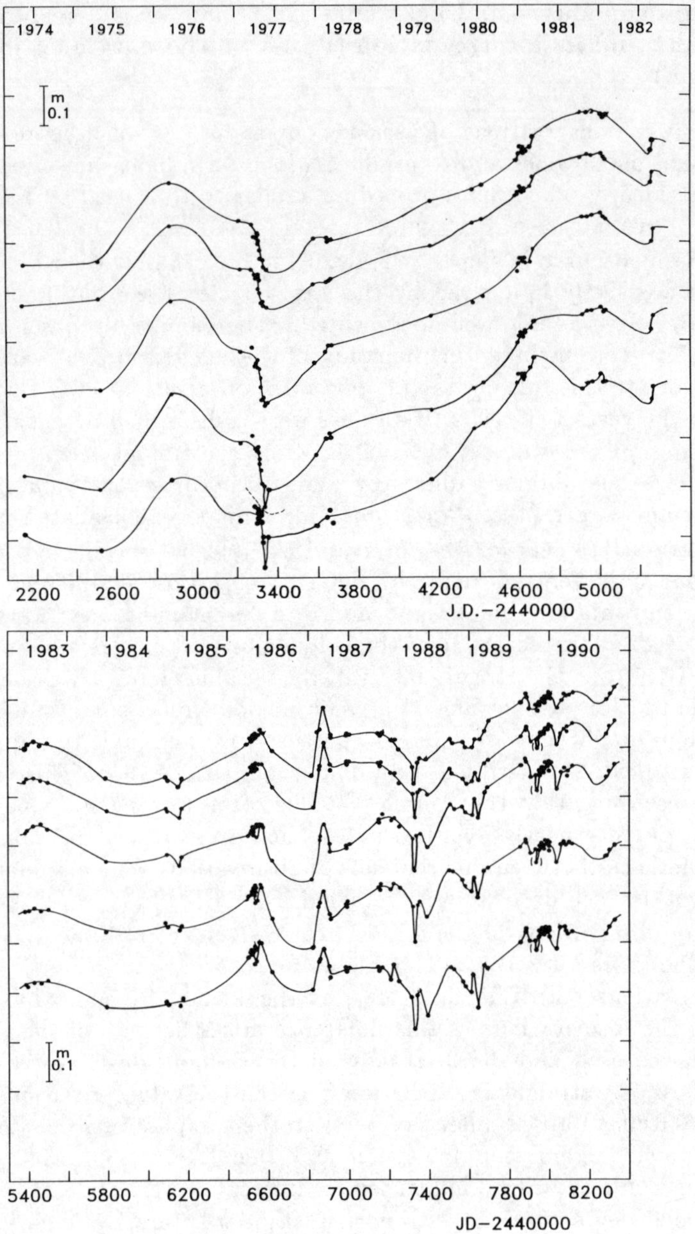

Fig. 1. The *VBLUW* (Walraven system) light-curves, 1974-1991. Note (i) the different kinds of long- and short- time-scale variations; (ii) the gradually rising mean brightness (secular light variation)

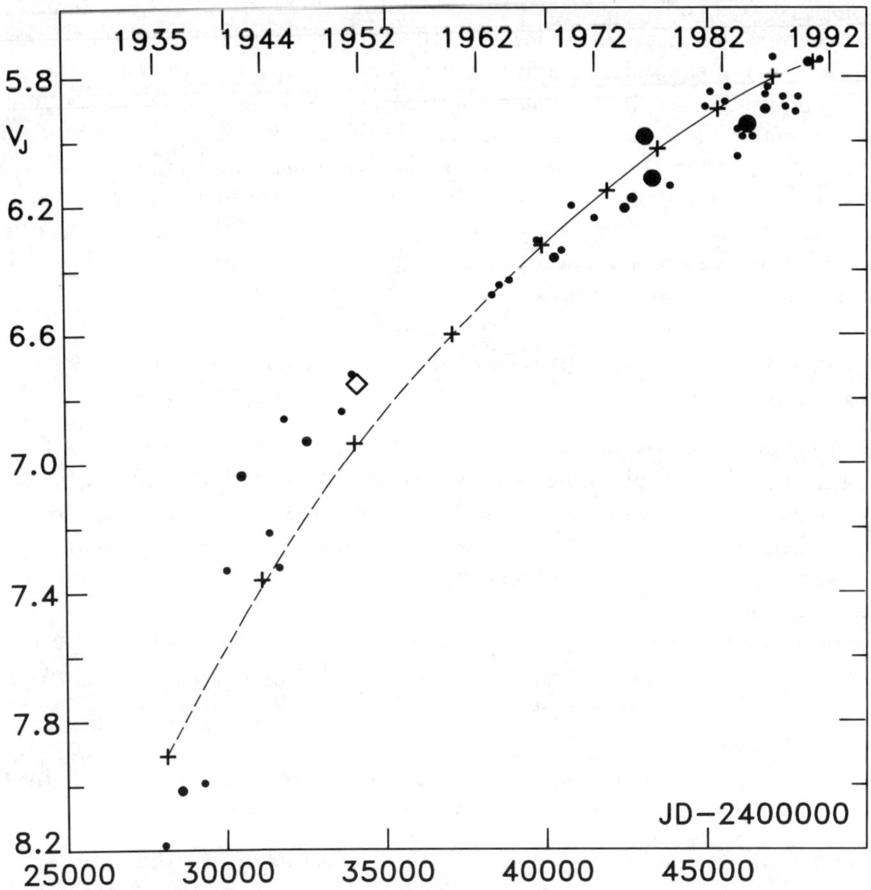

Fig. 2. The seasonal averages of the V_J magnitudes of η Carinae between 1935 and 1992. The dotted curve through the + signs is the computed brightness increase adopting that the secular light variation is due to the decrease of the circumstellar extinction by the expansion of the Homunculus.

ences between individual observers.

Figure 2 shows the so obtained light-curve for 1935-1992. The observations between 1935 and 1952 are from O'Connell (1956). The sizes of the dots reflect their weights. The visual observation (diamond) of de Vaucouleurs and Eggen (1952) fits this light curve. The dotted curve through the + signs represents the computed secular rise if it is only due to a decrease in circumstellar extinction by the expanding Homunculus (Table 3 in Paper III). We applied the same method as that used by Lamers and de Groot (1992) for P Cygni. This is done by means of an expression, which gives the column density $N(t)$ as a function of time t, adopting a homologously expanding shell. Then, by relating $N(t)$ to the optical depth and, thus, to the circumstellar extinction A_V, the expression mentioned above can be written in terms of the difference in brightness at any two given times. It appears that the computed curve fits the complete time interval 1935-1992 surprisingly well. The deviations are in most cases not much more than $0\overset{m}{.}1$ (Fig. 1).

It must be noted that two slightly different theoretical models by Menshchikov (1989), based on the structure of the dust envelope including dust-grain growth and destruction and employing radiative hydrodynamic calculations, revealed light curves which fit the secular rise in visual light also reasonably well. They show departures from the observed light curve by not much more than $0\overset{m}{.}2$.

4. The Secular variation 1971-1992 in the near-infrared (based on the data mentioned in 2d. and e.)

Our knowledge on the secular variation in the near-infrared during the last 22 yr is based on the *JHKL* photometry of Whitelock et al. (1983) and Carter et al. (1992). Since they were all done with the same instruments by the same observers, they should be homogeneous. Some other near-infrared photometry including the *M* pass-band, made by various observers between 1971 and 1987, appeared to be satisfactorily consistent with the ones mentioned above.

The opposite behaviour of η Car's secular brightness variation for $\lambda \leq 4$ μm (a rise) and for $\lambda \geq 4$ μm (a decrease) is noteworthy. We give a qualitative interpretation in Paper III (but see Fig. 3).

5. Conclusions from the analysis of nearly 60 yr of optical photometry and 20 yr of near-infrared photometry

In Paper III we have searched for astrophysical explanations of the various types of light and colour variations of η Car. The results can be summarized as follows:

1. The secular rise 1935-1992 amounting to $\sim 2\overset{m}{.}5$ in V, can largely be explained by a decrease of the circumstellar extinction (Fig. 1). The present circumstellar extinction due to the Homunculus, amounts to $A_{V_J} = 1\overset{m}{.}4 \pm 0\overset{m}{.}1$. Evolutionary causes to the secular variation are likely unimportant.

   Problems are, however, raised by the facts that:

   a. the colour variations in the *VBLUW* system deviate from the normal extinction law;

   b. the nett variation 1974 - 1992 between $\lambda = 0.32$ - 4 μm does not agree with the normal extinction law either, but shows bumps in the Balmer-line region and at ~ 2 μm and a dip longward of 4 μm.

   This indicates that other processes must be involved.

   Tentative explanations are proposed in terms of: i) the growth of a luminous disk (perhaps with a hot spot) in a possible binary system, showing Balmer emission, and/or a large disk or H II halo encompassing the core; ii) the growth of a warm dust shell around the core; iii) the gradual cooling of the cold dust component by the expansion of the Homunculus. Homunculus expansion is also responsible for the diminution of the circumstellar extinction mentioned above,

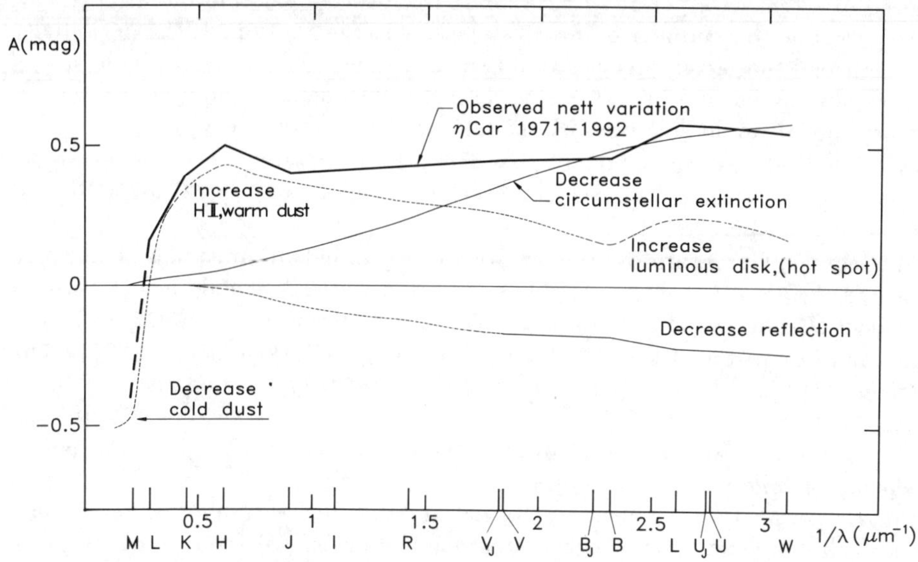

Fig. 3. An explanation for the observed nett light variation of η Car (fat line) between 1971 and 1992, as a function of the wavelenght, due to the changing physical conditions of the circumstellar material.

as well as for the decrease of the reflected optical light of the Honunculus in the optical region, of which the amount is estimated.

All these phenomena contribute, each in its own way, to the observed secular light variation. The resulting nett light variation 1971-1992, expressed by the amplitude as a function of $1/\lambda$, is graphically sketched in Fig. 3. It must be emphasized that only the decrease of the cirumstellar extinction and the reflected light are given by quantitative relations. The other two, i.e. the decrease of the cold dust luminosity and the overlapping relations for growing gaseous and dust regions, are qualitatively estimated, and found by subtracting the qualitative effects mentioned above from the observed nett variation (fat line).

2. The colour variations during the long-term light variations with a time-scale of the order of 1-3 yr are typical for hot S Doradus-type variables like S Dor, AG Car, HR Car, R71, etc. It is, therefore, almost certain that the main object in the core is an LBV.

3. An exciting feature of the monochromatic light curves is the presence of a number of minima with a maximum depth in the ultraviolet of $\sim 0^{m}_{.}1$ and likely of a short duration. The depths show a variable progression to the ultraviolet, suggesting atmospheric eclipeses of a hot source. A least-squares solution re-

sulted in a tentative period of $52^{\rm d}.4$, or $104^{\rm d}.8$ if secondary minima are present. However, due to a number of inconsistencies we urge caution. If the periodicity is confirmed the inconsistencies could then be explained by the variable size of the expanding pseudo-photoshere of the LBV with density inhomogeneities. A binary model would suit the disk model mentioned above very well.

Assuming that the maximum observed depths represent a total eclipse, the photometric parameters of the eclipsed body turn out to be typical for a B0.5 V star with M_{V_J} = -3.6.

A model for η Car as an LBV binary with its extended circumstellar structures, is presented in Fig. 4 (upper panel). The indicated warm dust halo has the size of the compact IR-Source for $\lambda < 4$ μm as observed by Bensammar et al. (1985) and the maximum diameter of the H II halo is according to Aitken et al. (1977). The lower panel explains the peculiar eclipses of the early B-type companion by blobs of ejected matter from the LBV.

We would never have arrived at most of these conclusions, if we had not had at our disposal such homogeneous photometric data, obtained by dedicated long-term photometric programmes. For many normal and peculiar variable stars such programmes (photometric as well as spectroscopic) are essential for making progress in understanding various aspects of stellar evolution and circumstellar material. Small telescopes (50-100 cm) are sufficient for these purposes and, therefore, any observatory should keep them in honour. There is a tendency nowadays to underestimate their use; however, an observatory without small telescopes is like a hand without its little finger.

References

Aithen D.K., Jones B., Bregman J.D., et al., 1977, *Astrophysical Journal*, 217, 103

Bensammar S., Letourneur N., Perrier F., et al., 1985, *Astronomy and Astrophysics*, 146, L1

Carter B.S., Marang F., Feast M.W., et al., 1992, *SAAO* Rep. No. 764

van Genderen A.M., Thé P.S., 1984, *Space Science Reviews*, 39, 317 (Paper I)

van Genderen A.M., de Groot M.J.H., Thé P.S., 1994, *Astronomy and Astrophysics*(in press, Paper III)

Lamers H.J.G.L.M., de Groot M.J.H., 1992, *Astronomy and Astrophysics*, 257, 137

Maeder A., Meynet G., 1987, *Astronomy and Astrophysics*, 182, 243

Manfroid J., Sterken C., Bruch A., et al., 1991, ESO Scientific Rep. No. 8

O'Connell, D.J.K., 1956, *Vistas in Astron.* 2, 1165

Sterken C., 1983, *The Messenger* 33, 10

Sterken C., Manfroid J., Anton K., et al., 1993, ESO Scientific Rep. No. 12

de Vaucouleurs G., Eggen O.J., 1952, *Publications of the ASP*, 64, 185

Whitelock P.A., Feast M.W., Carter B.S., et al., 1983, *Monthly Notices of the RAS*, 203, 385

Note added in proof

Another important conclusion on the internal visual extinction caused by the warm dust shell can be drawn by adopting the value of $1^{\rm m}.4$ as the total visual extinction by the Homunculus (see Section 5.1). To this end we correct η Car's present M_{V_J} of $-7^{\rm m}.8$, i.e. the value corrected only for foreground reddening, by a further $1^{\rm m}.4$, resulting in $M_{V_J} = -9^{\rm m}.2$. This result is to be compared with $M_{V_J} \sim -14^{\rm m}.8$ in 1843 (see Fig. 24 in Paper I). Thus, there should be an additional internal extinction amounting to at most $A_{V_J} \sim 5.6$. Since the present M_{bol} is about $2^{\rm m}.5$ fainter than

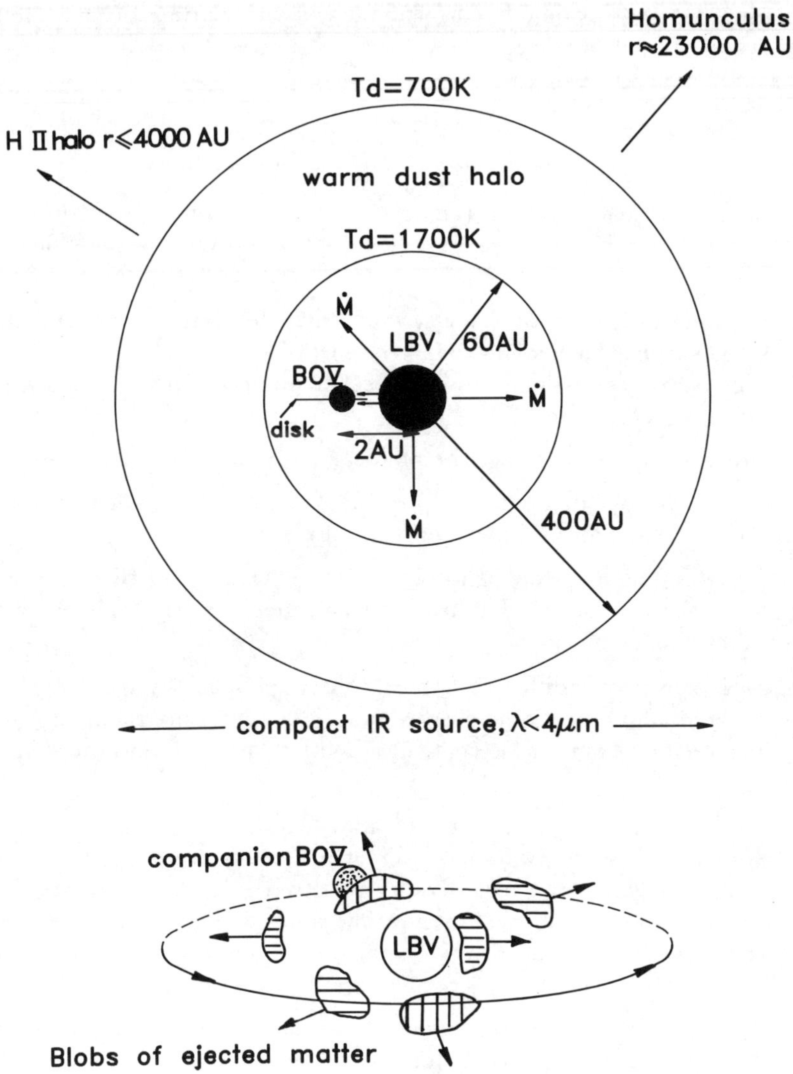

Fig. 4. A model for η Car as an LBV binary with various circumstellar structures (upper panel) and an explanation for the observed peculiar minima.

in 1843 (Paper I), perhaps due to the released thermal and rediative energy content of the ejected material (Davidson 1989, in: Physics of Luminous Blue Variables, K. Davidson, A.F.J. Moffatt and H.J.G.L.M. Lamers (Eds.), p. 101), the present value of $A_{V,J}$ for the warm dust shell will probably lie between 3^m and $5\overset{m}{.}6$. It is of interest to note that the infrared spectral study by Hamann et al. (1994, *Astrophysical Journal*, in press) suggests a total internal visual extinction of 5^m to 6^m.

We acknowledge the invaluable correspondence with R.B. Stothers about the luminosity excess of $\sim 2\overset{m}{.}5$.

DISCUSSION

Young: If there is a binary system, the orbital velocity should be of the order of 100 km s^{-1}, and should be detectable spectroscopically. What is the evidence for this?

van Genderen: Indeed, the orbital velocity should be a few hundreds of 100 km s^{-1}. However, no evidence has been found so far, because of the disturbing fact that there are hundreds of variable emission lines in the spectrum of η Carinae, caused by the various, fast moving, structures.

Burki: Is there any chance to detect the blue companion in the ultraviolet?

van Genderen: I do not think so, the possible B-type companion is about four magnitudes fainter than the LBV, which is also a blue star.

Hall: During systematic long-term photometry of η Carinae, are there any special requirements on the size of the diaphragm which must be used, i.e., how many seconds of arc in diameter?

Sterken: η Carinae is a very difficult object to observe, especially when it comes to centring it in the diaphragm. Some observers use different diaphragms during one observing run, others do automatic centring. Centring is an important element in the error budget.

van Genderen: Yes, diaphragm size should always be the same: for example, for the Walraven photometry we always used 16″, which is the size of the Homunculus. The observer always gets strict instructions on how to do the observations with respect to sky bakground, sequence, integration times, standard stars, etc. Thus, the observations are very homogeneous. As far as I know this is also true for the near-infrared observations of the SAAO observers.

Harmanec: Although it is clear that there is no direct evidence for duplicity of η Carinae, I wish to encourage you: for some less extreme Be stars which are known to be double-lined spectroscopic binaries (like CX Dra), systematic Hvar photometry shows very large cycle-to-cycle variations in their orbital light-curves, with modulation that is stronger in U and missing during some cycles.

van Genderen: Thank you!

Moffat: I have great respect for the famous Strömgren and Walraven systems; however, they were not designed with LBVs in mind, especially an exotic one like η Carinae. Could one not avoid many of the subtle problems of variability interpretation by using narrow-band filters that avoid at least the major emission lines?

van Genderen: No, that is not possible, there is hardly any undisturbed piece of continuum.

Kurtz: The problem with a special filter system is that *LTM* is then difficult in conjunction with other programmes.

Thé: In connection with the presence of Herbig-Haro objects I would like to mention that Herbig Ae/Be stars can also be about 10^6 year old when they are close to the main sequence, especially when the mass is about 3 or 4 $M_\odot$.

van Genderen: I am glad to hear that.

Le Contel: You mentioned that the object is "compact in the infrared". At which wavelength was the observation made?

van Genderen: Immediately below 4.6 μm, by Bensammar et al. (1985, *Astronomy and Astrophysics*, 146, L1).

Kurtz, Sterken

Friendly coffee

EVIDENCE FOR CIRCUMSTELLAR DISKS AROUND VARIABLE HERBIG Ae/Be STARS FROM LONG-TERM PHOTOMETRY

P.S THÉ
Astronomical Institute "Anton Pannekoek"
University of Amsterdam
Kruislaan 403
NL-1098 SJ Amsterdam
The Netherlands

ABSTRACT. In a classical paper George Herbig (1960) suggested that a number of Ae- and Be-type stars located in star-forming regions are massive counterparts of the then already known T Tauri pre-main-sequence objects. While T Tauri stars have masses of approximately one solar mass (or less), Herbig stars can have masses up to 15 times this value. Furthermore, their luminosities are about 100 to 1000 times larger than those of classical T Tauri stars. During the last five years, evidences have been accumulating, from polarimetric, spectroscopic and photometric observations, that Herbig Ae/Be stars are surrounded by circumstellar disks. In this paper the evidence for the presence of circumstellar disks around Herbig stars with large brightness variations, from long-term photometry, will be discussed. It is strongly believed that the brightness decreases are caused by obscuring dust clouds orbiting around the central star in the outer cooler regions of the disks seen edge-on. Employing a simple model, it is possible to estimate the properties of the dust clouds, such as mass, temperature, distance from the central star, orbital velocity and the period of revolution around the central star. The results of these calculations show that the above parameters are similar to those in our planetary system. Therefore, we believe that most probably, a) the formation of planetesimals in the circumstellar disks of Herbig Ae/Be stars is on-going, and b) the obscuring clouds will in the long run probably become planet-like objects.

Keywords: Herbig Ae/Be stars – variable stars

1. Introduction

In this section the general properties of Herbig Ae/Be (HAeBe) stars will be described briefly. In general, HAeBe stars can be found in star-forming regions containing gas and dust. However, there are also stars with properties similar to HAeBe objects, situated outside such regions; they are called **"isolated HAeBe stars"**. For inclusion of a star in the HAeBe stellar group Herbig restricted himself to stars of spectral type A or earlier with Hα emission lines. This limitation comes entirely from the analogy with the lower-mass T Tauri stars. Today, it is known that there are stars without emission lines that are intimately associated with T Tauri stars in star-forming regions. Furthermore, emission-line stars of type F, often also show pre-main-sequence (PMS) characteristics.

C. Sterken and M. de Groot (eds.), The Impact of Long-Term Monitoring on Variable Star Research, 31–40.

One of the more recently established PMS characteristics of HAeBe stars is the presence of excess infrared radiation due to thermal re-emission of dust grains. The important consequence of the presence of dust grains in the outer cooler regions of the circumstellar disk is that the extinction law towards the central star is very often anomalous, even when this star is in its maximum brightness, not obscured by a dust cloud. Another consequence is the variability of the stars in this group. Most HAeBe stars are variable in brightness; stars of spectral type earlier than A0 exhibit, in general, much lower variation (about $0\overset{m}{.}1^m$) than the later-type stars, which can have deep **Algol-type minima** up to 3^m below maximum brightness. The latter are believed to be caused by dust clouds.

Variability of the emission lines of HAeBe stars is a common phenomenon. A double-peaked Hα emission line can change to a P Cygni-type profile within a couple of months. Or from a normal to a reversed P Cygni profile, indicating mass accretion. The measured outflow of matter amounts to about $10^{-8} M_{\odot} \mathrm{yr}^{-1}$. On the other hand, from visual as well as ultraviolet (IUE) studies, clumpy accretion rates of $10^{-4} M_{\odot} \mathrm{yr}^{-1}$ were measured, sometimes with velocities up to 500 km s^{-1} or more. Very large velocities have also been detected from Lyα IUE observations; e.g. the average infall speed is about 1000 km s^{-1} for the star HR 5999. These very high accretion velocities should create shock waves in the gas close to the surface of the star, and will perhaps be the origin of bipolar flows perpendicular to the accretion disk.

From combined visual, infrared, sub-mm and mm data it has been shown that the excess infrared radiation of many HAeBe stars can be well fit by assuming that the excess emission above photospheric levels arises in a geometrically flat, optically-thick circumstellar accretion disk, with a hole in the inner parts and containing dust in the outer parts. However, a dissenting view and interpretation of the infrared emission in these objects has also been presented, in which the disk geometry is questioned and justifications for spherical dust shells around these objects are offered. From recent radio observations of HAeBe stars it has been concluded that the continuum emission from most of these objects is predominantly thermal and in many cases wind-related.

Recently, a new *Catalogue of members and candidate members of the Herbig Ae/Be (HAeBe) stellar group* has been published by Thé, de Winter and Pérez (1994), where the above briefly-described properties of the HAeBe objects are discussed more extensively. In the present paper I will pay attention exclusively to the evidences of circumstellar disks from long-term photometry of HAeBe stars, supported by some spectroscopic and polarimetric observations. In addition, I will also discuss the properties of dust clouds confined to these disks in connection with the possibility of planet-formation from dust grains in these clouds.

2. The Irregular Variability

Right from the beginning the Amsterdam working group of Herbig Ae/Be stars has joined the ESO group for Long Term Photometry of Variables (LTPV) for monitoring the brightness variations of HAeBe stars. Studies on individual stars using these LTPV data have been published by several authors of our group in

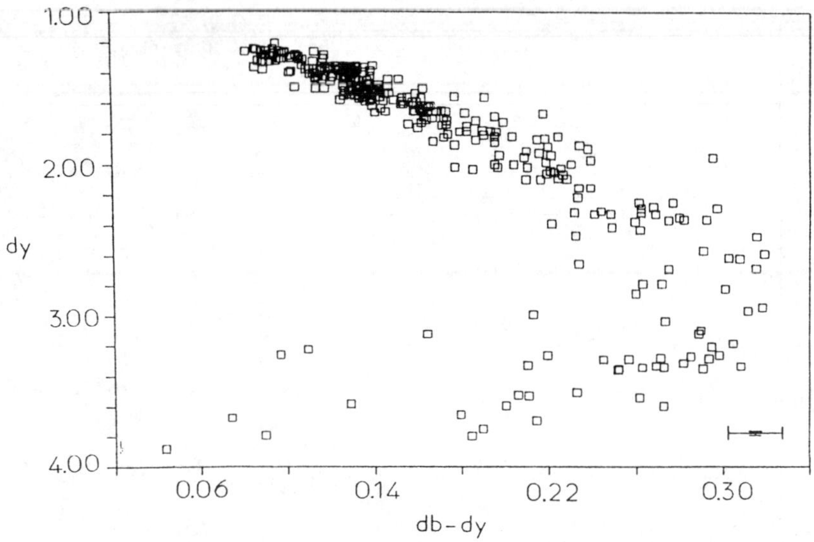

Fig. 1. The colour-magnitude diagram of UX Ori showing the blueing effect

a number of papers. The results of a general study of HAeBe stars based on the data obtained by the continuous LTPV observations in the Strömgren system were published by Bibo and Thé (1991). A continuation of this study, using more recently obtained data, is under way. From the study of Bibo and Thé (1991) it is clear that about 25% of the stars in their sample show very large brightness variations up to 3^m. They are mostly of spectral type A0 or later. These large brightness changes are irregular and completely unpredictable. Resemblance to minima of eclipsing binaries is the reason that the lowest brightnesses reached by HAeBe stars are called Algol-type minima. In the paper by Bibo and Thé (1991) many typical light-curves of HAeBe objects are shown.

The large and irregular variability of HAeBe stars is generally assumed to be caused by the obscuring effect of a dust cloud of large optical thickness orbiting in the outer region of the circumstellar disk. This disk is oriented edge on with respect to the observer. Rapid and large variations during one or two nights were observed frequently. We believe that this is caused by the fact that the screening cloud does not have a homogeneous optical thickness. The different transparencies in the cloud are causing the rapid fluctuations observed in the light curves of HAeBe stars. Furthermore, because there are many clouds of different densities orbiting around the central star, and because these densities may change in time, it is almost impossible to find periodicities in the above-mentioned light curves. It should be remarked, finally, that the small variations in brightness observed for diverse HAeBe

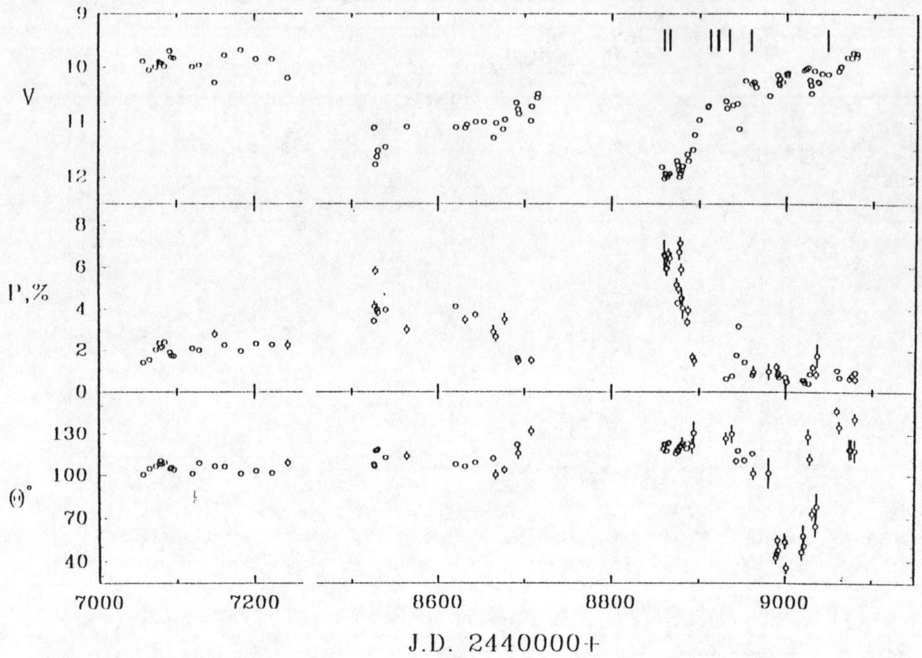

Fig. 2. Perfect correlation of brightness decrease and increase of linear polarization of the star UX Ori

stars, is most probably due to the non edge-on orientation of the disk of the stars.

From the discussion in this section it can be conluded that Herbig Ae/Be stars have circumstellar matter which is confined in a disk, and that dust grains are orbiting around the central star in the outer regions of the disk. The dust particles are forming clouds which are causing the irregular brightness variations of the HAeBe objects

3. The Blueing Effect

During the brightness variations of several HAeBe stars, it has been found for quite some time, that in the colour-magnitude relation a colour reversal occurs. When the star decreases in brightness it first becomes redder, which is a normal process, but, after reaching a certain magnitude, it reverses its redward trend and turns bluer when it dims further. In Fig. 1 the colour-magnitude diagram of UX Ori, a typical star which exhibits this effect, is shown as an example. This so-called **blueing effect** was already observed long time ago by Wenzel (1969), followed by Zatsjeva (1973) and many more authors. In the Strömgren photometric system it was observed recently by Bibo and Thé (1991).

A mechanism suggested to explain the blueing effect is based on the hypothesis that the development of a strong magnetic flux near the surface of the star is causing the large brightness drops, due to lowering of the temperature. Subsequently, the

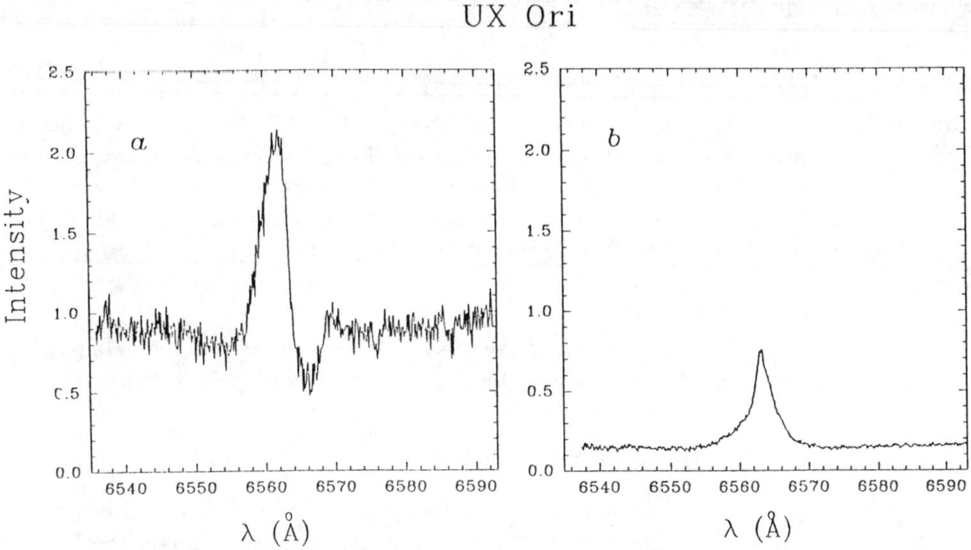

Fig. 3. The change of double-peaked Hα-emission profile into single-peaked in the spectrum of UX Ori

chromospheric regions around the star become more apparent, so that the star becomes bluer (see Herbst et al. 1983). However, there are strong objections against this mechanism, and at present it is believed that the above effect can be explained by the influence of dust clouds discussed in the previous section.

In this explanation it is accepted that the extended atmosphere of the star is indeed distributed in a disk-like structure, in which the dust particles are located on the cooler outer edge of the disk. The disk is oriented nearly edge-on to the observer. At the deepest Algol-type brightness decrease, the central star's blue light scattered by small dust particles becomes visible. The existence of the small particles in the neighbourhood of hot stars, of which the radiation tends to blow away such small dust grains, is possible when these small particles are overwhelmingly situated in and at the back of the dust cloud as seen from the central star. The masking effect of the dust cloud reduces the strength of the stellar radiation so that the small dust grains can survive in the neighbourhood of the central star. Furthermore, the gravitational pull of the dust cloud helps to protect the small dust grains from being blown away. A discussion of this matter has been published by Thé and Molster (1994).

From this discussion it is concluded that the observed blueing effect is supporting the belief that the circumstellar matter of HAeBe stars is distributed in a disk.

4. Polarimetric Observations

A direct proof that in the above explanation of the blueing effect we are dealing with scattered light, comes from polarimetric observations of isolated HAeBe objects. Since scattered light is linearly polarized, it can be expected that during brightness minima observations towards the star should show a large increase in linear polarization. Simultaneous photometric and polarimetric monitoring of several isolated HAeBe stars shows that this is indeed the case (Grinin et al. 1991). The observations show a perfect correlation between deep brightness decrease and increase of linear polarization. In Fig. 2 the results of the simultaneous observations of UX Ori is shown as an example.

Thus, it can be concluded from polarimetric observations that the blueing effect is indeed caused by small dust particles located in the disks of HAeBe stars.

5. Supporting Spectroscopic Observations

Spectroscopic observations made recently at La Silla (ESO) and at Kitt Peak (McMath Solar Observatory) have strengthened our belief that the matter in the extended atmosphere of HAeBe stars must be distributed in a disk. It is well known that at maximum brightness, when the star is not obscured by a dust cloud, the Hα emission in the spectrum of the star UX Ori clearly has a reversed P Cygni profile, indicating accretion of matter towards the stellar surface. The absorption dip in this profile is caused by the matter located between the star and the observer. When a dust cloud is obscuring the star, this absorbing matter will not be seen by the observer. The result is that, when the Hα-emission line is observed during minimum brightness, the double-peaked P Cygni profile should turn into a single-peaked one. This is indeed the case for the star UX Ori (see Fig. 3).

The observation of the Hα-emission profile during maximum and minimum brightness supports the belief that the circumstellar matter around HAeBe stars is distributed in a flat disk.

6. The Infrared Excess

A strong indication of the presence of dust grains in the outer regions of the disk comes from the near and far infrared observations. From these observations it is clear that Herbig Ae/Be stars possess excess of near and far infrared radiation. It is generally accepted now that this excess radiation is thermal re-emission of dust grains, absorbing radiation from the central star. An example of such infrared excess radiation is given for the star HD 45677 in Fig. 4.

7. Protoplanetary Discussion

In the above sections we have not yet discussed the properties of the dust clouds and the possible connection with the formation of planet-like objects from these clouds. We will now discuss the properties of the dust clouds most probably orbiting 6 well-known HAeBe objects listed in Table 1. In this table the astrophysical properties,

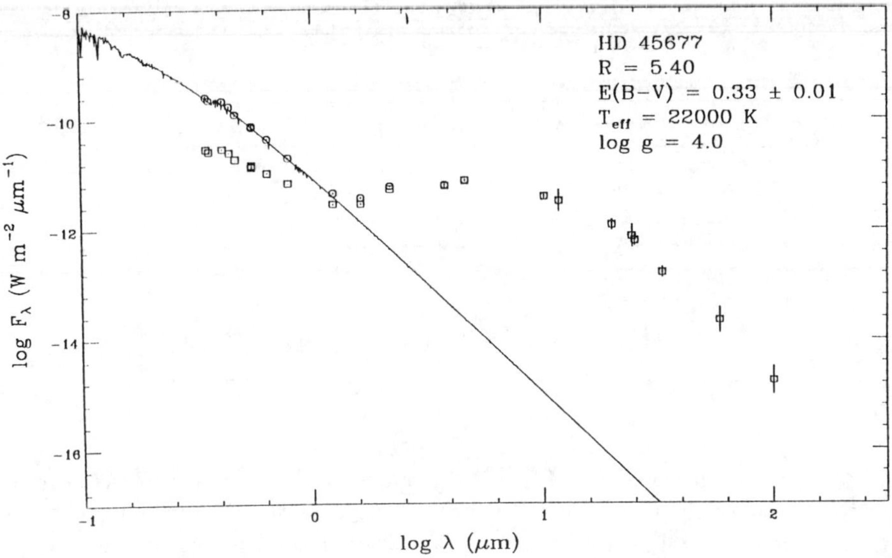

Fig. 4. The infrared excess radiation of the star HD 45677

TABLE I
Physical parameters of several HAEBE stars.

Name	SP.T.	Range in V [mag]	M_* [$M_\odot$]	R_* [$R_\odot$]	L_* [$L_\odot$]	T_{eff} [K]
UX Ori	A2 IIIe	9.6-12.3	2.8	3.3	65	9000
NX Pup	A0 IIIe	9.3-11.0	3.2	3.4	106	10100
HR5999	A7 IIIe	6.7- 8.5	2.3	3.0	29	7700
KK Oph	A5 Ve	10.3-12.0	1.9	1.9	14	8200
R CrA	B8 Ve	9.4-13.8	3.5	3.1	180	12000
T CrA	F0 Ve	11.7-14.3	1.6	1.6	6.5	7200

collected from different sources, are also given. Note that the maximum decrease of the brightness of all these stars is larger than 1^m.

Using Wien's displacement law, it is possible to estimate the average temperature of the dust grains from the maximum flux of the near-IR excess. This is the adopted temperature, T_d, of the dust cloud. Next, considering the dust particles as black bodies, it is possible to calculate, from the balance of absorbed and emitted

TABLE II
Physical parameters of the dust cloud.

Name	T_d [K]	r_d [AU]	M_{cl} [$M_\oplus$]	θ_{cl} [km/s]	P_{cl} [days]
UX Ori	1490	1.4	3.7	42	225
NX Pup	1360	2.4	1.0	34	764
HR5999	1400	1.0	6.3	45	243
KK Oph	1250	1.0	1.3	41	266
R CrA	1370	3.3	0.3	29	1237
T CrA	1310	0.5	4.4	27	185

radiation, the average distance, r_d, of the dust cloud from the central star. Assuming now that the shape of the dust cloud is spherically symmetric, a minimum critical mass M_{cl} for such a cloud, stable against gravitational tidal forces of the central star, can be calculated using Chandrasekhar's (1943) formula. Furthermore, if it is now assumed that the cloud is revolving in a circular orbit around the star, the orbital speed, θ_{cl}, and the period, P_{cl}, of the cloud can be calculated easily. The results of our estimates for the stars in Table 1 are listed in Table 2. From these results it is believed that most probably the dust clouds around Herbig Ae/Be stars are precursers of planet-like objects, such as in our own Solar System. For a general discussion of this matter see Thé and Molster (1994).

8. Concluding Remarks

The irregular and often very large variations of the brightness of Herbig Ae/Be stars indicate that they should be caused by the screening effect of dust clouds, orbiting in the outer regions of the circumstellar disk of these stars. Especially, when the disk is oriented edge-on to the observer the so-called blueing effect will take place. This is believed to be due to scattering of the light of the central star by small dust particles located at the back side of the dust cloud as seen from the central star. Direct support for this hypothesis comes from polarimetric observations, which show that the Algol-type brightness minima correlate perfectly with the large increases in linear polarization of the light of the central star.

Spectroscopic observations are also supporting the disk hypothesis. When the central star is obscured by a dust cloud the following change in the P Cygni profile of the Hα line has been observed. For the star UX Ori the double-peaked reversed P Cygni profile of its Hα line, changes to a single-peaked profile. The reason is that the obscuring cloud is not only screening the central star, but also the whole region between star and observer where the matter is located that causes the absorption dip in the Hα profile.

Simple considerations of the properties of the above-mentioned clouds obscuring the central star are also given in order to show that most probably these clouds are precursers of planet-like objects, such as in our own Solar System.

References

Bibo, E.A., Thé, P.S., 1991, *Astronomy and Astrophysics*, 89, 319
Chandrasekhar, S., 1943, *Principles of Stellar Dynamics*, Dover Publ., New York, p. 220
Grinin, V.P., Kiselev, N.N., Minikhulov, N.H., Chernova, G.P., Voshchinnikov, N.V., 1991, *Astrophysics and Space Science*, 186, 283
Herbig, G.H., 1960, *Astrophysical Journal, Supplement Series*, 4, 337
Herbst, W., Holtzman, J.A., Klasky, R.S., 1983, *Astronomical Journal*, 88, 1648
Thé, P.S., de Winter, D., Pérez, M.R., 1994, *Astronomy and Astrophysics, Supplement Series*, in press
Thé, P.S., Molster, F., 1994, Proceedings of the conference on *Planetary Systems: Formation, Evolution and Detection*, in press
Wenzel, W., 1969, in *Non-periodic Phenomena in Variable Stars*, p. 61, ed. L. Detre, IAU Colloquium No. 4
Zatsjeva, G.V., 1973, *Variable Stars*, Moscow, 19, 63

DISCUSSION

Waelkens: Are the disk masses you mention just dust masses or total—that is, gas plus dust—masses?

Thé: I talked about the dust masses, without the gas.

Delgado: Is there in the light curve some correlation between depth of the minima (or height of the maxima) and the time interval between them?

Thé: Since no periodicities in the brightness variations have been found, we can not answer your question.

Young: If the brightness variations are due to dust obscuration, the reddening near maximum light should tell something about the size distribution of the dust particles. What is the ratio of total-to-selective absorption?

Thé: The ratio of total-to-selective extinction varies from object to object. For example, in HR 5999, $R = 5.5$.

Kurtz: Has anyone tested for pulsational variability during times of maximum brightness when we might be seeing through to the photosphere?

Thé: No one has tested for this.

Harmanec: How sure are you that Herbig Ae/Be stars are not just normal Be stars which have especially strong and extended envelopes?

Thé: Herbig Ae/Be stars do have characteristics which are not typical for Be stars. For example, strong variations in the emission lines, excess infrared radiation due to thermal re-emission by dust, etc.

Harmanec: Is it not possible that your mass estimates of Be/Ae stars are incorrect because the spectral classes refer in fact to dense shells, not to the stars themselves?

Thé: It is possible, but there are other effects which make that the spectral types are in error. For instance, many of the lines are in emission. Our spectral types usually are just estimates from photometry.

Waelkens: Herbig Be stars and classical Be stars have very different energy distributions: IR excesses in the former are caused by free-free emission, in the latter by dust. There may be a generic relation, but both kinds of objects can easily be distinguished observationally.

Chochol: Do you have any indications that some of the Herbig Ae/Be stars are binaries?

Thé: Yes, there are binaries found among Herbig Ae/Be stars. Those which are certainly physically related are: AK Sco, a spectroscopic binary, and TY CrA, which is an eclipsing binary. HR 5999 has a companion, but perhaps it is not physically related.

Thanks for a nice Dinner – Waitress, Hall, Sterken

TEN YEARS OF MONITORING SOUTHERN Be STARS: RESULTS FROM THE ESO LTPV PROJECT

N. VOGT*, R. MENNICKENT
Astrophysics Group
Universidad Catolica
Casilla 104
Santiago 22
Chile

and

C. STERKEN**
Faculty of Sciences
University of Brussels
Pleinlaan 2
B-1050 Brussels
Belgium

ABSTRACT. A brief description of the brightness and colour variations of 15 southern Be stars, observed between 1982 and 1991 by means of differential photoelectric Strömgren photometry, is given. Seven stars of the sample revealed pronounced long-term variations on time scales of years to decades with amplitudes larger than $0^{m}.2$. In two of them, relatively sudden fadings by up to 1 mag in a few weeks were detected; they are interpreted as possible obscuration events by circumstellar dust clouds. Six other stars are characterized by variations on shorter time-scales with periods between 4.6 and 88 days, among them one massive eclipsing binary and one newly-detected ellipsoidal variable. Most stars of our sample show only marginal variations—or no variations at all—in the $b-y$ colour index; the UV dominated c_1 index is redder when the star is faint, and bluer when the star is bright. This implies that the amplitude of photometric variablilty is usually largest in the u band. The amplitudes of short-term and long-term variations are correlated: stars with stronger surface activities ("short-term") tend to display also major changes in their circumstellar envelopes ("long-term").

Keywords: Be stars – variable stars – photoelectric photometry

1. Introduction

The LTPV project (*Long-Term Photometry of Variables*) was initiated more than ten years ago at the European Southern Observatory, La Silla, Chile, to monitor stellar variability on time-scales of months to years (Sterken, 1983). The LTPV project was organized as an international cooperation. The measuring procedure

* also at Max-Planck-Institut für Astrophysik, Karl-Schwarzschild-Str. 1, D-85740 Garching
** Belgian Fund for Scientific Research (NFWO)

C. Sterken and M. de Groot (eds.), The Impact of Long-Term Monitoring on Variable Star Research, 41–54.

consists of differential *uvby* photometry of a programme star—using two comparison stars—with one of the smaller ESO telescopes. Each star is being monitored with a typical frequency between 2 and 14 days for about 6 months per year. All data obtained untill 1990 have been published in two data catalogues (Manfroid et al. 1991; Sterken et al. 1993), which also contain the identifications of variable and comparison stars. For further details on the general strategy of this campaign, see Sterken (1994).

Since Be stars were known to exhibit photometric variability at long time-scales, they were included in the observing list of the LTPV programme. Dietrich Baade selected 15 Be stars of special interest which have now been monitored for more than a decade. Here we present a brief summary of the main results. More detailed publications on the behaviour of each individual star are in press (Vogt et al. 1994; Vogt and Sterken, 1994), or in preparation.

2. An Overview

Although each star is characterized by its individual signature in its light curve on all time-scales covered, it is possible to classify the 15 stars roughly into five types of variablity, according to the predominant behaviour of each star:

1. Seven stars with remarkable long-term variablilty (amplitude $> 0^{m}_{.}2$, time-scales of years to decades): 27 CMa, ω CMa, HR 3237, 48 Lib, V 923 Aql, V 1294 Aql and ϵ Cap;
2. Four stars with only marginal or no long-term trends, but with random and/or periodic variations at shorter time-scales: 10 CMa (P=88 days), HR 2855 (P=72 days), HR 4074 (P = $4^{d}_{.}66$) and HD 173219;
3. One newly-detected ellipsoidal variable: HD 50123 (P=$28^{d}_{.}6$);
4. One known massive eclipsing binary with variable light curve: V 505 Mon (P = $53^{d}_{.}77$);
5. Two nearly-constant stars: λ Eri and HR 2142

3. Long-Term Variations, Obscuration

Most Be stars are variable on time-scales of years to decades with amplitudes up to $\sim$ 1 mag (Feinstein, 1975). Seven stars of our sample also display such characteristics:

3.1. HD 56014 = HR 2745 = 27 CMA = EW CMA

Photometric variations up to $0^{m}_{.}3$ at time-scales of decades (Feinstein, 1975), as well as spectacular radial-velocity variations up to 230 km s^{-1} (Baade, 1981) were reported for this star. Our photometry (Fig. 1) shows slow irregular variablility with increasing brightness towards the end of the time interval covered. The most remarkable feature in the light curve, however, is a relatively sudden colour change in February 1988 with a subsequent drop—especially in the u and v band—in March/April 1988 ($\sim$ JD 244 7250). An examination of spectrogrammes obtained at the Manuel Foster Observatory in Santiago in December 1987, and in January, March, April and October 1988 reveals weak emission at the violet flank of the

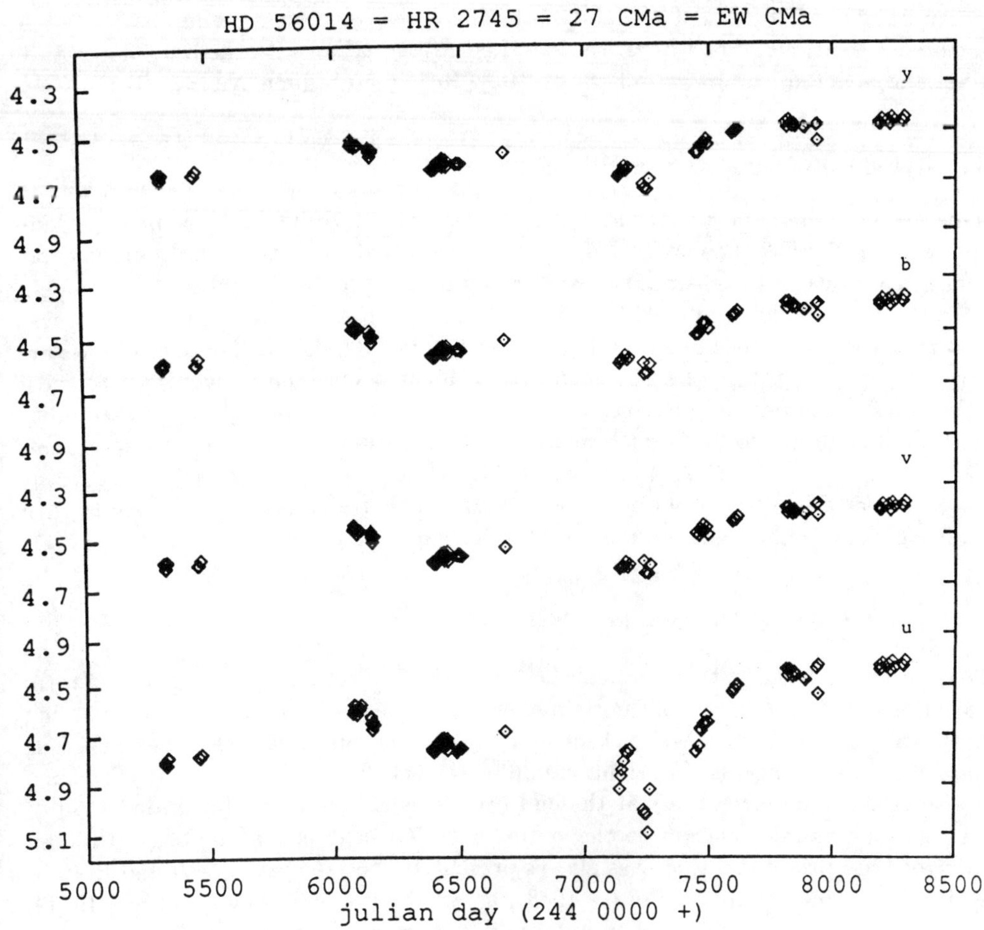

Fig. 1. Overall light curve of 27 CMa in the y, b, v and u bands (top to bottom). Note the obscuration event near JD 244 7300

Hβ absorption; the HeI and Balmer absorptions are strongly displaced to the red during the "obscuration event" that occurred in March 1988.

3.2. HD 56139 = HR 2749 = 28 CMA = ω CMA

Long-term variability with an amplitude of the order $0\overset{m}{.}3$ seems to be typical (Feinstein, 1975). Our data demonstrate that, in the bright state ($V \sim 3\overset{m}{.}6$), the activity at short time-scales (days to weeks) is much stronger than in the faint state ($V \sim 4\overset{m}{.}0$). Untill early 1986 the star was in the bright state; afterwards, within a 35-day interval (JD 244 6444–6479: January/February 1986), the star faded by $0\overset{m}{.}16$, $0\overset{m}{.}88$, $1\overset{m}{.}02$ and $0\overset{m}{.}45$ in the y, b, v, and u bands, respectively, displaying between JD 244 6479 and 6499 a faint, but constant, level. After an unfortunate gap in the observations of 635 days, the star had recovered an intermediate magnitude level and was fading slowly and continously during subsequent years.

3.3. HD 68980 = HR 3237 = MX PUP

This suspected pole-on Be star is characterized by a slowly variable V magnitude in the range $4\overset{\mathrm{m}}{.}6$ - $4\overset{\mathrm{m}}{.}9$ (Feinstein, 1975). Our data reveal variability on different time-scales with rather large amplitudes: a general tendency of brightening is present between JD 244 5200 and 6500 ($\sim 0\overset{\mathrm{m}}{.}13$ in 1300 days), and a subsequent fading was observed between JD 244 7100 and 8300 ($\sim 0\overset{\mathrm{m}}{.}22$ in 1200 days). These gradients are stronger in y than in u, the star seems to be bluer when faint, in contrast to most other Be stars in our sample. In addition, there is an apparently random variability with amplitude up to $0\overset{\mathrm{m}}{.}15$ on time-scales of 20–50 days. Since the star was faint at the beginning (1982) and at the end (1991) of the time interval covered, one may suspect a long-term period of $\sim$ 10 years. This result is also compatible with earlier observations: another faint state was seen in $\sim$ 1969–70 (Feinstein, 1975).

3.4. HD 142938 = HR 5941 = 48 LIB = FX LIB

During recent decades, this classical shell star revealed rather pronounced long-term variations of the V/R ratio of the Balmer emission lines with variable period ranging between 7 and 12 years (Mennickent and Vogt, 1991, and references therein). Our data show a slow increase in brightness up to JD 244 7200, and a subsequent decline, reaching the same brightness at the end (1991) as it had at the beginning (1983). There is a possible relation to the periodic V/R variations in the sense that an intermediate brightness level was always present at the extrema $V > R$ and $V < R$, while the brightness maximum ($\sim$ 1988), as well as possible minima (1982, 1991), coincide with symmetric emission-line profiles ($V = R$). The amplitude of the long-term brightness variations is about $0\overset{\mathrm{m}}{.}2$ in v, b and y and $0\overset{\mathrm{m}}{.}5$ in u. Apparently there is a strong variability in the Balmer jump. Finally, we observed four sudden fadings of up to $0\overset{\mathrm{m}}{.}5$ lasting 1–2 days; the significance and interpretation of these fadings is still under study.

3.5. HD 183656 = HR 7415 = V 923 AQL

This shell star is a spectroscopic binary with an orbital period of $214\overset{\mathrm{d}}{.}75$ ($K_1 =$ 6.2 km s^{-1}) which reveals long-term radial-velocity variations in the range -60 to $+20$km s^{-1} with a quasi-period of 5.8 years (Koubsky et al. 1989). Our data give indications for quasi-periodic photometric variations with a period of about 7 years (1.1 cycles covered) and an amplitude of $0\overset{\mathrm{m}}{.}07$ in y and $0\overset{\mathrm{m}}{.}25$ in u (Fig. 2). Note the remarkable phase shift between both band passes: the minimum in y was reached at JD 244 5600, while that in u was attained about 1000 days later. Both curves coincide in their maximum at JD 244 7800. In addition to this trend, there is some short-term variability of up to $0\overset{\mathrm{m}}{.}1$ in a few days time.

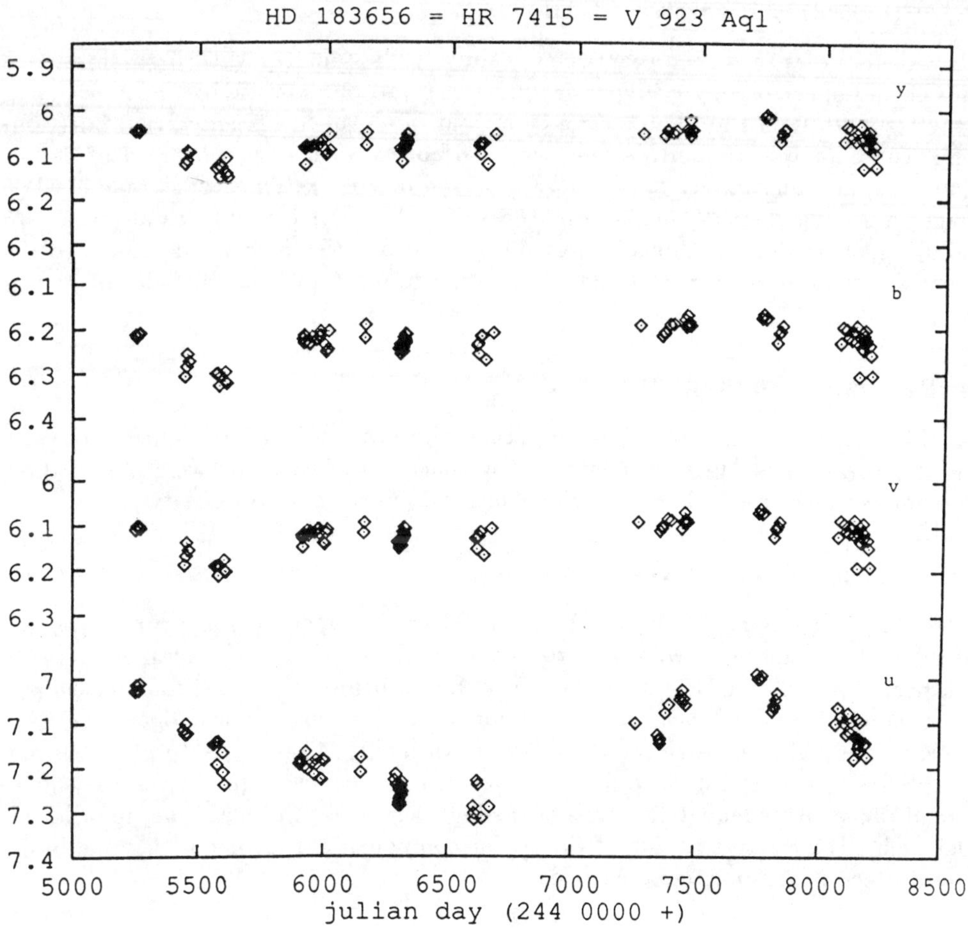

Fig. 2. Overall light curve of V 923 Aql in the y, b, v and u bands (top to bottom). Note the different form of the light curve in the u band

3.6. HD 184279 = V 1294 AQL

A very complete discussion of this shell star's spectroscopic and photometric behaviour over several decades is given by Ballereau and Chauville (1987). Short-term and mid-term variability of radial velocity, V/R ratio and V magnitude are present with a typical quasi-period of 5.5 years. Our data confirm the cyclic variability of this star with a period of about 5.9 years, covering a total of 1.4 cycles, with two relatively-sharp minima at approximately JD 244 5600 and 7800. The amplitude of this cyclic variation is $\sim 0\overset{m}{.}2$ in y and $\sim 0\overset{m}{.}35$ in u. In addition, a secular brightening of about $0\overset{m}{.}15$ (u) and $0\overset{m}{.}10$ (y) over 8 years is present. Short-term variations of up to a few hundredth of a magnitude in 10 days are occasionally present, but the star also shows periods of rather constant light.

3.7. HD 205637 = HR 8260 = ϵ CAP

Previous studies in recent years refer mainly to photometry with high time resolution and spectroscopy (Cuypers et al. 1989). To our knowledge, no systematic long-term study of the photometric behaviour of ϵ Cap has been carried out. Our light curve shows pronounced long-term variability with a quasi-period of 780 or 240 days, and an amplitude of $\sim 0\overset{m}{.}3$ in u and $0\overset{m}{.}1$ in y. The star is redder when faint. A secular decrease of the brightness of $\sim 0\overset{m}{.}1$ over 8 years was also seen. In addition, there are pronounced short-term variations which seem to show incoherent wave structures with $\sim$ 30 days cycle length and typical amplitudes of $0\overset{m}{.}1$ in all bands.

4. Periodic Variations, Binaries

In this section we describe those Be stars whose prevailing variability occurs at shorter time-scales (days to months). For nearly all of them we were able to find periodic variations which are—perhaps in most cases—due to binarity.

4.1. HD 48914 = V 505 MON

This eclipsing binary (P = $53\overset{d}{.}77$) with a B4 or B5 supergiant as primary, shows Hα emission, sometimes with a narrow central shell absorption (Boulon et al. 1974; Turner, 1976). The LTPV data cover a total of 9 primary and 6 secondary eclipse minima which enabled us to derive an improved eclipse ephemeris (Vogt & Sterken, 1994). The light curve shows considerable variability from cycle to cycle during all phases. In particular, width and shape of the eclipse light-curves vary in a broad range. An adequate analysis of the light-curves is in preparation, in order to determine the relative shapes of the stellar components, together with some basic parameters of the emitting envelope.

4.2. HD 48917 = HR 2492 = 10 CMA = FT CMA

This is an example of prominent spectroscopic and photometric variations with a period of $2\overset{d}{.}63$ days (Baade 1984, Balona et al. 1992). Our data reveal a coherent oscillation with a period of $88\overset{d}{.}5 \pm 0\overset{d}{.}3$, present throughout the entire 9-year observing time (Fig. 3). Its amplitude of about $0\overset{m}{.}08$ is nearly identical in all four bands. In addition, random short-term variability is superimposed on the nearly sinusoidal 88-day periodicity and causes most of the scatter present in the light-curves of Fig. 3.

4.3. HD 50123 = HR 2545

Although recognized as of composite spectrum and mentioned among the "bright southern stars of astrophysical interest" by Andersen and Nordström (1977), no systematic photometric study of this star has been published so far. The LTPV data revealed that it is an ellipsoidal variable with an orbital period P = $28\overset{d}{.}601$ and a radial velocity semi-amplitude $K_1 = 25 \pm 6$ km s^{-1}. The secondary seems to

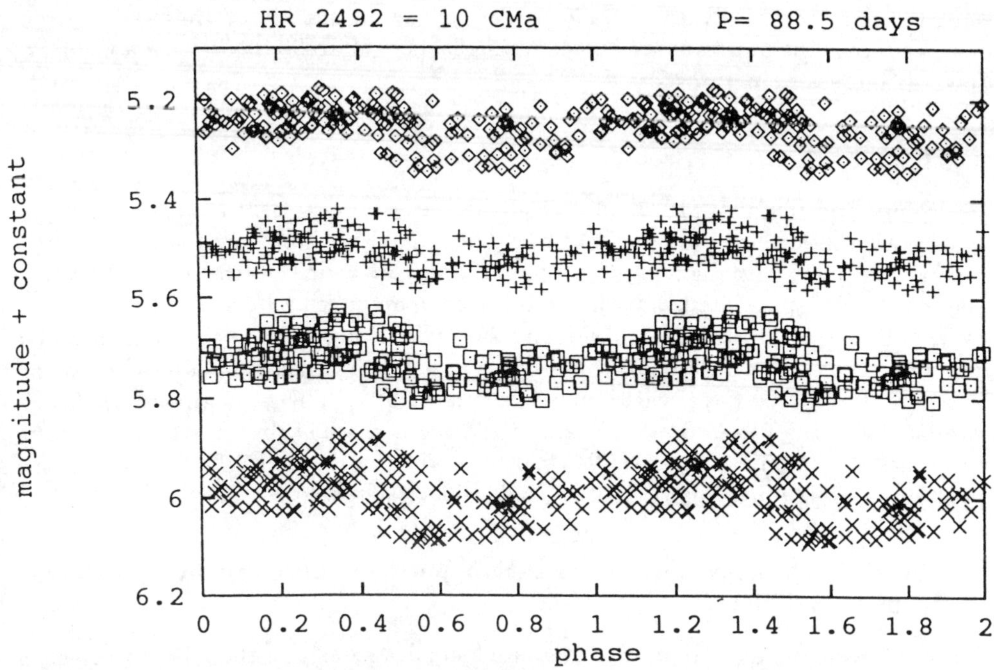

Fig. 3. Light-curve of 10 CMa, phased with the coherent oscillation period of $88\overset{\mathrm{d}}{.}5$, in the y, b, v and u bands (top to bottom). The scatter is mainly due to random variability on other time scales

be a K giant with a mass of about $1.5 M_\odot$. For details on this analysis see Vogt et al. (1994).

4.4. HD 58978 = HR 2855 = FY CMA

Only marginal long-term photometric variability within $< 0\overset{\mathrm{m}}{.}1$ was reported (Feinstein, 1975; Adelman, 1992). Our data also contain mainly erratic variations within a total range of $\sim 0\overset{\mathrm{m}}{.}1$, part of them due to coherent periodic variations with $P = 72\overset{\mathrm{d}}{.}8 \pm 0\overset{\mathrm{d}}{.}3$ and an amplitude of about $0\overset{\mathrm{m}}{.}05$ in u.

4.5. HD 89890 = HR 4074

Rapid changes in the line profiles of the HeI 4471 and MgII 4481 absorption line profiles, with periodic radial-velocity variations with $P = 2\overset{\mathrm{d}}{.}25$ (Baade 1984), as well as secular variations of these lines in their equivalent widths (Vogt et al. 1990) were reported. In contrast to these quite remarkable changes in the spectrum, no long-term variability seems to be present in our photometry. However, we detected incoherent oscillations with a period of about $4\overset{\mathrm{d}}{.}66$ days, just twice as long as the spectroscopic period given by Baade (1984). The amplitude of these oscillations reaches some $0\overset{m}{.}05$, and subsequent damping lasts $\sim$ 30 to 50 days, with a tendency towards increasing periods.

4.6. HD 173219

Hutchings and Redman (1973) identified this early-type Be star as a spectroscopic binary with an orbital period of $53\overset{d}{.}395$. They could not find any trace of the secondary spectrum although its mass should be in the range between 8 and $24 M_{\odot}$; therefore, they speculated that the invisible companion may be a black hole. However, subsequent searches for X-ray emission (Hutchings, 1981) failed to reveal a significant flux. Our data show that erratic variability is present on all time scales between 10 and 100 days, with amplitudes up to $0\overset{m}{.}1$. The y bandpass displays a secular decrease of $\sim 0\overset{m}{.}04$ in 8 years, while the u channel displays an increase in brightness by $0\overset{m}{.}05$ between JD 244 5200 and 7500, appearing $0\overset{m}{.}09$ fainter in the next season. Significant periodicities could not be detected.

5. The Colour Behaviour; Correlation between Long-Term and Short-Term Variability

Since the *uvby* bands were always measured simultaneously in the LTPV project, a detailed analysis of the colour variations as a function of brightness is feasible. As already mentioned, the amplitudes of the variations in the u channel exceed those in the remaining Strömgren bands for most stars: they appear bluer in ultraviolet bands at the time when the star is bright. In Figs. 4 and 5 we give two examples for this colour behaviour: normally, the temperature-sensitive index $b - y$ is nearly constant while the c_1 index (the difference between $u - v$ and $v - b$, a measure of the size of the Balmer jump) is related to the brightness in a roughly linear manner, with some intrinsic scatter (e.g. ϵ Cap in Fig. 4). Nearly all stars of our sample behave similar to this example. The only remarkable exception is HR 3237 (Fig. 5), where we find a small—but significant—variability of the $b - y$ index which becomes redder when the star varies from the brightest to an intermediate level ($V \sim 4\overset{m}{.}65$), and becomes bluer from this in-between level to the faintest states. Anti-correlated to this is the c_1 index, which is bluer from the brightest to intermediate magnitudes, with subsequent reddening when going to the faintest states. We do not have any obvious explanation for this anomalous behaviour of HR 3237.

Our data allow us to consider another interesting question: is there a correlation between long-term and short-term variability? As a measure of the long-term amplitude we used the total range of variation observed within the 8-year time interval covered. As short-term activity we define the standard deviation calculated from 4 subsequent data points, excluding gaps of larger than 10 days. This method was chosen in order to quantify variations in time-scales of hours to days. Not considered in both amplitudes (LTV and STV) were the obscuration events and the sudden dips observed for 48 Lib. Fig. 6 shows the correlation for the y band. The linear relation is significant to a 4.5σ level. The remaining bands give similar slopes, with slightly larger scatter. This result implies that stars with strong

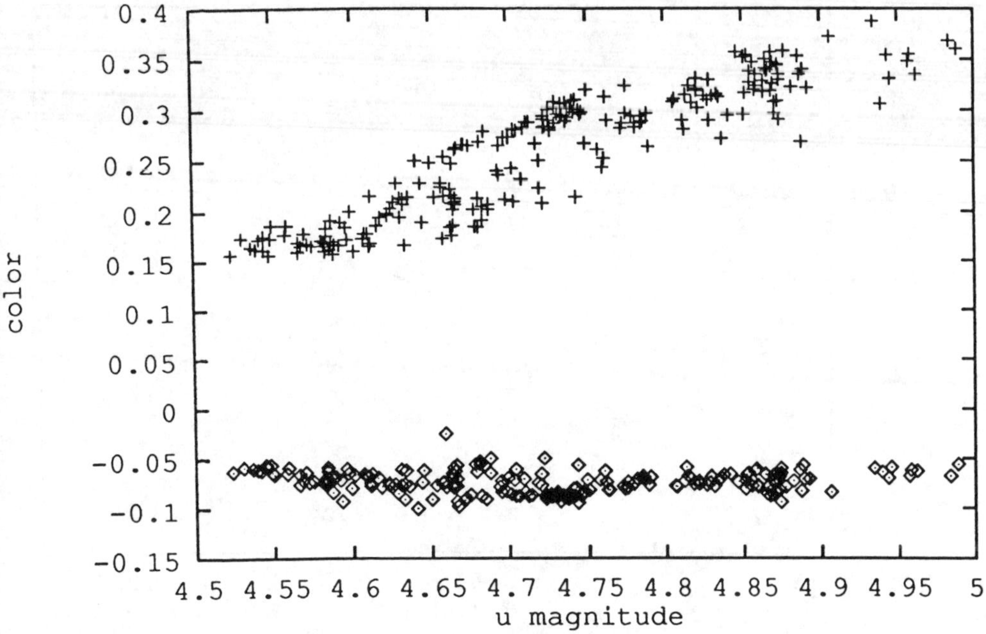

Fig. 4. $b-y$ (◇) and c_1 (+) colour indices *vs.* u magnitude for ϵ Cap. c_1 increases when the star becomes fainter

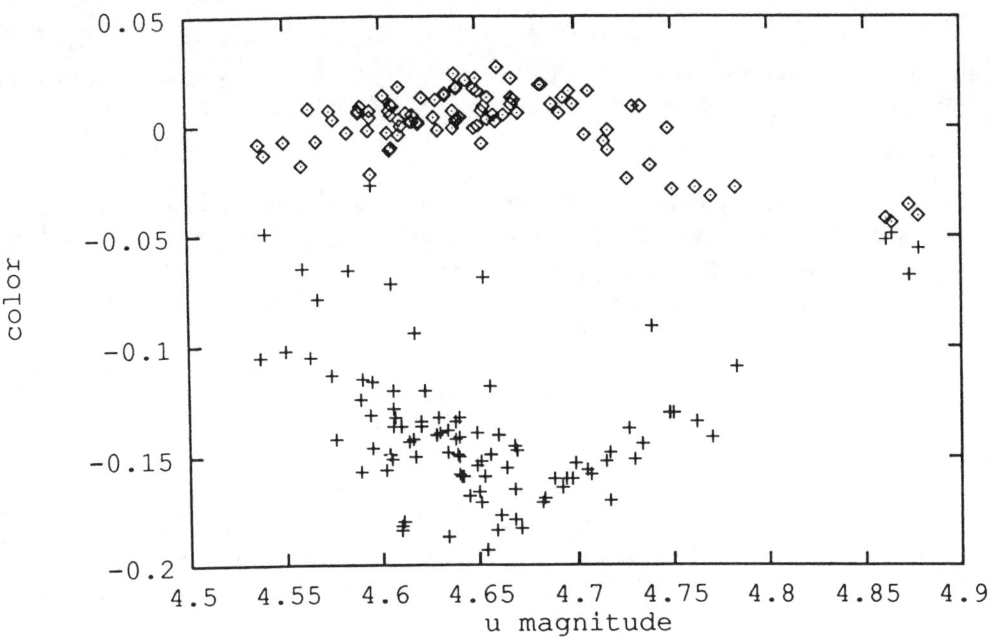

Fig. 5. $b-y$ (◇) and c_1 (+) colour indices *vs.* u magnitude for HR 3237. Note the anomalous variability in both colours, compared to other Be stars like ϵ Cap

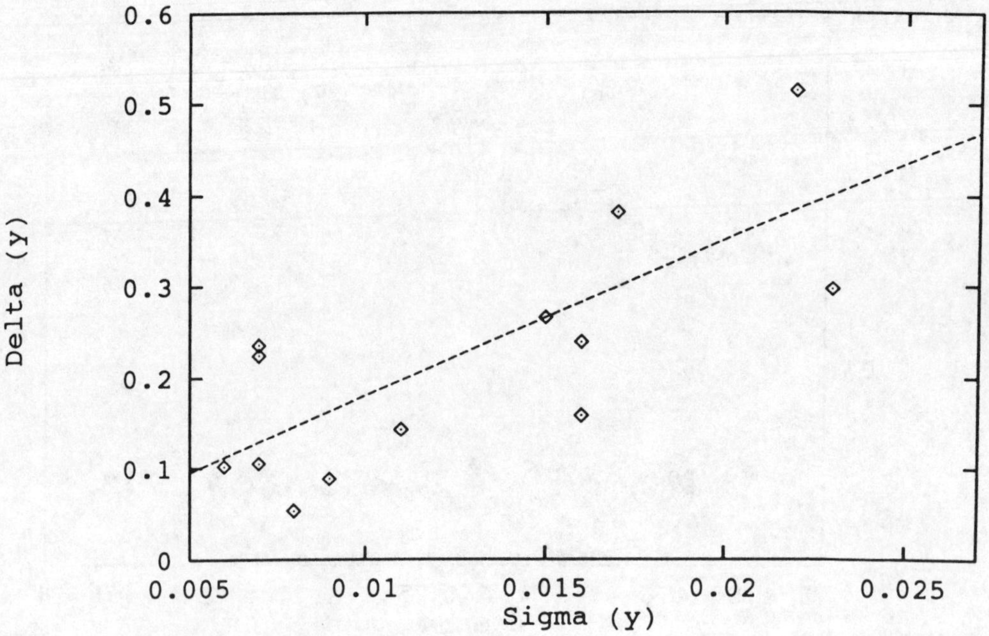

Fig. 6. Correlation between the total amplitude of the long-term variability (Δy) and the mean standard deviation of data points within intervals less than 10 days as a measure of short-term variability (σ_y) in the y band

long-term variations also tend to be more active at short time-scales, and vice versa.

6. Conclusions

Three main types of photometric variability can be distinguished in Be stars:

1. STV: Short-term variability on time-scales of hours to days, often periodic, with amplitudes of a few percent. They refer to activities on the stellar surface, such as non-radial pulsations (Baade, 1987), or rotation (for stars with inhomogeneous temperature distribution on their surface, see, for example, Balona, 1990). We had only one possible example for this kind of variation: HR 4074 with a $4\overset{d}{.}66$ transient incoherent oscillation.
2. MTV: Mid-term variability (10–100 days) which, in the case of Be stars, is often related to a binary period. There are two certain examples in our sample (V 505 Mon and HD 50123) and we suppose that also the other stable periods found in this range for 10 CMa and HR 2855 (P = $88\overset{d}{.}5$ and $72\overset{d}{.}8$ resp.) may be related to the orbital motion of a binary configuration. On the other hand, at least in two cases, (HR 2142 and HD 173219) known binary periods (81 and

58 days resp.) could not be recognized in the photometric data.

3. LTV: Long-term variability on time-scales of several hundred days to decades. Photometric and spectroscopic variations are often related, especially the V/R ratio of the Balmer emission and the radial velocity of II and HeI absorption lines. We had examples of a rather close relation between photometry and spectroscopy (48 Lib, V 1294 Aql), but in most other cases this correlation is not very significant. The physical interpretation of the long-term variations is still rather uncertain: probably processes in the circumstellar envelope and/or interactions between envelope and star are involved (mass loss, stellar winds). Correlated variations will give valuable hints on changes in structure, size or other parameters of the circumstellar disk.

There is a significant correlation in amplitudes of STV and LTV: relatively strong surface activities in Be stars imply also larger variations in their circumstellar disks.

Photometric variations on all time scales (LTV and STV) are, in most cases, strongest in the u band, as compared to longer wavelengths. Most stars are bluer when bright and redder when faint. Apparently, the height of the Balmer jump is varying together with the overall photometric flux. Only HR 3237 has shown a peculiar colour behaviour of unknown origin.

In two cases (27 and ω CMa) we were able to observe sudden fadings of up to 1 mag at time-scales of a few weeks. Both stars are characterized by low $v \sin i$ values and, likely, are viewed nearly pole-on. We interpret these behaviours as obscuration events, and speculate that part of the ejected material may condense into dust which agglomarates in circumstellar clouds—even outside the equatorial plane—and occasionally moves into the line of sight causing these obscuration and reddening events. The simultaneous displacement of the Balmer absorption lines towards the red (in 27 CMa) may indicate that a large fraction of the ejected gas is falling back onto the stellar surface rather than escaping from the star.

Our long-term monitoring approach has increased considerably our knowledge of Be-star variablilty; however, there are still significant gaps in the coverage. In order to resolve the rather complex variability of Be stars, one should strive to get a more homogeneous coverage in time than we have acquired so far.

Acknowledgements

We are indebted to Dr. Dietrich Baade for his collaboration in the early stages of the LTPV project, and especially for selecting the target stars and for supervising the first observations. This research has made use of the Simbad data base, operated at CDS, Strasbourg, France. N.V. and R.M. acknowledge the support by the Chilean "Fondo Nacional de Ciencia y Tecnologia" (Grant FONDECYT 637/91). C.S. acknowledges financial support from the Belgian Fund of Scientific Research (NFWO).

References

Adelman, S. J., 1992, *Publications of the ASP*, 104, 392
Andersen, J., Nordström, B., 1977, *Astronomy and Astrophysics, Supplement Series*, 29, 309
Baade, D., 1981, in *Workshop on Pulsating Stars*, Eds. G.E.V.O.N. and C. Sterken, Nice Observatory, 267
Baade, D., 1984, *Astronomy and Astrophysics*134, 105
Baade, D., 1987, in *Physics of Be stars*, IAU Coll. 92, Eds. A. Slettebak and T. P. Snow, Cambridge University Press, 361
Balona, L. A., 1990, *Monthly Notices of the RAS*, 245, 92
Balona, L. A., Cuypers, J., Marang, F., 1992, *Astronomy and Astrophysics, Supplement Series*, 92, 533
Ballereau, D., Chauville, J., 1987, *Astronomy and Astrophysics, Supplement Series*, 70, 229
Boulon, J., Doazan, V., Letourneur, N., 1974, *Astronomy and Astrophysics*, 33, 473
Cuypers, J., Balona, L. A., Marang, F., 1989, *Astronomy and Astrophysics, Supplement Series*81, 151
Feinstein, A., 1975, *Publications of the ASP*, 87, 603
Hutchings, J. B., 1981, *Publications of the ASP*93, 55
Hutchings, J. B., Redman, R. O., 1973, *Monthly Notices of the RAS*, 219
Koubsky, P., Gulliver, A. F., Harmanec, P., Ballereau, D., Chauville, J., Graf, T., Horn, J., Iliev, L. H., Lyons, R. W., 1989, *Bulletin of the Astronomical Institutes of Czechoslovakia*, 40, 31
Mennickent, R. E., Vogt, N., 1991, *Astronomy and Astrophysics*, 241, 159
Sterken, C., 1983, *The Messenger*, 33, 10
Sterken, C., 1994, this conference
Manfroid, J., Sterken, C., Bruch, A., Burger, M., de Groot, M., Duerbeck, H.W., Duemmler, R., Figer, A., Hageman, T., Hensberge, H., Jorissen, A., Madejsky, R., Mandel, H., Ott, H.A., Reitermann, A., Schulte-Ladbeck, R.E., Stahl, O., Steenman, H., vander Linden, D. and Zickgraf, F.-J.: 1991, *Astronomy and Astrophysics, Supplement Series*, 87, 481.
Sterken, C., 1983, *The Messenger*, 33, 10
Sterken, C., 1994, this conference
Sterken, C., Manfroid, J., Anton, K., Barzewski, A., Bibo, E., Bruch, A., Burger, M., Duerbeck, H.W., Duemmler, R., Heck, A., Hensberge, H., Hiesgen, M., Inklaar, F., Jorissen, A., Juettner, A., Kinkel, U., Liu Zongli, Mekkaden, M.V., Ng, Y.K., Niarchos, P., Püttmann, M., Szeifert, T., Spiller, F., van Dijk, R., Vogt, N., and Wanders, I. 1993, *Astronomy and Astrophysics, Supplement Series*, 102, 79
Turner, D. G., 1976, *Astrophysical Journal*210, 65
Vogt, N., Sterken, C., 1994, *Information Bulletin on Variable Stars*, 3958
Vogt, N., Barrera, L. H., Navarro, M., 1990, *Astrophysics and Space Science*, 173, 145
Vogt, N., Mennickent, R. E., Sterken, C., 1994, *Astronomy and Astrophysics*, in press

DISCUSSION

Sterken: 27 CMa is a Be star where β Cephei-type pulsations have been observed recently. In how far these pulsations have anything to do with the shifts in the light curves that you showed is not clear to me. One could analyse the data in the instrumental system, and be less dependent on the transformation errors. I wonder whether the scatter seen in the $\Delta y/\Delta(b-y)$ diagram shown by Thé for the reddest colours is not due to transformation errors?

Burki: Unfortunately, your observations are affected by gaps in time. The complete description of the variability of these stars will only be achieved by organising long-term, high-precision **and** continuous monitoring. Thus, we have to select preferentially **circumpolar** stars.

Hric: Is the double wave in the light curve of HD 50123 also present in the other passbands?

Vogt: It is only present in the y and b bands with amplitudes $0^m_.08$ and $0^m_.05$, respectively; it may be present in v with a ~ $0^m_.01$ amplitude. In u it is absent.

Chochol: I should like to discuss photoelectric photometry of the Be star V505 Mon, the object which you have mentioned in your talk. We were able to discover the eclipses in this binary ($P \sim 54$ days) because we have observed it also at the time when the Full Moon was located in the sky just a few degrees from this object. The eclipses occurred at the time of Full Moon. Normally, there are not that many photoelectric observations under such conditions.

Kurtz: Is the ellipsoidal variability of HD 50123 plausible theoretically, given the $28^d_.6$ period?

Vogt: Yes, it just fills a gap between the ellipsoidal variable with the longest period known (227 days; R CrB) and all other cases with periods shorter than 6 days. There should be more cases like this, but they are difficult to detect, unless one does LTM.

Hall: Actually, there is not a gap in orbital period among the ellipsoidal variables. Somewhat recently I published a paper which analysed the light curve of 13 ellipsoidal variables (1990, *Astronomical Journal*, 100, 554), all 13 of which have orbital periods falling within the "gap" between your longest period and your next-to-longest period.

Vogt: Yes, I agree that this "gap" should not exist in reality. My statement was based on an older paper by Beech (1985, *Astrophysics and Space Science*, 117, 69).

Demircan: The short-term variations from hours to days with amplitudes of ~ $0^m_.01$ you mentioned for Be stars are observed in most of the standard stars. Such variations cannot be eliminated by simple differential observing. The origin of such variations could well be the variable extinction of our own atmosphere. Did you consider this point?

Vogt: Yes, I agree, but all amplitudes that we consider as significant variations are larger (at least $0^m_.03 - 0^m_.05$). At the same time, our comparison stars show a long-term stability of less than $0^m_.01$.

Hensberge, Harmanec, Delgado, Le Contel, Breger, Szabados

LONG-TERM MONITORING OF THE COMPLEX VARIATIONS OF Be STARS

P. HARMANEC
Astronomical Institute
Academy of Sciences
251 65 Ondřejov
Czech Republic

ABSTRACT. This paper discusses several systematic long-term monitoring programmes which should help to find answers for the long-standing questions: *What causes the Be phenomenon?* and, *Why are Be stars so variable on several different timescales?* Present knowledge and some of the new findings about the long-term variations and secular changes of periodic rapid variations are discussed in detail. It is argued that, since the amplitudes of all known types of variations are rather small, the utmost care must be paid to standardization of observing procedures at different observatories and to careful and uniform data reduction. Some practical hints to achieve these goals are discussed.

Keywords Be stars – variable stars

1. Introduction

This review is *not* intended to give an overview of the complex problem of Be-star research and its recent achievements. The reader interested in a deeper insight into it is referred to the following conference proceedings and books: Slettebak (1976), Jaschek and Groth (1982), Underhill and Doazan (1982), Kogure and Hirata (1982), Harmanec and Pavlovski (1983), Slettebak and Snow (1986), Baade (1991), and Balona et al. (1994). Here, I shall only discuss a few interesting problems which could—and should—be addressed by systematic long-term spectral, photometric and polarimetric observing programmes. The choice of topics naturally reflects my personal preferences and puts a special emphasis on some of the more technical aspects of the problem.

Be stars belong to the most variable objects among hot stars and the history of their spectral investigation is as old as stellar spectroscopy itself. There is now a rather general consensus that the Balmer emission in the spectra of Be stars (which defines them as a class) originates in extended "mantles" around them, which are large in comparison to the B-star radii. However, the consensus stops right here. In various current concepts, the mantles are presented as everything from spherical envelopes to very thin and flattened equatorial disks, helmet-type loops maintained by local magnetic fields or even accretion disks in binaries.

C. Sterken and M. de Groot (eds.), The Impact of Long-Term Monitoring on Variable Star Research, 55–72.

The cause of the Be phenomenon is not known. All existing single-star models assume that the Be envelopes originate from material which somehow flows out from the star while some of the binary models assume that the Be envelope is formed by material flowing into the Be star from a companion.

What has made the Be stars so attractive for observers is their remarkable—and still mostly unpredictable—variability. It is this variability which—at the same time—represents a challenge even for the most successful static structural models of Be envelopes. The most pronounced spectral and light variations of Be stars are long-term, with characteristic time-scales of decades or at least several years. This, of course, also complicates the situation since it is conceivable that some characteristic patterns of variability—or even the regularity of some events—could only be recognized by future astronomers a few centuries later—if they still care...

In my view, the unsolved problems concerning Be stars include:

• origin of the envelopes, their geometry, (in)homogeneity, dynamics and associated variability;

• co-existence of hot and cool gas (relative to photospheres of underlying stars) with highly correlated variability patterns; and

• nature of periodic light and line-profile variations which occur on a time scale of $0\overset{d}{.}1$–1 d and their location (in or above the stellar photospheres).

Here, I shall try to indicate how systematic long-term monitoring programmes can contribute substantially to the ultimate solution of some of these problems.

2. Long-term spectral, light and colour changes and their interrelation

2.1. OBSERVATIONAL EVIDENCE

2.1.1. Data collection

Because of the associated time-scale, any serious investigation of the long-term variations of Be stars requires dedicated systematic observations lasting many years.

Spectroscopy: During the first half of this century, several such spectroscopic programmes were started and led indeed to the recognition of certain characteristic patterns of spectral changes. We owe very much to people like Drs. Maury, Merrill, McLaughlin, Curtiss, Struve, Gerasimovič, Heard, Miczaika and Burwell—just to mention some of the most active researchers. Merrill and Burwell (1933, 1943, 1949, 1950) provided the community with an invaluable information service and bibliography of all known Be stars. (This was very much a pioneering undertaking at that time. The bibliographic service was continued by Jaschek et al. (1971) and since 1980 through the Be Star Newsletters. The Jascheks later initiated the creation of the Strasbourg Data Center. The Be Star Newsletter will soon be replaced by an electronic news bulletin accessible via e-mail.)

After World War II, systematic patrol observations of many northern Be stars were continued at Haute Provence Observatory (following an initiative of Dr. Herman, see especially Hubert and Hubert 1979), David Dunlap Observatory (Copeland and Heard 1963), Perkins Observatory (Slettebak 1982) and at the Crimea (cf., e.g., Boyarchuk 1957). In 1972, systematic high-dispersion spectroscopic studies of Be stars became the main observing programme of the 2-m telescope in Ondřejov.

Kogure and Hirata began such studies at the Okayama Observatory and Dachs and his collaborators at La Silla. Coordinated optical and far-UV studies of long-term changes have been promoted especially by Doazan and her collaborators, and more recently by Peters and Bjorkman. For several years, Barker monitored spectral changes at Hα with an intensifier-dissector-scanner at the University of Western Ontario. Peters has taken over this patrol service, using a CCD detector at KPNO. In recent years, the number of multi-site multi-wavelength observing campaigns has been growing steadily. With apologies for all possible inadvertible omissions, I have to refer the readers to the *Be Star Newsletters* for references and details of all of these, and many other, observing efforts.

Photometry: Long-term light and colour variations of Be stars have not been documented so well. The reason is that their amplitude at optical wavelength only rarely exceeds $0^{\mathrm{m}}.1$-$0^{\mathrm{m}}.2$, which requires photoelectric measurements. Early, more systematic, studies of Be-star light variability were carried out by Lynds (1959, 1960), Hill (1967), Feinstein (1968, 1970, 1975), Sharov and Lyutyj (1972, 1976) and North and Olofsson (1977).

In 1972, a 0.65-m reflector equipped with a *UBV* photometer was put into service through a collaborative effort of Czech and Croatian stellar astronomers at Hvar Observatory. Systematic monitoring of light and colour variations of bright Be stars has been chosen as the main observing programme for this telescope. The first results were published by Harmanec et al. (1980). Soon afterwards, Harmanec, Horn and Koubský proposed and organized a long-term campaign on systematic photoelectric observations of all bright northern Be stars (cf. Harmanec et al. 1981, 1982 and *Progress Reports* in the Be Star Newsletters). They compiled and five times updated a *Be Star Observing Program* which contained a list of all Be stars and their **obligatory** comparison and check stars, together with their coordinates, *UBV* magnitudes, spectral types and available knowledge about their variability. These lists were distributed among several tens of interested observers and, obviously, have been used quite widely (no matter that their usage was not always acknowledged). A similar observing programme for southern Be stars was later prepared by Stagg (1983).

In the mid-seventies, photoelectric monitoring of Be stars was also started by Percy at Kitt Peak and Toronto (cf., Percy and Lane 1977, Percy et al. 1981) and for southern stars by Dachs at La Silla (Dachs 1982), among others. Percy later became a very active participant in the observing campaign. Although his main interest was in rapid changes, he contributed very significantly also to monitoring of the long-term changes and to promoting such studies among other observers (see Percy et al. 1988, Percy and Attard 1992 for summaries of Toronto results). A large body of first-quality photoelectric observations was accumulated in the eighties by Balona and collaborators in Sutherland (see, e.g., Balona and Engelbrecht 1986, Cuypers et al. 1989, Balona et al. 1992). Regrettably—since Balona's main interest was once more in rapid changes—most of his observations were limited to the *b* filter. Consequently, to get information about long-term changes from his data is not always straighforward. A very important contribution came from studies by Alvarez and Schuster (1982) and Schuster and Alvarez (1983) who published and analyzed 13-colour photometry of many Be stars. Further data have been contributed through a collaborative effort of French and Mexican astronomers (Alvarez, Ballereau, and Sareyan with collaborators) and in the course of the ESO long-term campaign (Vogt, this volume).

A number of detailed studies of individual Be stars resulted from the Hvar photometry of Be stars. However, the publication of the whole body of data was delayed for reasons mentioned later, and has only now become available (Harmanec et al. 1994, Pavlovski et al. 1994).

2.1.2. Long-term spectral changes

These were classified already by McLaughlin (1961) into two categories: E/C and V/R & RV changes. The E/C (emission-to-continuum) changes have, in fully developed cases, the character of cyclic transitions

$$B - Be - Be(H shell) - Be(H + metallic\, shell) - Be - B$$

and occur on a time-scale of several years to decades. The V/R (violet-to-red) variations are slow variations in the relative intensity of the double emission peaks of the Balmer lines which *sometimes* occur during the Be (Be shell) phases. They are usually accompanied by radial-velocity (RV hereafter) variations of *the whole emission line* for several tens of $km\ s^{-1}$ which are in phase with the V/R changes. Copeland and Heard (1963) concluded that a typical time-scale of V/R changes is about 6 years. In certain cases the V/R and RV variations were so regular over a number of cycles that the respective stars were misinterpreted as spectroscopic binaries. More typically, however, the V/R variations die out after 3-4 cycles of unequal length and amplitude.

Well-observed cases of long-term spectral changes include γ Cas (Doazan et al. 1983, 1984a, Horaguchi et al. 1993), BU Tau (Gulliver 1977, Kogure and Hirata 1982), V832 Cyg (=59 Cyg; Barker 1982, Doazan et al. 1985, 1989), ζ Tau (Deplace 1970, Harmanec 1984a, Mon et al. 1992), V744 Her (=88 Her; Doazan et al. 1982a,b), 4 Her (Hubert 1971, Koubský et al. 1994), V923 Aql (Koubský et al. 1989), V1294 Aql (Horn et al. 1982, Ballereau and Chauville 1987), and o And (Gulliver et al. 1980), among others.

A more recent, and certainly important, finding is that strong resonance *absorption* lines of C IV observed in the far-UV spectra of Be stars also exhibit pronounced long-term changes (Doazan et al. 1984b, 1985, 1989).

The percentage of Be stars among B stars used to be quoted as 10% in the textbooks. However, this figure has been growing steadily, and may now increase quite dramatically in connection with the extensive use of electronic detectors for stellar spectroscopy. Coté and van Kerkwijk (1993) reported the discovery of six new Be stars. Waters et al. (1993) found clear evidence of hydrogen emission in the IR Brα - Brγ lines of the known "strong-wind" star τ Sco where Hα appeared as an absorption line. Many similar findings will, no doubt, come in the near future. *This naturally raises the question of whether or not the Be phenomenon could be a more pronounced manifestation of a normal state of the dynamical atmosphere of any normal B star.*

2.1.3. Long-term light variations

Systematic photometric studies of Be stars soon showed that most of them—if not all—are light variables. Yet, until recently, several bright Be stars were on the list of recommended photometric standards, published, e.g., in the Nautical Almanac (for ζ Oph it is still so). Curiously, the light variations of V1294 Aql are well documented exactly because of its inclusion among photometric standards (cf., e.g., Horn et al. 1982).

Ten years ago I pointed out that the observed long-term variations of Be stars can, in principle, be classified into two categories (Harmanec 1983):

- light curves reminiscent of mild cases of nova outburst (e.g., γ Cas; Doazan et al. 1983), and
- light curves which are the inverse of nova light curves, i.e. reminiscent of light curves of R CrB stars (e.g., BU Tau: Sharov and Lyutyj 1992; V744 Her: Harmanec et al. 1978).

In each observed case, the initial "departure" from a "normal" brightness was much steeper than the subsequent (often non-monotonic) recovery. There is no obvious correlation between the spectral type of the underlying star and the type of the light variation. However, the overall character of variation *always remains the same* for particular stars for which more than one cycle of such light changes were observed, despite cycle-to-cycle variations in the shape and duration of the light curve. The first type of light curve occurs more frequently than the other.

2.1.4. Correlation between light, colour and spectral changes Thanks to all observational efforts (only partly mentioned here) we can say with confidence that the long-term spectral and photometric variations of certain Be stars are correlated and apparently reflect changes in the circumstellar envelopes. The understanding of the character of this correlation was greatly improved by the studies of North and Olofsson (1977), Dachs (1982), and Hirata (1982), among others. Here, I would like to elaborate on a scheme which I proposed earlier (Harmanec 1983):

Depending on the type of light variations (as described above), one can recognize two types of correlation: *positive correlation* and *inverse correlation* between the brightness and emission strength.

The positive correlation means that the transition from the B to the Be phase is accompanied by a rather steep increase in the optical brightness of the object. At the same time, the accompanying colour variations are indicative of a change of the luminosity class of the object from V to I in the $U - B/B - V$ diagram.

The inverse correlation means that the initial increase in the emission strength is accompanied by a light decrease; in the $U - B/B - V$ diagram, the object moves down along the main sequence towards later spectral subclasses.

For both types of correlation, however, the subsequent development of the shell spectrum—if it occurs during the E/C variations—is accompanied by another light decrease or at least by a "standstill" in the light curve. This is again reminiscent of the humps seen in the light curves of novae in the phase of development of the nebular spectrum.

Typical examples of the light curves corresponding to positive and inverse cor-

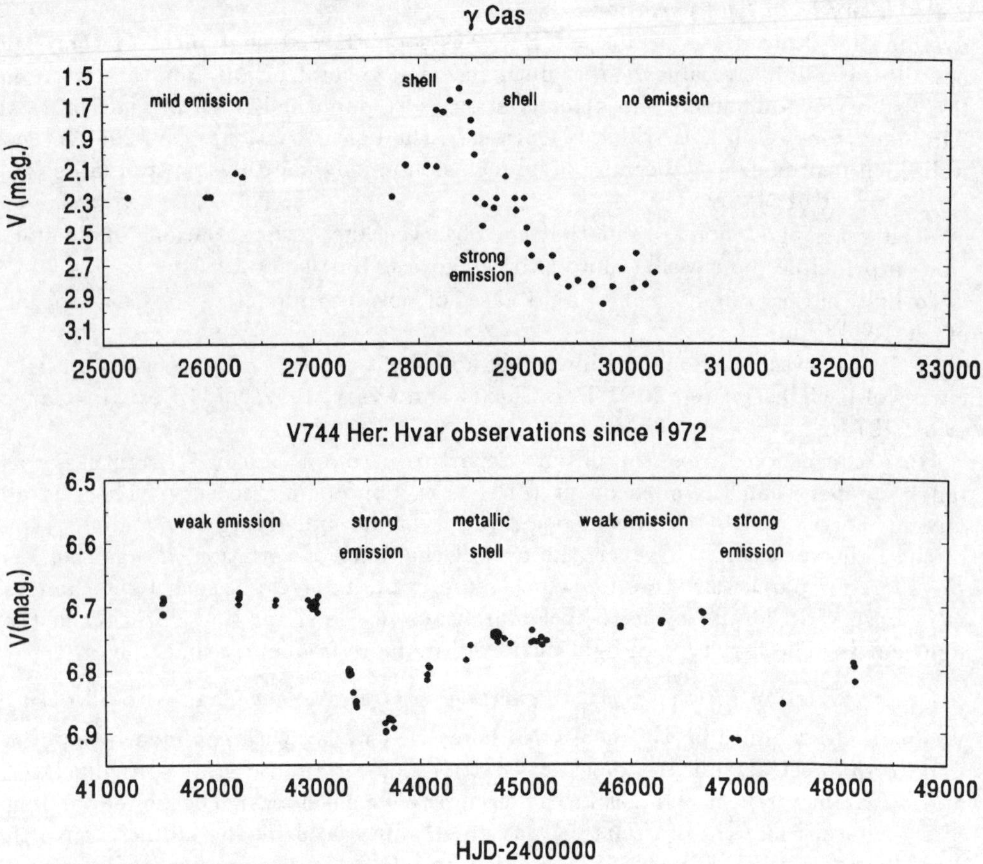

Fig. 1. Long-term light variation of γ Cas (after Doazan et al. 1983) and V744 Her (based on Hvar observations: Harmanec et al. 1978 and Pavlovski et al. 1994). The corresponding spectral changes are also indicated. These two stars are typical examples of positive and inverse correlation, respectively, between the emission strength and brightness—see text

relations are shown in Fig. 1.

Horn et al. (1982) have demonstrated that for three different stars which show the inverse correlation, BU Tau, V744 Her and V1294 Aql, the timescale of the associated light variations is remarkably similar—in spite of the fact that the spectral types of these objects are B8, B6, and B0.5, respectively. This fact certainly indicates something important about the process of formation of the Be envelopes.

I also wish to add that the above-mentioned two types of correlation represent "pure" cases. Thus, for instance, Juza et al. (1994), investigating long-term spectral and photometric variations of κ Dra, found that the strenghtening of the emission started as an essentially positive correlation, being accompanied by an increase in brightness. However, at a certain moment, the light began to drop to a local minimum which roughly correlated with the maximum emission, as if the star "switched" to an inverse correlation. Does this observation invalidate the very existence of the two types of correlation just described? The answer is *no* as I shall try to explain below.

In passing, I want to add that for κ Dra an increase of the emission strength is also accompanied by an increase of continuum polarization (Alvarez et al. 1990).

2.2. INTERPRETATION

When calling attention to the two types of correlation, I also suggested their very simple interpretation, namely an aspect effect of flattened Be envelopes of variable dimensions and/or optical thickness.

Let us assume that a Be envelope is a disk-like structure, which has the largest radius and optical thickness in the equatorial plane of the star and which extends continuously from the stellar photosphere. Let us assume further that for some unspecified reason, the dimensions and/or optical thickness of this envelope vary slowly with time. This also means that the inner parts of the disk, close to the stellar surface, may become optically thick for continuum radiation and act as a "pseudophotosphere" which is, however, cooler than the photosphere of the underlying star.

Let us consider the situation when the optical thickness and dimensions of the envelope are increasing with time. One may expect that the Balmer emission will become stronger at the same time. If one observes such an envelope roughly equator-on, then its densest equatorial parts will partly shield the stellar radiation, thus causing a light decrease and reddening of both $B-V$ and $U-B$ indices due to an increasing contribution from the cooler pseudophotosphere of the disk. *This is exactly what one observes for the inverse correlation.*

On the other hand, if the observer is well above the equatorial plane, an increase of the optical thickness of the disk leads to a more extended pseudophotosphere. This increases the observed effective radius of the object and, consequently, its brightness and photometric luminosity class. *One then observes the positive correlation between emission and brightness.*

An important thing to realize is that if the optical thickness and extent of the disk seen under an inclination, say, $30° - 60°$, continue to increase, the disk may become optically thick also along the line of sight and the positive correlation may change into the inverse one. This naturally explains more complicated cases like κ Dra and also the temporal appearance of shell phases for stars which are not observed equator-on.

I would like to stress that this simple-minded qualitative model postulates flattened envelopes, and it is conceivable that, as soon as theoreticians construct satisfactory quantitative models of Be-star disks, the comparison of the theoretical predictions with observations along the lines outlined above could lead to a new method of estimating the inclination of the equatorial planes of individual Be stars.

A very promising development in the theory of the formation of Be-star disks is represented by the model of a wind-compressed disk suggested by Bjorkman and Cassinelli (1993) and its development through 2-D hydrodynamical calculations by Owocki et al. (1994). These calculations have demonstrated that a radiation-driven wind from rapidly rotating Be stars can indeed lead to the formation of disks around them. The model disks are flattened towards the equatorial plane of the star, but *are not detached* from the stellar photosphere and may become spheroidal enough close to the star to extend even over its polar regions. In other words, the model disks are reasonably similar to the disks postulated here to explain the two observed types of correlation between brightness and Balmer emission. Further development of the theory must, however, find some mechanism to increase the optical thickness of the disk (Owocki, priv.com.).

There is also another interesting aspect of the new model: Owocki et al. found that a part of the material in the inner parts of the disk is falling back to the star. This indicates that a firm observational distinction between *accretion disks* and *disks created by an outflow* may be even more complicated observationally than thought so far. This fact should also be kept in mind since a number of Be stars are known binaries (cf., e.g., Harmanec 1987).

3. Rapid light and spectral variations of Be stars

3.1. OUTLINE OF THE PROBLEM

After a decade of concentrated effort, many investigators have come to the conclusion that systematic long-term studies are inevitable if we really want to understand the nature of the rapid variations of Be stars.

The story started with the discovery by Walker, Yang and Fahlman (1979) of line-profile changes in the form of bumps of apparent emission and/or absorption, carried over the line profiles of the rapidly rotating O9.5e star ζ Oph from blue to red, obviously by rotation and possibly by additional velocity fields. Baade (1982) and Bolton (1982) reported periodically variable line asymmetries for two other Be stars, ω CMa and λ Eri, and interpreted them as evidence for photospheric non-radial pulsations (NRP hereafter). Bolton (1982) also reported accompanying periodic light variations of λ Eri. Harmanec (1983, 1984b) pointed out that the known periodic rapid variations (periods between 0.4 and 2 days) generally have *non-sinusoidal* light curves and suggested that the periods of these variations give a good estimate of the *rotational* periods of respective stars. This was fully confirmed by a number of studies (see, e.g., Percy 1987, Balona 1991, Harmanec 1991 and references therein). Vogt and Penrod (1983) modelled an extended series of line profiles of ζ Oph and concluded that the observed changes can be well described by two models: tilted "spokes" in a circumstellar disk and sectorial NRPs. They preferred the latter model which has now been rather generally adopted. Yet, a few investigators still object against it (for details and further references consult, e.g., Balona et al. 1994, Baade 1987, 1991 and Harmanec 1989).

The present observational evidence indicates that:

- in cases where an attempt to derive the bump-recurrence periods was carried out, they were found to be indistinguishable from the photometric period;
- in all known cases, these periods are close to the expected rotational periods of the stars concerned; however, the lack of accurate knowledge of the equatorial radii and inclinations of particular Be stars makes it impossible to prove or disprove the identity of both periods;

• multiple periods were reported from analyses of some more extended spectroscopic observations (although in some cases it was possible to show that all reported periods were in fact submultiples of one period, comparable in length to the rotational period) but there is no really conclusive proof of multiperiodicity of light variations;

• the periodic light curves vary secularly both in amplitude and shape, the range of variations decreasing even below the detection threshold at some epochs, and there are also reports of long-term variability in the amplitude of spectral changes;

• there is at least a suspicion of possible correlations between the long-term variations of the Balmer emission on the one hand, and the amplitude of either light or spectral changes on the other.

3.2. REASONS FOR LONG-TERM MONITORING OF RAPID CHANGES

The importance of the problem of rapid variations lies in the fact that, if the variations are due to NRP, they could open the way to studies of stellar interiors via stellar seismology, while, if they are due to some corotating structures, astronomers would get a powerful method of measuring the true rotational periods of Be stars. Quite possibly, these two hopes are **not** *mutually exclusive.*

It is my belief that without further—and principally new—evidence the discussions between the proponents of various concepts can go on endlessly. We especially need to derive accurate radii of individual Be stars or concentrate on objects with known radii. On the observational side, the progress may come from:

• accurate parallaxes (HIPPARCOS?);
• long-baseline interferometry when it resolves the Be-star dimensions;
• systematic long-term studies and proper analyses.

Here, I shall only remark on the last point.

1. In cases where rapid light variations of a particular Be star have been observed for a long enough period, it is possible to derive a very accurate linear ephemeris describing the variations: see, e.g., EM Cep (Rachkovskaya 1975, Breinhorst and Karimie 1980), LQ And (Harmanec et al. 1991), or η Cen (Štefl et al. 1994). The same could be carried out for many other well-observed Be stars. In contrast to this, period analyses of the rapid spectroscopic variations are usually based on observing runs of a few days only. In the best available studies, certain periods were reported which were repeatedly detected in various short data strings. However, the large gaps between individual spectral series do not allow a unique determination of period(s) of spectral changes, comparably accurate to the photometric ones. However, this is just what is needed since there is as yet no firm proof that these two are *exactly* equal to each other.

2. Our recent analysis of six years of photometry of η Cen (Štefl et al. 1994) indicated the presence of a single period of $0\overset{d}{.}6424241$, yet the light curve shows a great deal of scatter. We did our best to reveal another coherent periodicity in the residuals but we failed. The plot of the residuals versus time (see Fig. 2) indicates, however, large systematic changes in the amplitude and shape of the the $0\overset{d}{.}64$ light curve. Similar studies for other Be stars are needed since, if such

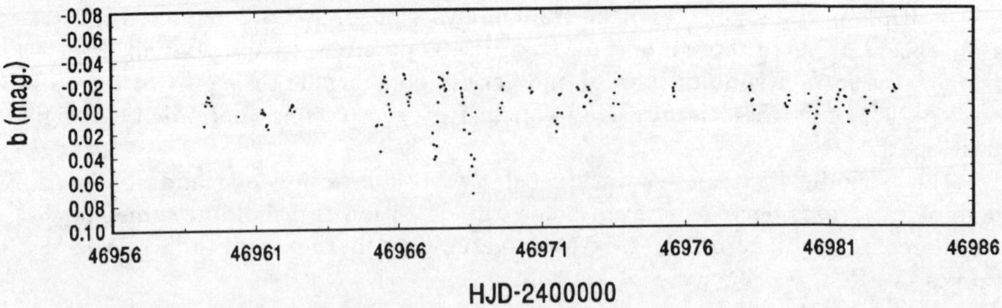

Fig. 2. The O-C deviations from the mean $0\overset{d}{.}6424$ light curve of η Cen versus time (adapted from Štefl et al. 1994).

behaviour is typical, any model attempting to interprete the rapid changes should try to explain it. In passing, I would like to mention that, if we had restricted the period analysis to a shorter string of data (say, one season), we could also describe the variations formally by several periods. That is why I suspect that the few reports of multiperiodic light variations of Be stars—for instance Pavlovski et al. (1993) for EW Lac— represent a Fourier decomposition of a single-periodic variation with variable amplitude rather than a real detection of multiperiodicity.

3. Kambe et al. (1993) found that the velocity range over which the travelling bumps were observable in the line profiles of λ Eri varied with the long-term changes of the Balmer emission. Also this finding should be verified for more Be stars since it may have important consequences for the interpretation.

4. Practical aspects of long-term observing projects

4.1. COMPLETENESS VERSUS QUALITY AND HOMOGENEITY OF DATA

Before we can make our observations from space or the Moon, it will remain very hard to obtain a good coverage of long-term variations from a single site. Long-term monitoring of *rapid changes* from one site is hopeless since their periods are of the order of one day and—apart from acute aliasing problems—some phases of ever changing light curves or profile variations remain inevitably hidden to the observer. That is why the combination of data from several observatories becomes a must.

Unfortunately, it is not always easy to combine the data obtained with different instruments and reduced by independent observers. Personally, I strongly favour the reduction of all data by one person, or at least with the same software, and along exactly defined rules. Below, I mention a few things which should be kept under control during spectral and photometric observations and data reduction.

4.1.1. Spectroscopic observations and reductions

Photographic spectroscopy: It is too late to discuss the observational strategy since the "golden age" of photographic spectroscopy has come to its end. I shall, therefore, only make two comments on the reduction.

1. Perhaps the most problematic step in the data reduction is the transformation of photographic densities into intensities. Textbooks recommend taking into account the wavelength dependence of the calibration. This is certainly true for spectrogrammes covering much more than, say, 1000 Å. Our experience at Ondřejov shows, however, that—at least for calibration spectra with, say, less than 10 steps—it is much better to neglect this wavelength dependence. Instead, we record the densities of the calibration steps at several different wavelengths in such a way as to cover a much wider range of densities and *combine the individual calibration curves—with proper shifts—numerically* by a least-squares method. In this way, one is able to obtain a more accurate mean calibration, less affected by accidental errors of the individual calibration steps. Even the errors of a mean calibration for one particular package of emulsion due to wavelength dependence are usually smaller than the errors of a single calibration based on, say, eight calibration steps. A standard and uniform processing of the plates is assumed. (We warn that standing waves, occurring in some automatic developing machines, may produce very unpleasant systematic transparency variations accross the plate.)

2. It is our experience that spectral variability is usually more clearly seen in time (or phase) plots of line intensities or half-widths than of equivalent widhts. The reason for this may lie in the fact that the equivalent width is affected by errors in both, calibration and continuum definition, and in wavelength scale. The above observation is only true, however, if one compares the results from reasonably similar dispersions. Thus, for instance, the line intensities from 4 Å mm^{-1}, and 8 Å mm^{-1} spectra can usually be compared directly while those from 10 Å mm^{-1} and 20 Å mm^{-1} spectra differ systematically.

Spectroscopy with electronic detectors This is a very rapidly developing field which was discussed by much more qualified persons at a number of recent meetings. I shall only stress a few important things:

3. Background and comparison spectra should always be obtained before and after each stellar spectrum and used in the reductions.

4. Flat-fielding is a non-trivial problem and may cause troubles. In some stable coudé spectrographs it is possible to take the flat-field spectra only once per night, but this should be tested for each particular instrument. For spectrographs which move with the telescope, frequent flat-fielding is always recommended.

5. Reticons seem still better than CCDs for studies in which an extremely high S/N ratio (say, over 1000) is needed (Walker 1987).

6. Interference can be a problem, especially for thinned CCDs. It should be avoided, even at the expense of a slightly different wavelength range used. Flat-field exposures closely mimicking the conditions of the stellar exposures may help.

7. Rectification of the spectra, too, is a non-trivial problem, common also to photographic spectrogrammes. If it is carried out interactively on a computer screen, it is always better if the whole series of spectrogrammes is rectified by one person. Yang and Walker (priv. com.) recommend first rectifying the mean spectrum of a series of spectra, approximate the deviations of individual continua from this mean by a low-degree polynomial and to use the mean continuum corrected for/by this polynomial approximation to rectify individual spectra. They claim

this is the best method to ensure a uniform rectification if one wants to trace very small details in the line profiles. If a detector with a small number of pixels ($\approx$500) is used for high-dispersion spectroscopy, care must be exercised when observing strong broad lines that a reasonable portion of continuum will be recorded on each side of the line. This problem is acute for rapidly-rotating Be stars.

4.1.2. Photometric observations and reductions

Because of the relatively small variations involved, a meaningful observational project to monitor long-term light and colour changes of Be stars must ensure a reliable transformation of data from different observatories to some standard system. In the case of the campaign which we organized via the Be Star Newsletter, most observations, including the Hvar data, were secured in the Johnson UBV system. We soon found that the usual linear transformation formulae do not give good results in $U - B$ and replaced them with a bilinear dependence on both $(U - B)$ and $(B - V)$ indices (Harmanec et al. 1977). Yet, we still obtained small but systematic colour-dependent deviations of all-sky data from season to season. We did many experiments with various transformation schemes until we finally succeeded with nonlinear transformation equations designed in such a way that each of the three extraterrestrial instrumental U, B, V magnitudes is made a cubic function of the stardard $(B - V)$, and a linear function of $(U - B)$. Harmanec, Horn and Juza (1994) re-reduced over 46000 Hvar (altitude 260 m) and about 5000 Skalnaté Pleso (altitude 1780 m) all-sky UBV observations by this technique. The seasonal UBV all-sky values for stars of all spectral types, and from both observatories, were reproduced with a precision of $0\overset{m}{.}01$. See their paper also for the details of the reduction technique. Below, I stress some important points:

8. Carefully selected and tested comparison and check stars should be obligatory for all observers participating in the observations.

9. The check star should always be observed *as frequently as* the variable and its observations should be published, at least in graphical form, since this is the best way to judge the real accuracy of each given data set and detect possible observational problems. More generally, the whole scheme of observations should be agreed upon, and then strictly followed by all observers. We, for instance, always observe the check star also before the first, and after the last observation of the variable since this allows an easier re-reduction of the data in cases when the comparison is later found to be slightly variable.

10. Since the colour properties of each telescope with a photometer (and, therefore, its instrumental colour system) vary slowly with time, it is impossible to study secular low-amplitude variations using one passpand only. Note, however, that the use of more complicated transformation formulae, which depend on two colour indices, implies that the accuracy of each standard magnitude depends on the errors of measurements in all three bandpasses (although certainly not to the same extent).

11. There is no such thing as a "neutral density" filter. If one has to attenuate the light of bright stars in the photometer, then the ideal solution is to observe *all programme and standard stars* through the same grey filter and derive the nightly and seasonal transformation coefficients for such data. If this is not possible be-

cause of the dynamical range of the photometer and/or low brightness of some of the observed stars, then it is necessary to calibrate the filter through repeated observations of suitably bright stars *at least once per night* in all individual passbands of the photometric system used. I also recommend testing possible dependence of the filter transparency (due to the band width used) on the colour of stars. I had to use this method during my recent observations with the 0.5-m reflector and *uvby* photometer at Sutherland. I found that the attenuating filter No.4 attenuates for about $2^{m}_{.}5$ in y and b, $3^{m}_{.}0$ in v and $8^{m}_{.}0$ in u. The attenuation in the two UV filters was, therefore, strongly colour-dependent and, moreover, I found slight transparency changes of the filter of about $0^{m}_{.}01$ from one night to another.

12. Drifts of the instrumental zero-point of magnitudes occur frequenly, even in such excellent sites like Sutherland or La Silla. My reduction software, therefore, also includes the terms which describe either/both linear or quadratic drift of the zero point during the night. I found that it is very easy to separate the effects of the zero-point drifts and linear extinction during the computer reduction as long as one avoids observations for which the air mass of all observed stars changes more or less monotonically with time. If this happen, a strong correlation between linear extiction and zero-point drifts occurs and it is then better to fix the extinction coefficients in the reduction of such data.

13. My software is based on multi-night reduction and the colour (second-order) extinction coefficients are included among *seasonal* extinction coefficients where they actually belong. Determination of the colour extinction coefficients each night along with linear extinction and zero-point coefficients is unwise and leads to much less accurate values.

14. Given the fact that there is a strong correlation between $B-V$ and $U-B$ for all stars on and near the main-sequence, it is necessary to use a carefully selected mixture of standard stars. Always include stars of all colours, some reddened stars, some stars of higher luminosity classes and some A and F stars. Thanks to a large number of Hvar and Skalnaté Pleso all-sky UBV observations, spanning nearly 20 years, we were also able to replace the original Johnson values (which contain accidental errors of up to a few hundreds of a magnitude) by more accurate mean values that followed from our reduction. In this way, we also derived reliable UBV values for most comparison, check and red standard stars selected for the Be-star observations. Thanks to this, both nightly and seasonal transformations become a natural by-product of our regular differential observations of Be stars, and almost no time is lost to special observations of transformation standards: virtually all all-sky observations of the obligatory comparisons and checks on good nights can now be used to transform the data. Since our comparison stars are rather evenly distributed over the northern sky, they can also be used in other observing programmes. Their new UBV values can be found in Harmanec, Horn and Juza (1994).

4.2. NEW DATA SOURCES: OBSERVING CAMPAIGNS/EXISTING ARCHIVES

I certainly do not wish to call for any "segregation" here. However, it is my experience that multi-site observing is usually more fruitful if based on a stable collaboration between observers who know each other and the advantages and limitations of the observing facilities used, than if a call for coordinated observations by anybody interested is announced in some newsbulletin.

The success of such observations also depends on a well-prepared observational strategy. The organizer of the campaign should unify in advance not only the observational scheme, program objects, needed frequency of observations, or wavelength region(s) used but also an obligatory list of all standard objects and/or calibration procedures. If this gets out of control, it is sometimes very hard to bring all data onto a comparable scale.

Obviously, the automatic photometric telescopes (APTs) will soon be very widely used for long-term programmes. I had the chance to analyze some Be-star observations from an APT currently operating in Phoenix. A realistic scatter belt of these data was something like $0\overset{m}{.}03$ but I am sure that this will improve significantly in the near future. The APTs may become especially useful in combination with CCD photometers since even some climatically less perfect sites could then be exploited for long-term monitoring of brightness changes.

We can also get a better insight into the long-term variations of Be stars through investigations of already existing archives.

1. There is a recent tendency to underestimate the quality of the old photographic data. However, I convinced myself several times that it is possible to find some variations, believed to be known only thanks to electronic detectors, in the old photographic spectra. Presently, we are collaborating with Dr. G. Scholz to employ the rich collection of Potsdam spectrogrammes of Be stars, secured at the very beginning of this century. The Ondřejov 5-channel microdensitometer, and the user-friendly software SPEFO, developed by my colleague Dr. J. Horn, permit obtaining new pieces of information (e.g., accurate emission-line velocities which could not be measured by the original investigators) from these, almost one century old, data.

2. Potentially, a great wealth of information on long-term light variations may come from the accurate photometry secured by Hipparcos. We should encourage the colleagues involved in that project to publish the data as soon as possible.

Acknowledgements

I thank the Organizing Committee and Dr. C. Sterken personally for inviting me to the workshop and for allocating funds which allowed me to participate. I am also greatly obliged to Drs. I. Hubeny, P. Koubský, J.R. Percy and G.A.H. Walker for their constructive criticism of the draft version of this manuscript.

References

Alvarez M., Schuster W.J. 1982, *Rev. Mex. Astron. Astrof.*, 5, 173

Alvarez M., Hubert-Delplace A-M., Arsenijević J., Ballereau D., Chauville J. 1990, *Rev. Mex. Astron. Astrof.*, 21, 398

Baade D. 1982, *Astronomy and Astrophysics*, 105, 65

Baade D. 1987, in Slettebak and Snow 1987, 361

Baade D. (ed.) 1991, *Rapid Variability of OB Stars: Nature and Diagnostic Value*, Garching, ESO Conf. Proc. No. 36

Ballereau D., Chauville J. 1987, *Astronomy and Astrophysics, Supplement Series*, 70, 229

Balona L.A. 1991, in Baade 1991, 249

Balona L.A., Engelbrecht C.A. 1986, *Monthly Notices of the RAS*, 219, 131
Balona L.A., Cuypers J., Marang F. 1992, *Astronomy and Astrophysics, Supplement Series*, 92, 533
Balona L., Henrichs H., LeContel J.M. (eds.) 1994, *Pulsation, Rotation and Mass Loss in Early-Type Stars*, IAU Symp. 162, Kluwer: Dordrecht (in press)
Barker P.K. 1982, *Astrophysical Journal, Supplement Series*, 49, 89
Bjorkman J.E., Cassinelli J.P. 1993, *Astrophysical Journal*, 409, 429
Bolton C.T. 1982, in Jaschek and Groth 1982, 181
Boyarchuk A.A. 1957, Izv. Krym. Astrophys. Obs. 17, 89
Breinhorst R.A., Karimie M.T. 1980, *Publications of the ASP*, 92, 432
Copeland J.A., Heard J.F. 1963, Publ. David Dunlap Obs. 2, 317
Coté J., van Kerkwijk M.H. 1993, *Astronomy and Astrophysics*, 274, 870
Cuypers J., Balona L.A., Marang F. 1989, *Astronomy and Astrophysics, Supplement Series*, 81, 151
Dachs J. 1982, in Jaschek and Groth 1982, 19
Delplace A.-M. 1970, *Astronomy and Astrophysics*, 7, 68
Doazan V., Franco M., Rusconi L., Sedmak G., Stalio R. 1983, *Astronomy and Astrophysics*, 128, 171
Doazan V., Franco M., Rusconi L., Sedmak G., Stalio R. 1984a, *Astronomy and Astrophysics, Supplement Series*, 55, 1
Doazan V., Morossi C., Stalio R., Thomas R.N., Willis A. 1984b, *Astronomy and Astrophysics*, 131, 210
Doazan V., Grady C.A., Snow T.P., Peters G.J., Marlborough J.M., Barker P.K., Bolton C.T., Bourdonneau B., Kuhi L.V., Lyons R.W., Polidan R.S., Stalio R., Thomas R.N. 1985, *Astronomy and Astrophysics*, 152, 182
Doazan V., Barylak M., Rusconi L., Sedmak G., Thomas R.N., Bourdonneau B. 1989, *Astronomy and Astrophysics*, 210, 249
Doazan V., Harmanec P., Koubský P., Krpata J., Ždárský F. 1982a *Astronomy and Astrophysics, Supplement Series*, 50, 481
Doazan V., Harmanec P., Koubský P., Krpata J., Ždárský F. 1982b *Astronomy and Astrophysics*, 115, 138
Feinstein A. 1968, *Zeitschrift fuer Astrophysik*, 68, 29
Feinstein A. 1970 *Publications of the ASP*, 82, 132
Feinstein A. 1975 *Publications of the ASP*, 87, 603
Gulliver A.F. 1977, *Astrophysical Journal*, s, 35, 441
Gulliver A.F., Bolton C.T., Poeckert R. 1980, *Publications of the ASP*, 92, 774
Harmanec P. 1983, *Hvar Obs.Bull.* , 7, 55
Harmanec P. 1984a, *Bulletin of the Astronomical Institutes of Czechoslovakia*, 35, 164
Harmanec P. 1984b, *Bulletin of the Astronomical Institutes of Czechoslovakia*, 35, 193
Harmanec P. 1987, in Slettebak and Snow 1987, 339
Harmanec P. 1989, *Bulletin of the Astronomical Institutes of Czechoslovakia*, 40, 201
Harmanec P. 1991, in Baade 1991, 263
Harmanec P., Pavlovski K. (eds.) 1983, *Workshop on rapid variability of early-type stars*, Hvar Obs. Bull. 7
Harmanec P., Grygar J., Horn J., Koubský P., Kříž S., Ždárský F., Mayer P., Ivanović Z., Pavlovski K. 1977, *Bulletin of the Astronomical Institutes of Czechoslovakia*, 28, 133
Harmanec P., Horn J., Koubský P., Kříž S., Ždárský F., Papoušek J., Doazan V., Bourdonneau B., Baldinelli L., Ghedini S., Pavlovski K. 1978, *Bulletin of the Astronomical Institutes of Czechoslovakia*, 29, 278
Harmanec P., Horn J., Koubský P., Ždárský F., Kříž S., Pavlovski K. 1980, *Bulletin of the Astronomical Institutes of Czechoslovakia*, 31, 144
Harmanec P., Horn J., Koubský P. 1981, in *Workshop on Pulsating B Stars*, Ed. by G.E.V.O.N. and C. Sterken, Nice Obs. Publ., 397
Harmanec P., Horn J., Koubský P. 1982, in Jaschek and Groth, 269
Harmanec P., Matthews J.M., Božić H., Pavlovski K., Huang L., Guo Z.-H., Percy J.R., Plume R., Ružić Ž, Wehlau W.H., Bohlender D.A., Horn J., Koubský P., Walker G.A.H., Yang S. 1991, *Bulletin of the Astronomical Institutes of Czechoslovakia*, 42, 1
Harmanec P., Horn J., Juza K. 1994, *Astronomy and Astrophysics, Supplement Series*, (in

press)
Hill G. 1967, *Astrophysical Journal, Supplement Series*, 14, 263 & 301
Hirata R. 1982, in Jaschek and Groth 1982, 41
Horaguchi T., Kogure T., Hirata R., Kawai N., Matsuoka M., Murakami T., Doazan V., Slettebak A., Huang C.C., Cao H., Guo Z.H., Huang L., Tsujita J., Ohshima O., Ito Y. 1993, *Publications of the ASJ*
Horn J., Božić H., Harmanec P., Koubský P., Pavlovski K., Žďárský F. 1982, *Bulletin of the Astronomical Institutes of Czechoslovakia*, 33, 308
Hubert H. 1971, *Astronomy and Astrophysics*, 11, 100
Hubert A.-M., Hubert H. 1979, *An Atlas of Be Stars*, Paris-Meudon Obs. Publ.
Jaschek M., Groth H.-G. (eds.) 1982, *Be Stars*, IAU Symp. 98, Reidel: Dordrecht
Jaschek C., Ferrer L., Jaschek M. 1971, *Catalogue and Bibliography of B type Emission-Line Stars*, Obs. Astron. Univ. Nac. La Plata, Argentina
Juza K., Harmanec P., Božić H., Pavlovski K., Žižňovský J., Tarasov A.E., Horn J., Koubský P. 1994, *Astronomy and Astrophysics, Supplement Series*, (submitted)
Kambe E., Ando H., Hirata R., Walker G.A.H., Kennelly, E. J., Matthews J. M., 1993, *Publications of the ASP*(November issue, in press)
Kogure T., Hirata R. 1982, *Bull. Astron. Soc. India* 10, 281
Koubský P., Gulliver A.F., Harmanec P., Ballereau D., Chauville J., Gráf T., Horn J., Iliev L.H., Lyons R.W. 1989, *Bulletin of the Astronomical Institutes of Czechoslovakia*, 40, 31
Koubský P., Harmanec P., Horn J., Hubert A.-M., Hubert H., Floquet M. 1994, in Balona et al. 1994
Lynds C.R. 1959, *Astrophysical Journal*, 130, 577 & 603
Lynds C.R. 1960, *Astrophysical Journal*, 131, 390
McLaughlin D.B. 1961, *Journal of the RAS of Canada*, 55, 13 & 73
Merrill P.W., Burwell C.G. 1933, *Astrophysical Journal*, 78, 87
Merrill P.W., Burwell C.G. 1943, *Astrophysical Journal*, 98, 153
Merrill P.W., Burwell C.G. 1949, *Astrophysical Journal*, 110, 387
Merrill P.W., Burwell C.G. 1950, *Astrophysical Journal*, 112, 72
Mon M., Kogure T., Suzuki M., Singh M. 1992, *Publications of the ASJ*, 44, 73
North H.L., Olofsson S.G. 1977, *Astronomy and Astrophysics*, 56, 117
Owocki S.P., Cranmer S.R., Blondin J.M. 1994, *Astrophysical Journal*, (April 1, in press)
Pavlovski K., Ružić Ž., Pavlović M., Jeong J.H., Nha I.-S. 1993, *Astrophysics and Space Science*200, 201
Pavlovski K., Božić H., Harmanec P., Horn J., Koubský P., Juza K., Hadrava P., Kříž S., Ružić Ž., Štefl S., Žďárský F. 1994, *Astronomy and Astrophysics*, (submitted)
Percy J.R. 1987, in Slettebak and Snow 1987, 49
Percy J.R., Lane M.C. 1977, *Astronomical Journal*, 82, 353
Percy J.R., Jakate S.M., Matthews J.M. 1981, *Astronomical Journal*, 86, 53
Percy J.R., Coffin B.L., Drukier G.A., Ford R.P., Plume R., Richer M.G., Spalding R. 1988, *Publications of the ASP*, 100, 1555
Percy J.R., Attard A. 1992, *Publications of the ASP*, 104, 1160
Rachkovskaya T.M. 1975, Izv. Krym. Astrophys. Obs. 53, 168
Schuster W.J., Alvarez M. 1983, *Publications of the ASP*, 95, 35
Sharov A.S., Lyutyj V.M. 1972, *Information Bulletin on Variable Stars*, 698
Sharov A.S., Lyutyj V.M. 1976, in Slettebak 1976, 105
Sharov A.S., Lyutyj V.M. 1992, *Astronomicheskii Zhurnal*, 69, 544
Slettebak A. (ed.) 1976, *Be and Shell Stars*, IAU Symp. 70, Reidel: Dordrecht
Slettebak A. 1982, *Astrophysical Journal, Supplement Series*, 50, 55
Slettebak A. and Snow T.P. (eds.) 1986, *Physics of Be stars*, IAU Col. 92, Cambridge Univ. Press: Cambridge
Stagg C.R. 1983, *Hvar Obs.Bull.* , 7, 143
Štefl S., Baade D., Harmanec P., Balona L.A. 1994, *Astronomy and Astrophysics*, (submitted)
Underhill A.B., Doazan V. 1982, *B and Be Stars*, NASA SP-456 Monograph, Washington, USA
Vogt S.S., Penrod G.D. 1983, *Astrophysical Journal*, 275, 661
Walker G.A.H. 1987, in *The impact of very high S/N spectroscopy on stellar physics*, IAU Symp. 132, Ed. by G. Cayrel de Strobel and M. Spite, Kluwer: Dordrecht, p. 39

Walker G.A.H., Yang S., Fahlman G. G. 1979, *Astrophysical Journal*, 233, 199
Waters L.B.F.M., Marlborough J.M., Geballe T.R., Oosterbroek T., Zaal P. 1993, *Astronomy and Astrophysics*, 272, L9

DISCUSSION

Sterken: I am a bit sceptical about the night-to-night transmission changes of your neutral-density filter. That such filters are not neutral is well-known, but they are basically like any other glass or interference filter in your optical system. Therefore, wouldn't there be night-to-night changes in the other filters as well?

Harmanec: This, obviously, could happen, but I think that what I found was another effect. It may be a temperature dependence—it was much cooler on some nights than on others. The difference was systematic for a number of calibration stars and when applied as a correction to the programme-star observations it also decreased their scatter.

Breger: A scatter of $\pm0\overset{\mathrm{m}}{.}002$ for single APT measurements sounds much better than a band of $0\overset{\mathrm{m}}{.}03$. Nevertheless, $\pm0\overset{\mathrm{m}}{.}002$ relative to comparison stars can be achieved for manual photometric measurements. I wonder if there is a reason why such APT measurements cannot be as accurate as the best manual observations?

Moffat: Surely, many of the colour-dependent extinction/transformation problems will disappear by going to narrow-band filters. If Strömgren filters are inappropriate, this means that one should use even narrower filters ($\leq$ 100Å)!

Harmanec: You do not avoid the need for (non-linear) transformation into a standard system by narrowing the filters, unless you can ensure that the total transparency curves (telescope + filter + detector) are identical for two different instruments. The need for transformation results from the fact that the transmission curves differ. It was nicely discussed by Cousins in 1966.

Moffat: UV spectroscopy of most O,B stars shows discrete (or narrow) absorption components (DACs, NACs) that move from $\sim {}^1\!/_2 v_\infty$ to v_∞ with time, getting narrower in the process. The propagation speed is slower than the wind, implying some kind of perturbation spiraling out in the wind. If those are related to the travelling bumps discovered by Walker et al., then indeed all O, B stars may show Be-type phenomena at some level.

Harmanec: Yes, I agree. Actually, I have argued on several occasions (Harmaneč 1989, *Bulletin of the Astronomical Institutes of Czechoslovakia*, 40, 201, see also Harmaneč 1991, ESO Workshop on *Rapid Variability of OB-Stars*, Ed. D. Baade, ESO Conference and Workshop Proceedings No. 36, p. 265) that there may be a causal relationship between the narrow absorption components in the UV lines and the travelling bumps seen in the optical profiles. I also called attention to the fact that one cannot expect a simple correlation of the two because the rapid rotation leads to local spiralling out of the matter so that you do not observe the optical and UV feature projected against the stellar disk at the same time. Actually, Prinja, Henrichs and their collaborators have demonstrated a clear rotational modulation of the development of narrow components (see, for example, IAU Symp. 162).

Chochol: During the active phase of Be stars emission lines are present. If they are located on the edge of the filter's passbands, the small differences in the width of the U, B, V filters could have an influence on the scatter of observations when you collect the data from different observatories. Did you find such an effect during your long-term campaign on Be stars?

Harmanec: In UBV photometry, this is only critical for emission-line stars with very strong lines, I think. Usually, the strong emission is in Hα and in Hβ only. I can say that when we reduced the UBV observations of five rapidly-variable Be stars observed during the 1983 minicampaign from six different observatories with my HEC22 reduction program, we could put all these data together without any additional zero-point shifts (see Stagg et al. 1988, *Monthly Notices of the RAS*, 234, 1021).

Sterken: I agree with your statement that old plates contain useful photometric information. But when doing such archaeophotometry—I have some doubt that you will find funding for such a project, since plate vaults are as much in danger as are small and medium-size telescopes—you must use the same software and reduction procedures as in other photometry. We may not forget that the only difference is that the detector is different, and that the data have been taken in the past.

Harmanec, Cramer

DISCUSSION No 1: Be STARS

Young: Passband width really is not effective in reducing transformation difficulties until you resolve line profiles. If you make the band half as wide, a given spectral feature becomes twice as important. Narrower filters decrease the colour term in the extinction, but *not* the transformation problems (see Young 1992, *Astronomy and Astrophysics*, 257, 366).

Hall: (to Harmanec): You mentioned that APT photometry typically shows a "scatter band" of $0^{\mathrm{m}}.03$. That sounds like terrible photometry. I just want to clarify that photometric accuracy (or precision) is normally quantified by the r.m.s deviation or a standard deviation. Since the total width of a "scatter band" is approximately plus *and* minus 2σ, a scatter band of $0^{\mathrm{m}}.03$ is equivalent to a standard deviation of about 7 millimags. That puts it in a better perspective.

Harmanec: I agree, and I have had no intention to underestimate such photometric measurements! This was just a plain statement of what one sees in, say, phase diagrams.

Hall: Did I hear correctly that your neutral-density filter absorbed 2 mag in V, 3 mag in B, and 5 mag in U? If so, then that is a horrible filter, not a neutral-density filter at all.

Harmanec: This was my point: people are using dense grey filters which are not neutral at all. In the specific case of filter No. 4 of the 0.5-m SAAO telescope, the attenuation amounts to $2^{\mathrm{m}}.5$ in y and b, $3^{\mathrm{m}}.0$ in v and $7^{\mathrm{m}}.9$ in u.

Young: The large variation of "neutral" filters with wavelength is quite typical of coloured glass or dyed gelatin filters. Only the evaporated-metal-film filters are approximately neutral. However, these are not very neutral at high densities.

Kurtz: (to Harmanec): Petr, why should guiding *increase* errors? It should *decrease* them.

Harmanec: In principle, yes. However, I can imagine situations when it will not be so. One would need to know, for instance, the range in coordinates and the frequency of action of the guiding mechanism. It could happen that this frequency interferes more inconveniently with a 20 sec than with a 10 sec integration. I read this information just on the train coming to the meeting. I shall certainly make inquiries to the people operating the Phoenix10 APT when I get back to Ondrejov.

C. Sterken and M. de Groot (eds.), The Impact of Long-Term Monitoring on Variable Star Research, 73–74.

Hall: (to Harmanec): When my photometry (APT or otherwise) uses a neutral-density filter, our practice is to observe the variable *and* the comparison star, *both* through the *same* filter, and the *same* spot on the filter. The power of the differential process thereby protects you greatly from many potential dangers which might produce systematic errors.

Harmanec: Well, I need comparable data over a time scale of years, and from various stations. Therefore, I must transform the data and that requires a large magnitude and colour range of standard stars. Not all photometers have large enough dynamical range to permit observing all stars through the attenuating filter. There are situations where one has to accept the second method: to calibrate the filter in each colour and apply these corrections to the raw data before calculating the transformation. Again, this method worked very well at Sutherland and we were able to combine, without problems, the SAAO data with simultaneously observed La Silla data.

Thé: I have a question for the audience: what is the evolutionary status of B[e] stars? At present it is confusing that these so-called post-main-sequence stars are found to have properties like pre-main-sequence stars.

Moffat: The (luminous) B[e] stars may be related to the LBV (Luminous Blue Variable) phenomenon, probably with bipolar outflow (disk + fast polar wind).

Waelkens: I disagree. B[e] stars were originally defined as Be stars with forbidden lines in emission. But that is not necessarily a homogeneous group. Clearly, luminous B[e] stars in the Magellanic Clouds studied by Zickgraf et al. are evolved objects linked to LBVs, but it is likely that a star such as HD 45677 is much less luminous. There may be various different occasions in which circumstellar matter leads to forbidden emission lines in young as well as in old stars.

Wolf: B[e] stars of the Magellanic Clouds have high-velocity winds like B supergiants, as is evident from their UV spectra. They show slow winds from optical spectra. The fast wind favours the explanation that the B[e] supergiants are post-main-sequence objects.

Williams: We find the analogous position for the WR stars: stars with WR spectra may be either high-luminosity Population I or low-luminosity Population II objects; spectral classes like WR and B[e] are descriptions of spectra and not necessarily definitions of stars or predictors of their other properties.

Harmanec: My feeling is that it is conceivable that many, phenomenologically different, emission-line stars are essentially controlled by the same physical processes. If, during the Be phase, an extended enough Be envelope is built, its outer parts can produce some forbidden emission line. I think it is very important to have independent accurate determinations of radii and masses of individual emission-line stars.

LONG-TERM RADIO MONITORING OF WOLF-RAYET STARS WITH NON-THERMAL EMISSION

K. A. van der HUCHT
Space Research Organization Netherlands
P. M. WILLIAMS
Royal Observatory Edinburgh
T.A.Th. SPOELSTRA
Netherlands Foundation for Research in Astronomy

and

J.P. SWAANENVELT
Space Research Organization Netherlands

ABSTRACT. Radio observations of Wolf-Rayet stars over a period of up to 20 years show that some of them display variable non-thermal radio emission, some continuously, others at episodes. Two such episodes have been observed from the archetype, the WC7+O4-5 system WR 140 (HD 193793) with a binary period of 2900 d. The non-thermal emission from another WC7+O system, WR 125, faded in the mid-1980s, while continuous, variable non-thermal emission is found in WR 146 (WC6) and WR 147 (WN8). In all four cases the origin of the non-thermal radio emission is thought to originate in the collision of the wind of the Wolf-Rayet star with the wind of a companion or in accretion on the companion itself.

Keywords: WR stars – Radio observations – Non-thermal emission

1. Introduction

The characteristic emission-line spectrum which defines the Wolf-Rayet (WR) spectral type results from mass loss of the order of $5 \times 10^{-5} M_{\odot} yr^{-1}$ in a dense stellar wind (for a review of the overall WR characteristics, see van der Hucht 1992). These dense, hot winds are also strong emitters of thermal *bremsstrahlung*, which places WR stars among the most luminous stellar radio sources. Radio observations currently provide the best method of determining wind densities and mass-loss rates. The first WR star detected at radio wavelengths was, not surprisingly, also the nearest: WR11 = γ^2 Velorum (Seaquist 1976). (WR designations refer to the Sixth Catalogue of Galactic Wolf-Rayet stars, van der Hucht et al. 1981).

Abbott et al. (1986) presented a sample of radio observations of 40 WR stars with radio flux densities $S_{6cm} \geq 0.3$ mJy within a distance of 3 kpc and north of declination $-47°$, and listed six non-thermal radio sources. Thus, like 24% of the luminous ($L \geq 10^6 L_{\odot}$) OB stars in the solar neighbourhood (Bieging et al. 1989), some 15% of the WR stars exhibit, superposed on their free-free radio emission, an

C. Sterken and M. de Groot (eds.), The Impact of Long-Term Monitoring on Variable Star Research, 75–84.

often variable non-thermal component. The first star to show a non-thermal radio spectrum was WR140 (HD 193793, Florkowski & Gottesman 1977). Subsequent observations show that this emission varies on a time-scale of years and does so periodically (Williams et al. 1990). The insights provided by these and theoretical studies have extended the need for long-term monitoring to a new wavelength region for a new type of variable star.

2. Radio emission of Wolf-Rayet stars

Radio observations of WR stars before 1992 are summarized by van der Hucht et al. (1992, Table 4). Because of their fast and dense stellar winds all OB and WR stars have thermal IR and radio free-free emission excesses over their photospheric radiation. The spectral index α, defined by $S_\nu \propto \nu^\alpha$, of a thermal source must lie between the extremes of $\alpha \approx -0.05$ for an optically thin source and $\alpha \approx +2.0$ for an optically thick source. In a stellar wind the density decreases sharply with radius, so the radiation at any frequency comes from a combination of optically thin and thick regions, giving a spectral index intermediate between these extremes, whose canonical value is $\alpha = 0.6$ (Wright & Barlow 1975; Abbott et al. 1984). However, some OB and WR stars have additional, sometimes variable, radio excesses which are in most cases non-thermal (Abbott et al. 1984, 1986; Bieging et al. 1989).

Attempts to explain this (e.g. White 1985) required small populations of relativistic electrons accelerated by Fermi processes in the strong shocks that are thought to permeate the radiation-driven stellar winds of *single* hot luminous stars. In those models weak surface magnetic fields ($\sim$ a few Gauss) are necessary to produce optically thin synchrotron emission of relatively low brightness temperatures, but also *large* emitting volumes because, for it to be observable, non-thermal radio emission has to be generated outside the free-free radio photosphere at the observed wavelength. WR radio photospheres have radii of the order of 100 R_{WR} (optical photosphere radius) at mm wavelengths and of the order of 1000 R_{WR} at cm wavelengths. White's 1985 model does not explain why only a limited fraction of the luminous OB stars and the WR stars exhibits excess non-thermal radio emission. Also, for the observed X-ray luminosties of OB and WR stars, models limited to *single* stars have been proposed (e.g. Lucy & White 1980; White & Long 1986; Chen & White 1991; White & Chen 1992; Usov & Melrose 1992), but again these models do not explain the observed variable excess X-ray luminosities of a sizable fraction of the OB and WR stars (for the WR stars, see van der Hucht 1992, Table 2.)

Any explanation for the excess X-ray and non-thermal radio emission must take account of the photoionization and free-free opacity of the Wolf-Rayet stellar wind. Consequently, early models based on binary interactions by Cherepashchuk (1976) and Prilutskii & Usov (1976) have not been favoured. For example, Stevens & Willis (1988) argued that, if WR6 (EZ CMa, $P = 3^{\mathrm{d}}.76$) were a WN5+c binary, its X-ray luminosity should be larger than is observed. In contrast is the recent identification of Cygnus X-3 ($P = 4^{\mathrm{h}}.79$) as a massive WR+c X-ray binary (van Kerkwijk et al. 1992, van Kerkwijk 1993). Here we discuss a class of WR binaries where excess X-ray and non-thermal radio emission from binary interactions can be observed if the wind is not homogeneous or if the separation in the binary system is large enough for the interaction region to be located far out in the WR stellar wind. Furthermore, if a binary orbit is eccentric, the separation of the components varies giving rise to the following two consequences: the intrinsic source will vary because of the varying wind density in the interaction region and the varying pathlength through the WR stellar wind will cause the extinction to vary. These effects, including reduced extinction due to inhomogeneity in the WR stellar wind, have been observed in the WC7+O4-5 binary WR140 (HD 193793, P = 7.94 yr, e = 0.85), whose X-ray and non-thermal radio variations are modelled in terms of phase-dependent circumstellar extinction by Williams et al. (1990).

The discovery that the variability at all wavelengths of WR140 is related to its long-period duplicity and caused by orbital modulation, has triggered many new IR and radio observations of this star and of related objects, and has stimulated a new look at existing X-ray observations of WR stars.

Ever since the second observed (1985) dust-emission event of WR140 (Williams et al. 1990), this object has been monitored by us at radio wavelengths with the *Westerbork Synthesis Radio Telescope*, and independently by observers at the *VLA*. In addition, WR146 (HM19-3, WC6) and WR147 (AS431, WN8), listed by Abbott et al. (1986) as non-thermal radio emitters (see also van der Hucht 1992, Table 4), are being monitored by us with the *WSRT*, and also the WC7+O9 object WR125 with the *VLA*, since its recent IR outburst which may have a recurrence interval of about 15 yr (Williams et al. 1992, 1994). Here we report on the *WSRT* monitoring of WR140, WR146 and WR147. The case of WR147 is discussed by Churchwell et al. (1992). On the basis of the variable non-thermal radio emission of those objects, it is suggested that all are long-period binaries with separations sufficiently wide for their WR winds to collide outside the WR radio photospheres with (winds of) companions.

3. Radio light-curves of the archetype, WR140 (WC7+O4-5)

Because of its large eccentricity (e = 0.85), the interaction of the stellar winds in the archetypal system, the WC7+O4-5 binary WR140 (HD 193793), reaches a maximum during periastron passage with a consequence that might never have been predicted if it had not been observed. Dust condenses in the stellar wind for about two months around periastron passage, causing a conspicuous change to the infrared flux. In the case of WR140, it was the analysis of the infrared observations (i.e. the IR period of 2900 d, the RV solution of optical spectra based on that period, and the resulting orbital elements, notably the orbital inclination) that led to the beginning of understanding the radio observations.

The spectral energy distributions of WR140, are shown in Fig. 1 of the paper *Long-term infrared photometry of WC stars making circumstellar dust* (Williams & van der Hucht, these Proceedings). The ultraviolet and visible fluxes for this system are apparently constant and a hot-star model is fitted to them. There are two sets of observations in the infrared and radio, representing the system at minimum and maximum flux levels. WR140 is also a strong and variable X-ray source and all these phenomena are ultimately caused by the collision of the fast ($\sim$ 3000 km

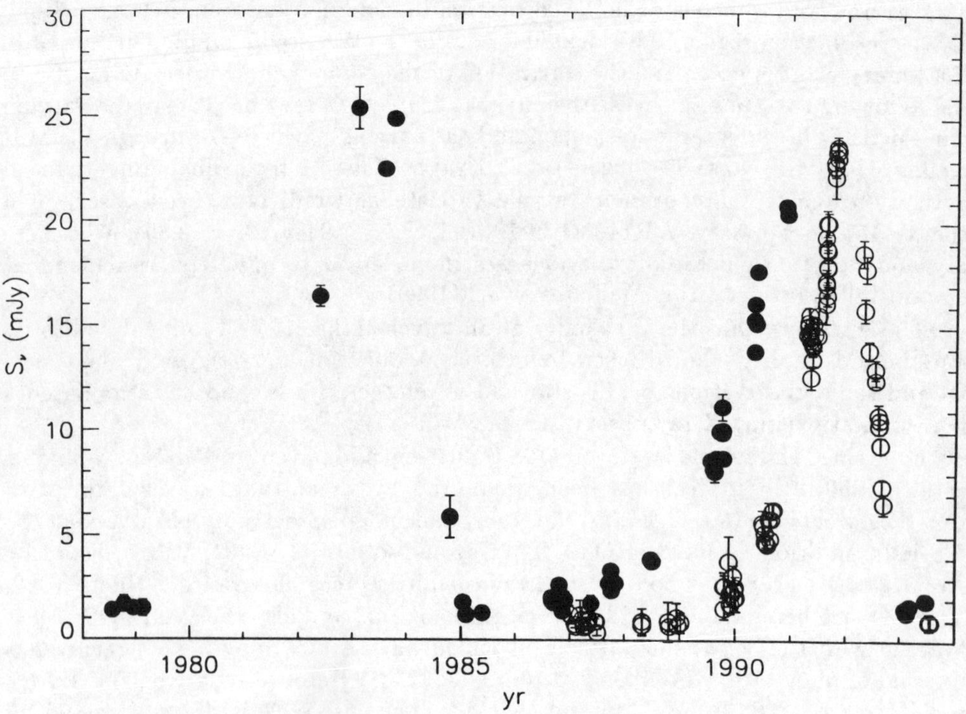

Fig. 1. Radio observations of WR140 at 6cm (filled symbols) and 21cm (open symbols). NROA data: before 1980, *VLA* data: 1982-1985, *WSRT* data: since 1986. Bars represent the background noise in the radio maps

s^{-1}) stellar winds from the WC7 and O4-5 components and the modulation of its effects by the binary motion (Williams et al. 1990). These processes have been investigated theoretically by Usov (1991, 1992), Stevens et al. (1992) and Eichler & Usov (1993).

Fig. 1 shows the radio flux from WR140 as observed at 6 cm and 21 cm with the *WSRT* in an attempt to map the anisotropic stellar wind of this interacting binary system. Around the time of the binary's periastron passage, the fluxes are consistent with free-free emission by the stellar winds of the WC7 and O4-5 components. During the orbital period, the spectrum changes and the flux levels rise by factors exceeding 20, reaching maxima around phase 0.8. The rise takes $\sim 0.4P$ and that at 21 cm lags $\sim 0.1P$ behind that at 6 cm. This is interpreted in terms of the unveiling of a non-thermal source deep in the stellar wind by decreasing circumstellar absorption, the lag of the 21 cm flux being attributable to the greater free-free absorption at this wavelength. In contrast, the fading of the 21 cm and 6 cm fluxes is faster and simultaneous, suggesting occultation of the non-thermal source by denser material. Since the radio maximum occurs, on the basis of the RV solution of Williams et al. (1990) and confirmed by Annuk (1991), when the O-type component is closer to us than the WR star, we can discuss the observed

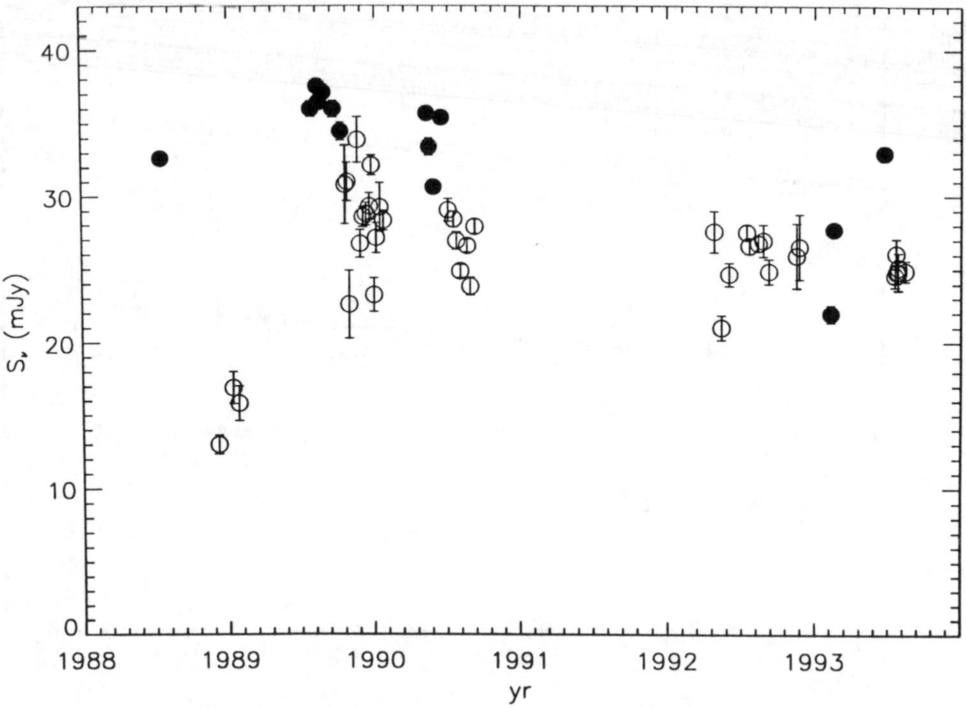

Fig. 2. *WSRT* observations of WR147 at 6 cm and 21 cm (symbols as in Fig.1)

light-curves in terms of the movement of a non-thermal source (associated with a standing shock at the collision surface of the two stellar winds) seen in a low-density cavity around the O-type component in an eccentric orbit through the dense Wolf-Rayet wind.

To constrain the properties of the non-thermal radio source in WR140 further, we observed it with global 6 cm *VLBI* in June 1990, and again in September 1991, i.e. close to the time of radio maximum. A preliminary analysis of the 6 cm data obtained in 1990.5 shows that the non-thermal radio source had a size of about 7 milliarcseconds (mas), while at that time the separation between the binary components was about 19 mas and the size of the free-free radio photosphere of the WC7 star in WR140 was about 62 mas. Thus the non-thermal source in WR140 had at that time at 6 cm a size of 11% of the size of the free-free radio photosphere. This demonstrates that the non-thermal radio emission does not originate in the entire WR wind sphere (as postulated by White 1985), but from a limited region which, in our model, is the low-density cone around the O-type star.

4. Radio monitoring of WR147 (AS431, WN8+?)

With *MERLIN* Moran et al. (1989) discovered, and with the *VLA* Churchwell et al. (1992) confirmed, that WR147 is a double radio source, with as south-

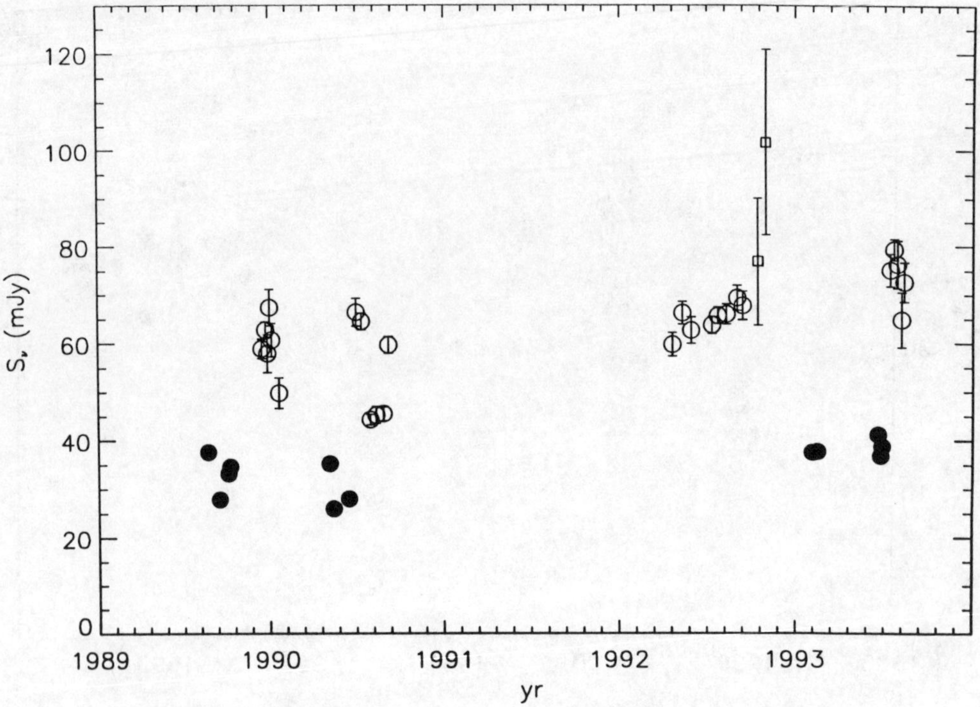

Fig. 3. *WSRT* observations of WR146 at 6 cm (filled circles), 21 cm (open circles) and 49 cm (open boxes)

ern component the WN8 star and at an angular separation of 580 mas the non-thermal northern component. The two components are connected by a well-resolved bridge of 2cm radio emission. WR147N is probably the interaction region of the WR147S WR stellar wind colliding with a companion star or compact object. From IR photometry Churchwell et al. (1992) derived a distance of WR147 of d = 0.63 kpc (implying a source separation of 364 AU $\approx$ 7820 R_{WR}), making it an object of ordinary WR X-ray luminosity (0.14 $L_\odot$) and ordinary WR mass-loss rate ($\dot{M} = 4.2 \times 10^{-5} M_\odot$ yr^{-1}), contrary to earlier belief (Caillault et al. 1985; Stickland et al. 1985), but retaining its peculiar non-thermal radio characteristics.

The *WSRT* radio light-curves (Fig. 2) do not show periodicity but variability of 25% at 6 cm and of 40% at 21 cm on a time-scale of weeks. The spectral index α_{6-21cm} is also variable with time, ranging between the thermal value of +0.7 and a value of +0.1. Since this modulation has a much shorter time-scale than the orbital period expected for the WR147S+WR147N system, a third component or disk orbiting WR147N suggests itself. *WSRT* radio monitoring will be continued to search for periodicity.

The large linear separation between WR147S and WR147N is consistent with the observed ordinary X-ray flux, as we cannot expect colliding wind interaction on the scale as seen in WR140 with its much smaller separation. We cannot discriminate whether thc X-ray flux originates in WR147N or in WR147S or in both. The observed X-ray luminosity of WR147 could just as well originate in the wind of the WN8 star WR147S, or be related to gravitational capture by a neutron star (WR147N) orbiting far out in the diluted wind of the WN8 star (Churchwell et al. 1992).

5. Radio monitoring of WR146 (HM19-3, WC6+?)

WR146 is a WC6 star (Smith et al. 1990), at a distance $d = 0.75$ kpc (van der Hucht et al., in preparation). Its ionization and C/He ratio resemble those of WR140 in appearing anomalous for its spectral subtype (Eenens & Williams 1992) suggesting wind interaction. We observed it with the *WSRT* since 1989 at 6, 21 and 49 cm. The data given in Fig. 3 reveal a variability of 60-100% on a time-scale of a few months. The object shows an average spectral index $\alpha_{6-21cm} = -0.7$ and $\alpha_{21-49cm} = -0.3$, i.e. the non-thermal radiation is more pronounced than in the case of WR147.

Here again, the observed variable non-thermal radio emission points to interaction between the WR wind and a companion, as in the case of WR147.

6. Conclusion: time-scales and the incidence of non-thermal radio emission

WR binaries with orbit sizes of the order of magnitude of the sizes of the WR radio photospheres or larger show their wind-collision interaction regions, notably at radio wavelengths. Because of the required binary component separations, minimum orbital periods of the order of 10 yr are required to reveal their non-thermal radio sources. Also, the inclination of the orbital plane with respect to the line-of-sight has to be favourable to make the non-thermal emission observable. In the case of long-period WC binaries the observability of dust formation is not inclination dependent, fortunately. Also the observability of colliding-winds induced excess X-rays should be independent of inclination. The combination of the three excess phenomena (X-ray and IR excess at periastron passage and radio excess at conjunction) in long-period WC binaries, and of two of those (no dust formation) in long-period WN binaries, make such objects important laboratories for high-energy astrophysics.

Regular radio, IR and X-ray monitoring of the WR stars discussed here and other hitherto alleged single WR stars is going to be a rewarding enterprise for decades to come.

Acknowledgements

We gratefully acknowledge the support of the staff of the *Westerbork Synthesis Radio Telescope.*

References

Abbott, D.C., Bieging, J.H., Churchwell, E. 1984, *Astrophysical Journal*, 280, 671

Abbott, D.C., Bieging, J.H., Churchwell, E., Torres, A.V. 1986, *Astrophysical Journal*, 303, 239

Annuk, K. 1991, in: K.A. van der Hucht & B. Hidayat (eds.), *Wolf-Rayet Stars and Interrelations with Other Massive Stars in Galaxies*, Proc. IAU Symposium 143 (Dordrecht: Kluwer), 245

Bieging, J.H., Abbott, D.C., Churchwell, E.B. 1989, *Astrophysical Journal*, 340, 518

Caillault, J.-P., Chanan, G.A., Hefland, D.J., Patterson, J., Nousek, J.A., Takalo, L.O., Bathun, G.D., Becker, R.H. 1985, *Nature*, 313, 376

Chen, W., White, R.L. 1991, *Astrophysical Journal*, 366, 512

Cherepashchuk, A.M. 1976, *PPis'ma Astron. Zh.*, 2, 356 (= *Soviet Astronomy Letters*, 2, 138).

Churchwell, E., Bieging, J.H., van der Hucht, K.A., Williams, P.M., Spoelstra, T.A.Th., Abbott, D.C. 1992, *Astrophysical Journal*, 393, 329

Eenens, P.R.J., Williams, P.M. 1992, *Monthly Notices of the RAS*, 255, 227

Eichler, D., Usov, V. 1993, *Astrophysical Journal*, 402, 271

Florkowski, D.R., Gottesman, S.T. 1977, *Monthly Notices of the RAS*, 179, 105

van der Hucht, K.A., Conti, P.S., Lundström, I., Stenholm, B. 1981, *Space Science Reviews*, 28, 227

van der Hucht, K.A., Williams, P.M., Spoelstra, T.A.Th., de Bruyn, A.G. 1992, in: L. Drissen, C. Leitherer, A. Nota (eds.), *Non-isotropic and Variable Outflows from Stars, Astronomical Society of the Pacific Conf. Ser.*, 22, 253.

van der Hucht, K.A. 1992, *Astronomy and Astrophysics Review*, 4, 123

van der Hucht, K.A., Williams, P.M., Setia Gunawan, D.Y.A., in preparation.

van Kerkwijk, M.H., Charles, P.A., Geballe, T.R., King, D.L., Miley, G.K., Molnar, L.A., van den Heuvel, E.P.J., van der Klis, M., van Paradijs, J. 1992, *Nature*, 355, 703

van Kerkwijk, M.H. 1993, *Astronomy and Astrophysics* (Letters), 276, L9

Lucy, L.B., White, R.L. 1980, *Astrophysical Journal, Letters to the Editor*, 241, 300

Moran,J.P., Davis, R.J., Bode, M.F., Taylor, A.R., Spencer, R.E., Argue, A.N., Irwin, M.J., Shanklin, J.D. 1989, *Nature*, 340, 449

Prilutskii, O.F., Usov, V.V. 1976, *Astronomicheskii Zhurnal*, 53, 356 (= *Soviet Astronomy*, 20, 2)

Seaquist, E.R. 1976, *Astrophysical Journal*, 203, L35

Smith, L.F., Shara, M.M., Moffat, A.F.J. 1990, *Astrophysical Journal*, 358, 229

Stevens, I.R., Willis, A.J. 1988, *Monthly Notices of the RAS*, 234, 783

Stevens, I.R., Blondin, J.M., Pollock, A.M.T. 1992, *Astrophysical Journal*, 386, 265

Stickland, D.J., Lloyd, C., Willis, A.J. 1985, *Astronomy and Astrophysics* (Letters), 150, L9

Usov, V.V. 1991, *Monthly Notices of the RAS*, 252, 49

Usov, V.V. 1992, *Astrophysical Journal*, 389, 635

Usov, V.V., Melrose, D.B. 1992, *Astrophysical Journal*, 395, 575

White, R.L. 1985, *Astrophysical Journal*, 289, 698

White, R.L., Chen, R.L. 1992, in: L. Drissen, C. Leitherer & A. Nota (eds.), *Non-isotropic and Variable Outflows from Stars*, Proc. STScI Workshop, *Astronomical Society of the Pacific Conf. Ser.*, 22, 274

White, R.L., Long, K.S. 1986, *Astrophysical Journal*, 310, 832

Williams, P.M., van der Hucht, K.A., Pollock, A.M.T., Florkowski, D.R., van der Woerd, H., Wamsteker, W.M. 1990, *Monthly Notices of the RAS*, 243, 662

Williams, P.M., van der Hucht, K.A., Bouchet, P., Spoelstra, T.A.Th., Eenens, P.R.J., Geballe, T.R., Kidger, M.R., Churchwell, E. 1992, *Monthly Notices of the RAS*, 258, 461

Williams, P.M., van der Hucht, K.A., Kidger, M.R., Geballe, T.R., Bouchet, P. 1994, *Monthly Notices of the RAS*, 266, 247

Wright, A.E., Barlow, M.J. 1975, *Monthly Notices of the RAS*, 170, 41

DISCUSSION

Delgado: Is there some significant difference between the binary frequency among WC stars, and the binary frequency among WN stars?

van der Hucht: The binary frequency of WN and WC stars was discussed by van der Hucht et al. (1988, *Astronomy and Astrophysics*, 199, 217). In the volume-limited sample with $d < 2.5$ *kpc*, the number of WR stars, with 37% binaries, is rather small. No difference in frequency between WN and WC binaries is seen. However, there is a difference in period. All galactic WC binary periods are larger than seven days. Of the WN binaries, 50% have periods of less than seven days.

Grunting Bruch, learning Schimpke

Hensberge, Schwarzenberg-Czerny, Waelkens, Harmanec

LONG-TERM INFRARED PHOTOMETRY OF WC STARS MAKING CIRCUMSTELLAR DUST

P. M. WILLIAMS
Royal Observatory Edinburgh
Blackford Hill
Edinburgh UK

and

K. A. van der HUCHT
SRON Space Research Utrecht
Sorbonnelaan 2
Utrecht NL

ABSTRACT. Infrared photometry of WC type Wolf-Rayet stars over up to 20 years shows that some of them form dust in their winds episodically. Three such episodes have been observed from the archetype, the WC7+O4-5 system WR 140 (HD 193793), giving a period of 2900 d. A recurrence interval of $\sim 13\,y$ is estimated for another WC7+O system, WR 137 (HD 192641), while times in excess of $15\,y$ are indicated by observations of WR 125 (WC7+O9) and WR 48a (WC8). Two systems of later subtype, WR 70 (HD 137603) and WR 98a (IRAS 17380–3031) vary on shorter time-scales and one WC4 system, WR 19, was observed in the aftermath of a dust-formation episode. Both WR 125 and WR 140 are X-ray and non-thermal radio sources.

Keywords: infrared photometry – WC stars – circumstellar dust

1. Introduction

Wolf-Rayet (WR) stars emit most of their flux in the far ultraviolet, so it is reasonable to ask why they should be observed in the infrared and what can be learned from doing so over long time-scales. The answer to both questions is provided by the behaviour of a small group of WR stars which emit puffs of dust for a few months at intervals of about a decade. Dust within $\sim$ 100 AU of a star and absorbing some of its uv-visible radiation is sufficiently heated to re-radiate this energy in the infrared. In the case of WR stars with strong uv-radiation fields, even re-radiation by a small amount of dust, say $10^{-8} M_{\odot}$, is sufficient to cause a ten-fold rise in the near-infrared flux. Thus, infrared photometry is a very sensitive diagnostic of dust formation by hot stars.

This is illustrated by the spectral energy distributions of the archetypal system, the WC7+O4-5 binary WR 140 (HD 193793), shown in Fig. 1. The ultraviolet and visible fluxes for this system are apparently constant and a hot-star model is fitted to them. There are two sets of observations in the infrared and radio, representing

C. Sterken and M. de Groot (eds.), The Impact of Long-Term Monitoring on Variable Star Research, 85–100.

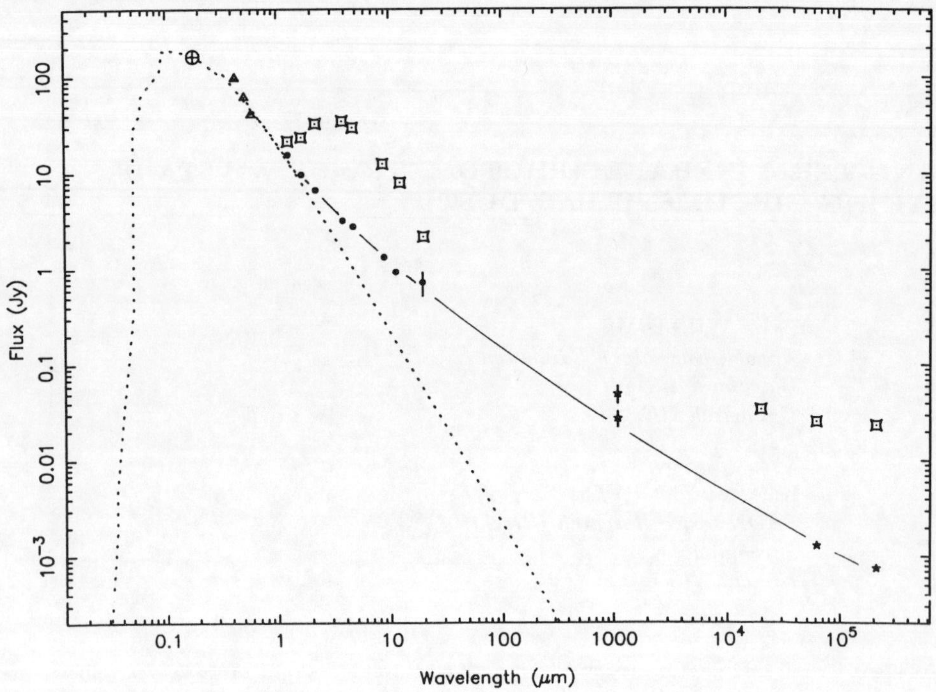

Fig. 1. Spectral energy distributions of WR 140 from observations with the IUE and ground-based optical, infrared, millimetre and radio telescopes. The broken line represents the spectrum of a hot-star model and the line through the lower sets of infrared and radio data points defines the system at quiescence.

the system at minimum and maximum flux levels. Because the radio maxima occur at different times from the infrared and by a different process, they will be discussed in a separate contribution (van der Hucht et al., these Proceedings). WR 140 is also a strong and variable X-ray source and all these phenomena are ultimately caused by the collision of the fast ($\sim 3000 km\ s^{-1}$) stellar winds from the WC7 and O4-5 components and the modulation of its effects by the binary motion (Williams et al. 1990a, hereafter W90). These processes have been investigated theoretically by Usov (1991), Stevens et al. (1992) and Eichler & Usov (1993).

The quiescent infrared–radio spectrum of a WR star is flatter than the photospheric spectrum of an ordinary hot star owing to free-free radiation from the dense stellar wind which carries the WR star's prodigious mass loss. This form of continuum is observed from all WR stars and provides another reason for observing them at long wavelengths: determination of their mass-loss rates. At maximum, the infrared spectrum of WR 140 is quite different from the power-law form at quiescence: it is that of heated dust (Fig. 1). Such maxima have been observed thrice in the last twenty years (Fig 2).

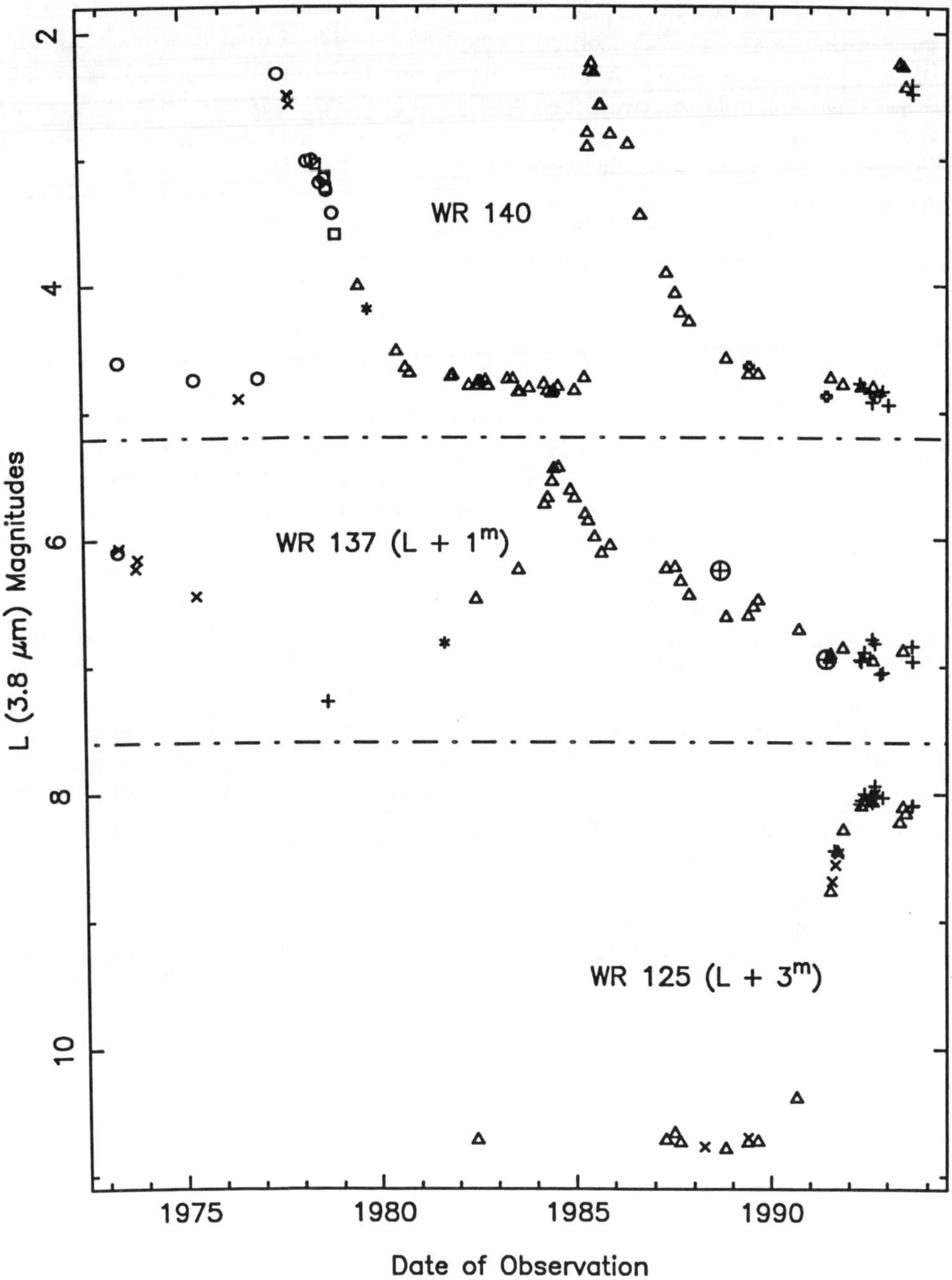

Fig. 2. Synoptic infrared light curves of WR 140, WR 137 and WR 125. Different symbols mark observations made at different observatories.

In this review, we begin with the infrared variations of WR 140 and their interpretation in terms of the formation and dissipation of dust in the stellar wind. We will then examine the infrared variations and dust-formation episodes of related stars. These include the two other WC7+OB systems, WR 137 (HD 192641) and WR 125, whose synoptic 3.8μ m light curves are also given in Fig. 2, and a fourth system for which we have long-term observations, WR 48a. We will describe two systems whose dust emission varies on a shorter time-scale than the other stars, the WC9+B0 binary WR 70 (HD 137603) and WR98a, discovered from its IRAS spectrum. The last system discussed is WR 19, which, unexpectedly for its early sub-type (WC4), was discovered in 1988 to be fading from a dust-formation episode.

2. IR light curves of the archetype, WR 140

Photometry of WR 140 at 3.8μ m presented in Fig. 2 shows three well-defined outbursts in 1977, 1985 and 1993, as well as a fading from what must have been another outburst in 1969. In this figure, as in most of the light-curves presented here, we plot observations made at different observatories with different symbols as a reminder of possible differences between their photometric systems. The 3.8μ m L filter (sometimes called L' to distinguish it from the original L) is now in widespread use and, where necessary, colour equations are used to transform observations made through the older 3.5μm or 3.6μ m filters to this system. This is the optimum wavelength for the study of newly-formed circumstellar dust because its emission peaks near here (cf. Fig. 1).

The first (1977) outburst was observed independently by Williams et al. (1978) and Hackwell et al. (1979). The dust was observed to cool as its emission faded and comparison of this with the earlier observations led Williams & Antonopoulou (1979) to suggest that the outbursts might be a recurrent phenomenon. The star was monitored in the infrared with UKIRT and, in the five weeks between 1985.21 and 1985.31, the infrared flux was observed to have begun its rise to maximum. The time of maximum was a function of wavelength; e.g., in 1985.36 at 2.2μ m, in 1985.41 at 3.8μ m and a year later at 12.5μ m. Comparison of these observations with the earlier ones allowed determination of a period of $2900 \pm 10d$ (W90) and solution of the radial-velocity orbit of the system. The orbital eccentricity was found to be very high and dust formation occurred during periastron passage when interaction of the WC7 and O4-5 stellar winds was at its greatest. The conditions under which the winds in the interaction region can be compressed and cool sufficiently to allow dust condensation have been discussed by Usov (1991).

The infrared photometry has continued and the next dust-formation episode was observed, as expected, in 1993 (Fig. 2). Because WR 140 was too close to the Sun for observing during this latest outburst, it was not been possible to define the fast rise more exactly or, yet, refine the period. Light curves at three wavelengths, 2.2μ m (K), 3.8μ m and 8.75μ m phased to the 2900d period are compared in Fig. 3. Apart from the sharp rises at all three wavelengths, the light curves are very different. The one at 2.2μ m shows the fastest decline, reaching quiescence around phase 0.4, whereas it is not obvious that the 8.75μ m flux has stopped fading before the next outburst occurs. The behaviour at 3.8μ m is intermediate;

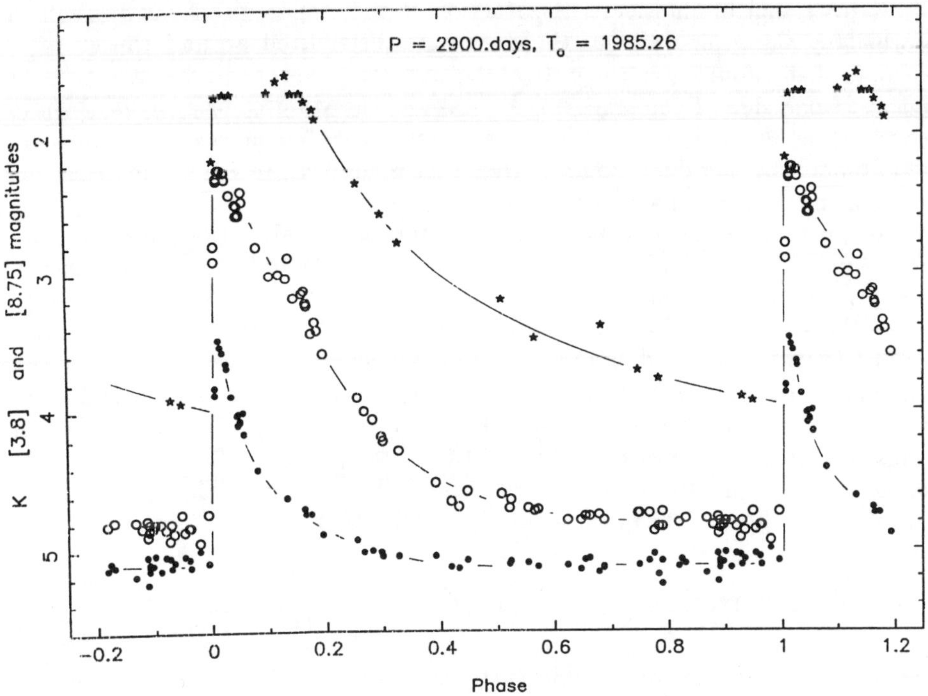

Fig. 3. Photometry of WR 140 phased to the photometric elements in W90

the observed flux at this wavelength just before outburst is probably from the stellar wind only. Observations at still shorter wavelengths (1.25 and 1.65μm) show even narrower maxima than at 2.2μ m while those at still longer wavelengths (12.5 and 19.5μm), though sparser, show an even slower decline than at 8.75μ m.

We do not have optical light curves to match the infrared but there is no evidence for significant variation in the overall luminosity of the WR 140 system during these infrared excursions. As can be seen from Fig. 1, the luminosity of the WR 140 dust shell is a tiny fraction of that in the uv and visible. Shortly after the 1977 outburst, at $\phi = 0.06$ on the ephemeris used for Fig. 3, Fernie (1978) observed that the UBV magnitudes of WR 140 were the same as those observed well before the outburst. Ultraviolet fluxes determined from low-resolution IUE spectra observed through a large aperture show scatter around the 10% level but no significant excursion around infrared maximum (W90).

3. Modelling the IR variations of WR 140

The luminosity of the dust cloud is about 0.3% that of the whole system, the exact percentage depending on the models used for the WC7 and O4-5 stars. The latter star is, of course, the primary and dominates the near-uv and visual flux output.

The 0.3% fractional luminosity is the fraction of the stellar uv-visual flux which is absorbed by the grains and, if the dust were distributed around the stars in an isotropic shell, would give the optical depth directly. More probably, the dust condenses to one side of the stars, in the wake of the wind-collision cone around them and on the other side of the O4-5 star from the WC7 star. Even so, the solid angle subtended by the dust cloud at the stars is unlikely to be as small as one percent of a sphere given that the cone has an opening angle $\sim 34°$ (Eichler & Usov 1993) and spreading in the azimuthal direction during orbital motion. We therefore expect the dust cloud to be optically thin in the uv-visible and, since grain opacity is roughly proportional to frequency, all the more so in the infrared. This makes modelling the spectrum very simple and we may decompose the observed spectral energy distribution into wind and dust-cloud components.

The dust is assumed to be heated by the geometrically diluted stellar radiation only and to be in radiative equilibrium. The temperature T_g of a spherical grain of radius a located a distance r from a star of temperature T_* and radius R_* is related to these quantities through the grain's Planck mean absorption cross-sections $\overline{Q}(a,T)$ appropriate to the grain and stellar temperatures:

$$4\pi a^2\ \overline{Q}(a,T_g)\ T_g^4 = \pi a^2\ \overline{Q}(a,T_*)\ \frac{4\pi R_*^2 T_*^4}{4\pi r^2}$$

The dust is taken to be amorphous carbon. The $8-13\mu$ m spectrum is smooth and does not show either the broad features associated with astronomical silicates or silicon carbide or the narrow graphite feature at 11.52μ m. For such grains, the Planck mean cross-sections are approximately proportional to aT provided that the grains are small ($a < 0.06\mu m$). This is likely to hold as the grains are condensing in an expanding medium whose density is falling so that grain growth is difficult. Consequently, the equilibrium grain temperature falls off with distance from the star as $r^{-0.4}$. The grains move with the wind wherein they condense, which will have reached its terminal velocity at this distance from the stars. Radiation pressure gives the grains a drift velocity relative to the wind but this is small compared with the terminal velocity and we neglect it. The cooling of the grains as they are carried away by the wind accounts for the forms of the light curves at different wavelengths in Fig. 3.

In modelling the evolution of the dust cloud we assume that grain formation occurs at a particular distance from the star determined by factors such as radiation field and wind density. Continuous dust formation for any length of time results in a cloud having a radial extent proportional to the wind velocity and to the time during which condensation occurred. The grains will occupy a range of radial distance and, hence, temperature. Because the dust is optically thin, the emission is proportional to the products of the grain opacities and Planck functions $\kappa_\lambda B(\lambda, T_g(r))$ for this range of grain temperatures weighted by the radial mass distribution $m_g(r)$. The flux F_λ observed at a distance d from the system is given by:

$$4\pi d^2 F_\lambda = \kappa_\lambda \int_{r_i}^{r_i+\Delta r} m_g(r)\ B(\lambda, T_r(r))\ dr$$

In the simplest models, dust formation is assumed to take place at a constant rate so that $m_g(r)$ is constant and the radial extent, Δr, is given by the wind velocity and age of the cloud, if dust is still forming, or length of time for which dust condensed. Such modelling of an observed spectral energy distribution allows only two free parameters: the distance to the inner edge of the dust cloud, r_i, and the total mass of dust to be fitted by observations at four or more wavelengths between 1 and 20μ m. Models fitted to the 1985 observations by W90 indicated that $2.8 \times 10^{-8} M_\odot$ of dust formed between 1985.21 and 1985.54, forming an expanding cloud. The value found for r_i was found to be similar for the first few observations, supporting the idea that dust formation was occurring at a particular distance from the star. The rate of dust formation of about $8 \times 10^{-8} M_\odot y^{-1}$ corresponds to 0.13% of a mass loss ($\sim 6 \times 10^{-5} M_\odot\ y^{-1}$) or 3–7% of the total carbon being lost by the WC7 star, the range quoted reflecting the uncertainty in the carbon abundance of the wind. Clearly, the modelling can be improved and it is hoped that observations following the 1993 outburst will strengthen the database to be fitted so that the effects of more processes can be examined.

4. The WR 140 look-alike, WR 125

The bottom light curve in Fig. 2 is that of WR 125. Although early observations of this star showed a stellar-wind spectrum only, we monitored it in the infrared for possible dust formation because it resembled WR 140 in three significant respects, all pointing to abnormal activity deep in its wind. It was a bright X-ray source (Pollock 1987), a strong non-thermal radio source when observed by Abbott et al. (1986) and its spectrum showed abnormally broad emission lines for its subtype (Torres et al. 1986). Our monitoring began in 1986 and we were rewarded by the discovery of dust formation in 1990.6 (Williams et al. 1992, hereafter W92).

Observations during 1991–93 show that dust formation by WR 125 was much more prolonged than that by WR 140, continuing for nearly three years until early 1993. Light curves at 8.75μ m (combined with some observations at 8.4μ m), 3.8μ m and 2.2μ m are given in Fig. 4. Modelling of the spectral energy distributions as above showed that the total amount of dust formed was $\sim 5 \times 10^{-7} M_\odot$, corresponding to a mean formation rate ($\sim 2 \times 10^{-7} M_\odot y^{-1}$), somewhat greater than that of WR 140.

The slower rise to maximum allowed examination of the dust-condensation rate through time-dependent modelling of the light curves (Williams et al. 1994, hereafter W94). It was assumed that dust condensed in the wind at the inner radius, r_i, determined from modelling the spectral energy distributions and that this was carried with the wind at its terminal velocity ($\sim 2900 km\ s^{-1}$, W92). It was not possible to fit both the single 1990 observation and the well-defined rise in 1991 with a constant rate of dust formation during condensation. We introduced a step in the condensation rate for the first nine months which produces a step in $m_g(r)$. The step function in dust-formation rate — to one-quarter of the maximum rate in 1990.65 to the maximum in 1991.4 — produces the saturating flux curves seen in Fig. 4 because of the expansion and cooling of the dust as soon as it is formed. By 1993, the first dust to form had cooled sufficiently to make no significant contribu-

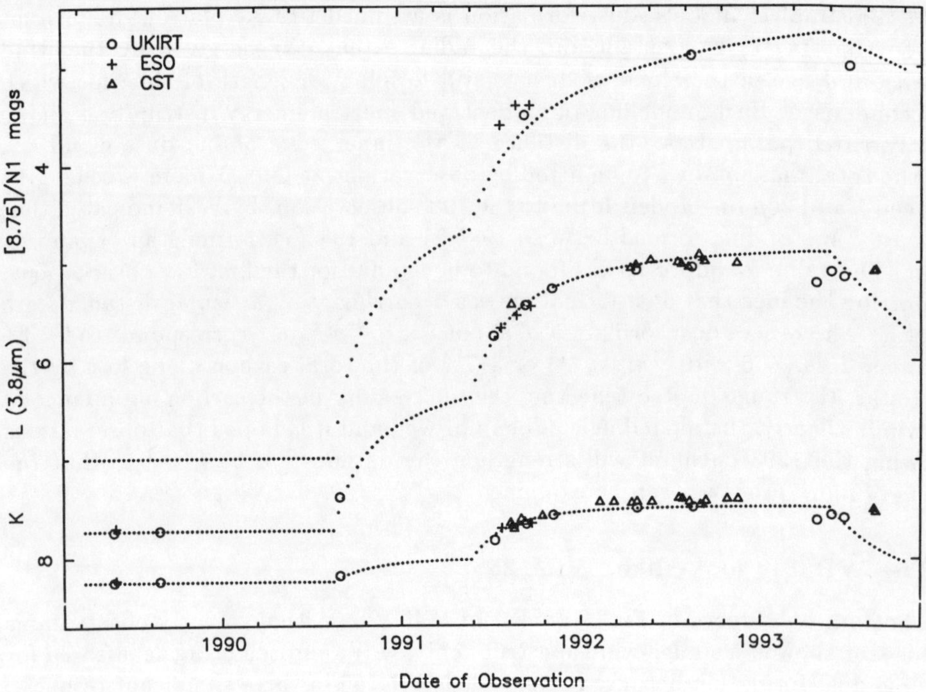

Fig. 4. Observed light curves of WR 125 from about one year before dust began. Superposed on the symbols marking the observations are dots marking magnitudes calculated at 10-day intervals from the model described in the text

tion to the 2.2μ m flux, which therefore reaches a steady state until dust formation ceases. If dust formation were to continue for long enough, flux levels at longer wavelengths would also saturate.

It is also evident from the most recent observations in 1993 that the fading is not steady, as if dust condensation did not cease abruptly, but sputtered out. Considering that dust formation requires density enhancements in the wind and the prevalence of instabilities in the wind interaction region (Stevens et al. 1992), this should not surprise us. As in the case of WR 140, the maximum luminosity of the dust cloud was less than 1% of the system.

Given that its terminal WR wind velocity is the same as that of WR 140 (W92), we expect the dust formed by WR 125 to move from the star at the same speed and the infrared flux to fade at the same rate. Because the amplitude at 3.8μ m was slightly larger than that of WR 140, we expect the flux from WR 125 at this wavelength to fade to its wind level in about five years. Together with the $\sim 2.5y$ rise time, this indicates that we could not have missed an outburst between our 1982 and 1987 observations and that the last maximum must have occurred some time before 1978. Accordingly, if outbursts by this star are recurrent, the period must be longer than 15y. It has not yet been demonstrated that WR 125 is a binary; the spectrum (O9) of the companion to the WC7 star has only recently been classified (W94) and radial- velocity observations over many years will be required to search for and establish an orbit.

5. The slightly erratic system WR 137

The middle synoptic light curve in Fig. 2 is that of another system containing a WC7 star, WR 137 (HD 192641). The rise to maximum in 1984.5 (Williams et al. 1985) is not well defined. It was certainly slower than that of WR 140 and appears to show the sort of step function introduced to explain the form of the rise of WR 125. The fading is much slower than that of WR 140: the 3.8μ m flux appears to have levelled off only recently but the fading time is hard to specify given the short-term variations superposed on the longer-term trend. The slower fading is a consequence of the fact that the wind velocity of WR 137 is only two-thirds of that of WR 140. However, the short-term variations are harder to understand. The comparative lack of variations at shorter wavelengths like 1.65μ m indicates that the stellar wind and photosphere were not varying while the dust emission varied, suggesting that there were several further minor episodes of dust formation after the 1984 maximum. Comparison of the fading after 1984 with earlier observations in 1973–75 suggests a recurrence time of about 13y for this system and it will be important to observe it frequently in the years leading up to the maximum expected in about 1997. Although this star has a composite spectrum, its status as a binary is uncertain. Annuk (1991) has determined an orbit from photographic spectra but Underhill (1992) did not find systematic motion in higher quality spectra observed between 1986.6 and 1991.6. Like WR 140, this is a relatively bright star, easily observed in the visible and it is to be hoped that the questions of the orbit and recurrence of dust formation can be solved in the next five years when we expect its activity to be greatest.

6. Slow fading by WR 48a

The fourth WR star for which a well-defined rise to maximum has been observed is WR 48a. It was discovered by Danks et al. (1983) on the rise to its maximum in 1979.5 and we have been observing it since 1982. Its fading is the slowest of the stars yet observed and, as expected from cooling dust, the fading is slowest at the longest wavelength (Fig 5). Thirteen years after outburst, the fluxes at 2.2μ m and longer wavelengths were still fading when last observed, although the shorter wavelength fluxes may have reached the stellar wind levels. It is not an easy system to observe, being about 7" northeast of a visually brighter star (CPD -62 3053) which can fall near the edge of photometric apertures and contribute significantly to the near-infrared flux observed from WR48a as it fades. It is also heavily reddened ($A_v \sim 8.3$, Williams et al. 1987a) and it has yet to be established whether there is a significant circumstellar component of this, as proposed by Danks et al., or whether it is all interstellar. We have little information on the visual magnitude: comparison of the images of the star on UKST short-exposure red plates taken at the time of infrared

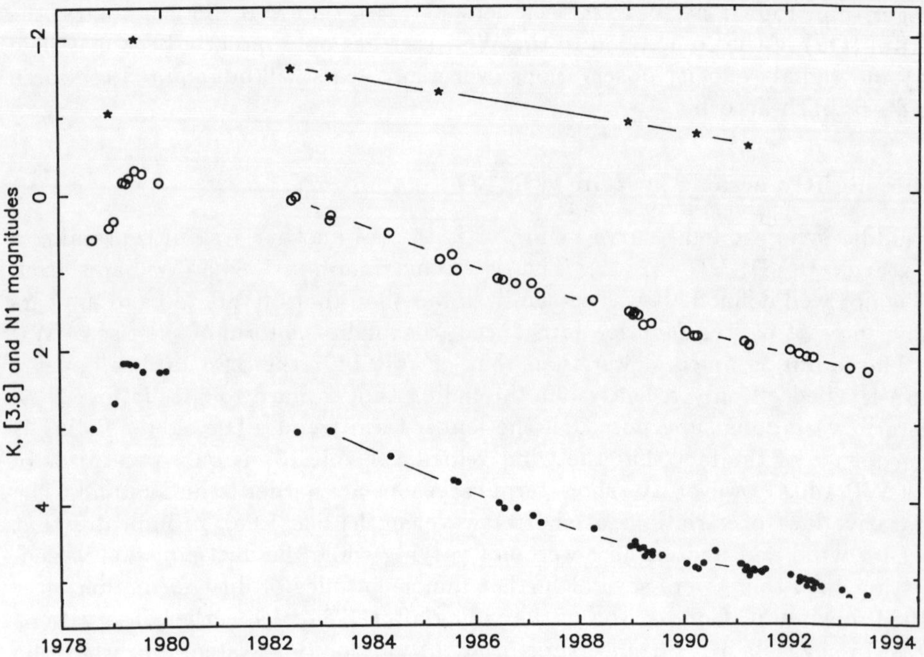

Fig. 5. Synoptic light curves of WR 48a since discovery.

maximum and a year earlier (1978.5) show no significant change in the brightness in this wavelength region. The slow cooling rate of the dust suggests a slow wind velocity, consistent with the late spectral subtype and emission-line widths. The terminal wind velocity will have to be determined from the blue-shifted absorption component of the 2.058μ m He I line as soon as the dust emission at this wavelength has faded sufficiently — but before the next dust formation episode!

7. Stars with variable dust emission: WR 70 and WR 98a

The infrared flux from the WC9+B0 binary WR 70 (HD 137603) varies on a shorter time scale and with smaller amplitude than the stars discussed so far (Fig. 6). After occasional observations in the early 1980s had shown WR 70 to be variable on a shorter time-scale than the stars like WR 140 and WR 48a, it was observed more frequently from 1989. It showed a well defined rise to maximum in 1989.25 attributable to an increase in the amount of heated dust but the subsequent behaviour is harder to understand. The broad minima look like partial eclipses but we do not expect dust clouds to be small enough to be eclipsed or opaque enough in the infrared to cause eclipses.

Cohen et al. (1991) showed that the source IRAS 17380–3031, whose low-

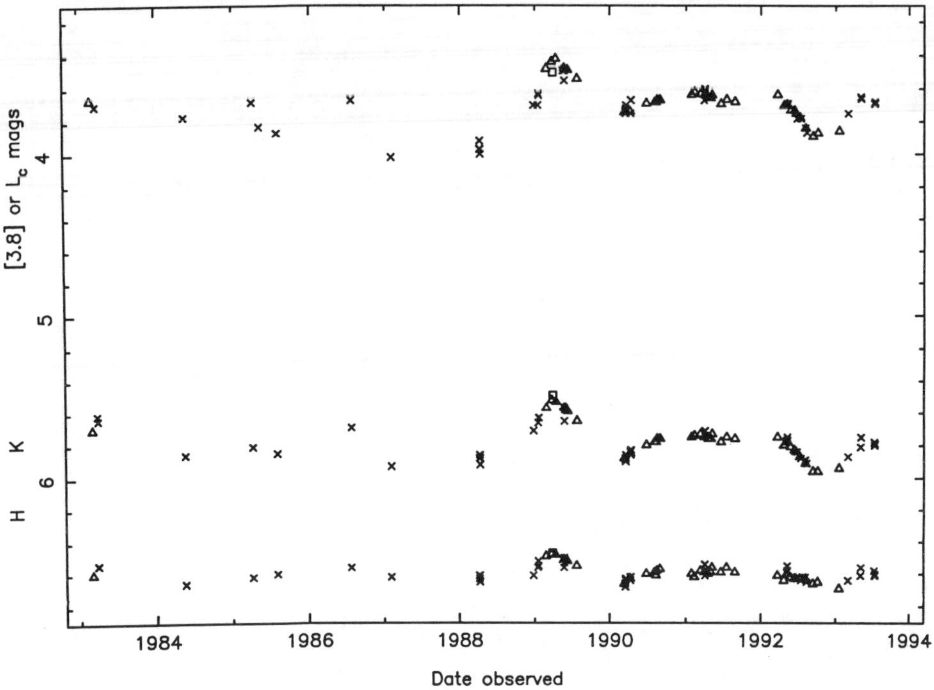

Fig. 6. Infrared light curves of WR 70 (HD 137603). As usual, different symbols mark observations made at different observatories and colour equations have been applied to the SAAO magnitudes to make them directly comparable with the ESO observations.

resolution IRAS spectrum resembled those of late subtype WC stars with strong circumstellar-dust emission, was indeed a late subtype WC star and we refer to it as WR 98a, interpolating in the *6th Catalogue* (van der Hucht et al. 1981). Our initial ground-based photometry confirmed the existence of a circumstellar dust shell, as expected from the IRAS data, and subsequent observation show that the infrared flux varies on a time-scale of months. At present we do not have enough data to characterize the variations and note that the system's heavy extinction ($A_v \sim 12.5$) makes it difficult to study at shorter wavelengths.

8. Dust emission by the WC4 system WR 19

The general pattern of the incidence of dust formation by WR stars is that no WN stars or WC stars earlier than WC7 make dust at all, and that dust formation only becomes common amongst the WC8 and later subtypes. For this reason, the observation in 1988 that the early (WC4) type star WR19 had a circumstellar dust shell, which faded over the next two years (Williams *et al.* 1990b), was most unexpected. We have continued observing this star, whose infrared spectrum is

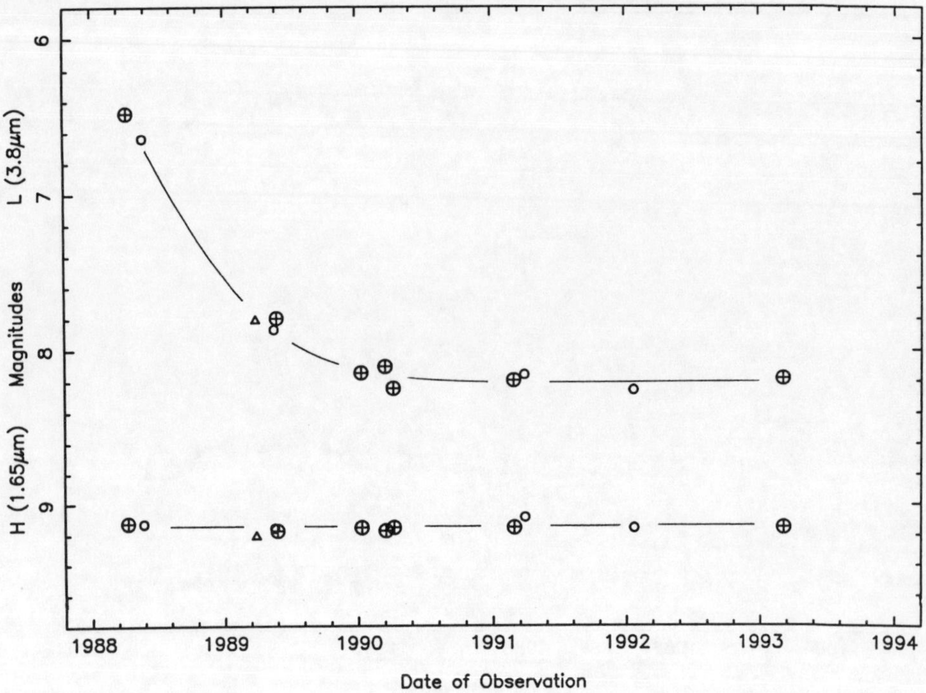

Fig. 7. Light curves of WR 19, discovered as it was fading from what we presume was a dust-formation episode in 1987 or the previous year.

now that of its stellar wind only. Light curves at 3.8μ m and 1.65μ m are given in Fig. 7. The behaviour at $2.2\mu m$ was intermediate, the flux reaching its wind level by 1989. The rate of fading at 3.8μ m, $\sim 0^m.9y^{-1}$, is very similar to that of WR 140, suggesting that the wind velocity of WR 19 is also $\sim 2900 km\ s^{-1}$. We do not know for certain that WR 19 is a member of a binary system; absorption lines are not evident in its spectrum but the dilution of the emission lines suggests that it has a companion $\sim 0^m.8$ brighter than the WC4 star in the visible (Smith et al. 1990).

9. Conclusion: time-scales and the incidence of episodic dust makers

We should distinguish the three time-scales of *flux fading*, which depends on the wind velocity, *flux rising*, which depends on how long conditions in the wind allow dust formation to continue, and *recurrence*, which is unrelated to the wind but, by analogy with WR 140, is the period of the binary assumed to trigger the episodes. In the case of WR 140, the sum of rising and fading times at 8.75μ m is comparable to the recurrence time. Consequently, at shorter wavelengths, we observe the stellar wind for part of each cycle and at longer wavelengths we always observe dust emission. Hence, the probability of observing an episodic dust maker in action depends on the interplay of wind velocity, wavelength of observation and period. A given system will exhibit excess emission over that expected from the stellar wind for longest at the longest wavelength but the accuracy of photometry also falls off with increasing wavelength. Fortunately, the longest wavelength at which one can work accurately with near-IR instruments, $3.8\mu m$, is most sensitive to newly-formed dust. Re-observation at intervals of 3–4 years of stars not currently known to be dust makers is one strategy for discovering more of these systems.

Stars of late spectral subtype, like WR 48a, have lower wind velocities and much longer fading times than WR 140. Unless their recurrence times are longer than their fading times at a particular wavelength, we will not observe the wind at that or longer wavelengths; only the episodic replenishment of the dust giving amplitudes of variation smaller than those of the systems like WR 140 and WR 125. A photometric survey of galactic WC 8–10 stars by Williams et al. 1987b) included stars previously known to have circumstellar dust emission and found convincing evidence for variation of only one star, WR 70, not previously known to be a variable. However, because the earlier observations had been made in different systems, conservative criteria were adopted for assessing variability as ??? 11 of the stars had no previous infrared photometry. Re-observation of stars currently known as continuous dust makers to search for low amplitude variability might reveal periodic replenishment of some of the dust clouds.

Apart from the search for new systems, the study of those described here is a long-term undertaking. In the next five years, we plan to observe and define the fading of WR 140 and WR 125 from their maxima this year, search for the rise to maximum of WR 137 expected in about 1997 and characterize the shorter time-scale variations of WR 70 and WR 98a. In the longer term, the monitoring of WR 19 and WR 48a will be continued at least until they show second outbursts — and thereafter to confirm that the light curves do repeat. We already know that we will have to observe WR 125 to *beyond the year 2008* to see the next infrared maximum of this system! Care will be needed to ensure homogeneity of the photometry done over such a long period over which instrumentation will undoubtedly change.

The infrared eruption of WR 140 in 1977 is a little less puzzling now than when first observed, but we still don't know *how* dust grains are formed in such a hostile environment. It may seem ironic that the principal contribution so far made by the infrared observations of this system has been in allowing interpretation of observations made in *other* wavelength regions, but we see this as an instance of the value of working across the spectrum. Further study of the other systems reviewed here is required to see how widely the WR 140 paradigm applies.

Acknowledgements

We gratefully acknowledge that most of the observations presented here were made at the European Southern Observatory (ESO), the United Kingdom Infrared Telescope (UKIRT), the South African Astronomical Observatory (SAAO) and the Carlos Sanchez Telescope (CST), with a few made at the Anglo-Australian Telescope (AAT) and at the San Pedro Mártir observatory—in Service Observing programmes, visitor allocations and by individual astronomers who have included one or more of these stars in their own observing programmes. We would like to thank Dave Anthony, Mike Barlow, Patrice Bouchet, John Caldwell, Brian Carter, Liz Conlon, Michael Feast, Alan Fitzsimmons, Tom Geballe, R. de Jong, Mark Kidger, Dave Laney, A. Leene, Fred Marang, Pieter Mulder, Hans Nieuwenhuizen, Greg Roberts, Frank Spaan, Mauricio Tapia, Diah Setia Gunawan, Pik Sin Thé, Patricia Whitelock, Hartmut Winkler and Dolf de Winter for their observations.

References

Abbott, D.C., Bieging, J.H., Churchwell, E. & Torres, A.V., 1986. *Astrophysical Journal*, 303, 239

Annuk, K, 1991. In: *Wolf-Rayet Stars and Interrelations with Other Massive Stars in Galaxies, IAU Symposium 143,* eds: van der Hucht, K.A. & Hidayat, B., (Dordrecht, Kluwer), p 245

Cohen, M., van der Hucht, K.A., Williams, P.M. & Thé, P.S., 1991. *Astrophysical Journal*, 378, 302

Danks, A.C., Dennefeld, M., Wamsteker, W.M. & Shaver, P.A., 1983. *Astronomy and Astrophysics*, 118, 301

Eichler, D. & Usov, V., 1993. *Astrophysical Journal*, 402, 271

Fernie, J.D., 1978. *Information Bulletin on Variable Stars*, 1377

Hackwell, J.A., Gehrz, R.D. & Grasdalen, G.L., 1979. *Astrophysical Journal*, 234, 133

van der Hucht, K.A., Conti, P.S., Lundström, I & Stenholm,B., 1981. *Space Science Reviews*, 28, 227

Pollock, A.M.T., 1987. *Astrophysical Journal*, 320, 283

Smith, L.F., Shara, M.M. & Moffat, A.F.J., 1990. *Astrophysical Journal*, 348, 471

Stevens, I.R., Blondin, J.M. & Pollock, A.M.T., 1992. *Astrophysical Journal*, 386, 265

Torres, A.V., Conti, P.S. & Massey, P., 1986. *Astrophysical Journal*, 300, 379

Underhill, A.B., 1992. *Astrophysical Journal*, 398, 636

Monthly Notices of the RAS, 252, 49

Williams, P.M. & Antonopoulou, E., 1979. *Monthly Notices of the RAS*, 187, 183

Williams, P.M., Beattie, D.H., Lee, T.J., Stewart, J.M. & Antonopoulou, E., 1978. *Monthly Notices of the RAS*, 185, 467

Williams, P.M., Longmore, A.J., van der Hucht, K.A., Talavera, A., Wamsteker, W.M., Abbott, D.C. & Telesco, C.M., 1985. *Monthly Notices of the RAS*, 215, 23p

Williams, P.M., van der Hucht, K.A. & Thé, P.S., 1987a. *Quarterly Journal of the RAS*, 28, 248

Williams, P.M., van der Hucht, K.A. & Thé, P.S., 1987b. *Astronomy and Astrophysics*, 182, 91

Williams, P.M., van der Hucht, K.A., Pollock, A.M.T., Florkowski, D.R., van der Woerd, H. & Wamsteker, W.M., 1990a *Monthly Notices of the RAS*, 243, 662. (W90)

Williams, P.M., van der Hucht, K.A., Thé, P.S., & Bouchet, P., 1990b. *Monthly Notices of the RAS*, 247, 18p

Williams, P.M., van der Hucht, K.A., Bouchet, P., Spoelstra, T.A.Th., Eenens, P.R.J., Geballe, T.R., Kidger, M.R. & Churchwell, E., 1992. *Monthly Notices of the RAS*, 258, 461. (W92)

Williams, P.M., van der Hucht, K.A., Kidger, M.R., Geballe, T.R. & Bouchet, P., 1994. *Monthly Notices of the RAS*, 266, 247

DISCUSSION

Kurtz: Why are the dust-formation episodes so regular?

Williams: Those of WR140 are triggered by the periastron passage of the binary companion; the eccentricity is very high (0.84).

Kurtz: What governs the dust-formation timescale?

Williams: I would speculate that stars with more prolonged episodes of dust formation than WR140 have orbits of lower eccentricity.

Kurtz: What happens in the blue?

Williams: The luminosity of the dust is less than 1% of that of the whole system. So, I don't expect to see a significant change in the blue.

Kurtz: Can you perhaps observe the next IR increase in WR140 in the L-filter during the day?

Williams: Yes, so long as we don't let sunlight fall on the telescope!

Sterken: Concerning your remark on the lack of homogeneity of IR photometric data: an old and—within the boundary conditions—reliable method to beat the atmosphere and instrumental zero-point shifts is differential photometry. So I would expect IR observers to select one or two comparison stars with a similar energy distribution (in the IR) as the target star. But I have never seen this in IR monitoring!

Williams: You are right; infrared observers must adopt more of the practices of optical astronomers. At least the prime standard for the WR140 $JHKLM$ photometry is close to it in the sky!

Le Contel: Are there any speckle observations of these stars in the infrared?

Williams: I don't believe so.

Moffat: What is the typical precision of your IR data?

Williams: Including calibrations, about $0\overset{m}{.}04$ in $JHKL$ and $0\overset{m}{.}1$ in the ten-micron region.

Moffat: One could do differential photometry simultaneously with an IR-array detector; but, alas, for WR140 there are no bright enough comparison stars within $1\overset{\prime}{.}$ to include on the chip.

Williams: Yes; but most of our targets are too bright for the infrared-array cameras. WR48A has now faded to be of comparable brightness to the nearby field stars. This affects aperture photometry at J and H and we should observe it with an array.

Moffat: Have you monitored a complete sample of WC (e.g. WC9) stars?

Williams: No, but we should as a means of detecting stars which top up their circumstellar dust reservoirs periodically.

Jorissen: There has been a lot of discussion in recent years about the origin of the SiC and graphite grains found in the Murchison meteorite. Some people are claiming that some of these grains come from WR stars, on the ground of their isotopic composition (e.g. $^{12}C/^{13}C$ ratio). Do your observations support the possibility of SiC or graphite-grain formation in the winds of WR stars?

Williams: I believe they do. The amorphous carbon made by the WR stars might be transformed to graphite elsewhere. Of course WR stars contribute very little dust to the interstellar medium compared with carbon stars and AGB stars.

Lloyd and Gosset study the details

PHOTOMETRIC MONITORING OF WOLF-RAYET STARS
Strömgren *uvby* and narrow-band photometry

E. GOSSET*, G. RAUW**, J. MANFROID***, and J.-M. VREUX
Institut d'Astrophysique – Université de Liège
avenue de Cointe 5
B-4000 Liège (Belgium)

and

C. STERKEN‡
Vrije Universiteit Brussel
Pleinlaan 2
B-1050 Brussel (Belgium)

ABSTRACT. We describe our on-going project of photometric monitoring of Wolf-Rayet stars. Our strategy is based on the complementarity between data issued from the *Long-Term Photometry of Variables* project and data issued from dedicated observing runs. This method is illustrated by a few concrete examples.

Keywords: WR stars – *uvby* photometry

1. Introduction

1.1. THE WOLF-RAYET STARS

For many years, Wolf-Rayet stars (hereafter WRs) have been considered as exotic objects. They only gained importance once it was realized that they correspond to an evolved stage of the most massive stars. Presently, they are a key point in the understanding of the upper part of the Hertzsprung-Russell (HR) diagram. The WRs are characterized by a strong mass loss ($\dot{M} \sim 10^{-5}$ $M_\odot$/year) and by large terminal velocities of the wind ($v_\infty \sim 2500$ kms^{-1}). In fact, their wind is so dense that the central object is usually not visible. The ejected material is partly constituted of processed matter (i.e. H- and He-burning products). Through this dense and energetic wind, WRs are also responsible for about 50 % of the kinetic energy injected into the interstellar medium. Altogether these characteristics make

* Chercheur Qualifié au Fonds National de la Recherche Scientifique
** Aspirant au Fonds National de la Recherche Scientifique
*** Directeur de Recherches au Fonds National de la Recherche Scientifique
‡ Senior Research Associate NFWO

C. Sterken and M. de Groot (eds.), The Impact of Long-Term Monitoring on Variable Star Research, 101–116.

the WRs key objects for our understanding of the ecology of the galaxies. For more details, we refer to the recent review by van der Hucht (1992).

Despite the importance of WRs, several gaps are still persisting in our knowledge of these objects. The agreement with the evolutionary theories is in fact only qualitative. Large uncertainties remain on their luminosities, effective temperatures and masses. Their exact localization in the HR diagram is therefore not well known. In addition, a few facts predicted by the theory, such as the dependence of the mass loss and of the terminal velocity on the metallicity, are not really observed. The true origin of the strong wind is not understood and what is called the momentum problem is still a matter of debate (Lucy & Abbott 1993; van der Hucht 1992). The origin of the variability observed in most of these objects is also unclear. Several mechanisms are invoked ranging from the effect of rotation to the presence of pulsations or of a binary system, not to speak of wind instabilities. The third possibility further underlines our ignorance of the exact nature of the central objects since some or all of them could even host a binary system.

We are convinced that variability studies could help in unveiling the true nature of WRs but the endeavour is not easy due to the wind which can act as a filter (Vreux 1987) and also as a source (Owocki et al. 1988).

1.2. HISTORICAL ASPECT OF VARIABILITY STUDIES

For a long time, the only way to form a WR (i.e., to unveil deep layers) was thought to be through mass transfer in a binary system (see, e.g., van den Heuvel, 1976) leading to the following evolutionary sequence:

$$\mathrm{O} + \mathrm{O} \Rightarrow \mathrm{WR} + \mathrm{O} \Rightarrow \text{compact} + \mathrm{O} \Rightarrow \text{compact} + \mathrm{WR} \Rightarrow \text{compact} + \text{compact}.$$

Observationally, mainly thanks to contributions from Moffat and collaborators, variability was shown to be quite common among WRs. Some photometric observations led to the discovery of apparently periodic variations with a period usually in the range of a few days. At that time, due to the theoretical context, there was a tendency to systematically interpret the variability as a proof of binarity and to identify these variable WRs with the evolutionary stage WR + compact. A list of proposed WR + compact candidates can be found, for example, in Cherepashchuk & Aslanov (1984). It is important to remember that, at that epoch, a typical observing run was 5 to 12 days long with one, perhaps two, measurements of the same object during the night.

Later, theoreticians such as Maeder (1983), following an early suggestion by Conti (1976), introduced the possibility to form a WR from a single star, provided that the initial mass is high enough ($M_{init} \geq 40\ M_{\odot}$) and that physical effects like mass loss and mixing of internal layers are taken into account. At about the same epoch, Vreux (1985) pointed out a few facts suggesting that the WR periods could effectively be shorter than the ones usually accepted. First, the variability within one night was sometimes of the same order of magnitude as the night to night variations. Second, the tendency to perform only one measurement per night was frequently introducing a sampling regular enough to produce a strong aliasing phenomenon leading to an almost total ambiguity between low and high frequencies.

Ever since then, other causes of variability than binarity were again suggested to explain the observations. They were ranging from rather classical ones like rotation and/or magnetic field (Underhill 1983; Poe et al. 1989), to rather new ones like radial pulsations (Maeder 1985) or non-radial pulsations (Vreux 1985), or even more recently to instabilities in the wind (Owocki et al. 1988). For more details, we refer the reader to the reviews by Vreux (1987) and by Moffat & Robert (1991).

It is in such a context that in the second half of the eighties we initiated a total reinvestigation of the variability of some WRs on the basis of runs of 6-12 days but with as many observations as possible within one single night. The rather unexpected and quite outstanding result of these campaigns is that almost none of the published WR + compact periods were recovered. This was clearly indicating a serious problem linked, either to the too small amount of data at our disposal, or to the fact that the variabilities were not purely monoperiodic. Consequently, in order to improve our knowledge of WRs, we had to extend our small, limited observing runs to actual monitoring.

2. Monitoring of Wolf-Rayet stars

In agreement with our dutch colleagues (K. van der Hucht and A. van Genderen) and in order to avoid vain duplications, we have decided to concentrate our efforts mainly on the WN stars. Among these, we selected a subset of late WN stars because they turned out to exhibit the largest variations statistically. In addition, Noels & Scuflaire (1986) had suggested that late WN stars with some H residue could be non-radial pulsators. The main targets we will discuss here can be divided into two categories: the new targets (WR 108, WR 78, WR 25, WR 24) and the old targets (WR22, WR16, WR40), the old ones having been observed over a longer time. All these stars have been inserted in the LTPV list (LTPV: Long-Term Photometry of Variables, see Sterken, these Proceedings) and observed in the Strömgren *uvby* system. Of course, other WRs are also observed in the framework of the LTPV programme (e.g. WR 6, WR 73, ...). These LTPV data were complemented by intensive campaigns dedicated to particular problems. Most of these dedicated campaigns were performed in the Strömgren *b* or *uvby* system which is rather convenient for WRs. A few non-LTPV targets were also included in these specific campaigns (e.g. WR 12, WR 66, ...). Throughout our studies, we routinely performed differential photometry against two comparison stars.

On the basis of a few concrete examples, we plan to present what each project brings as information and how we combine the various sources of data to produce a coherent monitoring. We will first consider the LTPV data.

2.1. LONG-TERM PHOTOMETRY OF VARIABLES

The LTPV project initiated by Sterken (1983, and also these Proceedings) has been shown to be an extremely valuable tool for studying the long-term variability of stars. It is based on the use of the ESO and Danish 50-cm telescopes which are located at the European Southern Observatory (ESO) at La Silla (Chile). Our main targets have been included in the LTPV list with a requested rate of one or two measurements per night. This is a rather heavy load, explaining the small number of accepted targets.

2.1.1. The new targets

WR 108: WR 108 is a tenth-magnitude star of type WN9-Ofpe. No previous variability study has been reported. We acquired 86 observations on this object. It is certainly variable at the level of a few percent. No clear periodicity is present. We are awaiting more data. For this object, we could be at the limit of what can be done with a 50-cm telescope because the standard deviation of the differential magnitudes of the two comparison stars is of the order of one percent, whereas for the other brighter objects, we routinely reach standard deviations of 3 millimag.
WR 78: WR 78 is a rather bright ($V \sim 6.5$) WN7 star. Schild *et al.* (1969) reported a tentative P = $3\overset{d}{.}3$ periodicity in the radial velocities. The star is known to be variable with a standard deviation of a few percent (Seggewiss & Moffat 1979; Lamontagne & Moffat 1987; Magain et al. 1987). Recently, Balona *et al.* (1989) suggested a P = $22\overset{d}{.}7$ periodicity on the basis of Johnson B photometry exhibiting a standard deviation of 13 millimag.

From the 223 measurements acquired on this star, we derive standard deviations which are slightly lower: $\sigma_{uvby} = (\sigma_u, \sigma_v, \sigma_b, \sigma_y) = (0.011, 0.010, 0.009, 0.010)$. The power spectrum of the v data exhibits its highest peak at $\nu = 0.123$ d^{-1} whereas the b data rather favour $\nu = 0.006$ d^{-1}. In addition, the two previously quoted periods (P = $3\overset{d}{.}3$ and P = $22\overset{d}{.}7$) are clearly not present in our data. Conversely, our two quoted frequencies are not present in Balona *et al.*'s (1989) data. This points toward a strongly epoch-dependent behaviour.
WR 25: WR 25 is an eighth-magnitude star of type WN7+abs. It is also an unusually strong X-ray source for a WR. On the basis of a series of photometric runs spanning in time from a few days to about 43 days (Moffat 1978; van Genderen et al. 1987; Lamontagne & Moffat 1987; Balona et al. 1989), the star is believed to be constant ($\sigma \leq 0.008$).

We acquired 156 measurements spread over a time-interval of 1551 nights and derived standard deviations σ_{uvby} = (0.012, 0.013, 0.011, 0.010). The power spectrum of the b data is shown in Fig. 1; it has been computed using Deeming's (1975) method. Clearly, we detected a variation of the star. Fig. 1 suggests that the power is located at very low frequencies. A variation $\Delta b \sim 0\overset{m}{.}02$ is observed on a crude time-scale of 250 d. We conclude that WR 25 is variable on a long-term basis and that it is a typical example of the unique efficiency of the LTPV project. Indeed such a variability would probably never have been discovered without the LTPV project because the latter guarantees sufficient homogeneity and accuracy on a long-term basis to allow the detection of such low-amplitude variations. The long time-scale is compatible with the hypothesis of a binary system, which in turn could be linked to the particularly strong X-ray emission if that is also found variable on a compatible time-scale. Of course, we need more data of that sort and, hence, wish a long life to the LTPV project.
WR 24: WR 24 also is a WN7+abs star but slightly brighter ($V \sim 6.5$) than WR 25. Previous studies reported either constancy at the $\sigma = 0.008$ level (Moffat

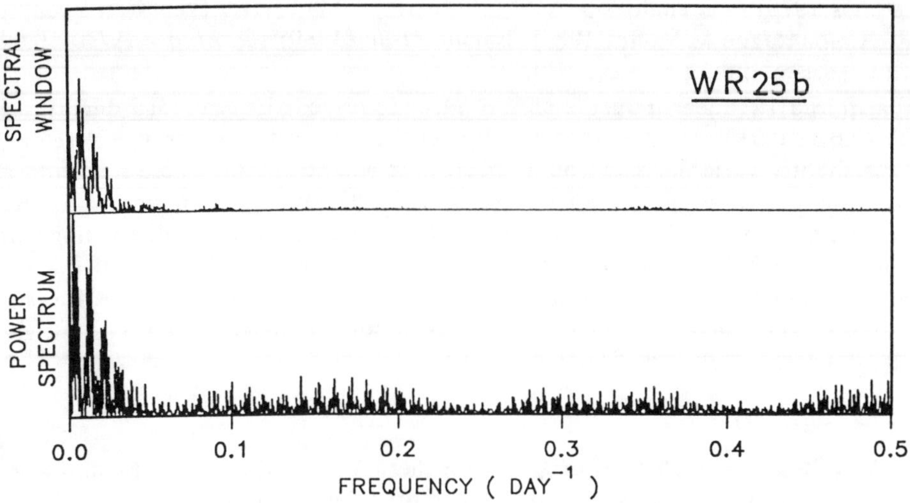

Fig. 1. Power spectrum of the *b* differential magnitudes of WR 25

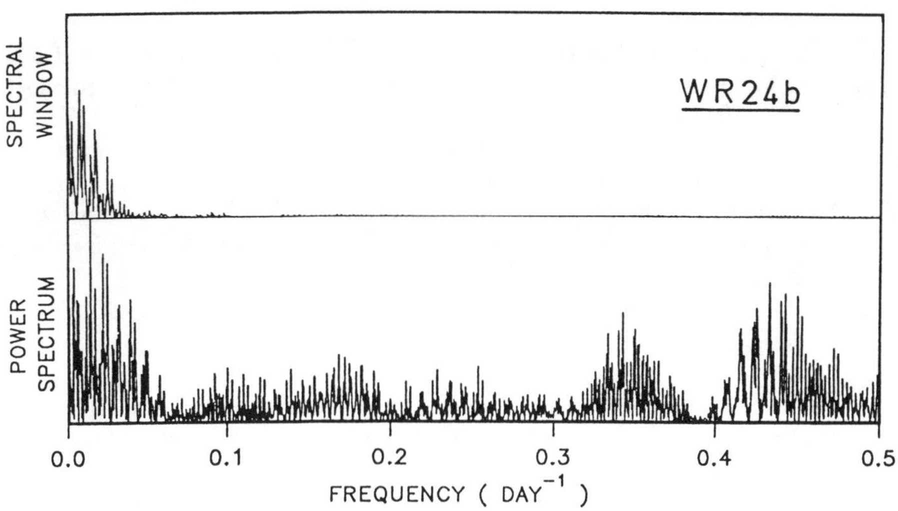

Fig. 2. Power spectrum of the *b* differential magnitudes of WR 24

& Seggewiss 1978) or a slight variability at the $\sigma = 0.009$ level (van Genderen et al. 1987; Lamontagne & Moffat 1987). Balona *et al.* (1989) reported a $\Delta B = 0.02$ amplitude corresponding to a $\sigma_B = 0.008$.

We acquired 181 measurements of WR 24 and derived the standard deviations σ_{uvby} = (0.011, 0.010, 0.010, 0.010), well above the observational error. Therefore, WR 24 is slightly variable. The power spectra associated to the whole set of data exhibit a main peak at $\nu = 0.43$ d^{-1} (P = 2.32 d) for the *u*, *v*, and *y* filters. The spectrum corresponding to the *b* filter is given in Fig. 2. A phase diagram with the purported period has a convincing amplitude of about 0.02 mag. However, the period is not present in all our data subsets and is absent from the Balona *et al.* (1989) data. This lack of ubiquity casts doubt on the detection of a single periodicity. Further data are clearly necessary.

2.1.2. The old targets

WR 16: WR 16 is one of the two brightest southern WN8 stars in the sky and also one of our old targets. This means that we acquired early data on the object: they have been published in Gosset *et al.* (1990). We refer the reader to that paper, although we will present here a brief summary of the characteristic results. Typically, the analysis of the 221 LTPV measurements results in the possibility to decompose the time series of the data in three to four components:

- two to three periodicities: $\nu = 0.110$ d^{-1} and $\nu = 0.066$ d^{-1} which are significant against the null-hypothesis of white-noise; and $\nu = 0.183$ d^{-1} which is not so markedly significant;
- a white-noise process of dispersion $\sigma = 0.01$.

However, nothing indicates that this decomposition transcends mathematical formalism and reaches an actual physical meaning. In addition, the quoted frequencies are not really ubiquitous in the different subsets of the data. Furthermore, Balona *et al.* (1989), on the basis of independent data, rather suggested the frequency ν = 0.058 d^{-1}. Therefore, we conclude that there is no clear, dominant, isolated and ubiquitous periodicity in the photometric variations of WR 16. As an alternative possibility, one often hears about pseudo-periodicities, a term which is a mixed bag of undefined ideas.

In fact, a deterministic multiperiodic process hidden in a white-noise process is a possible alternative. The complex beating phenomenon can exhibit successive domination of the basic periodicities or of their combination. In this hypothesis, we need more data in order to go beyond some recurrence time-scale (an objective only attainable through the LTPV project) and then have the power spectrum possibly gaining complete stability.

Alternatively, we could imagine that we have to deal with a dominantly stochastic process. Most of the stochastic processes can be mathematically modelled by linear Auto-Regressive Moving Average (ARMA) processes (see Scargle 1981; Box & Jenkins 1976). Any ARMA process with some positive autocorrelation at small lags (a few days) has a power spectrum dominated by low frequencies. Gosset *et al.* (1990) showed that WR 16 could be associated with such a case and that the existing photometric data are rather compatible with a simple ARMA process (see their Figs. 13 and 11, and their section 5.2). The three frequencies, products of the decomposition quoted above, even if they are significant against the white-noise null-hypothesis, are no longer significant against a null-hypothesis such as the adopted ARMA process. In such a context, the peaks in the power spectrum are just due to particular realizations of the stochastic process. Sometimes ARMA processes can transiently exhibit well-marked oscillations (see Figs. 8 and 9 of Scargle, 1981). Future studies based on new data will lead either to a similar decomposition (in which case the process will be effectively multiperiodic) or to other peaks, these being located under a similar envelope dominating at low frequencies (in this latter case, an ARMA process similar to the one approximated in Gosset et al. 1990, would be the right variability model).

WR 40: WR 40 is the other bright star of type WN8. Our analysis of the variability of WR 40, partly based on LTPV data, has been published in Gosset *et al.* (1989), where we reported a formal decomposition of the variability into a powerful (approximately) white-noise process ($\sigma = 0.02$), a significant periodicity $\nu \sim 0.16$ d^{-1} and, possibly, a further non-significant one ($\nu \sim 0.41$ d^{-1}). On the other hand, Balona *et al.* (1989) also concluded, from the analysis of their own observations, the existence of two periodicities in WR 40. Gosset and Vreux (1990) argued that it was not possible, on the basis of existing data, to distinguish between the two periodicities reported by Gosset *et al.* (1989) and the two reported by Balona *et al.* (1989). Therefore, contrary to the case of WR 16, it seems that WR 40 can already be recognized as the seat of a biperiodic process. At least, we can be sure that some determinism is present in the variability of WR 40. We still have to investigate the possibility of the presence of some chaos.

WR 22: A different case is the one of WR 22, a WN7 star ($V \sim 6.4$) which is reported as being an SB1 (single-line spectroscopic binary) with a period P = 80.35 d (van der Hucht *et al.* 1981). It has a rather well-determined orbit (Niemela 1979; Moffat & Seggewiss 1978; Conti et al. 1979) but no firm identification of the companion has been reported. The star is photometrically variable; although Moffat and Seggewiss (1978) have not reported any trace of an eclipse. However, Balona *et al.* (1989), investigating the variability of WR 22, observed a $\Delta B \sim 0.1$ mag dip on JD2447235.5.

At the same time WR 22 was observed at ESO in the framework of the LTPV project and this dip was also detected. A second dip was also noticed on JD2447476.8 i.e. three cycles later. Only a few measurements were available in each dip. Anyhow, thanks to the LTPV project, we serendipitously confirmed the discovery of an eclipse. It is interesting to note that the mentioned eclipse corresponds to the companion being behind the WR. Consequently, the companion must be luminous. There is no trace of the other eclipse which indicates that the inclination of the system is rather well constrained, making WR 22 a very interesting object.

We would now like to use the case of WR 22 to shift from LTPV to the subject of intensive campaigns.

2.2. INTENSIVE CAMPAIGNS

2.2.1. WR 22

In order to observe the eclipses of WR 22, dedicated campaigns are necessary since the eclipses are usually not coincident with LTPV runs. In addition, because the eclipses seem to have a width of about one day, we need a dense sampling within a one-day interval. This is completely impossible in the framework of the LTPV project. Such a campaign has been organized in March 1990. The results were published in Gosset *et al.* (1991). Essentially, we definitely confirmed the existence of the eclipse and measured more accurately its width. Unfortunately, the whole eclipse was not observable due to the daylight gap. In addition, the WR is intrinsically variable within the night (see Fig. 1 of Gosset et al. 1991). Accordingly, if we want to derive the shape of the eclipses, we need to observe several occurrences in order to fill the day gap progressively and reduce statistically the dispersion due to the intrinsic WR variability. Consequently, we are currently organizing several campaigns.

It may be interesting to point out that it is difficult at ESO, and often elsewhere, to obtain long runs around the epoch of eclipses. Due to the WR variability, it is difficult to fix accurately the luminosity outside eclipses. Consequently, we must preserve the knowledge acquired in the framework of the LTPV project in order to know precisely (statistically) this mean luminosity and be able to measure the depth of the eclipse. In that context, it is important to recall that WRs are objects whose photometry is particularly sensitive to conformity errors of bandpasses and to colour-transformation uncertainties (Manfroid & Sterken 1987; Manfroid & Sterken 1992; Manfroid et al. 1992). Hence, it is imperative to perform our intensive campaigns with the Danish 50-cm telescope equipped with its Strömgren photometer, i.e. to observe in the same instrumental system as the LTPV project.

2.2.2. WR 16

A typical reason to complement LTPV data by intensive campaigns is to investigate the variability within one night. A good example is the small campaign for studying (among others) WR 16 which is described in section 3.2 of Gosset *et al.* (1990). From the LTPV data, we get dispersions of the magnitudes of WR 16 such that $\sigma_{uvby} =$ (0.016, 0.014, 0.012, 0.015). But we do not know a priori which part of the variability comes from high frequencies ($\nu > 1$ d^{-1}). To answer this question we observed WR 16 during eight nights with the ESO 1-m telescope equipped with a one-channel photometer and a Strömgren b filter. We acquired several measurements per night. The total dispersion of these data is $\sigma = 0.0090$: this is less than for the LTPV data and is due to the small number of nights. The important point is that the dispersion of the nightly means is $\sigma = 0.0088$ whereas the highest dispersion within a single night is $\sigma = 0.005$. Clearly, most of the variability is at low frequencies: the ARMA analysis of section 2.1.2 which admits that point implicitly is therefore not invalidated.

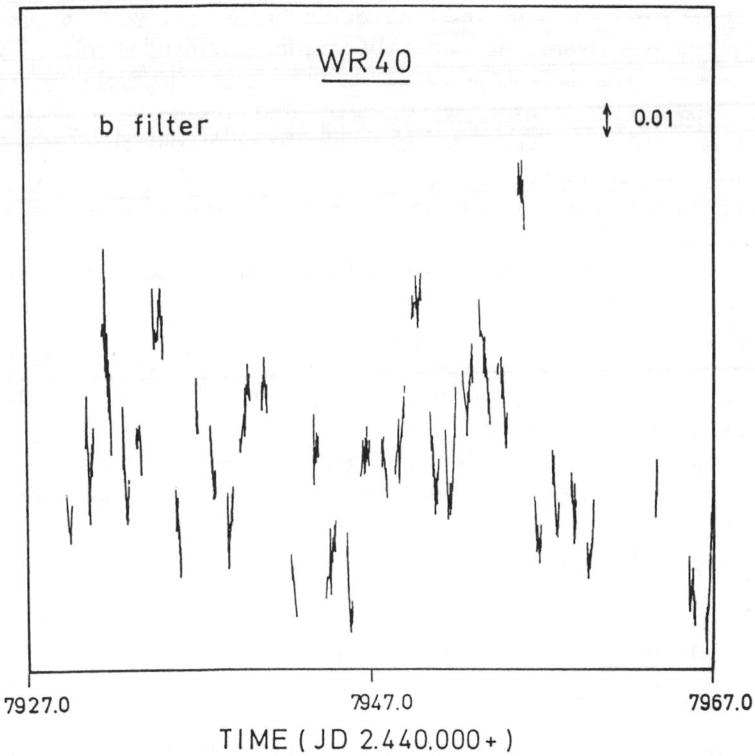

Fig. 3. Plot versus time of the *b* differential magnitudes of WR 40 measured during the March 1990 part of the campaign. The variability of the star within the night, although weaker than on longer time-scales, is well illustrated here

2.2.3. WR 40

The results for WR 40 derived from the LTPV data were sufficiently interesting to motivate a dedicated campaign. Therefore, from December 1989 to March 1990, we organized a big campaign. We merged several contiguous runs allocated to different observers (including an LTPV one) in order to gather a minimum of one or two measurements per night for the whole duration of the campaign. Altogether, we covered some one hundred nights. In addition, during the greater part of the month of March 1990, we studied the star with a time resolution of a few minutes. Moreover, the star was also observed from South Africa between JD2447894 and JD2448000 (by Luis Balona) in order to reduce the problems linked to the one-day aliasing. We present here preliminary and partial results deduced from this unprecedented campaign which, indeed, yielded much more information than a typical LTPV run.

A power spectrum of the photometric data corresponding to the whole campaign shows four families of aliases. Two families are related to the frequencies outlined by Gosset *et al.* (1989), Balona *et al.* (1989) and Gosset and Vreux (1990). The two other families are new and their presence requires further study of the problem. In Fig. 3, we present the Strömgren *b* filter differential magnitudes corresponding to the March subset of the data. One sees clearly the large variations from one day to the next. These are, of course, related to the four frequencies detected by our Fourier analysis but, in addition, we would like to point out that, within single nights, the star exhibits marked variations which are most probably not coherent since we do not notice them in the power spectrum. The typical time-scales of these variations are 3 to 12 hours.

Since we have at our disposal a considerable number of observations per night, we are also able to have a deep look at much shorter time-scales. Visual inspection of the data plots of individual nights suggests the existence of small dips and bumps in the lightcurve with typical time-scales of a few tens of minutes. In particular, on JD2447949, we notice a 45-min oscillation but only two cycles are clearly visible. The peak-to-peak amplitude is about 0.01 mag. The exact nature and origin of these variations is still unclear. This is definitely different from the behaviour observed during one night by Blecha *et al.* (1992) and interpreted as the signature of radial pulsations.

The first results for WR 40 (Gosset et al. 1989; Balona et al. 1989; Gosset & Vreux, 1990) justify by themselves the set up of such an observational campaign and illustrate the interest of complementing LTPV data by other observations. A detailed and fully documented study of these new data will be published elsewhere.

2.3. NARROW-BAND PHOTOMETRY

A criticism usually addressed at people studying the photometric variability of hot stars and in particular of WRs is that nobody knows what is actually varying. Is it the emission lines, the continuum or both ? It is possible to propose a first answer to these questions on the basis of the Strömgren photometry described above. The variability of almost all the WRs is lower in the *b* filter (see text above and also Table I). This is also the filter containing the He II λ4686 Å line (blended with a N III line), the most outstanding emission feature in the visible domain. Fig. 4 indeed exhibits the spectrum of WR 40 with, superimposed, the bandpasses associated with the Strömgren system. Consequently, it is natural to suggest (as discussed for example by van der Hucht et al. 1990, and van Genderen et al. 1989) that the variability is larger for the continuum than for the lines.

We decided to give a more straightforward answer to these questions and we organized a dedicated campaign. We performed differential photometry on three WRs (namely WR 22, WR 25, WR 40) simultaneously at the Danish 50 cm telescope equipped with the standard Strömgren photometer and at the ESO 1 m telescope equipped with a photometer and narrow-band filters designed for WR studies. The bandwidth of these particular filters is typically a few tens of Å. We will mention here five filters. The first one, at λ5021 Å, samples the continuum in a region free of strong emission lines; the same applies to the second one at λ6051 Å. The third filter is centred on the strong He II λ4686 Å line while the fourth filter is centred on the C IV λ5805 Å line and the last one on the He I λ5876 Å line. The localization of these narrow-band filters is illustrated in Fig. 5, to be compared with Fig. 4. As an example, we give in Table I the variability of WR 40 through the different filters.

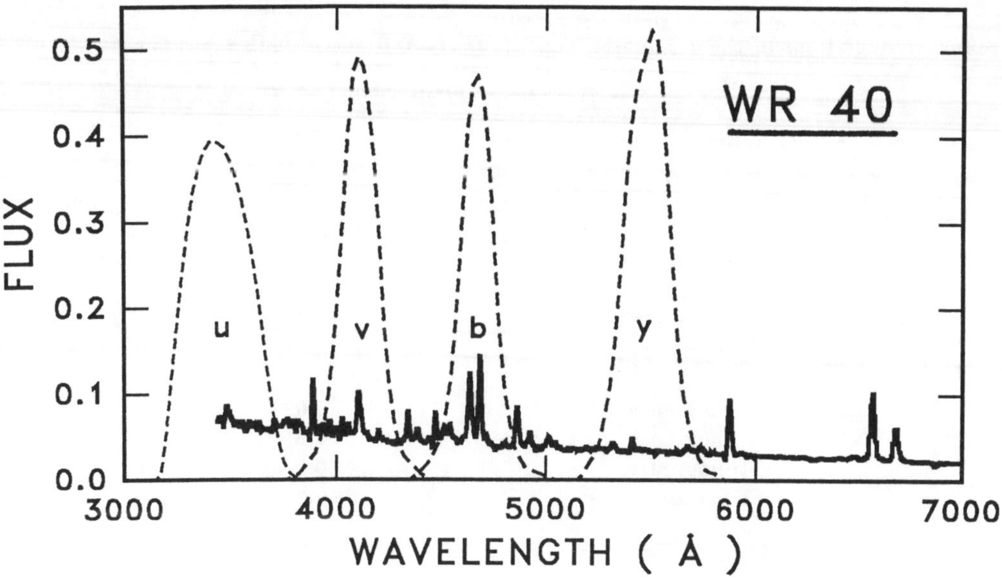

Fig. 4. Spectrum of WR 40 along with the four bandpasses of the Strömgren system

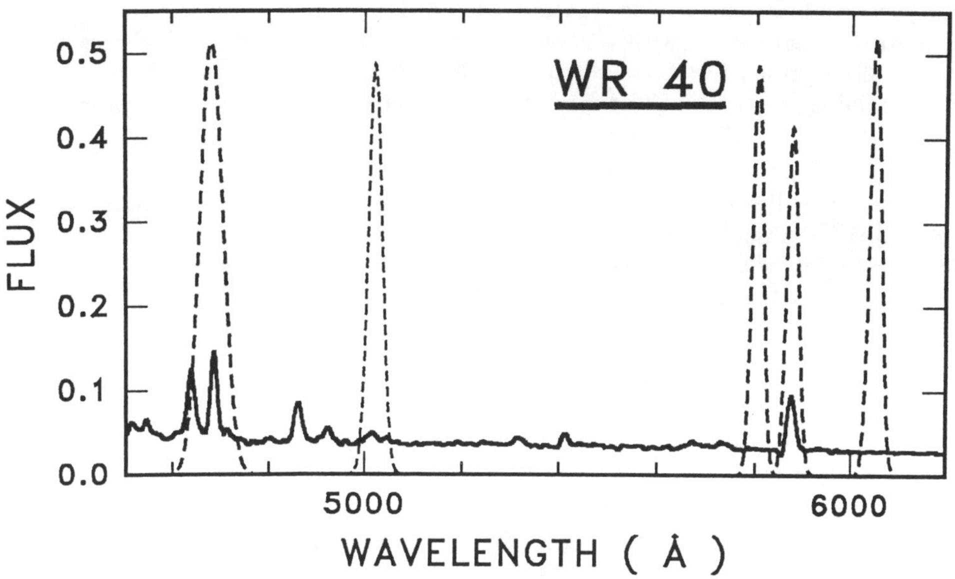

Fig. 5. Spectrum of WR 40 along with a few bandpasses of the particular narrow-band photometric system

TABLE I

Standard deviations of the differential magnitudes of WR 40 through the different filters

Filter	σ
Continuum λ5021 Å	0.022
Continuum λ6051 Å	0.022
He II λ4686 Å	0.018
C IV λ5805 Å	0.020
He I λ5878 Å	0.014
Strömgren *y*	0.023
Strömgren *b*	0.019
Strömgren *v*	0.020
Strömgren *u*	0.023

Clearly, the continuum varies with a dispersion $\sigma = 0.022$ in good agreement with what is observed in the Strömgren *u* and *y* filters. The variation in the fourth filter is almost as large because WR 40 is a WN8 star with accordingly weak carbon lines. The variability in the filter centred around He II λ4686 Å is somewhat lower ($\sigma = 0.018$) in agreement with the value ($\sigma = 0.019$) related to the Strömgren *b* filter. Finally, it is interesting to notice that the variability is the lowest in the fifth filter containing the He I λ5876 Å line. All this confirms the idea that the strongest variability is due to the continuum.

Although the above-mentioned example is a good illustration of the interest of performing dedicated narrow-band photometry of WRs, it is very difficult to organize a large campaign to investigate the variability of WRs through these filters. Indeed, these filters are not suited to cooler stars and it is impossible to conceive some kind of LTPV project using these too specific filters.

3. Conclusion

The aim of this contribution was to show how photometric investigations of the WR variability could be used to answer some important questions linked to the theoretical modelling of these objects, such as the existence (or not) of radial or non-radial pulsations, not to speak of the crucial question of the proportion of binaries. We described our strategy which consists in concentrating our efforts on a limited number of objects and in complementing the results of the LTPV project by the ones of dedicated campaigns of Strömgren *uvby* and particular narrow-band photometry. We hope we have shown the interest of the LTPV project for WR studies and that this will help to ensure its continuation.

Acknowledgements

We are greatly indebted to all the observers of the *Long-Term Photometry of Variables* project. Luis Balona and Gabriel Franco collaborated with us for our March 1990 campaign. This research was supported in part by contract ARC 90/94-140 "Action de recherche concertée de la Communauté Française" (Belgium) and by contract SC005 of the Belgian Programme Service Centres and Research Networks initiated by the Belgian State, Prime Minister's Office, Science Policy Programming. Our thanks also go to J.-P. Swings for having read and significantly improved a first draft of the present paper. C.S. acknowledges financial support from the Belgian Fund for Collective Basic Research (grant 2.0060.91).

References

Balona, L.A., Egan, J., Marang, F.: 1989, *Monthly Notices of the RAS*, 240, 103

Blecha, A., Schaller, G., Maeder, A.: 1992, *Nature*, 360, 320

Box, G.E.P., Jenkins, G.M.: 1976, Time Series Analysis, Forecasting and Control, revised edition, San Francisco: Holden-day

Cherepashchuk, A.M., Aslanov, A.A.: 1984, *Astrophysics and Space Science*, 102, 97

Conti, P.S.: 1976, *Mém. Soc. Royale Sci. Liège (Série 6)* 9, 193

Conti, P.S., Niemela, V.S., Walborn, N.R.: 1979, *Astrophysical Journal*, 228, 206

Deeming, T.J.: 1975, *Astrophysics and Space Science*, 36, 137

Gosset, E., Vreux, J.-M.: 1990, *Astronomy and Astrophysics*, 231, 100

Gosset, E., Remy, M., Manfroid, J., Vreux, J.-M., Balona, L.A., Sterken, C., Franco, G.A.P.: 1991, *Information Bulletin on Variable Stars*, 3571, 1

Gosset, E., Vreux, J.-M., Manfroid, J., Remy, M., Sterken, C.: 1990, *Astronomy and Astrophysics, Supplement Series*, 84, 377

Gosset, E., Vreux, J.-M., Manfroid, J., Sterken, C., Walker, E.N., Haefner, R.: 1989, *Monthly Notices of the RAS*, 238, 97

Lamontagne, R., Moffat, A.F.J.: 1987, *Astronomical Journal*, 94, 1008

Lucy, L.B., Abbott, D.C.: 1993, *Astrophysical Journal*, 405, 738

Maeder, A.: 1983, *Astronomy and Astrophysics*, 120, 113

Maeder, A.: 1985, *Astronomy and Astrophysics*, 147, 300

Magain, P., Vreux, J.-M., Manfroid, J.: 1987, *Information Bulletin on Variable Stars*, 3022, 1

Manfroid, J., Sterken, C.: 1987, *Astronomy and Astrophysics, Supplement Series*, 71, 539

Manfroid, J., Sterken, C.: 1992, *Astronomy and Astrophysics*, 258, 600

Manfroid, J., Sterken, C., Gosset, E.: 1992, *Astronomy and Astrophysics*, 264, 345

Moffat, A.F.J.: 1978, *Astronomy and Astrophysics*, 68, 41

Moffat, A.F.J., Robert, C.: 1991, in *Wolf-Rayet Stars and Interrelations with Other Massive Stars in Galaxies*, IAU symposium 143, eds. K.A. van der Hucht and B. Hidayat, Dordrecht: Kluwer Acad. Publ., 109

Moffat, A.F.J., Seggewiss, W.: 1978, *Astronomy and Astrophysics*, 70, 69

Niemela, V.S.: 1979, in *Mass Loss and Evolution of O Type Stars*, IAU symposium 83, eds. P.S. Conti and C.W.H. de Loore, Dordrecht: Reidel, 291

Noels, A., Scuflaire, R.: 1986, *Astronomy and Astrophysics*, 161, 125

Owocki, S.P., Castor, J.I., Rybicki, G.B.: 1988, *Astrophysical Journal*, 335, 914

Poe, C.H., Friend, D.B., Cassinelli, J.P.: 1989, *Astrophysical Journal*, 337, 888

Scargle, J.D.: 1981, *Astrophysical Journal, Supplement Series*, 45, 1

Schild, R.E., Hiltner, W.A., Sanduleak, N.: 1969, *Astrophysical Journal*, 156, 609

Seggewiss, W., Moffat, A.F.J.: 1979, *Astronomy and Astrophysics*, 72, 332

Sterken, C.: 1983, *The Messenger*, 33, 10

Underhill, A.B.: 1983, *Astrophysical Journal, Letters to the Editor*, 268, L127

van den Heuvel, E.P.J.: 1976, in *Structure and Evolution of Close Binary Systems*, IAU symposium 73, eds. P. Eggleton, S. Mitton and J. Whelan, Dordrecht: Reidel, 35

van der Hucht, K.A.: 1992, *Astronomy and Astrophysics Review*, 4, 123 van der Hucht, K.A., Conti, P.S., Lundström, I., Stenholm, B.: 1981, *Space Science Reviews*, 28, 227

van der Hucht, K.A., van Genderen, A.M., Bakker, P.R.: 1990, *Astronomy and Astrophysics*, 228, 108

van Genderen, A.M., van der Hucht, K.A., Bakker, P.R.: 1989, *Astronomy and Astrophysics*, 224, 125

van Genderen, A.M., van der Hucht, K.A., Steemers, W.J.G.: 1987, *Astronomy and Astrophysics*, 185, 131

Vreux, J.-M.: 1985, *Publications of the ASP*, 97, 274

Vreux, J.-M.: 1987, in *Instabilities in Luminous Early Type Stars*, eds. H. Lamers and C. de Loore, Dordrecht: Reidel, 81

DISCUSSION

Sterken: Looking at the spectra, it is tempting to believe that the right approach in observing such stars is by using scanner or spectrometer observations, and then singling out sections of the continuum that are free of emission lines—if one can calibrate the instrument, of course.

Gosset: The narrow-band filters we use are doing well, too, since their FWHM is only about 30 Å, i.e. of the order of the widths of the lines seen in WR spectra. Moreover, WR stars of Type WN do not have too many spectral lines. Of, course, spectrophotometry is the successor of the present-day photometry, but the precision is not yet what we need. Within the framework of the LTPV project, our differential photometry reaches a precision of $\sigma = 0\overset{m}{.}003$, a value far better than what spectrophotometry can yield today.

Breger: The periodicities of Wolf-Rayet stars you present, are very interesting. Periods in this range are quite difficult to determine and to verify. It might be easier for the audience to make a judgement, if you could present power spectra and argue the reality of these variations on the basis of the comparison of the power spectra between variable and comparison stars.

Gosset: This is certainly true; thanks for the suggestion, I shall take it into account.

Moffat: The near-absence or lack of variability in emission lines compared to continuum was already seen in WR40 in 1975–76 data (Moffat & Isserstedt 1980).

Gosset: Our special filters are narrower and, thus, allow more quantitative results. For example, they isolate the HeII λ4686 line from the NIII $\lambda4640 - -\lambda$ blend.

de Groot: Please comment on the reality of the 45-min "periodicity" in the viewgraph you showed during the discussion.

Gosset: First of all, I would not speak of periodicity before a more detailed analysis has been completed: the phenomenon stands out only during two cycles. However, the amplitude associated with these two cycles is well beyond the noise. This certainly deserves a detailed investigation.

Burki: A. Blecha discovered a period of about 10 minutes in a WR star. Did you derive any periodicity shorter than about half an hour in your data? Note that other WR stars have been observed with our CCD at the Swiss telescope on La Silla, and that several of them show short periods during some nights.

Gosset: Blecha and coworkers have observed the 10-min periodicity in WR40. Our sampling was not appropriate for detecting such a short period (we have roughly one measurement every 4–5 minutes). Taking into account the precision of our data, I am not sure whether we would be able to detect it if it is not persistent. In any case, we did not see it. As for the short-period variations of the other WR stars, I wish to stress that one should be very careful when using CCDs.

Kurtz: We have looked at WR40 for a total of about 25 hours on many nights with much higher accuracy than the CCD photometry of Blecha and coworkers. We see *no* evidence of the purported 10-min variability.

Gosset: Within a time span of one month, only during one night were such 45-min oscillations clearly seen. It could also be the case for the time scale suggested by the Swiss. One could imagine that the visibility of short periods (if they exist) is phase-dependent (I am speaking of a phase related to the longer periods described in Gosset & Vreux 1990, *Astronomy and Astrophysics*, 231, 100).

Kurtz: You must be very careful with power spectra at low frequencies because of the window pattern from negative frequencies spilling over into the positive frequencies. It is best to show the window pattern by generating a noise-free sine wave sampled at the times of the actual observations, and then looking at the power spectrum of that signal.

Gosset: Sure, you are right. We are routinely taking great care of all these effects but, for this talk, I have decided to put the emphasis on the results, not on the methods. We also use other methods of analysis than Fourier-related ones.

Le Contel: I would like to mention that I observed jumps or fluctuations at the level of $0\overset{\mathrm{m}}{.}01$ on a variable and two comparison stars on which the signal was at the millimag level. The effect was chromatic with a "pseudo periodicity" of 10 minutes, but with a time lag between the variable and the comparison stars depending on the distance of the comparisons to the variable. I interpreted this as an atmospheric effect, as it was season-dependent due to the closeness of the Sahara desert to the observing site (Granada). So, maybe, the 45-min variability you observed is real, maybe it is the signature of atmospheric-transparency variations.

Gosset: I am not completely convinced either, but such variations are usually not visible in our two-comparison-star differential magnitudes. We see no observational effect that could only affect the WR, except problems related to the presence of the nebula around WR40. We are investigating that aspect. The definitive answer will be in a forthcoming paper.

Williams, van Genderen and de Groot watch Moffat handling his last meal before his date with Alice in Wonderland (see p. 136)

VARIABILITY OF WOLF-RAYET STARS: CONSTRAINTS ON THE PROPERTIES OF HOT-STAR WINDS

Anthony F.J. MOFFAT
Département de Physique and Observatoire du Mont Mégantic
Université de Montréal
C.P. 6128; Succ. Centre Ville
Montréal QC H3C 3J7
Canada.

ABSTRACT. This paper describes how intense optical studies of the variability of Wolf-Rayet stars in photometry, polarimetry and spectroscopy are important—even crucial in some ways—to our understanding of the nature of hot-star winds. This applies as much to single WR stars as to binary systems containing WR stars.

Keywords: Wolf-Rayet stars – early-type stars – stellar variability – photometry – polarimetry – spectroscopy

1. Introduction

Stars of high mass evolve *internally* very rapidly. While the Sun will nucleosynthesize its central supply of hydrogen in some 9000 million years, a star 30 times more massive will do the same in a mere 5 million years (Maeder 1991a). Consequently, more massive stars will leave the main sequence at a much quickened pace.

Massive stars evolve more quickly also at their *surface*. Since the luminosity of a (main-sequence) star increases highly non-linearly with the mass ($L \sim M^3$), the increased radiation pressure (mainly on atoms, ions and electrons in hot-star winds, and on dust in cool-star winds) will drive more intense stellar winds in more massive stars. As an example, the Sun loses of order 10^{-14} of its mass per year while a $30M_\odot$ star will lose some $10^{-6}M_\odot$ per year. For massive stars, the mass-loss rates depend mainly on the luminosity and go something like $\dot{M} \sim L^{1.6}$ (e.g. Howarth & Prinja 1989).

As a consequence of these high mass-loss rates, stars of initial mass above about $30M_\odot$ will eventually lose a significant part of their outer envelopes at the same time as the products of CNO H-burning (enriched He and N) are mixed to the eroded surface. Such stars become hotter and develop increasingly strong winds; they are recognized as Wolf-Rayet stars of the N-sequence (i.e. WN). Later, when He-burning advances, a transition to a WC star can take place at the surface. The above scenario applies to the case of initial solar abundance. For lower metallicity

C. Sterken and M. de Groot (eds.), The Impact of Long-Term Monitoring on Variable Star Research, 117–140.

(e.g. LMC, where $Z \sim Z_{\odot}/3$, or especially SMC, where $Z \sim Z_{\odot}/10$), the minimum initial mass needed to lead to WR stars is higher, since the radiation pressure which drives the mass loss depends quite sensitively on the presence of heavy ions (Maeder 1991a).

WR stars are notorious for their high mass-loss rates, typically some $10^{-5} M_{\odot} yr^{-1}$ with a range of about a factor three on either side of this. These rates are about an order of magnitude higher than those of their predecessor O-stars. The physical origin of these strong WR winds is unknown, in large part due to the fact that the high wind density itself prevents us from seeing down to the surface of the stellar core and probing its structure. One is thus obliged to learn about WR stars on the basis of the (processed) radiation that emerges from their winds. Recent reviews of WR stars are given by van der Hucht (1992) and Maeder & Conti (1994); see also the proceedings of IAU Symposium No. 143 (van der Hucht & Hidayat 1991).

WR stars represent an important link through which all massive stars above a certain limiting mass appear to pass on their way from the main sequence to the neutron star/black hole graveyard (unless they explode before reaching the WR stage, e.g. at lower Z; or they evaporate previously and do not explode as supernovae; or the supernova explosion disperses all the mass). WR stars exhibit the strongest sustained winds known among hot stars. The ejection of large quantities of nuclear processed material at high speed (typical terminal wind speeds are $\sim$ 2000 km s^{-1}, within a factor $\sim$ 2) has an important impact on the ISM, even if the number of WR stars appears to be small (about 1200 total estimated in the Galaxy: Maeder & Lequeux 1983).

Although considerable progress has been made over the past decade or so, our fundamental understanding of WR stars is still considerably compromised by their very dense winds. At the same time, this makes them very interesting! In particular, snap-shot models of the winds (spectral emission lines) are generally limited to the assumption of homogeneity and spherical symmetry (e.g. Hillier 1991; Hamann 1991). That life may not be so simple is indicated by recent studies involving *time* variability. Such (mainly observational) studies (e.g. Moffat & Robert 1992) can provide crucial diagnostics e.g. on rotation, pulsation, rotating magnetic fields and magnetically generated flares, clumping, wind expansion speed (e.g. via clumps as tracers propagating in the wind), binary frequency, wind collisions in binaries, and the existence of WR + c systems (c = compact companion).

In this review, I will discuss recent advances in WR-star research based on photometric, polarimetric and spectroscopic monitoring on various time-scales in the optical. This is complemented by other work reported in these Proceedings, such as radio and IR variability of WR stars by van der Hucht and Williams, and the ESO LTPV photometric-monitoring project of WR stars (mainly WNL, i.e. late type, cool WN stars) by Gosset and collaborators. The study of UV variability has been somewhat limited by the low signal-to-noise ratio of IUE spectra, although they certainly deserve mention (St-Louis 1990; see also Prinja & Smith 1992).

2. Data Sources

The nature of monitoring programmes is set mainly by the expected time-scales

TABLE I
Typical expected time-scales and feasible techniques for optical monitoring of WR stars

Phenomenon	Timescale	Technique
pulsars in WR + c	msec - sec	phot
flares, pulsations, clumps	min - hours	phot, pol, sp
rotation, binaries, precession	days - years	phot, pol, sp

Note: Techniques refer to photometry and polarimetry, both preferable in the continuum, and spectroscopy

and the techniques available. In the case of WR stars these are summarized for optical work in Table 1. Note that this list is almost certainly biased by my own current ideas; one should always remain open to other previously unknown or unthought-of time-scales. It is my view that wherever possible, photometry and polarimetry are best done in the continuum, leaving studies of the emission lines to spectroscopic—even spectropolarimetric—monitoring. This means abandoning the temptation (because it is easier and often more expedient) to observe photometrically or polarimetrically in *standard* medium or broad-band filters (e.g. UBV, Strömgren, Walraven,...). Such systems were certainly not designed with WR stars in mind! The ambiguity and hence confusion that often result as to whether the variations occur in the continuum or in the (often very strong) emission lines or both, can considerably reduce the value of such data.

It should be noted that it can be of great advantage to obtain time-resolved data simultaneously in all three modes (continuum photometry and polarimetry, and spectroscopy), as argued e.g. in the case of wind-clump studies by Brown et al. (1994). For WR stars this has never really been attempted yet! Furthermore, the technique of spectropolarimetry and its time variability is beginning to look very promising for WR stars (e.g. Schmidt 1988; Schulte-Ladbeck et al. 1992). Besides revealing potential wind asymmetries via (variable) line-depolarization effects, spectropolarimetry should also be used to search directly for magnetic fields and their modulation by rotation and/or by intrinsic growth or decay.

Most work on variability has been done in the optical, where ground-based techniques are wel known and developed, leading to the highest precision. Other areas are gradually being exploited (radio, IR, UV, X, even gamma!), but none has *yet* attained the same high relative precision that is possible in the optical. The increasing possibilities of working outside the Earth's atmosphere, however, promise to change this during the next decade or so.

In Table 2, I list as an example the rather wide range of optical telescopes that I and my collaborators have used repeatedly in the last few years, to obtain time-resolved data of WR stars in any of the three modes: photometry, polarimetry and spectroscopy. Note the obvious trend epitomized by the two extremes: photon-intense studies involving high-resolution, time-resolved spectroscopy require the largest telescopes, while broad-band photometry can be done well on small telescopes (for the same stars). For best results, time-resolved, high-resolution spec-

TABLE II

Optical telescopes I have repeatedly used for monitoring WR stars

Size/site	Purpose	Time-step	Typical Run
3.6m CFHT, ESO	High-res. sp	< 1 hr	3-4 days
2m MPI(ESO), Leoncito (Arg.)	Broad + med-band pol	~ 1 d, (hrs)	~ 2 wks
1.6m Mont Mégantic (Québec)	Med-res. sp	~ 1 d, (hrs)	~ 1 wk
0.6m UTSO, CTIO (Chile)	Med-res. sp Broad-band pol Broad + med-band phot	~ 1 d, (hrs) ~ 1 d, (hrs) msec - 1d	several wks
0.25m APT (Arizona)	Broad-band phot	~ 1 d	yrs

Note: The Mégantic telescope can be considered as a "backyard" telescope, available for several ~ week-long runs every year, although not in a good climate

tropolarimetry will ultimately require the use of 8+ metre class telescopes.

Clearly, it is much easier to get more, continuous (or more precisely in practice, contiguous) time coverage (essential to minimize the adverse effects of aliasing and spurious harmonics) on smaller telescopes. I summarize current contiguous-night assigments I have enjoyed on optical telescopes of different size classes in Table 3. For many situations, a WET (whole Earth telescope) configuration would be extremely beneficial for any size class. Curiously, I personally have never used more than 2 (small) telescopes at (unfortunately only slightly) different longitudes, although I have finally succeeded with other collaborators in arranging for a joint run on three 4-metre class telescopes (1994 June) at widely different longitudes (CFHT, Hawaii; WHT, La Palma, Spain; and the 6m Zelenchukskaya telescope in Russia) to obtain nearly contiguous high-resolution spectroscopy of one key WR star over 3 days.

A final remark about data quality and, to some extent, quantity as well: As one goes from self-obtained through shared to service observing and automatic telescopes, it is normal that the amount of drive and devotion to a given project generally decreases. Nevertheless, in the long run, one will have to find ways to optimize the APT mode, preferably in WET configuration (short of using a devoted continuous pointing telescope in space or from the Poles), which is the only truly viable mode of proper attack on many scientific problems related to variability, not only of WR stars, but of many kinds of variable objects. This approach would, however, be highly suited to WR stars, in which most of the important time scales lie in the hours-days range.

TABLE III
Contiguous night runs on telescopes of different size classes

Diameter(m)	No. Contiguous Nights	
	Typical	Maximum
8	(1?)	?
4	~ 3	9
2	~ 5	21
1	~ 7	12
0.5	~14	90

Note: The 8-metre value is hypothetical. The 9 nights maximum on 4m class telescopes actually refer to three nights on three 4m class telescopes simultaneously

3. Science Highlights

I will now depart from categories depending on technique to the scientifically more interesting classification according to theme. My experience has shown that it makes good sense to divide the results on variability into three catagories: single WR stars, binary WR stars and ambiguous cases. I will not try to be complete, rather I will concentrate on recent highlights.

3.1. SINGLE WR STARS

In my view, one of the most important results to emerge lately is the recognition of stochastic variability in the winds of virtually all WR stars that have been monitored appropriately, regardless of subtype. In fact, all hot-star winds may show the same basic structure, which is most readily seen in the strong winds of WR stars. The most likely interpretation is that we are seeing the consequences of (possibly full-scale) turbulence of a compressible gas filled with supersonic shocks, much like what is seen in GMCs (cf. Moffat et al. 1994). The idea of such shocks originally stemmed from the presence of very high ionization (e.g. O VI) lines, as well as significant X-ray fluxes. This led to shock theories in O-star winds involving either forward shocks (Lucy & White 1980) or reverse shocks (Owocki 1991), depending on the speed of the density enhancements relative to the rest of the wind.

Observationally, what led me and my collaborators to this important discovery in WR stars was a series of circumstances starting with intense photometric monitoring of a large sample of mainly single WR stars (Moffat & Shara 1986; Lamontagne & Moffat 1987). Although, generally, only one data point was obtained per night over runs of typically 2 weeks, these data showed that most WR stars are stochastically variable at some (low) level. WR stars with slower winds (late type WN or WC) tend to be more variable, for a reason that was only appreciated later

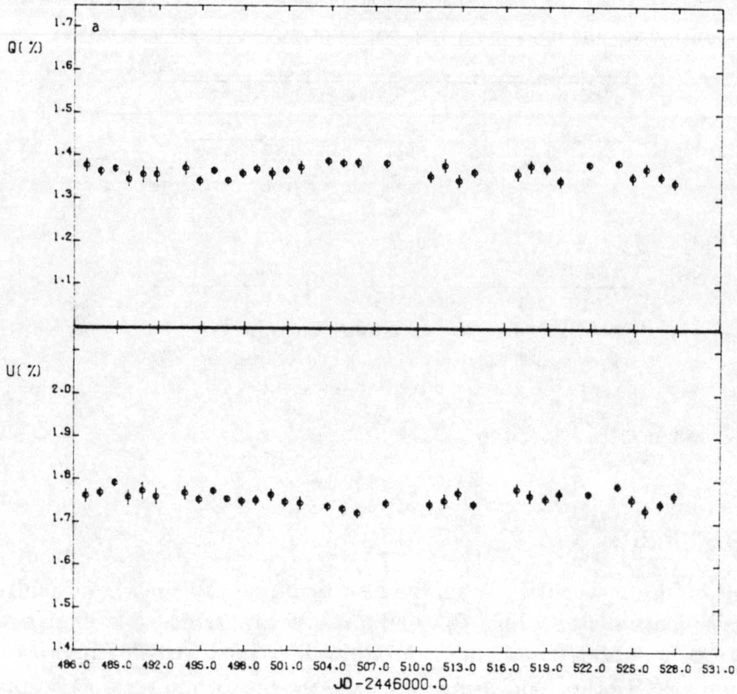

Fig. 1. Stokes' linear polarization parameters $Q = Pcos2\theta, U = Psin2\theta$, versus time (~ 6 week interval) for (a) the non-variable single WC7 star HD 156385 = WR 90 and (b) the highly variable WN8 star HD 96548 = WR 40. The random character of the Q,U plot for WR 40 in (c) shows the stochastic nature of the variations, without any preferred direction such as in a flattened disk. Taken from Drissen et al. (1987) and St-Louis et al. (1987)

during spectroscopic work (Robert 1992).

The photometric work was followed up by polarimetric monitoring, also mainly on a one-day basis, over several weeks (St-Louis et al. 1987; Drissen et al. 1987, 1993; Robert et al. 1989). If the photometric variability had been due to non-radial or radial pulsations in a spherical smooth wind, one would have expected the intrinsic continuum polarization to be extremely small or even zero and hence the net polarization (allowing for interstellar polarization, which is often quite high) to be almost constant in time. However, significant polarization variability *was* found in many single WR stars, with the same trend as in photometry: slow-wind WR stars tend to be more variable in polarization (cf. Fig. 1). These results strongly suggest the presence of a stochastically variable, inhomogeneous wind, on time-scales somewhat less than a day. The nature of the inhomogeneities could not, however, be ascertained from these data alone.

This led us to look at optical spectral lines for evidence of some kind of clumpy structure in the winds. Many optical emission lines arise via subordinate recombi-

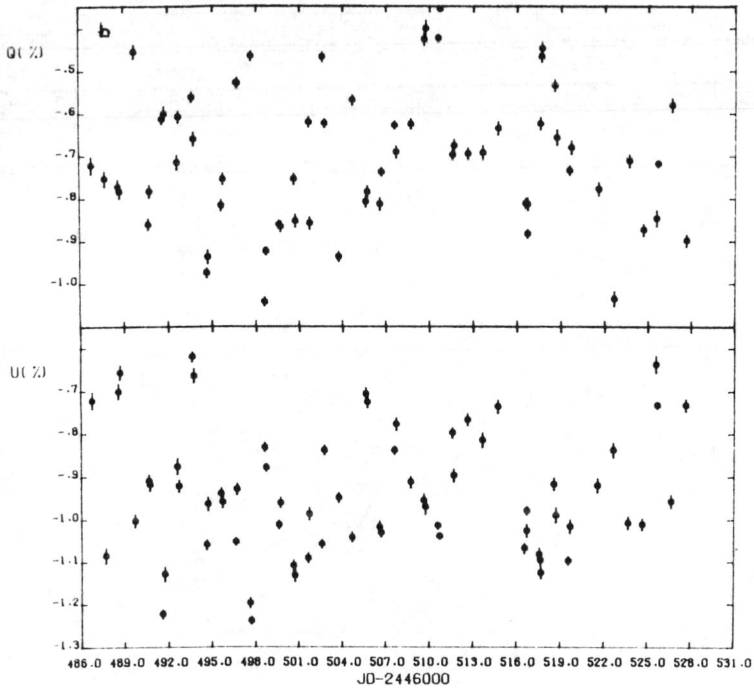

Fig. 1. b

nation transitions throughout the wind, thus allowing one to probe essentially the whole wind via Doppler expansion of emitting plasma. Extensive monitoring of some 9 WR stars of various subclasses (WN5-8; WC5-9) led Robert (1992) in her Ph.D. thesis (cf. also the summary in Moffat & Robert 1992) to the detection of emission substructures superposed on many of the broad emission lines (see Fig.2 for an example). After subtracting out the mean or minimum (to give non-negative values) spectrum, these substructures were seen to be quasi bell-shaped, so multi-gaussian fits were first made to them. This allowed the substructures on a given line on each spectrum to be characterized by three independent quantities: flux (f), width (σ_v, expressed in km s^{-1}) and central position (projected radial-velocity component). Later, the much more objective and powerful technique of wavelet analysis was applied (Moffat et al. 1994), which allows more rigorous statistical tests to be made as to their reality (Lépine 1994).

The first results to emerge concerning the substructures were the following:
- The number of substructures detected to a given threshold limit for a given emissin line is directly proportional to the wind terminal velocity of the star.
- Line variability at different Doppler velocities across an emission line is directly proportional to the underlying line flux.
- Lines of high-ionization species (formed in hotter regions close to the central core)

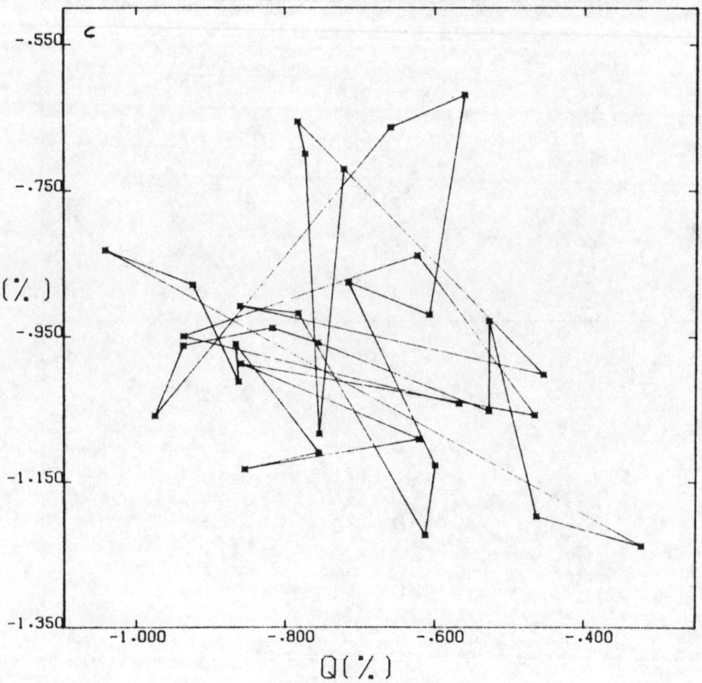

Fig. 1. c

tend to be *less* variable.

- The substructures tend to move away from the line centre with time.

These already implied some kind of density enhancements or clumps in the wind, that form, grow, then decay stochastically as they propagate outwards in an otherwise spherical wind. A hint of just what these changes physically are was revealed by the detection of:

- scaling laws: $f(\sigma_v), \tau(f)$, where τ is the lifetime,
- flux spectrum $N(f)$ converted, assuming a density-squared recombination-emission law, into a mass spectrum: $\mathcal{N}(m) \sim m^{-1.5}$,
- anisotropy, i.e. clumps seen coming out towards or away from the observer tend to show ~ 3 times higher velocity dispersion than those moving out perpendicularly to the line of sight: this implies a radial component of velocity dispersion that is greater than the tangential component;
- fractal dimension D ~ 2.3.

All of these strongly imply that we are seeing (shock-induced) eddies in a compressible, supersonic turbulent medium, much like what is seen in the ISM (especially GMCs: Henriksen 1991, 1994; Moffat et al. 1994; Lépine 1994), or even in Lyα clouds of the IGM (Giallongo et al. 1993). For example, in GMCs, one finds $\mathcal{N}(m) \sim m^{-1.5}$ on average (Williams & Blitz 1993) and non-spherical clumps

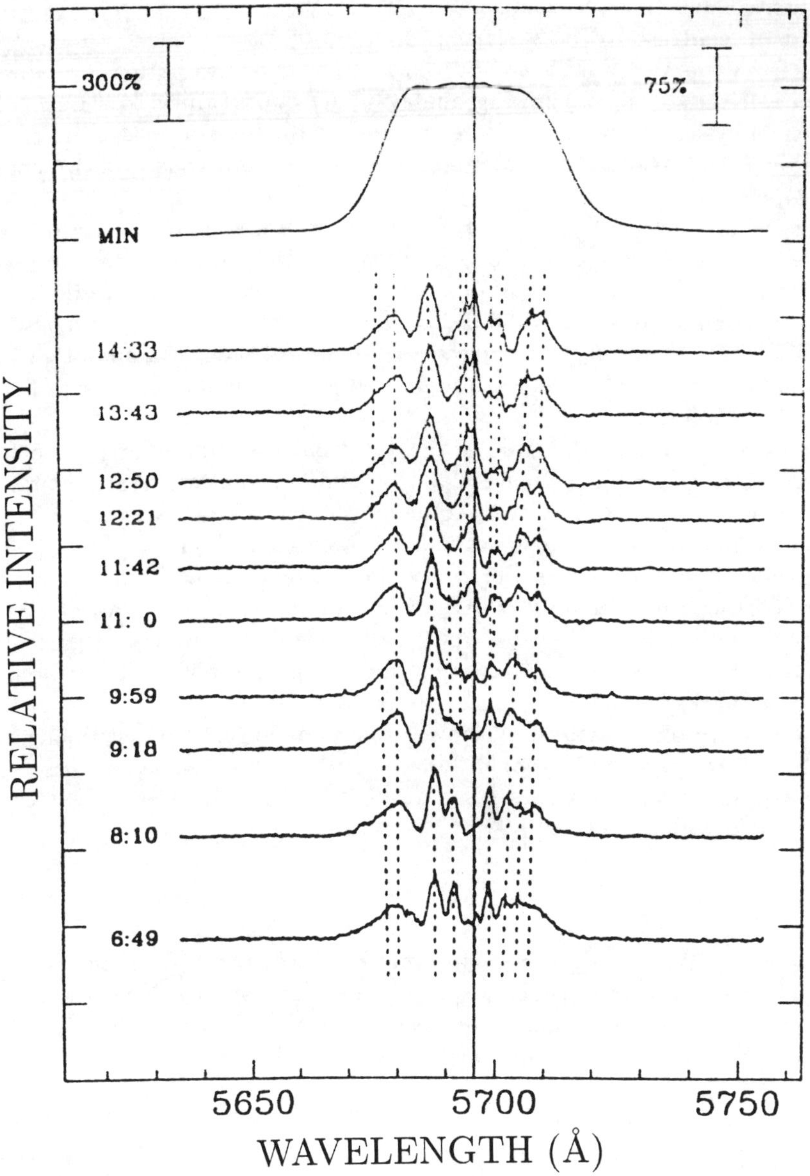

Fig. 2. Montage of substructures on a spectral line (C III 5696) from one night's data of the WC8 star HD 192103 = WR 135 (Moffat & Robert 1992). The top spectrum is the slightly smoothed minimum spectrum of four nights' data. Below this, the series of spectra was obtained by subtracting this minimum spectrum from each individual spectrum and displaying it on a 4x enlarged scale. UT times are indicated. Dashed straight lines are shown as a guide to trace substructures on successive spectra. Vertical bars indicate the scales in continuum units. Note the immediate impression of a stochastic superposition of a few large quasi-gaussian shaped substructures along with increasing numbers of smaller, narrower ones down to the instrumental noise limit

randomly oriented due to fluctuating gravitational fields (Fleck 1992). The latter property is in contrast to the preferred direction of higher velocity dispersion pointing out from the WR star, since outward directed net radiation pressure is the turbulence driver. Besides allowing one a unique opportunity to study time-dependent astrophysical turbulence in a compressible medium directly, the consequences of global wind clumping for hot-star science are numerous and important (Moffat 1994), e.g.:

- If the whole wind shows a hierarchical, turbulent, clumpy structure right down to the smallest scale, persisting out to large distances from the star, the density-squared dependence of the best traditional mass-loss indicators (i.e. radio/IR f-f emission or subordinate emission-line fluxes) will require reduction of the mass-loss rates by a factor of ~ 3 or more to correct for the assumption of a smooth wind, possibly for all hot stars (e.g. Moffat & Robert 1994). This would have a significant impact on stellar evolution.
- If the clumps are really an integral part of the wind (as suggested by the fact that the dispersion spectrum of line variability generally follows the line profile itself), then they can be used as tracers to define the wind kinematics, and hence lead to viable studies of the wind dynamics. However, it should be noted that the accelerations of the WR clumps are very slow compared to the accelerations obtained for O-star winds, mainly from UV-line fitting (e.g. a β-law: $v(r) = v_\infty(1 - R_\star/r)^\beta$, with β index close to unity), thus leading some investigators to believe that the clumps are shocks that are propagating at lower speeds compared to some kind of "background" wind (e.g. Owocki et al. 1988).
- The clumpy nature of the winds will have to be allowed for when studying the dynamics of wind interactions, either in binaries or with the ISM. Generally, the interactions will be less energetic, since the mean free path will be enhanced when there are "spaces" in the wind.

3.2. BINARY WR STARS

It is useful to study WR + O binaries, not only to obtain the masses, but also to probe the winds by eclipse and variable-transparency effects and by their interaction through collision. To do this properly, we need (1) spectroscopy to provide (a) RV orbits from well-behaved lines (usually weak lines of high ionization that are formed closest to the WR core where they are least likely to be perturbed), (b) phase-dependent profile changes related to occultation effects, and (c) information on the excess emission from the shock cone that wraps around the O-star, which generally will have the lower mass-loss rate. We also need either (2) light-curves or (3) time-dependent polarization curves (both preferably in the continuum), or both, to yield information on the orbital inclination and mass-loss rate primarily of the WR star (e.g. St-Louis et al. 1988, 1993a; Moffat et al. 1990; Lamontagne et al. 1991, 1994). The inclination is used to convert the spectroscopic values of $M sin^3 i$ into actual masses M. All of these techniques require monitoring. In fact, if the orbital period is close to a small number of integral days, it may take many weeks of data to piece together a full orbit from single-site observations. Even if this is not the case, it is often wise to observe several orbital cycles to check for consistency and

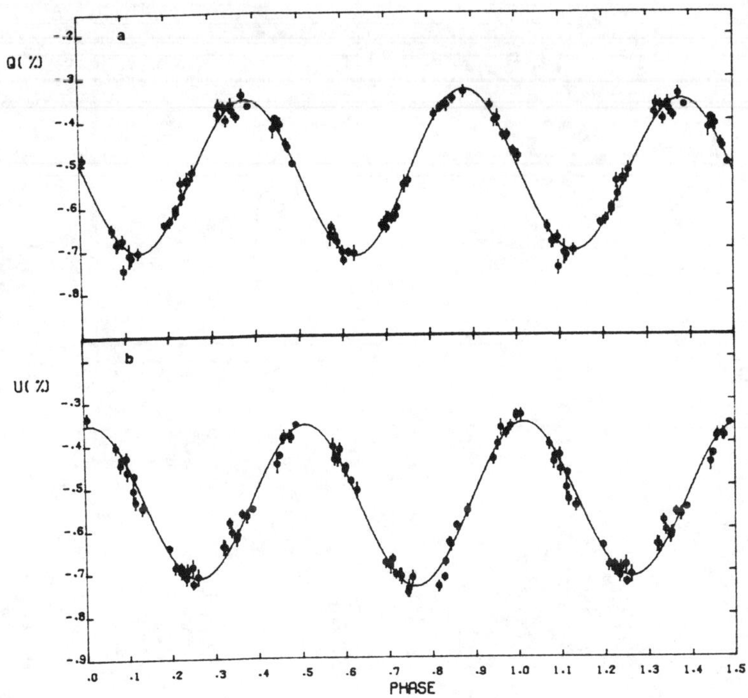

Fig. 3. Phase-dependent Stokes' parameters Q, U observed for HD 97152 = WR 42, WC7 + O5-7, P = $7\overset{d}{.}886$, fitted to a theoretical curve for a circular orbit (St-Louis et al. 1987)

sometimes simply to beat down any intrinsic noise. This is illustrated in Fig. 3 for the polarization orbit of a non-eclipsing WR + O system with P $\sim$ 8d, and in Fig. 4 for light-curves of several non core-eclipsing systems, including the same star as in Fig. 3. Both techniques (i.e. polarization and light curves) are sensitive to free electron scattering and yield inclinations and mass-loss rates, but probe different parts of (mainly) the WR wind. A consistency check between i and $\dot{M}$ from the two methods shows generally good agreement, but with errors significantly larger than formal errors calculated by the principle of propagating independent errors (Lamontagne et al. 1994).

It is interesting to note that essentially all WR + O non core-eclipsing systems (i.e. those showing only one dip per orbit; except those of long period or low inclination) show relatively shallow, v-shaped light curve dips, with apex centred at phase zero (WR star in front). The model is based on the assumption of a spherically symmetric wind and fits well in all cases. Certainly, highly flattened winds (e.g. Underhill & Nemec 1989) will not fit the data. It is also worthy of note that the mass-loss rates obtained by photometric light-curves and by polarimetry depend *linearly* on the electron density and are thus impervious to wind clumping. This explains why photometric and polarimetric mass-loss rates are systematically lower,

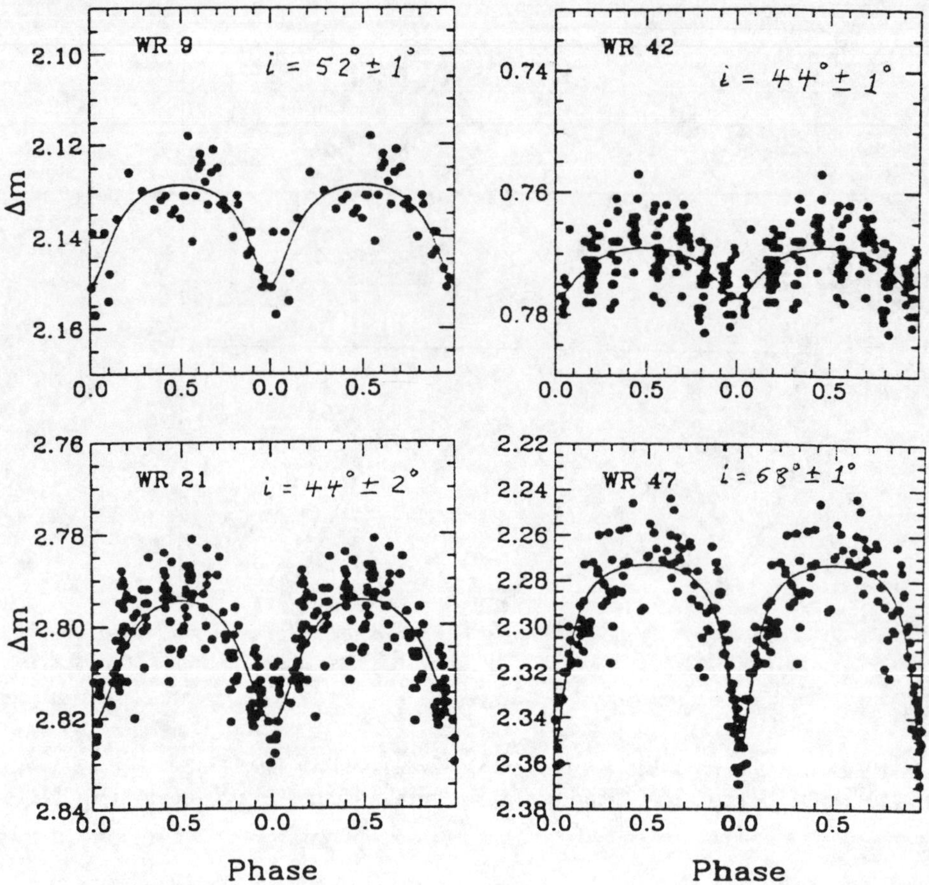

Fig. 4. Phased visual light curves of four non-core eclipsing WR + O binaries. The fitted curves are models of atmospheric transparency variations due to electron scattering of WR wind material along the line of sight to the O-star, which acts like a probe in a circular orbit (see Lamontagne et al. 1991, 1994). Note that the $O-C$ noise is significantly greater than the instrumental noise in some cases (especially for WR 47)

but more reliable than those from radio/IR, when based on the usual approximation of a smooth wind (St-Louis et al. 1993a; Moffat & Robert 1994).

So far there have been relatively few theoretical (e.g. Eichler & Usov 1993; Stevens et al. 1992) or observational (Shore & Brown 1988 and St-Louis et al. 1993b in the UV; Lewis et al. 1993, Marchenko et al. 1994a,b, Luehrs 1993 in the optical) studies of WR + O colliding winds. Part of the reason for this is that, again, one requires a large bank of data (i.e. good monitoring to cover all the phases and check for repeatability, but especially around eclipses at phase 0.5, when the O-star is in front). This is illustrated in Fig. 5 for V444 Cyg, WN5 + O6 (Marchenko et al. 1994a), which shows the phase-dependent behaviour of the weak, low-ionization He I line at 4471 Å. Stronger lines, although easier to observe, are often of higher ionization level and are perturbed more by eclipse effects, making the collision interpretation less straightforward. In the case of the He I variations in Fig. 5, based on some 50 high-quality phase-binned CCD spectra, a model of He I excess, formed downstream after cooling in the shock cone, is illustrated in Fig. 6.

Other recent interesting results of long-term monitoring of very massive binaries include: (a) the unique discovery of a slow, time-dependent (periodic?) transition of one of the WR stars in the unique 19-day WR + WR SMC binary HD 5980, from WN3-4 to WN6 over a decade (Koenigsberger et al. 1994), and (b) the 90-day photometric monitoring of the elliptical-orbit 6-day binary HD 93205, O3V + O8V, containing the earliest type (thus most massive?) star known in a binary (Antokhina et al. 1994): although there are no eclipses, the light curve does show an amplitude of ~ 0.02 magnitude, due to a combination of distortion and wind-collision effects. Work is in progress in an attempt to extract the orbital inclination, and hence deduce the mass of the potentially most massive star known in a binary, which, like other massive stars in the Carina Nebula, is expected to evolve into a WR star eventually. The work of Gies and collaborators (e.g. Gies et al. 1993, and references therein) relative to other O + O systems should also be mentioned.

3.3. AMBIGUOUS WR STARS

Some WR stars show no obvious, direct signs of a companion in their spectra, yet show periodic variations in light, polarization and/or RV. The (global line) RV variations are typically of low amplitude (~ 10 km s^{-1}), making it unlikely (unless the orbital inclination is implausibly low) that these stars have a normal OB-type companion, as commonly seen among WN6-7 stars classified as single-line, massive binaries (SB1). Among the some 50 galactic (non WN8; WN8 are all variable - see below) WR stars for which sufficient monitoring is available, two stars (only, so far?) clearly fall into this catagory: HD 50896 = EZ CMa = WR6, WN5, P = $3^{d}.76$ (Firmani et al. 1980), confirmed recently by Antokhin et al. (1994a) as truly the sole period below ~ 90 d, and HD 191765 = WR134, WN6, P = $2^{d}.34$ (Robert 1992; McCandliss et al. 1994). [Plus WR46, WN3p, which may have a 6.8-hour period: van Genderen et al. (1991 and references therein), Niemela et al.(1994)] Both of these stars are also unique in showing strong (and variable) line depolarization, implying a global asymmetry in the wind. (The continuum is most strongly polarized intrinsically, with lines of low ionization formed furthest out in the wind being least polarized.) These stars could be single and rapidly rotating, with some kind of global inhomogeneity (e.g. magnetic field loops, oblique rotator with spots, jets, etc...) or low-mass (most likely WR + c) binary stars (a WR + O binary with low inclination is improbable). The long-awaited proof that WR + c stars do exist was recently revealed in a spectacular fashion from the IR spectral monitoring of the strong X-ray source Cyg X-3 (van Kerkwijk 1993), now believed to contain a WN7 star.

We illustrate the above results for the star EZ CMa in Fig. 7 (possible detection of a flare: Matthews et al. 1993) and Fig. 8 (a model of an oblique rotator, based on

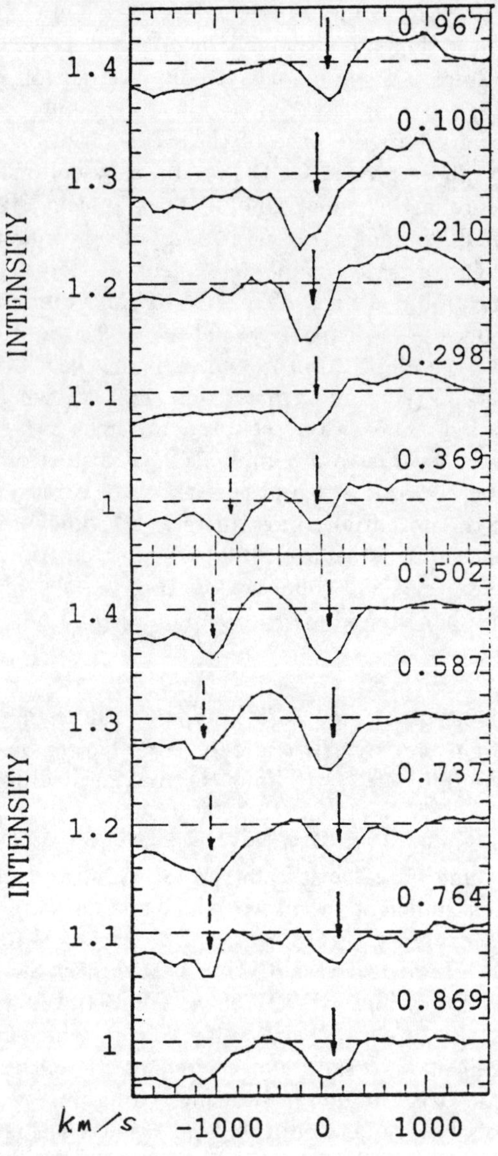

Fig. 5. Phase dependent variations of He I 4471 in V444 Cygni (WN5 + O6): Marchenko et al. (1994a). Note that each spectrum shown is binned from several spectra with mean phase as indicated. The WR star is in front at phase zero. Note the sine-wave orbital modulation of He I 4471 absorption arising in the O-companion, compared to the P Cygni enhanced absorption of the same line at large negative RV, as the shock wake behind the O-star swings around towards the observer and is thus seen superposed on at least one of the bright star disks (cf. Fig. 6), between phases ~ 0.3 and ~ 0.8. Broad excess emission is seen at all phases, coming from the same shock cone

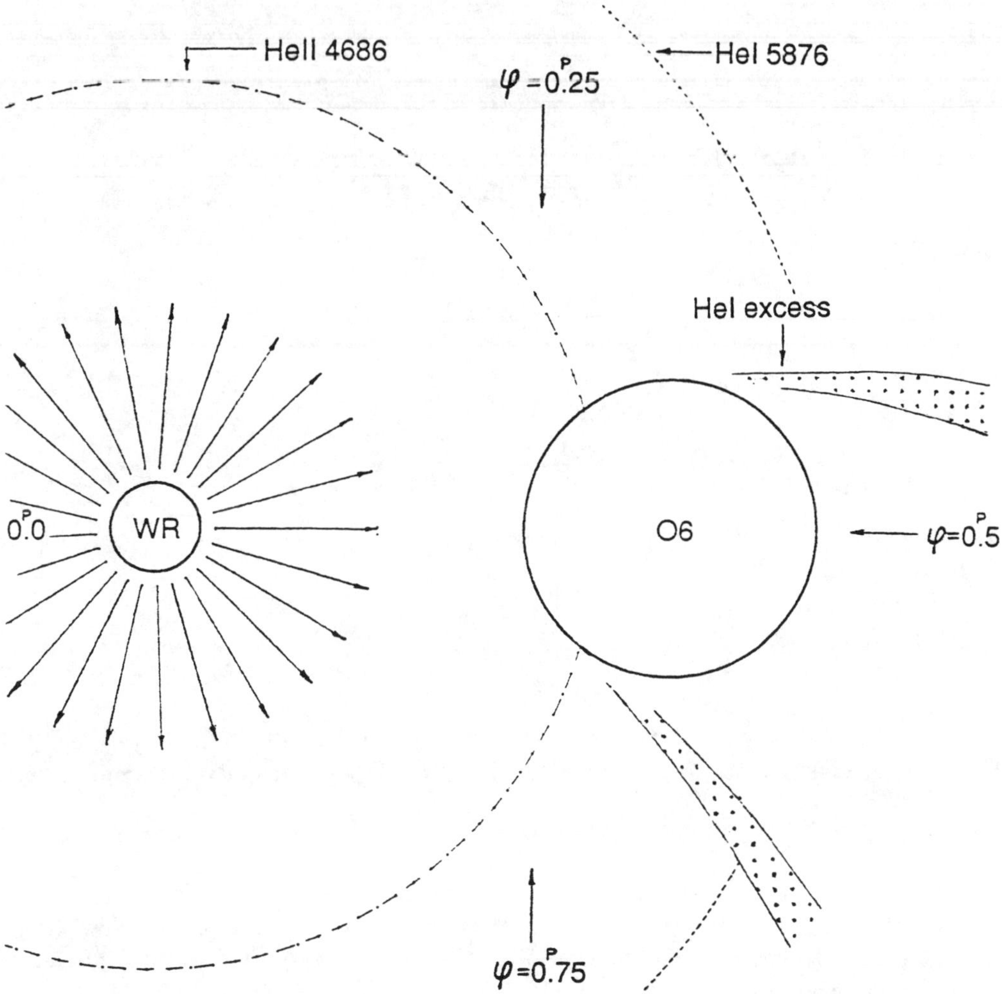

Fig. 6. Cartoon for the wind interaction leading to He I excess in V444 Cygni (Marchenko et al. 1994a), used to explain qualitatively the observed variations in Fig. 5

simultaneous continuum photometry, polarimetry and spectroscopy: Matthews et al. 1992). Similar flares have also been claimed from 75 hours of rapid monitoring of WR 134 (Marchenko et al. 1994d). Despite long-term searches, as yet no longer periodicity (e.g. due to precession, as might be expected for a massive star with a compact companion, by analogy with SS 433) has been seen in EZ CMa (Robert et al. 1992). On the other hand, such a longer (precession?) period has been claimed in WR 134 from two years of APT photometry by Moffat & Marchenko (1993).

Another group of WR stars shows the highest degree of intrinsic variation among

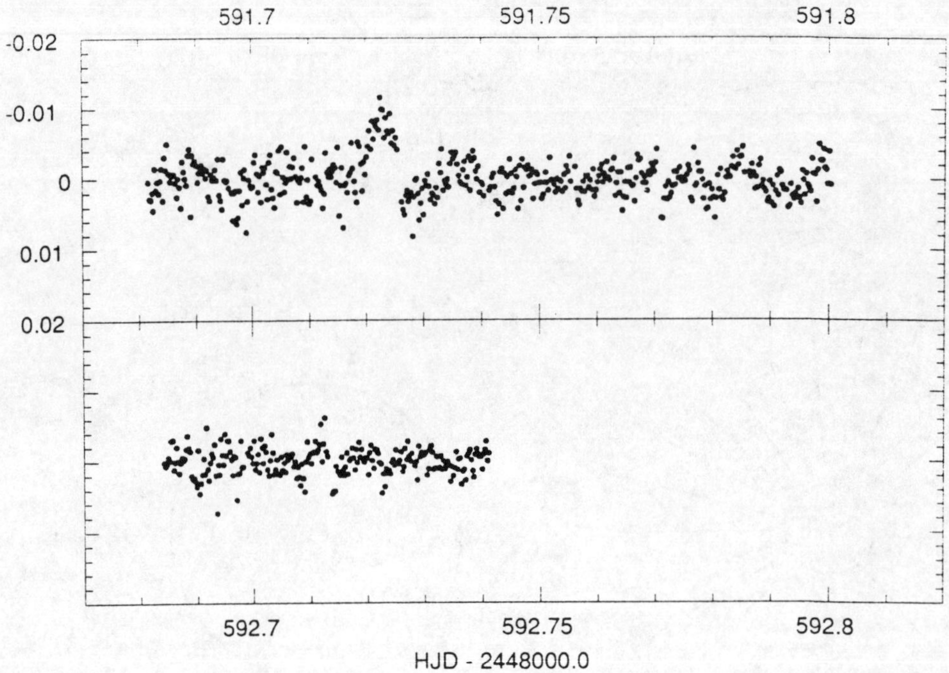

Fig. 7. Detection of a possible 1% flare of ~ 10 minute duration in Strömgren b rapid photometry of EZ CMa (see Matthews et al. 1993). Two nights of data (magnitudes vs. time) are shown

all WR subtypes: the WN8 stars (possibly along with the WC9 stars). They also have the narrowest emission lines, often with P Cygni profiles, implying slow winds and relatively large core radii (see Moffat 1989). The WN8 stars also have the following properties:

- no evidence of massive companions (i.e. no WR + OB),
- often (always?) runaway status: e.g. WR124, with a peculiar space velocity of over 150 km s^{-1} (Moffat et al. 1982), or WR123 = HD 177230, which is some 800 pc from the galactic plane (Lamontagne et al. 1983).

These properties suggest that WN8 stars might all be WR + c systems in which any X-ray emission is smothered or not produced in the first place in detectable quantity.

The two brightest WN8 stars in the sky, WR 40 and WR 16, have been monitored the most frequently. In particular, WR 40 was recently monitored contiguously for one month in truly continuum filters by Matthews & Moffat (1994). This study shows that the variations then as well as all those published before, can be explained by two periods (~ 12 d possibly from non-synchronously rotating spots or some other kind of asymmetry in the wind and ~ 17 d from an orbiting companion).

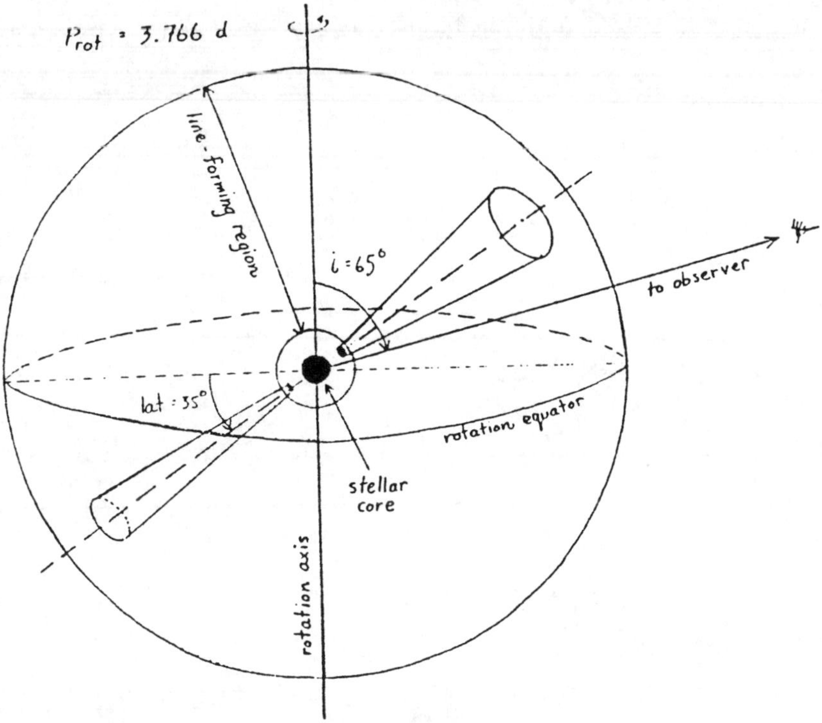

Fig. 8. Oblique-rotator model for EZ CMa = WR 6 based on simultaneous continuum photometry and polarimetry, along with spectroscopy (Matthews et al. 1992)

The two modulations have varying epoch-dependent amplitudes and there is also a component of truly random variability. Essentially all of the prevously claimed periods for this star can be explained as aliases or harmonics of these two periods. Whether this is true, or whether we are seeing some kind of NRP (Gosset et al. 1994), will have to await yet more observations.

In fact, we (Antokhin et al. 1994b) have obtained an unprecedented set of almost 90 contiguous one-site continuum photometry for 4 southern WN8 stars (WR 16, 40, 66, 82), covering various-time scales from msec to the full 90-day interval. These results basically confirm those of Matthews & Moffat (1994) in the case of WR 40, with the additional constraints that:

- variability occurs on a nightly basis, but always gradually during the night
- no periods are seen in the range of msec to days (except for WR 66 - see below): Marchenko et al. (1994c)
- no sign of a 10.5-minute periodicity, as claimed by Blecha et al. (1993), was seen in WR 40. This result is supported also by the photometric data of Schneider et al.

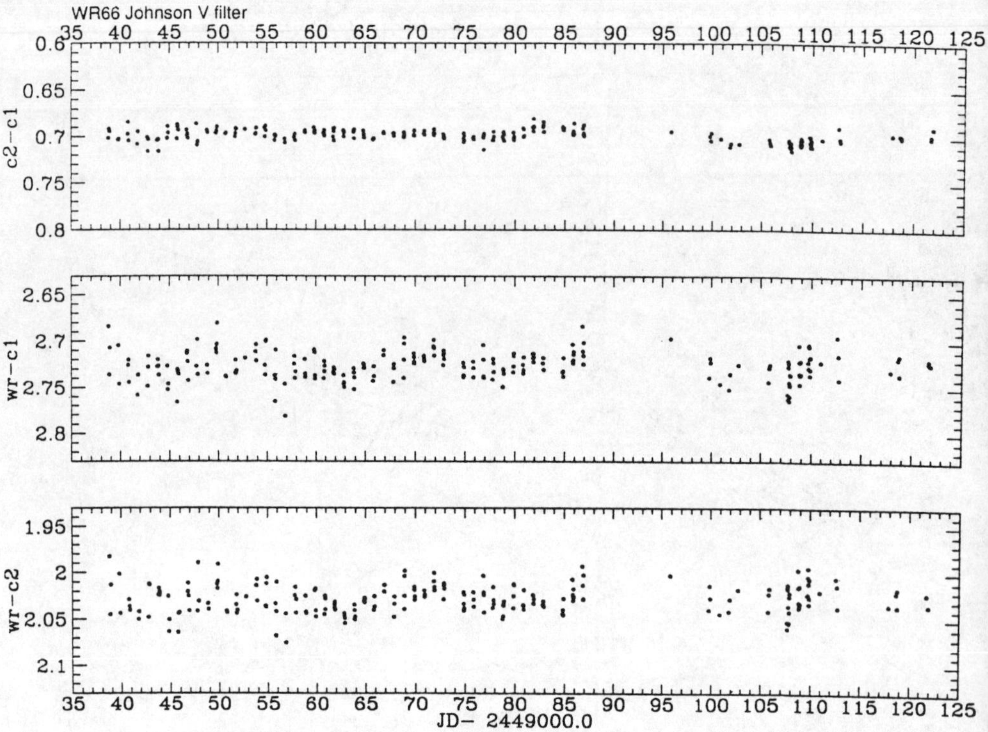

Fig. 9. Photometry of the WN8 star WR 66 and its two comparison stars (Antokhin et al. 1994b). Note that the WR star shows relatively large variations, while one of the comparison stars (C1 as it turns out) shows low-amplitude periodic variations of P > 2 days (cf. Fig. 10)

(1994) and Martinez et al. (1994)
- WR 66 shows strong evidence for a 3.51-hour sine-wave modulation with peak-to-peak amplitude of $\sim$ 0.02 magnitude, along with random noise (cf. Figs. 9 and 10). The presence of a *single* period in WR 66 would seem to make it unlikely to be due to NRP (e.g. Vreux 1985), while the period is too long to be due to radial pulsation, expected to lie in the range of 15 - 60 minutes for WR stars (Maeder 1985, 1991b).

My own speculation is that WN8 stars possibly represent some kind of puffed up WN7 star, as a result of the swallowing of a compact companion. This would explain their lack of OB companions, high runaway speed (accelerated after a supernova explosion, leading to the formation of a bound, compact companion), and strong variability. Note that one refers to hypothetical RSGs that have swallowed compact companions via rapid spiral-in as Thorne-Zytkov (1977) objets. In the case of WR 66, the compact star may not yet have reached the centre of the WN8 core. WN8 stars could thus be preceeded by a Cyg X-3 (WN7 + c) stage, whose rarity is a consequence of the rapid spiral-in decay of the orbit (see Cherepashchuk & Moffat

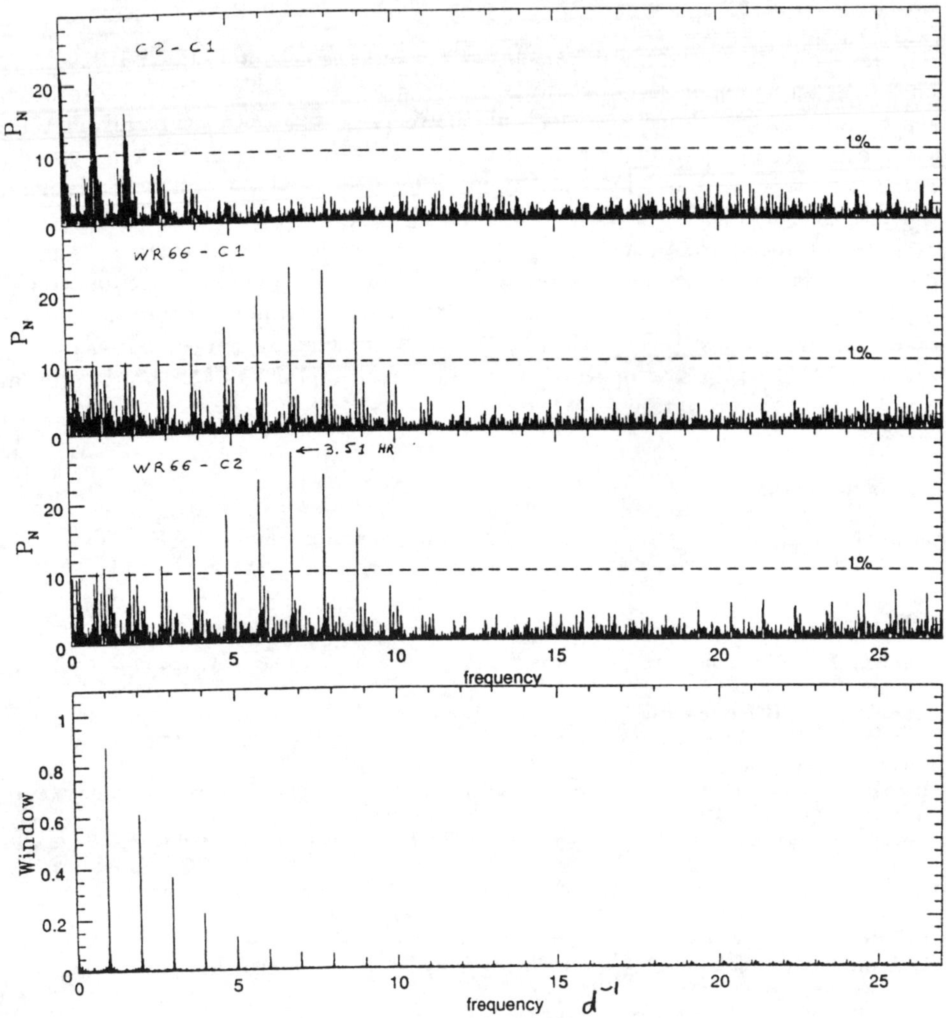

Fig. 10. Normalized power spectra of the data in Fig. 9, along with the window function. Note the prime period of 3.51 hours in WR 66, with its one-day aliases, all less significant, while C1 appears to show a period of several days (see Antokhin et al. 1994b)

1994). Clearly, more monitoring of WN8 stars (e.g. Northern Milky Way and the LMC) is needed.

I conclude this section by noting that time variability of WR stars is an important diagnostic, not only to probe the structure of the wind itself but also, in the cases of WR + O and WR + c binaries, to provide constraints on the WR star itself.

4. Future Work

WR stars show their strongest variability on time scales of hours to days. As is known by most experts who study variable stars from single-site ground-based observatories, such systems are very difficult to piece together, especially if there exists more than one period and epoch-dependent variations, often along with intrinsic noise. WR stars are thus ideally suited for dedicated variability study using Whole Earth Telescopes. A plea for a WET-type network of APT telescopes, at least for photometry, as proposed by D. Crawford some 10 years ago, is strongly endorsed. This would increase the uninterupted time-base and eliminate nasty time gaps, leading to a clearer scientific interpretation. Also, simultaneous multi-mode observations would best be exploited by carrying out multi-site spectropolarimetry, i.e. getting I, Q, U, V as a function of lambda and time. I think I should stop at this point, as I have a date with Alice in Wonderland...!

Acknowledgements

This work was financially supported by NSERC (Canada) and FCAR (Québec).

References

Antokhin, I.I., Bertrand, J.-F., Lamontagne, R., Moffat, A.F.J. 1994a, *Astronomical Journal*, submitted

Antokhin, I.I., Bertrand, J.-F., Lamontagne, R., Moffat, A.F.J. 1994b, in preparation

Antokhina, E. et al. 1994, in preparation

Blecha, A., Schaller, G., Maeder, A. 1992, *Nature*, 360, 320

Brown, J.C., Robert, C., Moffat, A.F.J., St-Louis, N., Antokhin, I.I. 1994, *Astronomy and Astrophysics*, submitted

Cherepashchuk, A.M., Moffat, A.F.J. 1994, *Astrophysical Journal, Letters to the Editor*, in press

Drissen, L., St-Louis, N., Moffat, A.F.J., Bastien, P. 1987, *Astrophysical Journal*, 322, 888

Drissen, L., Robert, C., Moffat, A.F.J. 1993, *Astrophysical Journal*, 386, 288

Eichler, D., Usov, V. 1993, *Astrophysical Journal*, 402, 271

Firmani, C., Koenigsberger, G., Bisiacchi, G.F., Moffat, A.F.J., Isserstedt, J. 1980, *Astrophysical Journal*, 239, 607

Fleck, C.F. 1992, *Astrophysical Journal*, 401, 146

Giallongo, E., Cristiani, S., Fontana, A., Travese, D. 1993, *Astrophysical Journal*, 416, 137

Gies, D.G., Wiggs, M.S., Bagnuolo, W.G. 1993, *Astrophysical Journal*, 403, 752

Gosset, E., Rauw, G., Manfroid, J., Vreux, J.-M., Sterken, C. 1994, these proceedings

Hamann, W.-R. 1991, IAU Symp. No. 143, 81

Henriksen, R.N. 1991, *Astrophysical Journal*, 377, 500

Henriksen, R.N. 1994, in *Instability and Variability of Hot-star Winds*, eds. A.F.J. Moffat et al. (Dordrecht: Kluwer), in press

Hillier, D.J. 1991, IAU Symp. No. 143, 59

Howarth, I.D., Prinja, R.K. 1989, *Astrophysical Journal, Supplement Series*, 69, 527

Koenigsberger, G., Moffat, A.F.J., St-Louis, N., Auer, L.H., Drissen, L., Seggewiss, W. 1994, *Astrophysical Journal*, submitted

Lamontagne, R., Moffat, A.F.J., Seggewiss, W. 1983, *Astrophysical Journal*, 269, 596

Lamontagne, R., Moffat, A.F.J. 1987, *Astronomical Journal*, 94, 1008

Lamontagne, R., Robert, C., Grandchamps, A., Lapierre, N., Moffat, A.F.J., Drissen, L. 1991, IAU Symp. No. 143, 316

Lamontagne, R., Moffat, A.F.J., Drissen, L., Robert, C., Matthews, J.M., Grandchamps, A., Lapierre, N. 1994, in preparation

Lépine, S. 1994, in *Instability and Variability of Hot-star Winds*, eds. A.F.J. Moffat et al. (Dordrecht: Kluwer), in press
Lewis, D., Moffat, A.F.J., Matthews, J.M., Robert, C., Marchenko, S.V. 1993, *Astrophysical Journal*, 405, 312
Lucy, L.R., White, R.L. 1980, *Astrophysical Journal*, 241, 300
Luehrs, S. 1993, Ph.D. Thesis, Univ. of Muenster, Germany
Maeder, A. 1985, *Astronomy and Astrophysics*, 147, 300
Maeder, A. 1991a, *Astronomy and Astrophysics*, 242, 93
Maeder, A. 1991b, IAU Symp. No. 143, 167
Maeder, A., Lequeux, J. 1982, *Astronomy and Astrophysics*, 114, 409
Maeder, A., Conti, P.S. 1994, *Annual Review of Astronomy and Astrophysics*, in press
Marchenko, S.V., Moffat, A.F.J., Koenigsberger, G. 1994a, *Astrophysical Journal*, in press
Marchenko, S.V., Moffat, A.F.J., Koenigsberger, G. 1994b, in preparation
Marchenko, S.V., Antokhin, I.I., Bertrand, J.-F., Lamontagne, R., Moffat, A.F.J., Piceno, A., Matthews, J.M. 1994c, in preparation
Marchenko, S.V. et al. 1994d, in preparation
Martinez, P., Kurtz, D., Ashley, R., Tripe, P. 1994, *Nature*, in press
Matthews, J.M., St-Louis, N., Moffat, A.F.J., Drissen, L., Koenigsberger, G., Cardona, O., Niemela, V.S. 1992, *Astrophysics and Space Science*, 22, 130
Matthews, J.M., Moffat, A.F.J., Marchenko, S.V. 1993, *Astronomy and Astrophysics*, 266, 409
Matthews, J.M., Moffat, A.F.J. 1994, *Astronomy and Astrophysics*, in press
McCandliss, S.R., Bohannan, B., Robert, C., Moffat, A.F.J. 1994, in *Instability and Variability of Hot-star Winds*, eds. A.F.J. Moffat et al. (Dordrecht: Kluwer), in press
Moffat, A.F.J., Lamontagne, R., Seggewiss, W. 1982, *Astronomy and Astrophysics*, 114, 135
Moffat, A.F.J., Shara, M.M. 1986, *Astronomical Journal*, 92, 952
Moffat, A.F.J. 1989, *Astrophysical Journal*, 347, 373
Moffat, A.F.J., Robert, C., Drissen, L., Lamontagne, R., Coziol, R., Mousseau, N., Niemela, V.S., Cerruti, M.A., Seggewiss, W., Van Wieren, N. 1990, *Astrophysical Journal*, 350, 767
Moffat, A.F.J., Robert, C. 1992, *Astrophysics and Space Science*, 22, 203
Moffat, A.F.J., Marchenko, S.V. 1993, *Astronomical Journal*, 105, 339
Moffat, A.F.J., Robert, C. 1994, *Astrophysical Journal*, 421, 310
Moffat, A.F.J., Lépine, S., Henriksen, R.N., Robert, C. 1994, Ap.Sp.Sci.Lib., in press
Moffat, A.F.J. 1994, in *Instability and Variability of Hot-star Winds*, eds. A.F.J. Moffat et al. (Dordrecht: Kluwer), in press
Niemela, V.S. et al. 1994, IAU Symp. No. 163, in preparation
Owocki, S.P., Castor, J.I., Rybicki, G.B. 1988, *Astrophysical Journal*, 335, 914
Owocki, S.P. 1991, IAU Symp. No. 143, 155
Prinja, R.K., Smith, L.J. 1992, *Astronomy and Astrophysics*, 266, 377
Robert, C., Moffat, A.F.J., Bastien, P., Drissen, L., St-Louis, N. 1989, *Astrophysical Journal*, 347, 1034
Robert, C. 1992, Ph.D. Thesis, Univ. de Montreal
Robert, C. et al. 1992, *Astrophysical Journal*, 397, 277
St-Louis, N., Drissen, L., Moffat, A.F.J., Bastien, P., Tapia, S. 1987, *Astrophysical Journal*, 322, 870
St-Louis, N., Moffat, A.F.J., Moffat, A.F.J., Drissen, L., Bastien, P., Robert, C. 1988, *Astrophysical Journal*, 330, 286
St-Louis, N. 1990, Ph.D. Thesis, UC London
St-Louis, N., Moffat, A.F.J., Lapointe, L., Efimov, Yu.S., Shakhovskoy, N.M., Fox, G.K., Piirola, V. 1993a, *Astrophysical Journal*, 410, 342
St-Louis, N., Willis, A.J., Stevens, I.R. 1993b, *Astrophysical Journal*, 415, 298
Schmidt, G.D. 1988, in Polarized Radiation of Circumstellar Origin, eds. G.V. Coyne et al. (Vatican City State: Vatican Obs.), p.641
Schneider et al. 1994, IAU Symp. No. 162, in press
Schulte-Ladbeck, R.E., Meade, M.R., Hillier, D.J. 1992, *Astrophysics and Space Science*, 22, 118
Shore, S.N., Brown, D.A. 1988, *Astrophysical Journal*, 334, 1021
Stevens, I.R., Blondin, I.M., Pollock, A.M.T. 1992, *Astrophysical Journal*, 386, 265
Thorne, K., Zytkov, A.N. 1977, *Astrophysical Journal*, 212, 832

Underhill, A.B., Nemec, F.L. 1989, *Astrophysical Journal*, 345, 1008
van der Hucht, K.A., Hidayat, B. 1991, eds. IAU Symp. No. 143 (Dordrecht: Kluwer)
van der Hucht, K.A. 1992, *Astronomy and Astrophysics Review*, 4, 123
van Genderen, A.M. et al. 1991, IAU Symp. No. 143, 129
van Kerkwijk, M.H. 1993, *Astronomy and Astrophysics*, 276, L9
Vreux, J.-M. 1985, *Publications of the ASP*, 97, 589
Williams, J.P., Blitz, L. 1993, *Astrophysical Journal*, 405, L75

DISCUSSION

Kurtz: What is the spectral type of your $3^{\mathrm{h}}.5$ variable comparison star?

Moffat: We tried to find blue comparison stars of similar magnitude as the variable. I do not have detailed information with me, but it will soon be published elsewhere with all the data.

Kurtz: Could you give a physical explanation for the very interesting brightness variation at periastron in the case of the $O3 + O8$ binary?

Moffat: E. Antokhina (Sternberg Astronomical Institute, Moscow) is modelling the light curve of the $O3V + O8V$ binary HD 93205. With a total amplitude of about $\sim 0^{\mathrm{m}}.02$, it cannot be due to eclipses. Rather, it must be due to variable heating, reflection and distortion effects, which are most pronounced near periastron. The details will be available soon (we hope) in published form.

Waelkens: You, as well as P.S. Thé earlier today, showed the power of polarimetry, and nicely demonstrated that this technique is most efficient when it is operated on a long-term basis. At ESO, the only polarimeter (PISCO) has recently be decommissioned with the argument that there were not enough "high-quality proposals". But the instrument was offered at a large telescope and, so, could never be operated on a long-term basis. Clearly, polarimetry offers a nice example of the fundamental importance of long-term projects in astronomy. [1]

Moffat: I agree! In fact, I have used PISCO on two occasions (1986 for 6 ESO nights, and in 1988 for 14 MPI nights). It is a very sophisticated instrument which provided very good data, although, if I remember correctly, its efficiency is somewhat compromised by light loss at the 2 m $f/15$ focus since PISCO was designed for an $f/8$ beam. I am sorry I did not apply again, especially now that H. Schwarz at ESO recently wrote a nice manual for PISCO.

van der Hucht: You mentioned that the WR + c possibility is still a missing link in the binary-evolution scenario. Would you care to comment on the claim of van Kerkwijk (1992, *Nature*, 355, 703; 1993, *Astronomy and Astrophysics*, 276, L9) that Cyg X-3 harbours a WN companion?

Moffat: You are right; I should have said *previously* missing link. The IR spectroscopic discovery last year that Cyg X-3 contains a massive Population I $\sim$ WN7 star with a compact (neutron star or black hole) companion in a tight $4^{\mathrm{h}}.8$ orbit, provided the first evidence for a viable WR + c system. It is probably a very lucky

[1] Note from the Editors: It should be stressed that the SOC of this ARW did invite key speakers on this topic but, unfortunately, these speakers could not attend the meeting for various reasons

find for such a close system, whose orbit must be decaying rapidly on spiral-in, maybe leading to a Thorne-Zytkov type bloated star with a compact object at its centre.

Demircan: Concerning the space velocities of WR binaries, I remember from Batten's catalogue that most of the gamma velocities are more than 50 km s^{-1}, but we know that the space velocities of the progenitor OB systems are not more than 15 km s^{-1}. Therefore, do you think the WR stars are runaway stars? Are there any statistics about this?

Moffat: Among WR + OB binaries, it is unwise to use the emission lines (formed in an expanding wind) to obtain the systemic velocity; it is best to use well-defined OB absorption lines. In that case, there is (as far as I remember) no evidence for excessive velocities. In the case of (apparently) simple WR stars (especially WN8), some of which may have been accelerated by a supernova explosion yielding a WR + c binary, there is some evidence for runaway motion, but not based on the emission lines (cf. earlier work, e.g. Moffat and Isserstedt 1980, *Astronomy and Astrophysics*, 85, 201).

A Belgian reunion – Hensberge, Burger, Gosset, Jorissen

DISCUSSION No 2: WR STARS

Vogt: (to Moffat): You distinguish between single WR, binary WR and ambiguous WR stars. How do you know that the so-called single WR stars are really single? Is it possible to exclude binarity?

Moffat: No, of course not (as for any kind of stars)! Labelling a WR star as single normally is taken to mean that it shows no periodic radial-velocity variations (none are close enough to have a visual orbit). If the orbital inclination is low, and the period very long (¿¿ a decade), we may mis-label the star as a single to stay on the safe side. The current $WR+OB$ binary-frequency estimate is around 45% and climbing!

van der Hucht: First of all, a binary needs a radial-velocity solution.

Williams: There is a severe selection effect: a companion has to be about $0\overset{m}{.}7$ mag brighter than the WR star for its absorption lines to be seen in the composite spectrum. Because the WR continuum is flatter than that of hot stars, the violet region of the spectrum is the best place to look for such absorption lines.

Kurtz: Since many of the WR stars have O companions, why don't we see more red supergiants with WR secondaries? Is it because of lifetimes, or is it an observational effect?

Moffat: Some people would say that rapid Roche-lobe overflow prevents a companion, that as a single star would become a red supergiant (RSG), from becoming such a star. However, it may be that even as single stars, WR and RSGs do not correlate spatially, suggesting that the RSGs ($M \leq M_{\odot}$) do not evolve into WR stars ($M_i \geq 30M_{\odot}$). In binaries, the original secondary could, in principle, become a RSG (depending on its mass) while the primary has become a WR star, but the timing is a problem: the WR star will normally have exploded before the secondary becomes a RSG. On the other hand, a few $WR+OB$ binaries do contain O-type supergiants.

Harmanec: (to Kurtz): Don, I think that it is very illustrative to note that *not a single* O supergiant now known to be the primary of a massive X-ray binary, was known as a spectroscopic binary before the boom of X-ray observations started—although, for the brighter ones among them the data were there. It is also useful to realise that a total eclipse of such a supergiant by a main-sequence star would

C. Sterken and M. de Groot (eds.), The Impact of Long-Term Monitoring on Variable Star Research, 141–142.

be only a few hundreths of a magnitude deep and could be misinterpreted as its intrinsic variability.

Niarchos: (to Moffat): The inclinations you have found in your examples seem to be quite small. It means that the systems show partial eclipses. Is there any evidence for this?

Moffat: The plot I showed refers to non-(core) eclipsing $WR + O$ systems: the number distribution in $\cos i$ is uniform (given the small numbers), suggesting a random spatial orientation, rather than favouring small (or large) inclinations. The eclipses involve variable transmission (due to electron scattering) of light from the O-star which orbits in the WR wind.

Williams: (to Moffat): Do the variations of the WN8 stars you observe come from their cores or from their electron-scattering wind?

Moffat: Our recent photometric data are in optical narrowband continuum windows. Thus, we are probing regions of the wind interior out to where the lines are formed. However, the $\tau_{es} \sim 1$ surface in most (all?) cases is still located in the wind, not at the surface of the star (defined as where the wind expansion velocity reaches the sound speed in the medium).

Mantel explaining extinction to Poretti

LONG-TERM VARIABILITY OF BARIUM AND S STARS

A. JORISSEN*
Institut d'Astronomie et d'Astrophysique
Université Libre de Bruxelles
Belgium

ABSTRACT. Chemically-peculiar red giants are ideal targets for long-term monitorings, since these binary stars have periods ranging from a few months to more than 10 years. The results of the monitoring of a sample of barium and S stars in the framework of the Long-Term Photometry of Variables program carried out at ESO for about 10 years are described. The two barium stars with the shortest orbital periods in the sample, HD 46407 (P = 458 d) and HD 121447 (P = 185 d), exhibit long-term variations of very low amplitude. Ellipsoidal variations are observed for HD 121447, whereas dust scattering and obscuration is invoked in the case of the eclipse-like variations recorded for HD 46407. A similar behaviour has been observed for the Tc-poor S star HD 35155 (P = 642 d), which is an interacting binary system exhibiting symbiotic-like features in the UV. We show that our long-term photometric and radial-velocity monitoring of HD 35155 sheds light on the variations observed with IUE for this system.

Keywords: red giants – K stars – barium stars – microvariability – eclipsing binaries – ellipsoidal stars variables—stars: HD 46407, HD 35155, HD 121447

1. Preamble

Barium and S stars, with their orbital periods ranging from a few months to more than ten years, are probably among the best examples for illustrating the unique capabilities of long-term monitoring. The Long-Term Photometry of Variables (LTPV) programme operating at the *European Southern Observatory* (Sterken, this volume) started the monitoring of barium stars in 1984, shortly after the discovery by McClure and collaborators (McClure et al. 1980, McClure 1983) of the binary nature of these chemically-peculiar red giants (PRG). Thanks to almost one decade of monitoring, quite unexpected results were obtained, unravelling some of the mysteries locked in these binary systems. It should be stressed that the LTPV programme, thanks to its policy of merging targets of quite different types, offers the rare opportunity to include stars like barium stars which *a priori* were not expected to show up as variable. But the long-term photometric monitoring of these stars has been quite rewarding, as will be shown in this paper.

* Research Associate FNRS Belgium

C. Sterken and M. de Groot (eds.), The Impact of Long-Term Monitoring on Variable Star Research, 143–162.

2. The Binary Nature of Barium and S Stars

Barium stars were defined as a class of PRG stars by Bidelman and Keenan (1951). They represent about 1% of G and K giants, and their characteristic spectral features are strong Ba II $\lambda 4554$ and SrII $\lambda 4077$ lines, as well as strong CN and CH bands. Their location in the HR diagram indicates that some are on the lower part of the first giant branch (corresponding to the shell H-burning stage in the evolution of low-mass stars) and others are in the He clump (corresponding to central He-burning; see McClure 1984 for a review). Detailed abundance analyses indicated that carbon and s-process elements (i.e. elements heavier than iron synthesized by slow neutron captures starting on iron-group nuclei) are overabundant in barium stars with respect to normal giants. The exact origin of these chemical anomalies long remained a mystery, since apart from the controversial He-flash, barium stars were believed not yet to have reached an evolutionary stage where these heavy elements can be synthesized by internal nucleosynthesis processes and brought to the surface. Current stellar evolution models predict such processes to occur only on the asymptotic giant branch (AGB), populated by luminous and cool stars of low and intermediate mass having exhausted their central He fuel (see Iben and Renzini 1983 for a review). The discovery of the binary nature of barium stars (McClure et al. 1980, McClure 1983) opened a new channel to investigate the origin of the chemical pecularities of these stars. Webbink (1986) and McClure and Woodsworth (1990) showed that the mass-function distribution of these spectroscopic binaries with one observable spectrum is compatible with a white dwarf (WD) companion. Although their statistical argument is quite convincing, direct evidence for a hot companion using the *International Ultraviolet Explorer* (IUE) satellite has been rather scarce (Dominy and Lambert 1983, Böhm-Vitense et al. 1984), indicating that the WD must be rather cool ($T_{\rm eff} < 10000$ K). In a former state of the binary system, the WD must have been an AGB star where heavy-element synthesis was taking place. Mass transfer from this star towards its less-evolved companion may then be the key to the chemical peculiarities exhibited by barium stars (see e.g. Jorissen and Boffin 1992).

S stars were defined by Merrill in 1922 as a class of cool giants exhibiting ZrO bands. They are cooler and more luminous than barium stars (e.g. Scalo 1976), and are often presented as being an intermediate phase between M giants (having $C/O < 1$ in their envelope) and C stars ($C/O > 1$) on the AGB (e.g. Iben and Renzini 1983). Recent evidences of a dichotomy in the family of S stars led to the revision of this traditional evolutionary scheme (Johnson 1992, Jorissen et al. 1993). There appears to be an excellent correlation indeed between the absence of Tc lines in an S star spectrum, the absence of IR excesses in the IRAS 12 and 25 μm bands, the moderate light variability and the binary nature of the S star as shown by both radial-velocity variations and strong UV continua. Technetium lines were first observed in S star spectra by Merrill (1952). They constitute a clear signature of on-going s-process nucleosynthesis in these stars, since Tc has no stable isotopes. The isotope ^{99}Tc involved in the s-process chain has a half-life against β-decay of about $3\ 10^5$ y, which is of the same order of magnitude as the lifetime of the star on the AGB. Tc-rich S stars are the genuine mass-losing AGB stars involved in the traditional M–S–C evolutionary sequence. Tc-poor S stars, however, are likely the cool counterparts of barium stars, and are not yet as evolved as Tc-rich S stars, as indicated by the absence of Tc and of IR excess. Tc-poor S stars are located on the upper part of the first giant branch or in the early AGB phase. Their orbital elements are very similar to those of barium stars (Jorissen and Mayor 1992b). A WD companion is thus expected for these systems as well, and here direct evidence has indeed been found on IUE spectra which often exhibit symbiotic-like features (Johnson et al. 1993).

3. Long-term Radial Velocity and Photometric Monitoring of Barium and S Stars

In 1984, orbital elements were known for only 7 barium stars (Griffin and Griffin 1981, McClure 1983) and one S star (Griffin 1984). Thanks to the continuing efforts of McClure at Victoria, Griffin at Cambridge and the Geneva team operating the CORAVEL spectrovelocimeters at ESO and Haute Provence Observatory, ten years later the number of known orbits amounts to 45 for barium stars (McClure and Woodsworth 1990, Griffin and Keenan 1992, Jorissen and Mayor, 1992a and in preparation) and 6 for S stars (Jorissen and Mayor 1992b). The orbital periods involved, ranging from a few months to more than 15 years (the longest period known to-date has been obtained by Griffin and Keenan for ζ Cyg and amounts to 17.7 y), really require this long-term effort. Information on the dynamics of mass transfer can now be gained from the analysis of a large and representative sample of orbital elements. Another achievement of our CORAVEL monitoring (Jorissen and Mayor 1992a) is the proof that binarity really is a *necessary condition* for a star to become a PRG. All 27 known southern barium stars with strong peculiarities (Ba4 or Ba5 in the usual scale of Warner 1965) were monitored since 1984, and 26 turned out to be binaries (Jorissen and Mayor 1992a). The star without radial-velocity variations (HD 19014) at the level of 0.5 km s^{-1} may well be a pole-on system (0.1/27 stars with a radial-velocity amplitude smaller than 0.6 km s^{-1} are predicted if $P = 5$ y).

Simultaneously to this radial-velocity monitoring, a photometric monitoring of several barium and S stars was undertaken in 1984 in the framework of the LTPV, as described in Jorissen et al. (1992a; Paper I). It was hoped that the binary nature of these systems might trigger some light variability, in the form of *ellipsoidal variables* for example. As Sect. 6 will show, this hope appeared to be justified, since one such variable was found among barium stars. There was *a priori* not much hope to find *eclipsing binaries*, since the companions to barium stars are cool WDs, as indicated in Sect. 2. A puzzling eclipse-like behaviour has nevertheless been found in two cases (Sects. 7 and 8). Our photometric monitoring was further motivated by the results of Landolt (1983) who observed 17 barium stars for about 12 years at the Cerro Tololo Inter-American Observatory and at Kitt Peak National Observatory. The precision of Landolt's non-differential UBV photometry, as derived from the r.m.s. deviation of standard stars, is about $\sigma_V = 0\overset{m}{.}007$. Variations at or above the 3σ level were found for 6 barium stars. However, no light-curves were provided, since less than a dozen observations per star were made, and no special effort was made to ensure that the instrumental system remained stable over the time-span involved. In these matters, significant improvements could be provided by the LTPV programme. The frequency of observations is one or two per week for about 3 months per year and per star. The observations are made differentially with respect to two nearby G or K giants from the Bright Star Catalogue, and changes in the instrumental systems are accounted for by the use of different colour transformation matrices. The variability was analyzed separately in the different instrumental systems. Our sample of 20 barium stars (as listed in Table 1 of Paper I, plus HD 84678) includes several stars in common with Landolt (1983), as well as stars randomly selected from the CORAVEL sample before their orbital elements were known. A few S stars monitored with CORAVEL (Jorissen and Mayor 1992b) are also included in the photometric sample. The main aim here was to disentangle radial-velocity variations due to orbital motion from those due to pulsations.

4. A Few Words About Precision

Since the expected variability level is low, special care had to be taken during data acquisition and reduction. First, in a project like the LTPV which relies largely on some kind of service observing (see Sterken, this volume), a strict observing policy had to be given to the observers. They were asked to integrate each star in the sequence APBPBPA (where P, A and B stand for programme star, first and second comparison stars, respectively) not less than 40s and not more than 240s, the latter requirement to prevent the whole sequence from becoming excessively long. This requirement ensures that the photon noise is smaller than $\sim 0\overset{m}{.}003$ for stars brighter than magnitude ~ 7. The sky measurement was done at the same location (150" N whenever possible). The APBPBPA integrating sequence also allows one to perform *internal* consistency checks (see also Hensberge, this volume). Within a given night sequence, the standard r.m.s. deviations were computed for the three samples of differences P-A, P-B and A-B involving contiguous stars in the sequence. These standard deviations should reflect just the quality of the measurements, and are by no means related to the intrinsic variations in these stars, which occur on much longer time-scales. The distribution of these standard deviations was then constructed from all the nightly sequences available for a given APB triplet, and a threshold corresponding to 3 times the median value of the distribution was defined. Nights with a standard deviation in excess of this threshold were carefully examined, and the measurement(s) responsible for the outlying pair(s) was(were) then removed. It was found that the use of this objective criterion to clean the data substantially improved the light-curves presented in Figs. 2 and 6, for example. Typically, about 10% of the data was removed in this way.

The "multi-night" reduction algorithm used by J. Manfroid (Manfroid 1985) allows to make optimal use of all constant stars (comparisons and standards) to derive the extinction. Possible slow variations in the extinction coefficients during a night can be handled by the algorithm, and sometimes a night can even be reduced in two or more separate parts. A crucial feature of the reduction procedure is the fact that the colour transformation is computed only once for all measurements in a given instrumental system, thus merging measurements from different observing runs. All measurements are thus reprocessed after each observing run. This procedure prevents the inter-run offsets that would occur for programme stars with such extreme colours as ours, if a different colour transformation were to be used for each run (see Manfroid and Sterken 1992).

The pair HR 8204 (= ζ Cap, G5IIBa2, y = 3.8, $b - y$ = 0.60) + HR 8213 (G5III, y = 4.51, $b - y$ = 0.54) illustrates the best precision that can be achieved in such conditions, since $\sigma_y = 2.5\ 10^{-3}$ mag for 32 measurements at the Danish 50 cm telescope spread over 1189 days. The high stability of G giants has also been emphasized by Grenon (1993). It is also noteworthy that a higher precision can be achieved on the Danish 50 cm telescope than on the ESO 50 cm telescope, because the former is equipped with a 4-channel photometer, contrary to the ESO 50 cm telescope where the 4 channels have to be measured sequentially. The same pair observed 47 times during the same period at the ESO 50 cm telescope yields $\sigma_y = 3.3\ 10^{-3}$ mag. This somewhat lower precision is a characteristic feature of the measurements performed at the ESO 50 cm telescope, and can probably be attributed to the fact that the whole observing sequence takes more time on the ESO 50 cm telescope than on the Danish 50 cm telescope.

5. Stability of Red Giants

5.1. NORMAL GIANTS

In order to identify the light variability specific to the PRG's, it is necessary to consider first the variability associated with normal giants. Not very much is known on that topic, however, as stressed by various recent reviews (Wing 1985, Johnson and Querci 1986). An important by-product of our LTPV monitoring is certainly that it provides information on the variability of red giants over time-scales of a few days to several years, down to amplitudes as small as $0\overset{m}{.}010$ in y.

Figure 1 presents the standard deviation σ_y in the y band as a function of $b - y$ for all the pairs (i.e. A-B, P-A and P-B) having more than 20 measurements spread over at least 1000 days from the Danish 50 cm telescope. It updates Fig. 5 of Paper I. There is a very clear tendency for K giants to be increasingly variable as they are redder. Examination of the individual measurements indicates that it is a night-to-night jitter which is causing the larger standard deviations of the reddest K giants. A period analysis using the method of Lafler and Kinman (1965) shows no significant periodicity in the range 1 to 100 days.

The greatest caution should be exercised, however, before drawing conclusions about the reality of such a tendency for σ_y to be larger for redder $b - y$ indices, because several instrumental effects may possibly be causing it, namely:

- second-order effects in the extinction coefficients. Neglecting these effects could translate into a residual (i.e. the deviation of the magnitude with respect to its average value over the whole monitoring) correlated with the airmass of the observation. The residuals of the three reddest barium stars exhibit no correlation at all with the airmass. Second-order effects on the atmospheric-extinction coefficients are indeed negligible in the b and y bands,

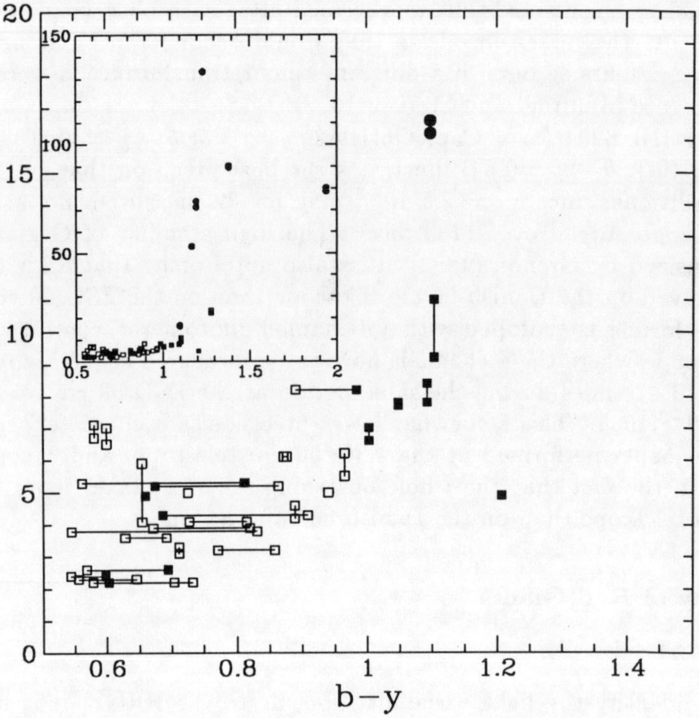

Fig. 1. The stability of red giants in the y band, expressed as σ_y (in 10^{-3} mag, from measurements performed at the Danish 50-cm telescope only) vs $b - y$. The standard deviation σ_y has been corrected for photon noise as described in the text, and refers to a pair P-A, P-B or A-B. The two pairs involving the variable star in the triplet (not always P!) were connected by a vertical line, drawn at the $b - y$ index of the variable star. The σ_y of the constant pair then defines the horizontal line connecting the $b - y$ indices of the two constant stars in the triplet. Barium stars and S stars are represented by filled squares or circles, respectively, whereas normal giants are represented by open squares. The insert extends the scale so as to include M giants and S stars, and show how the variability suddenly "explodes" for giants redder than $b - y = 1.1$

as shown by Tautvaišienė & Straižys (1985);

• programme and comparison stars too far away on the sky. Our programme and comparison stars are usually close enough (within 5°) for the differential extinction to be negligible for pairs observed not too far from the zenith, and the residuals show no correlation with the airmass anyway, as indicated above;

• photon noise, even though the standard deviation σ_y of pair ij (where i and j stand for P, A or B) plotted in Fig. 1 has been corrected for the photon noise, through $\sigma_y^2 = \sigma_{obs}^2 - \bar{\epsilon}_{ij}^2$, where σ_{obs} is the observed standard deviation and $\bar{\epsilon}_{ij} = (\bar{\epsilon}_i^2 + \bar{\epsilon}_j^2)^{1/2}$ is the average photon noise on pair ij. Moreover, there is one case where the reddest star in the APB triplet is also the brightest one;

- a faint star close to the location where the sky is measured, being on some nights inside the diaphragm and on other nights outside it, could be reponsible for such a jitter, although it is not clear how such an effect could introduce a correlation with the $b - y$ index;
- extrapolation of the colour transformation to the extreme values characterizing our stars. It has to be emphasized that a single colour transformation has been applied to these measurements which were all performed in the Danish 50 instrumental system. Therefore, this extrapolation is not expected to cause the night-to-night jitter that is observed.

Since none of the possible instrumental effects mentioned above appears to account convincingly for the observed tendency of σ_y to be larger for redder $b - y$ indices, we are tempted to consider it as an indication that the microvariability level of giant stars increases with later spectral types, a suggestion already expressed by Grenon (1993). The microvariability level of Fig. 1 (5 to 8 millimagnitudes for late K giants) is consistent with that deduced for K giants from the Geneva photometric catalogue (Table I of Grenon 1993). Schwarzschild (1975) actually predicted that giants with a deep convective envelope could exhibit some microvariability as a result of the fluctuations in the number of large convective cells visible at the surface.

5.2. BARIUM STARS

Barium stars spread all over the $(\sigma_y, b-y)$ plane of Fig. 1, but it should be noticed that the reddest K giants in our sample are in fact barium stars. As described in Sect. 5.1, the larger standard deviations of these stars are caused by a night-to-night jitter. In Fig. 1, HD 46407 tends to have a large σ_y ($0\overset{m}{.}0049$) given its $b - y$ index of 0.66. The index $R_y = \sum_{i=1}^{N-1}(y_{i+1} - y_i)^2/2(N-1)\sigma^2$, where the N y_i measurements are ordered in time, indicates, moreover, that there is a significant trend in the data, since $R_y = 0.29$ over 1394 d ($N = 99$, to be compared with $R_y = 0.78$ and $\sigma_y = 0\overset{m}{.}0035$ for the comparison pair; $R_y = 1$ if there is no trend in the data). Although HD 121447 has a σ_y of $0\overset{m}{.}010$ rather typical for its $b-y$ index of 1.10, its R_y index of 0.48 over 2337 d ($N = 45$, to be compared with $R_y = 0.89$ and $\sigma_y = 0\overset{m}{.}0060$ for the comparison pair) also reveals the presence of long-term variations. These long-term variations will be described in more detail in Sects. 6 and 8. It is noteworthy that these two stars are precisely those having the shortest orbital periods in our photometric sample ($P = 458$ d and 185 d for HD 46407 and HD 121447, respectively; McClure and Woodsworth 1990; Jorissen and Mayor, in preparation).

The comparison of the results of the LTPV monitoring with those of Landolt (1983) reveals that HD 121447 is the only variable barium star (the most variable, actually) in Landolt's sample that has been confirmed by the LTPV monitoring. Rather than the result of irregular variations, it is likely that these opposite conclusions are to be attributed to the better long-term stability of the photometric system achieved within the LTPV programme.

6. HD 121447: An Ellipsoidal Variable

This cool barium star (K7IIIBa5) exhibits the largest amplitude variations among barium stars, as seen on Fig. 1. The CORAVEL radial-velocity monitoring yields the second shortest orbital period among barium stars, with $P = 185^{\rm d}.7 \pm 0^{\rm d}.1$ in a circular orbit (Jorissen and Mayor, in preparation). Figure 2 shows that, when phased with this orbital period, the y light curve exhibits a double-sine pattern.

A period analysis of the y light curve by means of the Lafler and Kinman (1965) method reveals that the best photometric period is in fact $93^{\rm d}.0$ (see Fig. 3), which is exactly half of the orbital period. The good sampling of the radial-velocity data excludes the possibility that the spectroscopic period of $185^{\rm d}.7$ is an alias of the photometric period. It is thus likely that HD 121447 is an ellipsoidal variable. This is further suggested by the absence of variations in the colour indices within the experimental uncertainties (Fig. 2), as expected if the companion is a cool WD. From the absence of any far UV continuum on a deep IUE exposure of the HD 121447 system, Johnson et al. (1993) were able to put an upper limit of about $M_{\rm v} = 11$ for the absolute magnitude of the WD companion. It is also interesting to note that the orbital solution is characterized by residuals larger than expected (0.48 km s^{-1} for 22 measurements, to be compared with the uncertainty on one measurement amounting to 0.35 km s^{-1}). This may indicate that tidal interactions play a role in the system.

An analysis of the light-curve along the method outlined by Morris (1985) and Hall (1990) has been performed, yielding radii of the order of 50 to 100 $R_\odot$ for the barium star, depending on the adopted masses. More details will be given elsewhere.

7. HD 35155: An interacting, eclipsing-binary S star

In the course of an IUE investigation of S stars, Johnson and Ake (1983) discovered intense UV emission lines indicative of a state of high ionization in this otherwise ordinary S3/2 star (Keenan and Boeshaar 1980). This star was then monitored with IUE for more than 10 years by Ake et al. (1991). Spectral features typical of interacting binary stars were found, namely (see left insert in Fig. 5)

- a very broad CIV λ1549 emission line (FWHM ~ 2500 km s^{-1})
- narrower emission lines of SiIII], CIII], SiIV, NV, ... with variable intensities
- a λ^α power-law continuum for $\lambda < 2300$ Å (with $\alpha \sim -1$ to -1.5), whose intensity variations are uncorrelated with the line intensities
- longward of λ2300 Å there is an excess above the power-law continuum, disappearing when that continuum is the weakest
- absorption lines (of FeII, TiII,...?) whose intensity variations are anticorrelated with the brightness of the continuum

In the meantime, our CORAVEL monitoring (Jorissen and Mayor 1992b) confirmed the binary nature of that star, having a nearly circular orbit ($e = 0.07 \pm 0.02$) with a 642 d period, one of the shortest among binary S stars.

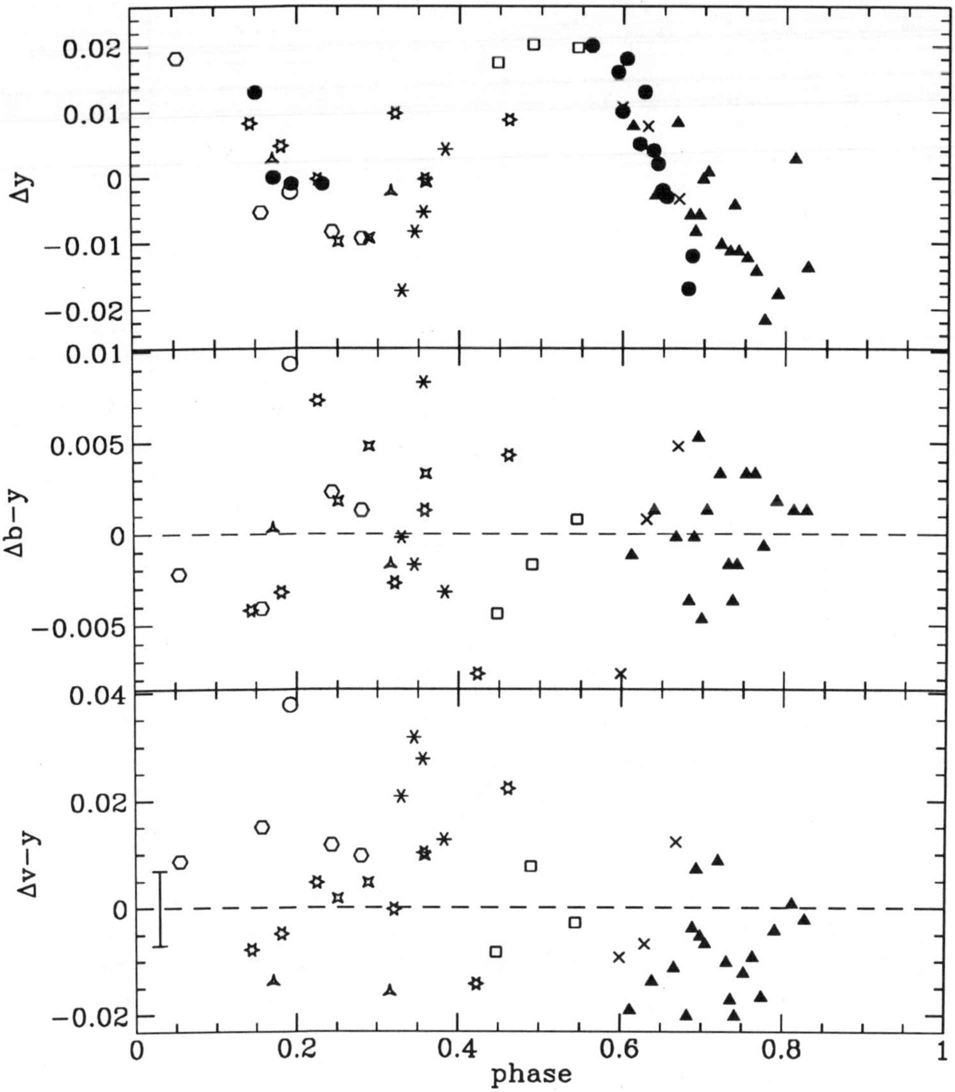

Fig. 2. Light curve of HD 121447 in the y, $b-y$ and $v-y$ colours (using HD 121699 and HD 117246 as comparison stars), phased with the orbital period of $185^{\mathrm{d}}.7$. The scatter in the $v-y$ colour is dominated by the photon noise, as indicated by the vertical bar corresponding to $2\bar{\epsilon}_{\mathrm{PB}}$ (see Sect. 5.1 for the notation). Different symbols refer to different cycles. The upper panel also contains data obtained in the Geneva V magnitude (black dots). The average y magnitudes from the LTPV observations have been used to compute the residuals in the Geneva system as well. The Geneva observations nicely fit in the trend derived from the LTPV data.

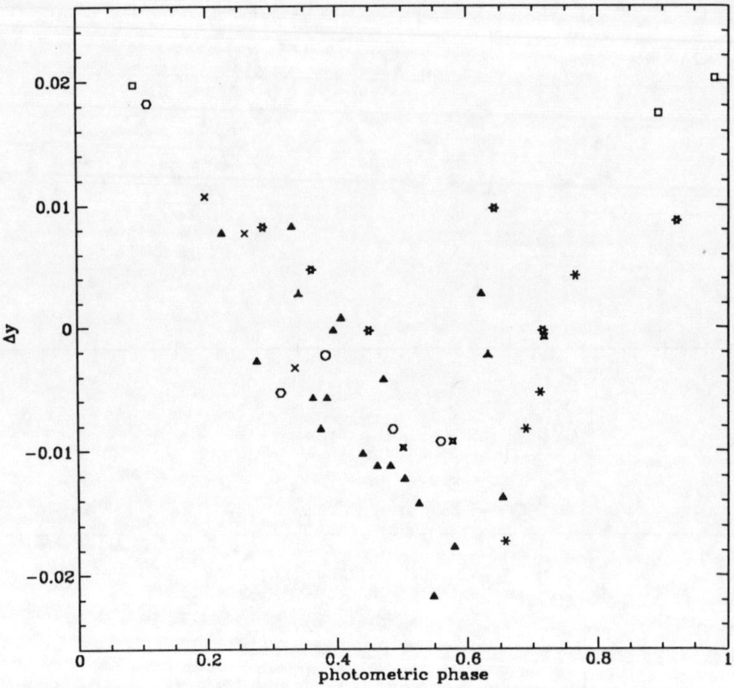

Fig. 3. Light curve of HD 121447 in the y band, phased with the photometric period of $93\overset{d}{.}0$

The light curve shows a clear dip in the y band at the time of the eclipse of the companion by the S star (Fig. 4), as well as smooth variations outside eclipse. The eclipse is not as clear in the v band. It is noteworthy that the observations during eclipse were obtained at a time when the orbital parameters were not yet known! It should be stressed that this serendipitous discovery was possible thanks to the substantial phase coverage achieved by the LTPV programme even for these long-period binaries. Similarly, it is quite remarkable that Ake et al. (1991) obtained by chance two IUE spectra (SWP 20615 and SWP 20878) exactly during the eclipse of the companion by the S star (see lower-left panel of Fig. 5). These spectra have, as expected, the weakest (though not completely absent) short-wavelength UV continuum and no excess longward of λ2300 Å.

Figure 5 shows the spectrum of HD 35155 all the way from the X-ray domain to the IR. The spectrum of the eclipsed light is also represented, as derived from the IUE spectra of Ake et al. (1991) and from the depth of the eclipse in the *uvby* colours. The spectrum of the eclipsed light component has a shape matching that of the system out of eclipse. This suggests that obscuration by dust and/or scattering processes may be involved.

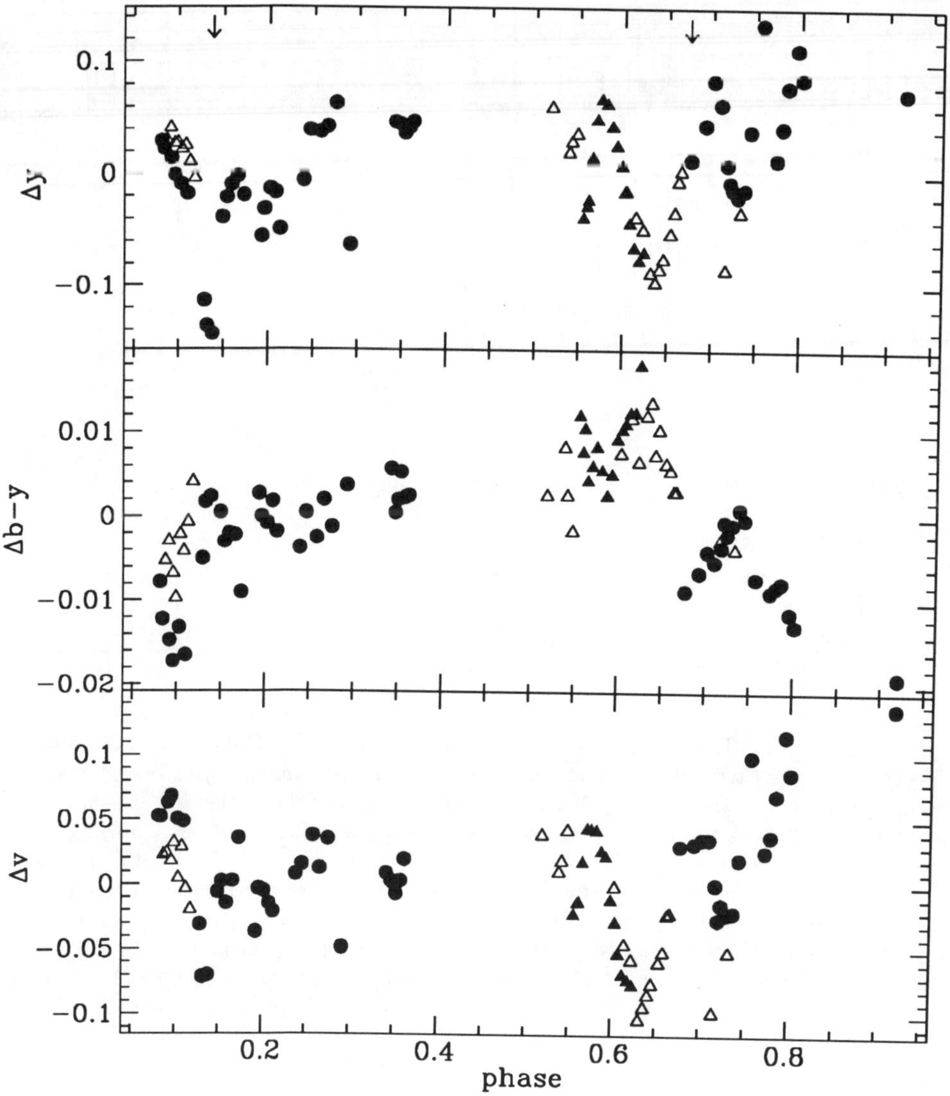

Fig. 4. The y, $b-y$ and v differential lightcurves [P-(A+B)/2] of HD35155 as a function of the orbital phase ($P_{\rm orb}$ = 642 d). The comparison stars are HR 1784 and HD 37828. The different symbols refer to the different cycles. The arrows in the upper panel correspond to the spectroscopic ephemeris for the eclipse of the companion by the S star (at phase 0.13) and for the transit of the companion (at phase 0.67)

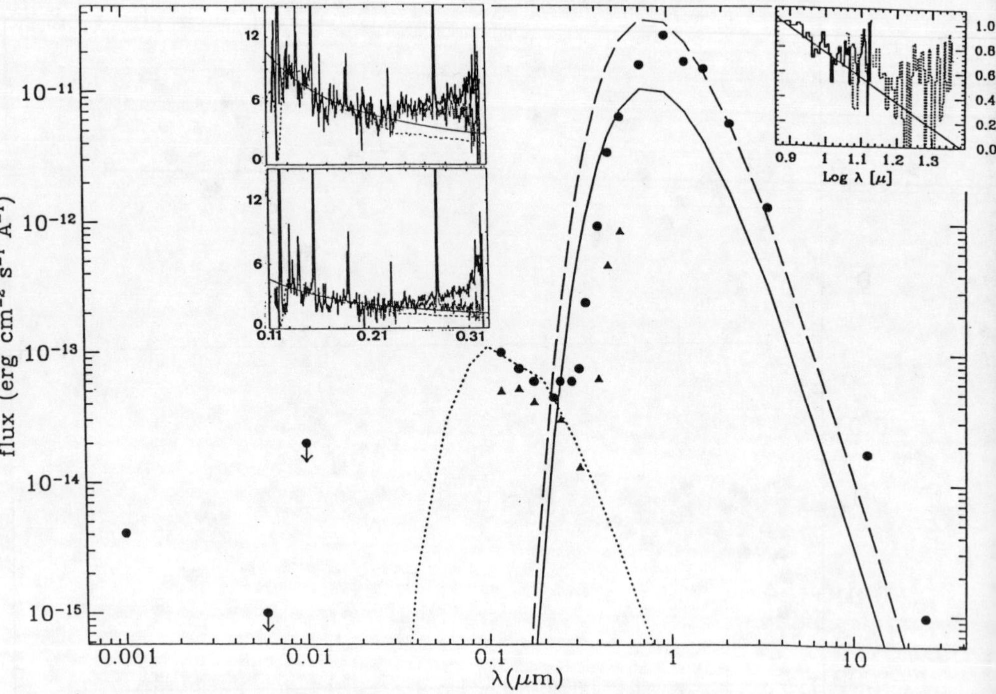

Fig. 5. Spectrum of HD 35155 from the X domain to the far IR. Dots refer to the observed spectrum whereas triangles refer to the eclipsed light component. Data in the X-ray domain are from Jorissen et al. (1994), from Ake et al. (1991) in the UV [the upper left panel is the UV spectrum out of eclipse (SWP22656), and the lower left is during eclipse (SWP20878)], from the LTPV for the *uvby* bands, from Eggen (1972) in the *RI* bands, from Catchpole et al. (1979) in the *JHKL* bands. The fluxes at 12 and 25 μm are from the IRAS catalog (IRAS Science Team 1988). The right panel is the IRAS low resolution spectrum from Groenewegen (1993), where an excess is clearly visible. Note that an upper limit of 2.3 10^{-26} erg cm^{-2} s^{-1} on the radio flux at 3.6 cm (Drake et al. 1991) has not been represented. Black-body curves with $T = 3600$ K (solid line, fitted to V, and long-dashed line, fitted to K) and $T = 25000$ K (short-dashed line) are represented

Ake et al. (1991) indeed note that periods of high obscuration occur when the overall UV flux drops and the FeII absorption spectrum deepens. The spectroscopic ephemeris derived from the CORAVEL monitoring indicates that these periods of high obscuration correspond to the passage of the companion behind the S star. At that time, the light curve in the optical domain also exhibits a clear dip. The y and $b-y$ light curves of Fig. 4 also clearly show that the light and colour variations in these bands are correlated in such a way that the system outside eclipse looks redder when fainter. Groenewegen (1993) points out that there is excess light longward of 15 μm in the IRAS low resolution spectrum (see Fig. 5), that is probably caused by dust emission. A deficit in the near UV is also suggested by the poor fit of the black-body curves to the observed fluxes. Whether obscuration and/or scattering by dust particles can consistently reproduce the variations observed in the different colours remains to be determined from a detailed model.

As was already the case for most of the barium stars monitored by Landolt (1983; Sect. 5.2), the conclusion drawn from the LTPV monitoring regarding the variability of HD 35155 disagrees with previous studies. Despite the rather large variations ($\Delta y \sim 0\overset{m}{.}2$) uncovered by the LTPV monitoring, Eggen (1972) found HD 35155 to be one of only two constant S stars (at the level of $0\overset{m}{.}1$ in V) among the sixteen studied. It is not quite clear whether the kind of variability found by the LTPV monitoring could have remained undetected by Eggen's observations, or whether these opposite conclusions indicate instead that HD 35155 may go through periods of quiescence. The latter hypothesis is actually supported by ROSAT observations of HD 35155 (Jorissen et al. 1994), which show that the X-ray flux from that system is strongly variable. A variable X-ray flux is also suggested by the fact that the HeII λ1640 line was missing in all IUE spectra obtained by Ake et al. (1991) but HeI λ10830 was observed by Brown et al. (1990). These lines are usually considered as being excited by X-rays (e.g. Zirin 1976).

Finally, the eclipsing nature of the system combined with the CORAVEL orbital parameters (Jorissen and Mayor 1992b) and the high resolution IUE spectrum (Ake et al. 1991) allow one to derive stringent constraints on the masses of the two stars. Ake et al. (1991) suggest that the CIII] and SiIII] UV lines are emitted in a region tied to the companion. The radial velocity of these lines can then be combined with the orbital solution for the giant to derive the mass ratio of the system, like in SB2 systems, yielding $M_1/M_2 \sim 3.0$, where M_1 and M_2 refer to the mass of the giant and of the companion, respectively (Jorissen et al. 1992b). Combining this value with the mass function then yields $M_2 \sin^3 i = 0.45$ $M_\odot$. The eclipsing nature of the system requires in turn that $\sin i \geq 0.9$ if the eclipse is total (given the orbital separation and the giant radius derived from the eclipse duration), so that $0.45 \leq M_2(M_\odot) \leq 0.6$. This analysis thus provides a strong direct indication in favour of a WD companion. Moreover, a mass $1.3 \leq M_1(M_\odot) \leq 1.8$ is then derived for the S star. Clearly, more IUE spectra are desirable in order to monitor the radial velocity of the CIII] and SiIII] lines and to confirm that they are tied to the WD companion. This again illustrates the importance of combined spectroscopic, photometric and radial-velocity monitoring in the field of PRG stars.

8. HD 46407: An eclipsing-like binary system

The star HD 46407 is a typical barium star of spectral type K0IIIBa3, whose chemical abundances were among the first to be determined by quantitative spectroscopic methods (Burbidge and Burbidge 1957). It was found to be a binary with a 458 d period by McClure and Woodsworth (1990).

As indicated in Sect. 5.2, HD46407 exhibits significant, albeit small, long-term photometric variations. Fig. 6 shows the y, $b - y$ and $v - b$ light curves phased with the orbital period. The y light curve exhibits a broad dip centred on the time of transit of the companion in front of the barium star as well as a narrower dip around the time of eclipse. Three such events were recorded, though with variable depths and phase lags (Fig. 8), suggesting that the eclipsed light does not come from a stellar companion.

The (y, $v - b$) diagram of Fig. 7 clearly shows that, like HD 35155 discussed in Sect. 7, the barium star HD 46407 looks redder when it is fainter and that the eclipse colours correspond to the average value over the cycle. It is, therefore, likely that some kind of dust scattering or obscuration process is involved, not unlike the situation encountered for HD 35155 and for the post-AGB star HR4049 (Waelkens et al. 1991). Indirect evidence for dust in the HD 46407 system is provided by excesses of about $0\overset{m}{.}1$ in the J and K bands (Feast and Catchpole 1977). A detailed analysis of the light curve is in progress, and should allow one to determine the exact geometry of the dust cloud reproducing the observed light and colour variations.

9. Conclusions

We have shown that long-term monitoring is essential to probe various processes related to the binarity of PRG stars, since their orbital periods may be as long as 10 y. Our photometric monitoring of barium and S stars within the LTPV programme yielded unexpected results, since two eclipsing-like stars were discovered (HD 46407, K0IIIBa3 and HD 35155, S3/2), despite the low luminosity of their WD companion. Obscuration and/or scattering by dust are probably involved in the light variations. Our ground-based LTPV and CORAVEL monitoring also proved to be of great help in interpreting the variations observed in the UV spectra of HD 35155 obtained by Ake et al. (1991) with the IUE satellite. In particular, the eclipsing behaviour observed in the *uvby* bands is present in the UV as well. Moreover, the combination of the spectroscopic orbit derived from the absorption spectrum of the giant with the UV emission lines associated with the companion makes it possible in principle to derive the mass ratio as in SB2 systems. The eclipsing nature of the system puts further constraints on the orbital inclination, so that the masses of the two components are within reach. In the same way, the combination of photometric and radial-velocity monitoring appears essential for the proper interpretation of the light variations observed for the cool barium star HD 121447 (K7IIIBa5). The ellipsoidal variations observed for this system set constraints on the radius of the barium star, and hence, on its position in the Hertzsprung-Russell diagram. Such a direct determination is the first one for a barium star, illustrating again the importance of long-term monitoring in the field of PRG stars.

In these times when super-large telescopes are threatening the further operation of their small brothers, it should not be forgotten that the modest telescopes used in the kind of long-term projects described during this conference can still provide fundamental data for our understanding of stellar physics.

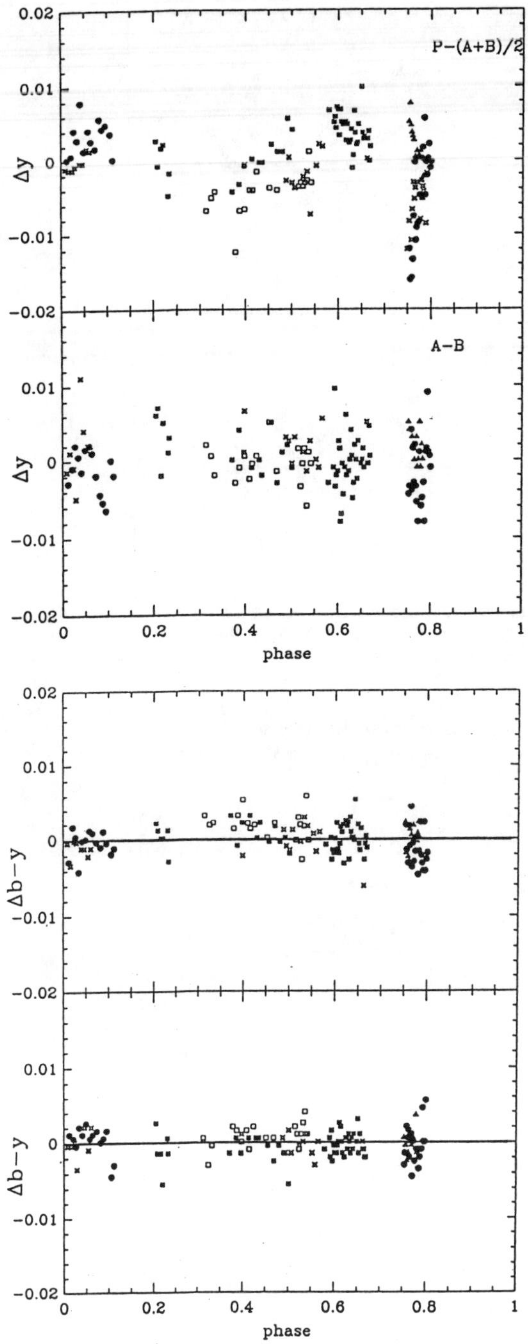

Fig. 6. a. Top: Light curve of HD 46407 in the y band. Top panel corresponds to the differential light curve of HD 46407 with respect to the average of the two comparison stars (HR 2379 and HR 2367). Bottom panel corresponds to the difference between the two comparison stars. Different symbols refer to different cycles.
b. Bottom: The same for the $b - y$ index

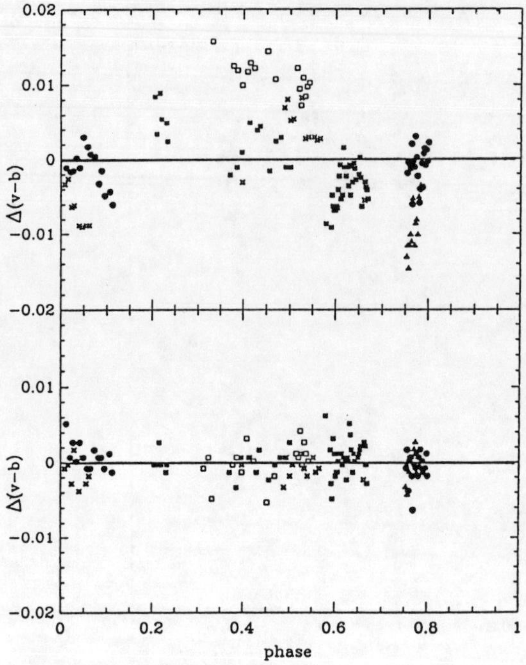

Fig. 6. c. The same for the $v - b$ index

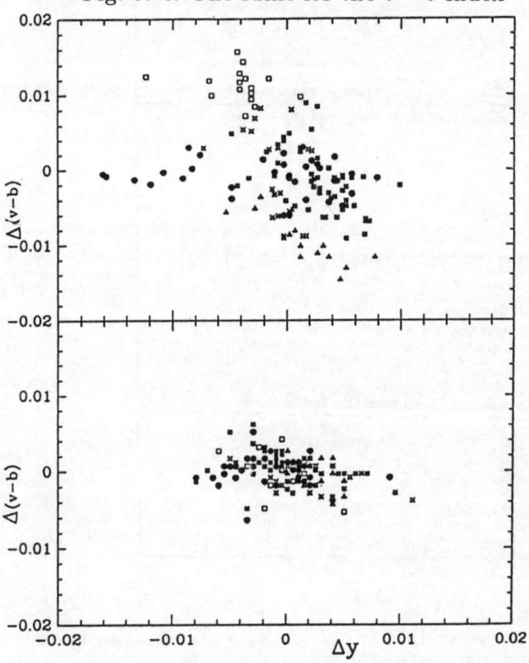

Fig. 7. The (y, $v-b$) plane for HD 46407 (upper panel) and its two comparison stars (lower panel). Symbols are as in Fig. 6. The plotted quantities are residuals with respect to the average values (brightness increases towards the right, and colour is redder upwards).

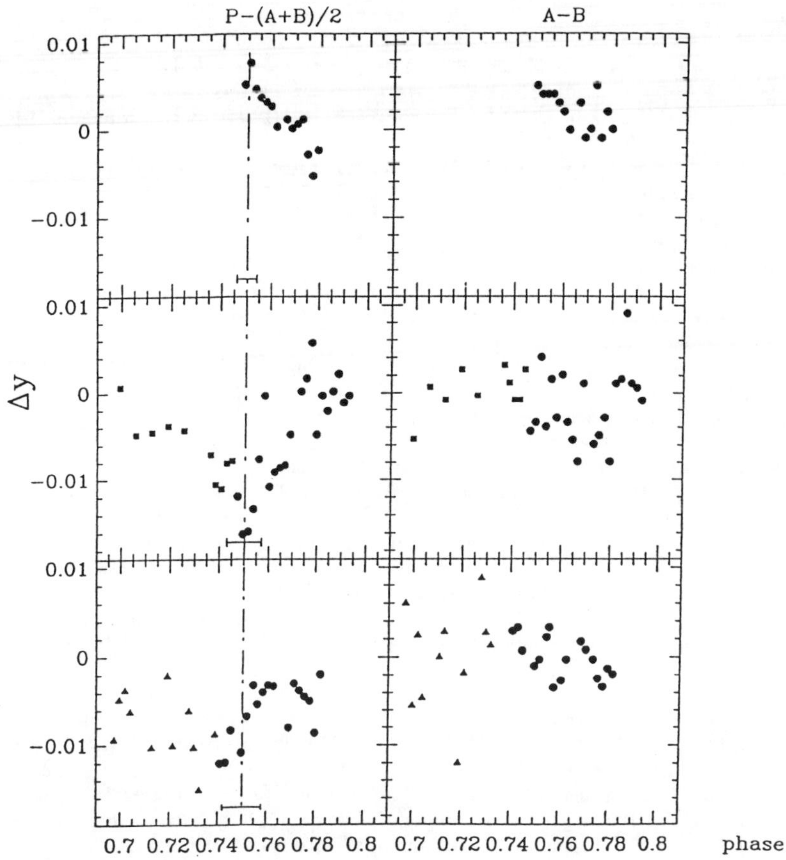

Fig. 8. Results of three observing campaigns around the predicted time of eclipse (at phase 0.75) of the companion by the barium star HD 46407: February 1985 (upper panel), November 1988 (middle panel), February 1990 (lower panel). Phases are computed according to the orbital elements from McClure and Woodsworth (1990). The uncertainty in the prediction of the time of mid- eclipse is given by the horizontal bar. The left panels present the differential magnitude of the programme star with respect to the average of the two comparison stars, whereas the right panel presents the difference between the two comparison stars. Dots refer to measurements from the Danish 50 cm telescope, whereas squares and triangles refer to measurements from slightly different systems at the ESO 50 cm telescope. The plotted magnitudes correspond to the residuals with respect to the average magnitude over the total time span of the monitoring in a given system.

Acknowledgements

Without the support of the European Southern Observatory to the LTPV, none of these results could have been obtained. Special thanks are also due to all the LTPV observers. The collaboration of M. Mayor and all CORAVEL observers has been essential for getting orbital parameters of many PRG stars. I also thank G. Burki and M. Burnet (Geneva Observatory) for making possible the observing campaign of HD 121447 on the Swiss telescope at La Silla.

References

Ake T.B., Johnson H.R., Ameen M.M., 1991, *Astrophysical Journal*, 383, 842

Bidelman W.P., Keenan P.C., 1951, *Astrophysical Journal*, 114, 473

Brown J.A., Smith V.V., Lambert D.L., Dutchover E.Jr., Hinkle K.H., Johnson H.R., 1990, *Astronomical Journal*, 99, 1930

Böhm-Vitense E., Nemec J., Proffitt Ch., 1984, *Astrophysical Journal*, 278, 726

Burbidge E.M., Burbidge G., 1957, *Astrophysical Journal*, 126, 357

Catchpole R.M., Robertson B.S.C., Lloyd Evans T.H.H., Feast M.W., Glass I.S., Carter B.S., 1979, *SAAO Circ.* 1, 61

Dominy J.F., Lambert D.L., 1983, *Astrophysical Journal*, 270, 180

Drake S.A., Johnson H.R., Brown A., 1991, *Astronomical Journal*, 101, 1483

Feast M.W., Catchpole R.M., 1977, *Monthly Notices of the RAS*, 180, 61P

Eggen O.J., 1972, *Astrophysical Journal*, 177, 489

Grenon M., 1993. In: *Inside the stars*, Weiss W.W., Baglin A. (eds.) Proc. IAU Coll. 137, *Astronomical Society of the Pacific Conf. Ser.*, 40, 693

Griffin R., Griffin R., 1981, *Monthly Notices of the RAS*, 193, 957

Griffin R.F., 1984, *The Observatory*, 104, 224

Griffin R.F., Keenan P.C., 1992, *The Observatory*, 112, 168

Groenewegen M.A.T., 1993, *Astronomy and Astrophysics*, 271, 180

Hall D.S., 1990, *Astronomical Journal*, 100, 554

Iben I.Jr., Renzini A., 1983, *Annual Review of Astronomy and Astrophysics*, 21, 271

IRAS Science Team, 1988, IRAS Point Source Catalogue, prepared by Beichman C.A., Neugebauer G., Habing H.J., Clegg P.E., Chester T.J., NASA RP-1190

Johnson, H.R., 1992, Binary S and MS stars. In: *Evolutionary Processes in Interacting Binary Stars*, Kondo Y., Sistero R.F., Polidan R.S. (eds.) Proc. IAU Symp. 151, Kluwer, Dordrecht, 157

Johnson H.R., Ake T.B., 1983. In: Baliunas S.L., Hartmann L. (eds.) *Proceedings of the Third Cambridge Conf. on Cool Stars, Stellar Systems and the Sun*, Springer, Heidelberg

Johnson H.R., Querci F.R., 1986, *The M Type Stars*, Monograph Series on Non-Thermal Phenomena in Stellar Atmospheres, CNRS-NASA SP-492

Johnson H.R., Ake T.B., Ameen M.M., 1993, *Astrophysical Journal*, 402, 667

Jorissen A., Mayor M., 1992a, Orbital elements of a new sample of barium stars. In: *Nuclei in the Cosmos*, Käppeler F., Wisshak K. (eds.) Inst. of Physics Publ., Bristol, 29

Jorissen A., Mayor M., 1992b, *Astronomy and Astrophysics*, 260, 115

Jorissen A., Boffin H.M.J., 1992. In: *Binaries as tracers of stellar formation*, Duquennoy A., Mayor M. (eds.) Cambridge Univ. Press, Cambridge, 110

Jorissen A., Manfroid J., Sterken C., 1992a, *Astronomy and Astrophysics*, 253, 407 (Paper I)

Jorissen A., Manfroid J., Mayor M., Sterken C., 1992b, *Information Bulletin on Variable Stars*, 3730

Jorissen A., Frayer D.T., Johnson H.R., Mayor M., Smith V.V., 1993, *Astronomy and Astrophysics*, 271, 463

Jorissen A., Schmitt J.H.M.M., Bickert K.F., 1994, in preparation

Keenan P.C., Boeshaar P.C., 1980, *Astrophysical Journal*, S 43, 379

Lafler J., Kinman T.D., 1965, *Astrophysical Journal*, S 11, 216

Landolt A.U., 1983, *Publications of the ASP*, 95, 644

Manfroid J., 1985, habilit. thesis, Univ. of Liège, Belgium

Manfroid J., Sterken C., 1992, *Astronomy and Astrophysics*, 258, 600

McClure R.D., 1983, *Astrophysical Journal*, 268, 264

McClure R.D., 1984, *Publications of the ASP*, 96, 117

McClure R.D., Fletcher J.M., Nemec J.M., 1980, *Astrophysical Journal*, 238, L35

McClure R.D., Woodsworth A.W., 1990, *Astrophysical Journal*, 352, 709

Merrill P.W., 1922, *Astrophysical Journal*, 56, 457

Merrill P.W., 1952, *Astrophysical Journal*, 116, 21

Morris S.L., 1985, *Astrophysical Journal*, 295, 143

Scalo J.M., 1976, *Astrophysical Journal*, 206, 474

Schwarzschild M., 1975, *Astrophysical Journal*, 195, 137

Tautvaišienė G., Straižys V., 1985, *Soviet Astronomy*, 29, 234

Waelkens C., Lamers H.J.G.L.M., Waters L.B.F.M., Rufener F., Trams N.R., Le Bertre T., Ferlet R., Vidal-Madjar A., 1991, *Astronomy and Astrophysics*, 242, 433
Warner B., 1965, *Monthly Notices of the RAS*, 129, 263
Webbink R.F., 1986. In: Leung K.C., Zhai D.S. (eds.) Critical Observations vs Physical Models for Close Binary Systems. Gordon and Breach, New York, 403
Wing R. F., 1985. In: *Cool Stars with Excesses of Heavy Elements*, Jaschek M., Keenan P.C. (eds.) Reidel, Dordrecht, 61
Zirin H., 1976, *Astrophysical Journal*, 208, 414

DISCUSSION

Delgado: Stars at the evolutionary phases occupied by Ba stars might develop deep convection zones, bringing up processed material to the surface. Is the binary mass-exchange the only possible model to account for such chemical peculiarities?

Jorissen: There is no way to synthesize the heavy elements observed at the surface of these barium stars during the shell H-burning phase on the red-giant branch along which many of these stars are now evolving. At least for these, the binary scenario seems to be required. Furthermore, our radial-velocity monitoring with CORAVEL at ESO has shown that 26 out of the 27 known barium stars with strong anomalies are binaries.

Hall: In evaluating the radius, mass ratio, inclination, etc. of your ellipsoidal variable, did you use the complete correct model? Specifically, I wonder whether you included the effect of gravity darkening? If this is neglected, then you will need a larger radius to produce the same amplitude. Conversely, if you include this, you will get a smaller radius, and I believe you wanted the radius to be smaller. A useful reference would be Hall 1990 (*Astronomical Journal*, 100, 554).

Jorissen: I only took into account purely geometric effects in this preliminary model. I fully agree that gravity darkening is important and has to be included for the final analysis.

Young: I am unhappy with your dust-cloud model, because dust is a forward-scattering rather than backward-scattering medium. So I would rather use the dust cloud around the other star.

Jorissen: If the dust cloud were to be around the barium star, there would be no way to illuminate it from behind since the WD companion is very cool (not seen with IUE).

Harmanec: Are you not concerned about the fact that the period of your ellipsoidal variable is so close to one half of a year? I actually noted that your light curve resulting from data folding indicates slightly different brightness levels for individual observing runs but not too much evidence of variations *within the given run*. I, therefore, suggest that you check for possible very small seasonal zero-point shifts in your data.

Jorissen: I think that the half-year orbital period is safe, because our *radial velocity* data are sufficiently well sampled. Also, there has been at least one photometric run during which significant variations have been found, and variability has since been confirmed from Geneva photometric data.

Harmanec: You showed us that the r.m.s errors of the magnitude differences of pairs of stars increase towards later spectral types of the objects involved. I would like to ask whether, when looking for possible systematic effects, you also investigated the $O - C$ deviations of your standard stars for possible colour-dependent small systematic effects?

Jorissen: No, especially because no sufficiently red standards exist in the *uvby* system. Therefore, extrapolation of the colour transformation toward such red stars may, perhaps, explain the observed effect. However, I emphasize that with the multi-night reduction algorithm used to reduce the LTPV data, the *same* colour matrix is applied to all the runs in the Danish 50 instrumental system. Remember also that the extinction coefficients are computed separately for every night from the comparison star measurements. And we checked that there are no second-order extinction effects in these, since the deviations Δy from the average value do not correlate with the airmass.

Kurtz: Can you tell me how you know the secondary stars are white dwarfs?

Jorissen: Statistical analysis of the mass-function distribution $f(M) = \frac{{M_1}^3}{(M_1+M_2)^2}\sin^3 i$ shows that it is compatible with the superposition of a random orbital inclination and a $\frac{{M_1}^3}{(M_1+M_2)^2}$ distribution peaked around $0.4 M_\odot$. This value indicates that, if the giant has a mass in the range 1 to 2 $M_\odot$, then the companion has a mass in the range 0.5 to 0.6 $M_\odot$ like white dwarfs have.

Niarchos, Hric, Chochol, Demircan

A LONG-TERM PHOTOMETRIC CAMPAIGN APPROACH TO THE STUDY OF SYMBIOTIC VARIABLES

L. HRIC, D. CHOCHOL, R. KOMZIK, A. SKOPAL, and Z. URBAN
Astronomical Institute
Slovak Academy of Sciences
059 60 Tatranská Lomnica
The Slovak Republic

and

P. NIARCHOS
Section of Astrophysics
Astronomy and Mechanics
Department of Physics
University of Athens
157 83 Zografos – Athens
Greece

ABSTRACT. This paper is based on a five-year experience with an international campaign on photometry of symbiotic stars. During this period, 1 475 photoelectric, 2 739 photographic and 56 872 visual observations were collected and published in five volumes of Contrib. Astron. Obs. Skalnaté Pleso. Special emphasis is put on specific problems related to collecting, processing and publishing large amounts of observational data. The list of observatories, astronomers, instruments and amateur groups of visual observers of variable stars collaborating within the framework of this campaign, as well as the most interesting results collected for the symbiotic stars UV Aur, T CrB, CH Cyg, V 1329 Cyg, AX Per and PU Vul are presented. Advantages and disadvantages of visual photometry are discussed.

Keywords: photometric campaign – symbiotic stars – light variability

1. Introduction

Symbiotic variables are the group of objects exhibiting composite spectra consisting of a cool-giant continuum and absorption-line characteristics, a hotter continuum in the blue part of the spectrum and many emission lines of a number of elements in various stages of ionization. During the periods of activity, changes of the hot continuum and emission lines occur.

Symbiotics are binary systems with strong interactions between the components. One component is the cool giant losing mass, by wind or by Roche-lobe overflow, to the more compact companion (white dwarf, main-sequence star or neutron star).

C. Sterken and M. de Groot (eds.), The Impact of Long-Term Monitoring on Variable Star Research, 163–176.

Activity stages are explained by nuclear burning of accreted material on the surface of the white dwarf or by the mass-transfer bursts and/or accretion-disc instabilities.

2. The light variability

The light curves of symbiotic variables are complex. We observe periodic, semi-regular and basically unpredictable irregular light variations.

1. **Periodic variations:**
   a) eclipses of the hot source by the red-giant companion in the binary orbit
   b) reflection effect (illumination of the red-giant hemisphere facing the hot component)
   c) pulsations of the cool giant
   d) rotation of the components.
2. **Semi-regular variations:**
   a) presence of complex structures of circumstellar matter in the symbiotic binary system with gaseous streams and hot spots.
3. **Irregular variations:**
   a) activity of symbiotic binaries caused by thermonuclear burning on the white dwarf, similar to the outbursts of classical novae
   b) activity of symbiotic binaries caused by an unstable increase of the mass loss from the cool giant or by instabilities in the accretion disc
   c) dust formation in the system.

Moreover, the brightness variations can be caused also by other mechanisms: e.g., winds in collision, jets and bipolar outflows, formation of strong nebulosity, eccentricity of the orbit, geometry of the components, precession of the accretion disc, and the presence of a third body. The light curve of a particular symbiotic star results from various combinations of the mechanisms mentioned above.

All brightness variations can be characterized by their time scale and amplitude. The time scales of periodic and semi-regular variations modulated by the binary motion or caused by Mira-like variations are mostly of the order of a hundred days. The relevant amplitudes are at most a few magnitudes in the visual. The irregular nova-like variations have amplitudes between 2 and 7 magnitudes on time scales of years to decades.

Only the long-term monitoring of symbiotic stars can reveal the full complexity of the light variations and clarify the physical nature of individual systems.

3. Campaign idea and coordination

The lack of UBV photometry for symbiotic stars has led us to the idea of launching an observational campaign for the photometry of selected symbiotic stars (Hric and Skopal, 1989). We prepared finding charts for 30 symbiotic stars with detailed instructions for photoelectric observations. The main goal of the proposed campaign is to collect photoelectric, photographic and visual observations and to describe the long-term variations in brightness. The detailed analysis should be done separately.

All contributions from individual observers and/or observing teams are gathered and subsequently published annually in Contrib. Astron. Obs. Skalnaté Pleso

TABLE I
The observatorics

	Observatory	Instrument	Type of observations	Astronomers
1	**Skalnaté Pleso** Astronomical Institute Slovak Academy of Sci. The Slovak Republic	0.6 m Cassegrain reflector	photoelectric	Chochol D. Hric L. Komžík R. Skopal A. Urban Z.
		0.3 m astrograph	photographic	Chochol D. Hric L.
2	**Stará Lesná** Astronomical Institute Slovak Academy of Sci. The Slovak Republic	0.6 m Cassegrain reflector	photoelectric	Chochol D. Hric L. Komžík R. Skopal A.
3	**Masaryk University Observatory**, Brno The Czech Republic	0.6 m Newton reflector	photoelectric	Papoušek J.
4	**Copernicus Obs. & Planetarium**, Brno The Czech Republic	0.4 m Nasmyth reflector	photoelectric	Hanžl D.
5	**Kryonerion Station** National Obs. of Athens, Greece	1.2 m Cassegrain reflector	photoelectric	Dapergolas A.
	University of Athens Greece			Niarchos P. Rovithis P. Rovithis-Livaniou H.
6	**Observatory of the Wroclaw Univ.** Poland	0.6 m Cassegrain reflector	photoelectric	Pigulski A.
7	**Rozhen Obs. &** Bulgarian Academy of Sci. Bulgaria	50/70/172 cm Schmidt telescope	photographic	Semkov E. Tsvetkova K.
8	**Catania Astrophysical Obs.**, Italy	0.91 m Cassegrain reflector	photoelectric	Blanco C.
9	**Sonneberg Obs.** Germany	0.6 m Cassegrain reflector	photoelectric	Luthardt R.
10	**Astron. Obs. of the Odessa State Univ.** Ukraine	0.2 m seven camera astrograph	photographic	Chinarova L.L.
11	**Crimea Station of the Sternberg State Astron. Inst.**, Moscow, Russia	0.4 m astrograph	photographic	Pikhun A.I.
12	**Private Observatory in Zweikirchen**, Germany	0.36 m automatic telescope	photoelectric	Agerer F.
13	**Private Observatory near Považská Bystrica** The Slovak Republic	4/300 mm Sonnar with 0.24 m guiding reflector	photographic	Velič Z.

(CAOSP) (deadline September 30). We have published five papers so far (Skopal et al. 1990; Hric et al. 1991; Skopal *et al.*, 1992b; Hric et al. 1993b and Hric et al. 1994, in press). The observations have been obtained at the observatories listed in Table 1. The observations were done during the years 1983 – 1993 (photoelectric), 1955 – 1993 (photographic), and 1973 – 1993 (visual). In summary, 36 symbiotic and symbiotic-like stars were observed in our campaign. The material consists of 1 475 photoelectric observations, 2 739 photographic plates and 56 872 visual-magnitude estimates. The list of amateur groups is given in Table 2.

TABLE II
The amateur groups

	Group, Abbreviation	**Type of observations**	**Coordinators**
1	Observatoire de Strasbourg, Association Francaise des Observateurs d'Etoiles Variables **AFOEV**, France	visual	Schweitzer E.
2	British Astronomical Association **BAA**, Cyprus	visual	Isles J.E.
3	Berliner Arbeitsgemeinschaft für Veränderliche Sterne **BAV**, Germany	visual	Korth S.

4. Campaign experiences and problems

The publication of such large amounts of observational material creates problems with the editorial board of the journal CAOSP. We are attempting to publish our photoelectric and photographic campaign data in the form of tables. Unfortunately, in cases of very large amounts of data (e.g. FG Sge in our latest paper of the campaign series), we are forced to publish only the figures; this applies especially to the case of visual observations. The small budget for CAOSP makes it difficult not only to publish but also to distribute our journal abroad.

We have good experience with observations carried out by technicians at the Skalnaté Pleso Observatory. They are fully able to obtain observations of high quality after 1 to 2 years of practice.

In a few cases the use of incorrect comparison stars resulted in a waste of observing time ($\leq 1\%$). More mistakes ($\sim 1\%$) were found in the treatment of data. Nevertheless, the constant supervision of incomming observational data by the principal investigator leads to rapid detection of errors and to subsequent correction. Our long-term monitoring also revealed a small-amplitude variability of the comparison star in one case (CH Cyg - S1: HD184786, M8); two other cases are suspect.

Electronic mail affects the efficiency of our campaign work. The deadline for publishing campaign results is only two weeks after receiving the latest observational data.

The accuracy of photoelectric observations is better than a few hundredths of magnitude for data obtained simultaneously at different observatories. During excellent nights at the Skalnaté Pleso Observatory we reach an accuracy of $0^{m}_{.}002$. Thanks to its high altitude the Skalnaté Pleso Observatory (1783 m) has relatively good photometric conditions compared with the other observatories listed in Table 1. As the symbiotic stars are objects with strong emission lines, the discrepancies between various instrumental systems, reflecting the differences in the widths of U, B, V filters, can spoil the agreement of observational data from different observatories. In many symbiotic stars, the emission lines of [O III] at 495.89 and 500.69 nm located just at the edge of the V passband, could cause these disagreements. We have described this effect in the paper on Nova Cygni 1992 (Chochol et al. 1993).

We would like to extend our observations to fainter symbiotic stars, i.e. down to 15 mag (at present, the limit is $\sim$ 12 mag in the V-band), using either a CCD guiding camera or a CCD detector for photometry.

The observational data from our campaign, as well as those published in the literature, are used for preparing the theses of university students. In general, new data collected within our campaign help to distinguish among the theories which explain the individual phenomena in symbiotic stars; e.g., accretion-disc instability versus thermonuclear-runaway models. The long-term monitoring of symbiotic stars, after detailed analysis of light curves, contributes new knowledge of fundamental data—radii, orbital periods and their changes, orbital inclinations, mass ratios (EG And, V 1329 Cyg, AX Per)—and reveal new interactions in some systems (CH Cyg, AX Per).

The annual publication of CAOSP puts the rigorously reduced photometric data at the disposal of the astronomical community in less than one year after the latest observations were obtained. We are now starting the more detailed quantitative analysis of the data accumulated over the five-year period of the campaign. However, through our campaign papers this approach is open to all researchers in the field of symbiotic stars. Moreover, spectroscopists will be able to make use of the information about the brightnesses of the stars of their interest at the times of their observations. We are, therefore, viewing the annual publication of our, mostly fresh, observational data as a service to the whole scientific community.

5. Campaign results

Basic observational characteristics of the light curves of symbiotic stars are published in our campaign papers. To show the fruitfullness of our campaign work we selected six symbiotic stars as examples of complex light variations and present some results of their more detailed analysis.

UV Aur is a symbiotic system which contains an evolved carbon star rather than an M-giant as its cool component. The period of the light variations was 365 days during the last five years. The older visual data in Fig. 1 support a period close to 400 days, caused by pulsations of the cool component (Khudyakova, 1985). Another long-term light change lasting over the entire period of observation is clearly present in Fig. 1. It supports the 14-year orbital period proposed also by Khudyakova (1985). In Fig. 2 we compare the photoelectric V and visual data; the agreement is very good.

CH Cyg. Long-term monitoring of the symbiotic binary CH Cyg (orbital period 5700 days) revealed the new outburst which began in February 1992 (Skopal et al. 1992a) shown in Fig. 3. This discovery started very extensive multi-frequency observations of this system [optical (Asiago, Ondřejov and Skalnaté Pleso Observatories), ultraviolet (IUE satellite), radio (Jodrell Bank - MERLIN, Hawai - JCMT, New Mexico - VLA)]. We revealed the fundamental nature of the CH Cyg outburst, which can now be understood in terms of an accretion event resulting from a direct impact of the material transferred from the pulsating giant component (Skopal et al. 1993 a, b). The fact that we observed extremely high terminal velocities of -1800 km s^{-1} in the H_α emission component, but a very small change in the star's brightness (bolometric luminosity much below the Eddington limit) contradicts some models for the CH Cyg outburst suggested previously: an obligue rotator model (Mikolajewski and Mikolajewska, 1988) and the triple-star model (Hinkle et al. 1993). The detailed behaviour in V during the last period is compared to the visual observations in Fig. 4.

V 1329 Cyg The outburst of the well-known symbiotic nova V 1329 Cyg started from $m_{pg} = 15$ in 1957 and reached a maximum of $m_{pg} = 11.5$ in 1966. According to Munari et al. (1988), the cause of the nova-like outburst was a non-degenerate H- burning flash on a luminous white dwarf. During our campaign we have gathered photographic data taken at three observatories (Fig. 5). The periodic sinusoidal variations with an amplitude of $1^m.5$ are clearly visible in the post-outburst light curve (Fig. 6) where we gathered photographic as well as B photoelectric data. Using the sinusoidal fitting method with an exponential decrease, we obtained a reliable value of the period (Hric at al.1993a). The following ephemeris is valid after the outburst:

$$JD_{min} = 2\,446\,784.9(\pm 2.4) + 963^d.306(\pm 0.772) \times E$$

While the pre-outburst orbital period of V 1329 Cyg was 951.9 days, its period increased by $11^d_{\cdot}4$ as a consequence of the outburst. The best explanation for this period increase is an accretion event.

AX Per The long-term monitoring of the eclipsing symbiotic binary AX Per revealed a new ephemeris for the primary minima,

$$JD_{min} = 2\,436\,673.3(\pm 0.6) + 679^d.9(\pm 1.2) \times E$$

(Skopal, 1991), which is more accurate than that previously determined by Kenyon (1982). The visual data collected in our campaign cover two cycles of AX Per's activity (Fig. 7). Detailed analysis of the UBV light curve during the primary minimum gave the basic parameters of the binary: $R_{giant} = 102 \pm 3\ R_\odot$, $R_{hot} = 36 \pm 3\ R_\odot$ (during the outburst), orbital inclination $i = 90° \pm 6°$, luminosities $L_{hot} = 340\ (d/2.5\ kpc)^2\ L_\odot$, $L_{giant} = 1600\ (d/2.5kpc)^2\ L_\odot$ (Skopal, 1993). An extensive investigation of this system revealed the existence of a collisionaly-heated region situated on the red-giant hemisphere as the consequence of a slow thermonuclear outburst of the hot component. This model contradicts the accretion-disc powered model, recently published by Mikolajewska and Kenyon (1992). Photoelectric V data together with visual observations during the last outburst are shown in Fig. 8.

PU Vul is a symbiotic nova which underwent an outburst in 1979. Visual observations (Fig. 9) cover the period of activity as well as the deep minimum in 1980 – 1981 caused by circumstellar dust formation (Belyakina et al. 1984, 1989). The gradual decrease of brightness which started in 1987 is well documented in photoelectric V data (Fig. 10).

T CrB is a well-known recurrent nova included in the group of symbiotic stars. Recently, Yudin and Munari (1993) presented IR observations which gave evidence for a pronounced ellipsoidal shape of the M3 giant in the system. The phase diagram of V data from our campaign, using the ephemeris $JD_{min} = 2\,435\,571 + 227.8 \times E$ (Hric et al. 1991), is given in Fig. 11. The double wave found by Yudin and Munari in the J-colour is also clearly visible in our V data. The variation is less significant

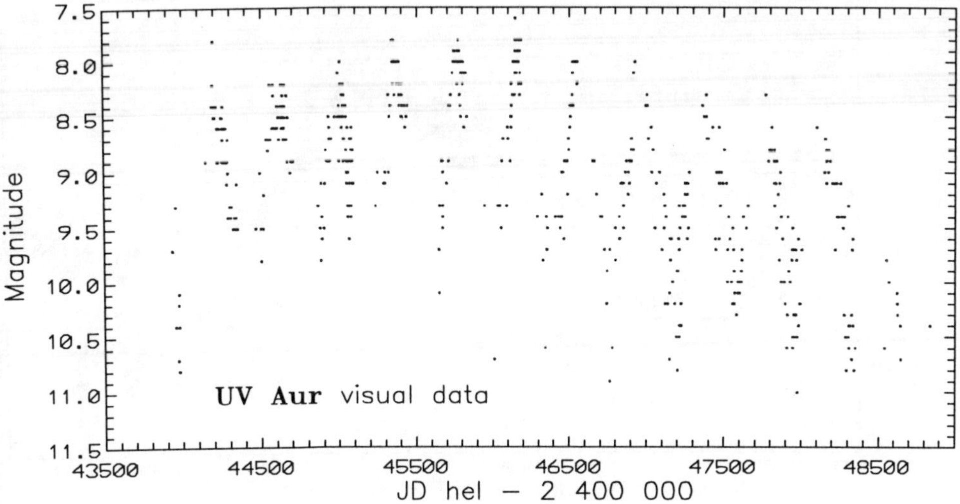

Fig. 1. UV Aur visual data

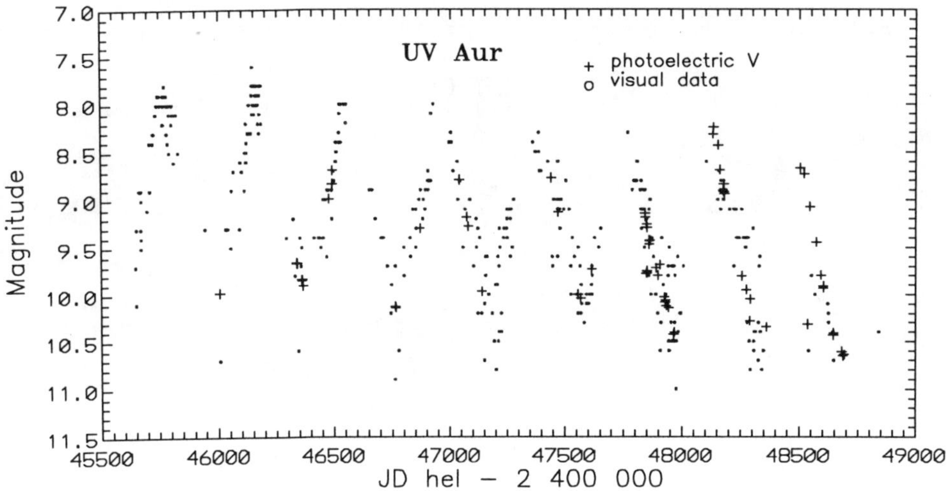

Fig. 2. A comparison of photelectric V and visual data for UV Aur

in the B-colour and not present in the U-colour.

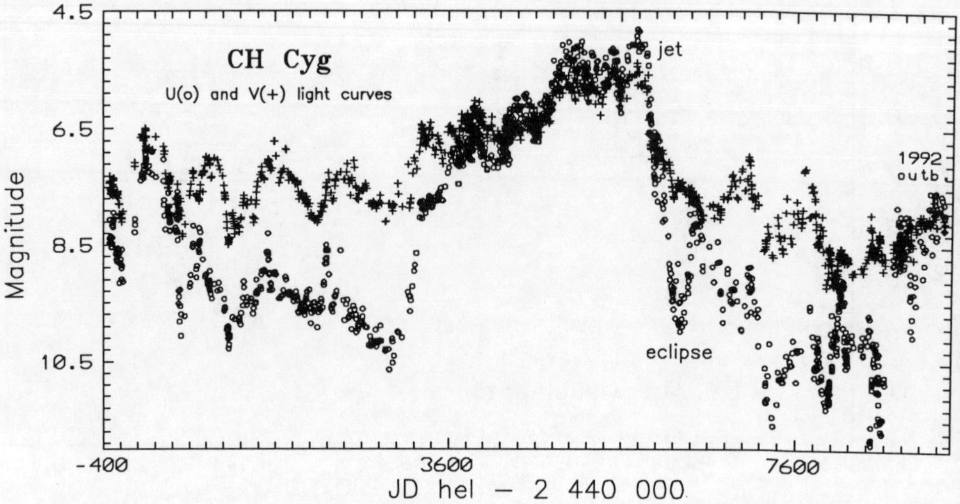

Fig. 3. A comparison of photelectric U and V light curves of CH Cyg

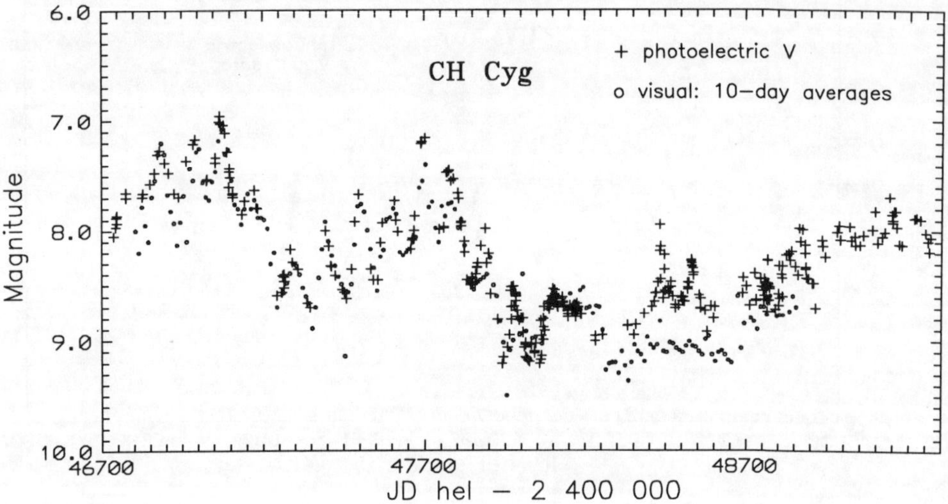

Fig. 4. A comparison of photelectric V and averaged visual data for CH Cyg

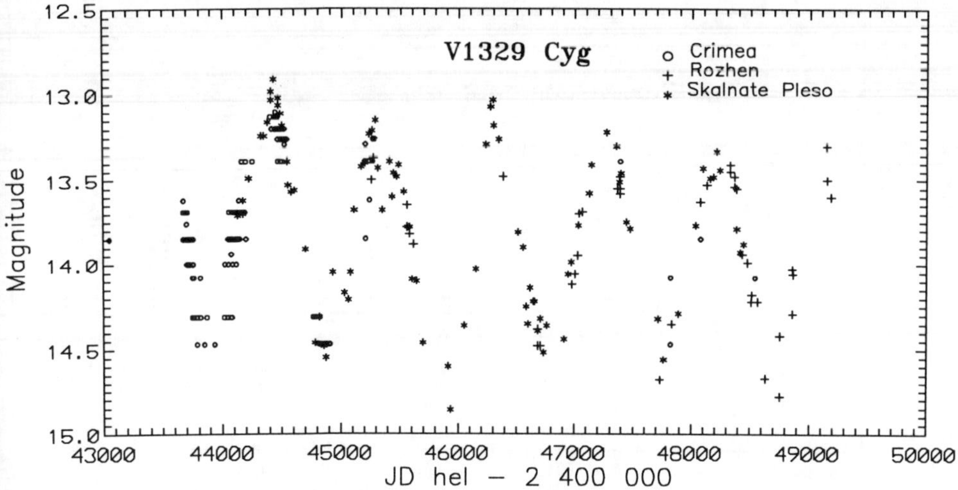

Fig. 5. A compilation of V 1329 Cyg photographic data

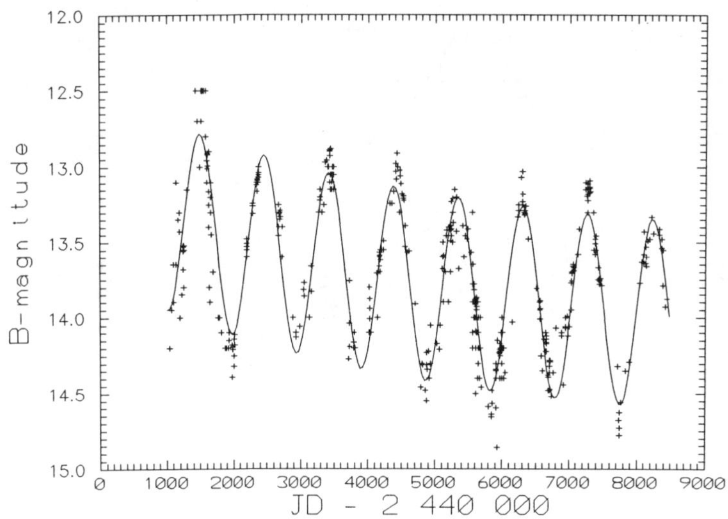

Fig. 6. Photographic and photoelectric B data of V 1329 Cyg. The sinusoidal fit with exponential decrease is shown.

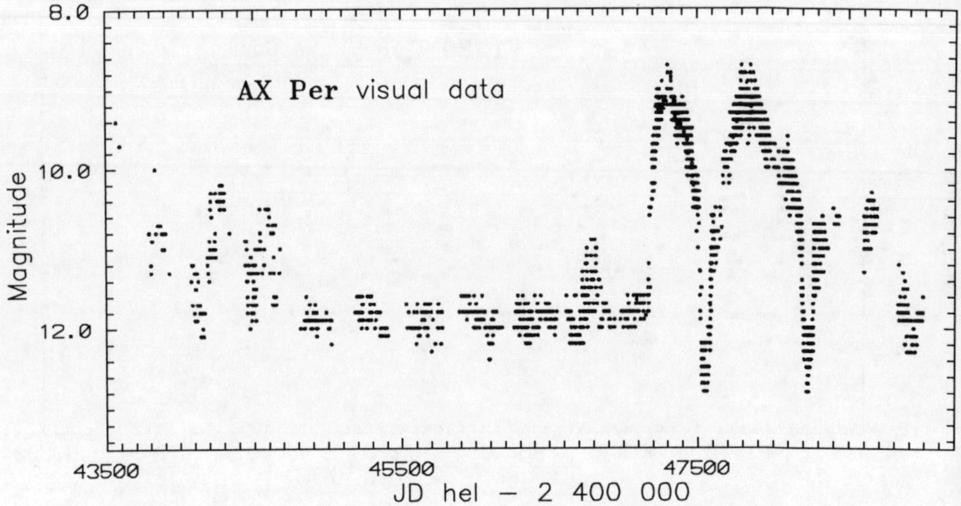

Fig. 7. Visual data for AX Per

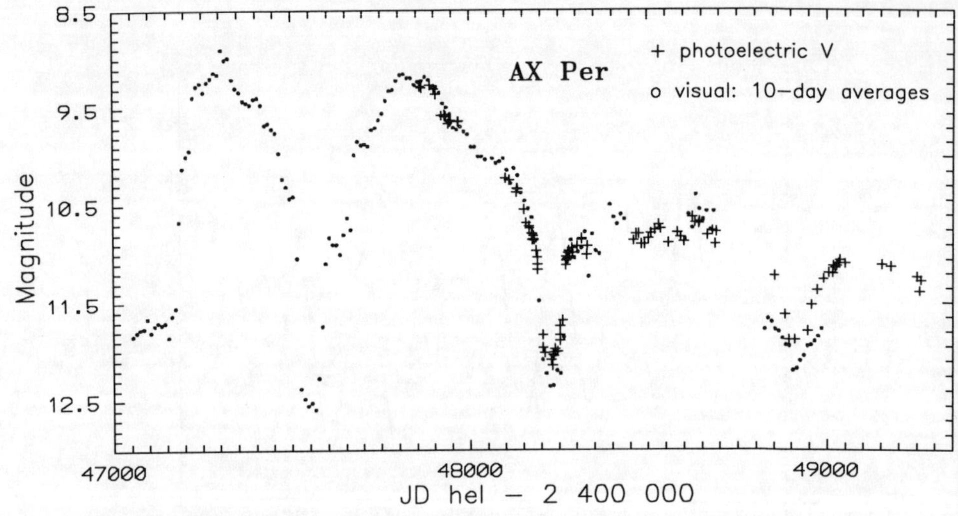

Fig. 8. A comparison of photelectric V and averaged visual data for AX Per

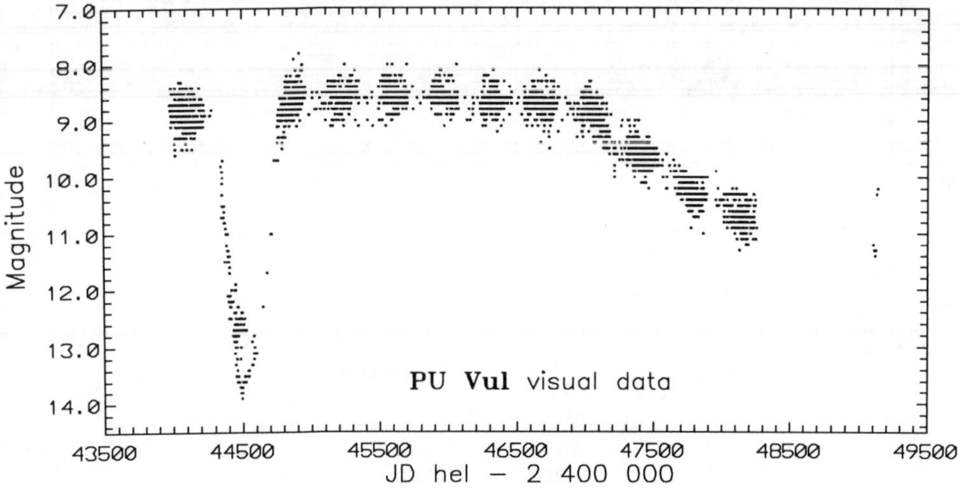

Fig. 9. Visual data for PU Vul

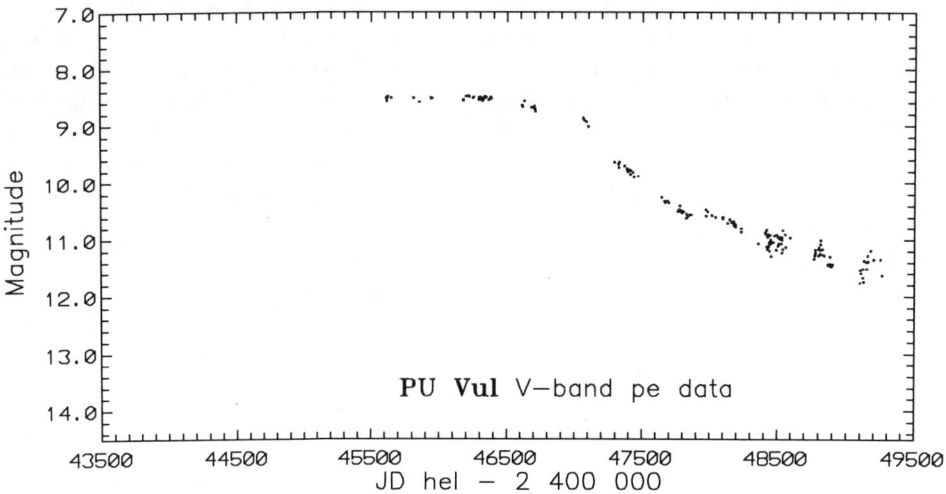

Fig. 10. Photelectric V data for PU Vul

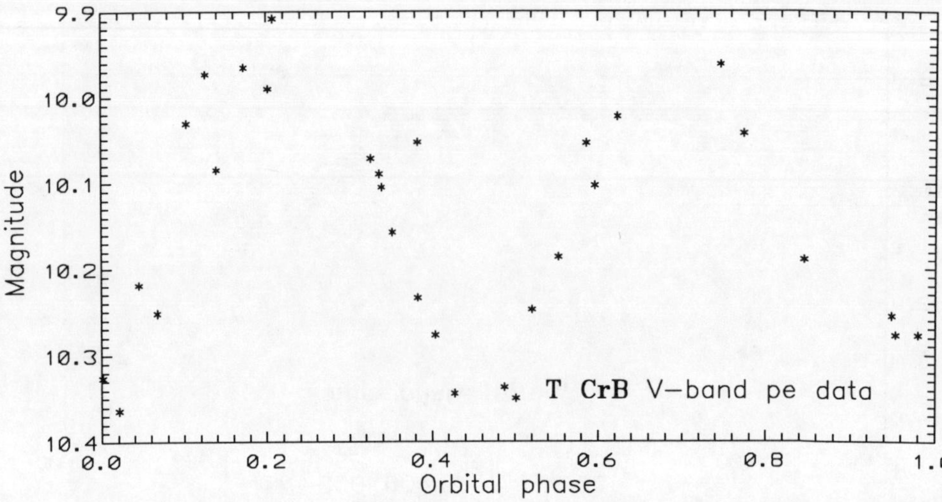

Fig. 11. Phase diagram for photoelectric V data of T CrB

6. Visual observations of symbiotic stars

Visual observations still represent an inevitable tool for studying long-term semi-regular as well as irregular brigtness changes of variable stars. Given the character of symbiotic-star variability, it is clear that such observations are also extremely usefull in this field. During our campaign we have cooperated with several groups of amateur astronomers, both within our two countries (The Slovak and Czech Republics) and abroad. These groups are well-organized and are behaving professionally as can be seen from some of our figures. In summary, our experience in this respect is as follows:

– **Advantages**

1. the human eye is a universally available detector
2. the inflow of data related to a rather large number of stars is much larger than any conventional professional approach could achieve
3. the resulting coverage of light curves greatly supersedes long-term photoelectric data which are rather fragmentary in many cases
4. the time-averaged mean values—summed over 5 days, 10 days or so—are fitting rather well with the photoelectric V data, as can be seen from the figures accompanying our discussion of several individual symbiotics (UV Aur, CH Cyg, AX Per and PU Vul)
5. this rather continuous observational coverage of a number of symbiotics assures quick alert service in the case of an unpredictable abrupt change.

– **Disadvantages**

1. the human eye's sensitivity is subject to the individual physiological capabilities in recognizing intensity changes as well as to differences in spectral sensitivity

2. in special cases, when non-typical phenomena occur (e.g. the sudden appearence of extraordinary emission lines), the natural observational scatter due to the above-mentioned spectral-sensitivity differences grows by a factor up to 3. This is similar to the divergence in various photoelectric instrumental systems (see our experience with Nova Cyg 1992 in Chochol *et al.*, 1993)
3. the visual observations of faint objects are sensitive to natural effects (mainly Moon phases) resulting in gaps in the observational coverage
4. there is a large scatter of individual observers' visual estimates (about $0^{\rm m}_{.}5$). This effect can be overcome by a larger number of observers and/or by introducing individual zero points. The first requirement is fulfilled in the case of almost all groups involved in the effort.

Acknowledgements

The results reported in this paper could be achieved only with the collaboration of all the people taking part in our campaign explicitly mentioned in Tables 1 and 2 as well as individual amateur astronomers from The Slovak and Czech Republics. We gratefully acknowledge the contribution of all of them.
This contribution was made possible through the NATO ARW support and the Slovak Academy of Sciences Grant No. 495/1993.

References

Belyakina, T. S. et al.: 1984, *Astronomy and Astrophysics*, 132, L12
Belyakina, T. S. et al.: 1989, *Astronomy and Astrophysics*, 223, 119
Chochol, D., Hric, L., Urban, Z., Komžík, R., Grygar, J., Papoušek,J.: 1993, *Astronomy and Astrophysics*, 277, 103
Hinkle, K. H., Fekel, F. C., Johnson, D. S., Scharlach, W. W. G.: 1993, *Astronomical Journal*, 105, 1074
Hric, L., Skopal, A.: 1989, *Information Bulletin on Variable Stars*, 3364
Hric, L., Skopal, A., Urban, Z., Dapergolas, A., Hanžl, D., Isles, J.E., Niarchos, P., Papoušek, J., Pigulski, A., Velič, Z.: 1991, *Contrib. Astron. Obs. Skalnaté Pleso* 21, 303 (Paper II)
Hric, L., Chochol, D., Komžík, R.: 1993a, *Astrophysics and Space Science*, 201, 107
Hric, L., Skopal, A., Urban, Z., Komžík, R., Luthardt, R., Papoušek, J., Hanžl, D., Blanco, C., Niarchos, P., Velič, Z. and Schweitzer, E.: 1993b, *Contrib. Astron. Obs. Skalnaté Pleso* 23, 73 (Paper IV)
Hric, L., Skopal, A., Chochol, D., Komžík, R., Urban, Z., Papoušek,J., F., Niarchos, P., Rovithis-Livaniou, H., Rovithis,P., Chinarova, L. L., Pikhun, A. I.,Tsvetkova, K., Semkov, E., Velič, Z. Schweitzer, E.: 1994, to appear in *Contrib. Astron. Obs. Skalnaté Pleso* 24 (Paper V)
Kenyon, S. J.: 1982, *Publications of the ASP*94, 165
Khudyakova, T. N.: 1985, *Pisma Astron. Zh.* 11, 623
Lund, L., Tomov, T., Vennik, J., Leedjärv, L.: 1986, *Pisma Astron. Zh.* 12, 870
Mikolajewska, J., Kenyon, S. J,: 1992, *Astronomical Journal*, 103, 579
Mikolajevski, M., Mikolajewska, J.: 1988, In: *'The Symbiotic Phenomenon'*, IAU Coll. No. 103, eds. Mikolajewska et al. Kluwer, Dordrecht, 233
Munari, U., Margoni, R., Mammano, A.: 1988, *Astronomy and Astrophysics*, 202, 83
Skopal, A.: 1991, *Information Bulletin on Variable Stars*, 3602
Skopal, A.: 1993, *Astronomy and Astrophysics*, (submitted)
Skopal, A., Hric L., Urban, Z.: 1990, *Contrib. Astron. Obs. Skalnaté Pleso* 19, 123 (Paper I)
Skopal, A., Hric, L., Komžík, R.: 1992a, *IAU Circ.* No. 5504

Skopal, A., Hric, L., Urban, Z., Pigulski, A., Blanco, C., Papoušek,J., Hanžl, D., Agerer, F., Niarchos, P., Rovithis-Livaniou, H., Rovithis,P., Tsvetkova, K., Semkov, E., Velič, Z., Michálek, F., Komačka,L', Schweitzer, E. and Korth, S.: 1992b, *Contrib. Astron. Obs. Skalnaté Pleso* 22, 131 (Paper III)

Skopal, A., Bode, M. F., Errico, L., Kenny, H. I., Ivision, R.: 1993a, *Monthly Notices of the RAS*, Letters, submitted

Skopal, A., Bode, M. F., Evans, A., Errico, L., Komžík, R., Hric, L., Urban, Z., Chochol, D., Vittone, A.: 1993b, *Monthly Notices of the RAS*, in press

Yudin, B., Munari, U.: 1993, *Astronomy and Astrophysics*, 270, 165

DISCUSSION

Duerbeck: There is no reason to publish visual estimates because they are available through anonymous ftp from CDS in Strasbourg. Visual observations of *many* observers usually show a scatter of $\pm 0\overset{m}{.}5$. This can be reduced by forming normal points or by concentrating on estimates of active, diligent, single observers, whose data sets may show a scatter of only $\pm 0\overset{m}{.}1$.

Chochol: We have published the visual observations from different groups of visual observers and some of these observations are not available from Strasbourg. Moreover, the goal of our campaign is not only to publish the pure observations, but also to give the preliminary analysis of light variations comparing the photoelectric, photographic and visual data.

Niarchos: I would like to make a few remarks concerning the contribution of the Greek group to that campaign. We have participated in this campaign for almost three years. To carry out the observations we use the 1.2 m Cassegrain reflector at Kryonerion Observatory. The altitude of the observatory is about 900 m above sea level, and the telescope is equipped with two photometers: a two-channel conventional photometer and a photon-counting one.

Hall: I think it is not necessary for you to "apologize" for having used visual estimates in your campaign. We all know that visually-estimated magnitudes are less precise and less accurate than photoelectric ones, but if the errors are much smaller than the variability being studied, which *is true* for the symbiotic variables, then that photometry is still useful. Serious damage is caused only by the total absence of data during a critical phase of the variable's light curve.

de Groot: I can see how you correct visual estimates made when the stars are faint and simultaneous or near-simultaneous photoelctric measurements have been made to bring them up to photooelectric standard. But what do you do when the stars are faint and an event is recorded by visual estimates only? Do you then have a method to correct the visual estimates and bring them up to photoelectric values?

Hall: The only meaningful correction I know of, to correct visually estimated magnitudes to Johnson V magnitude, is the following. If the estimate is made as a Δm with respect to a comparison star and the $\Delta(B-V)$ is known, then one can apply a "transformation correction" in the normal way, where the transformation coefficient in this case is approximately $E_V = 0.2$, but can be evaluated somewhat more exactly on an observer-by-observer basis.

LONG-TERM LIGHT AND PERIOD VARIATIONS OF SOME CONTACT BINARIES

O. DEMIRCAN
Ankara University Observatory
Science Faculty
06100 Tandogan Ankara
Turkey

ABSTRACT. Photoelectric differential *UBV* photometry of some contact binaries carried out at the Ankara University Observatory since 1987 was used together with earlier data in understanding long-term light and period variations of these systems. A typical error of our observations is about $0\overset{m}{.}01$ mag in *B* and *V*, and $0\overset{m}{.}02$ in *U*. The variable extinction effect is largely cancelled by the frequent comparison-star and sky measurements. Due to insufficient and unevenly distributed data, no significant long-term light variations could be detected. Seemingly irregular cycle-to-cycle light variations can be as large as $0\overset{m}{.}04$. The accumulated data reveal that the long-term orbital-period variations obtained by the analysis of eclipse-minima timing are formed by the beat phenomena of two or three variations caused by different mechanisms, and only the scatter in the $(O-C)$ diagrams could be related to the cycle-to-cycle light variations.

Keywords: long-term monitoring – period changes – contact binaries

1. Introduction

In order to gain insights into underlying physical mechanisms responsible for the light and orbital-period variations of interacting binaries, the nature of such variations should be displayed in the first step by long-term photometric observations. The period of long-term variations could well be many decades. However, the necessary data are accumulated faster in the case of short-period binaries. For this reason the photometric observations of some contact binaries of W UMa type (Table 1) are obtained in different observing seasons since 1987 at the Ankara University Observatory. A contact binary of W UMa type is formed by two late-type stars physically touching each other. The observations are used in diploma theses by fourth-year undergraduate students, in M.Sc. and Ph.D. theses by post graduate students, and in different research projects. I will present here general results about the long-term light and orbital-period variations of these systems obtained from long-term observations.

C. Sterken and M. de Groot (eds.), The Impact of Long-Term Monitoring on Variable Star Research, 177–184.

Table 1. The program stars of W UMa type observed at Ankara Observatory from 1987 to 1993. Stars denoted with * are stars for which long-term light and period variations were studied.

Name	87	88	89	90	91	92	93	Name	87	88	89	90	91	92	93
AB And*		+	+	+	+	+		AM Leo*		+	+	+	+	+	+
BX And*	+		+	+	+	+	+	AP Leo			+	+	+		+
OO Aql*	+	+	+	+	+	+	+	UV Leo			+	+	+		+
SS Ari*					+	+	+	XY Leo				+			+
BF Aur		+	+					V502 Oph*			+	+	+	+	+
44 i Boo			+	+	+	+	+	V566 Oph		+	+				+
CK Boo		+	+	+	+	+	+	V839 Oph		+	+	+	+	+	+
BO CVn					+	+	+	U Peg*			+	+	+	+	
KR Cyg		+	+					AW UMa*			+	+	+	+	+
V885 Cyg				+			+	AG Vir			+				
V1073 Cyg			+	+	+			AH Vir*		+	+	+		+	+
LS Del*	+	+	+	+	+			BH Vir			+	+		+	+
BV Eri	+	+	+					BD+17 2053			+	+			
YY Eri*		+	+	+	+	+	+	BD+31 849					+	+	+
AK Her*		+	+	+	+	+	+	BD+41 2447			+	+			
FG Hya		+	+	+	+		+	BD+53 0507							
SW Lac			+	+	+	+									

2. Observations

The photometry of the programme stars was obtained at the Ankara University Observatory by using a one-channel uncooled photometer with an EMI9789 QB photomultiplier until September 1991, and then a SSP-5 photometer with a R1414 side-on photomultiplier (from Hamamatsu Corp.) attached to the 30-cm Maksutov telescope. The observations comprise three five-second integrations through Johnson U, B and V filters in the sequence comparison-sky-variable-sky-variable-comparison-sky with occasional interruptions for background and check-star measurements. A constant field aperture of 1.0 mm is used during all observations. The eastern sky is affected by light pollution from the city. Therefore, the extinctions in eastern and western skies are different (Fig. 1). The extinction curves formed by the measurements of comparison stars (as in Fig. 1) display low-amplitude ($\sim 0^{m}.01$) variations on time-scales of $0^{h}.5$-$1^{h}.0$ which were introduced by sky-transparency variations. The mean linear extinction coefficients obtained from five-year observations are given in Table 2 for eastern and western skies separately (Gürol 1992). The variable sky-transparency effect is mostly (but not completely) eliminated in the differential observations, and introduces a limitation on the accuracy of our observations. The typical error in our individual differential data points is about $0^{m}.01$ in B and V and $0^{m}.02$ in U. The control of the photometer head, data acquisition and data reduction were carried out with a software system prepared by Müyesseroğlu (1992) at the Ankara University Observatory. The observers are research assistants and/or graduate and post-graduate students. Some undergraduate students assist the observers. One star is observed by many observers on different nights. In the accuracy of observations, no significant personal effect is detected.

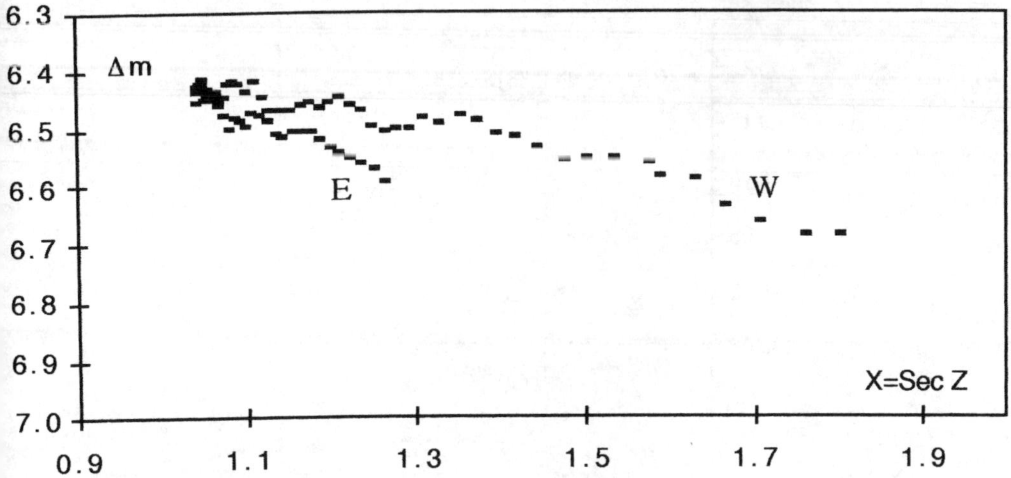

Fig. 1. A typical different extinction in eastern (E) end western (W) skies of Ankara University Observatory. A low-amplitude ($\sim 0\overset{m}{.}01$) periodic transparency variation is also apparent (after Gürol 1992).

Table 2. The mean linear extinction coefficients (MLEC) for eastern and western skies of Ankara University Observatory (after Gürol 1992).

	Eastern sky	Western sky
k_u	0.841 ± 0.027	0.798 ± 0.032
k_b	0.505 ± 0.015	0.471 ± 0.020
k_v	0.323 ± 0.013	0.303 ± 0.018

3. Long-term period variations

From our observations many new times of minimum light were calculated for each system by using the well-known method of Kwee and van Woerden (1956). Average values of the times of minimum were formed for the simultaneous U, B and V observations. We then used our data, together with times of minimum collected from the literature, to form $(O-C)$ curves. Knowing that the appearance of the $(O-C)$ diagrams depends very much on the choice of the light elements used in the calculation of the C-epochs, we formed many different $(O-C)$ diagrams by using different light elements. Such work is accomplished for a dozen systems from Table 1. As an example, three different appearances of the $(O-C)$ diagram of AB And are shown in Fig. 2. It was realized that, in general, the interpretation of $(O-C)$ diagrams is not an easy task. As it is seen in Fig. 2, the $(O-C)$ diagram of AB And could be considered to consist of either two straight lines , a parabola, or a part of a sine curve.

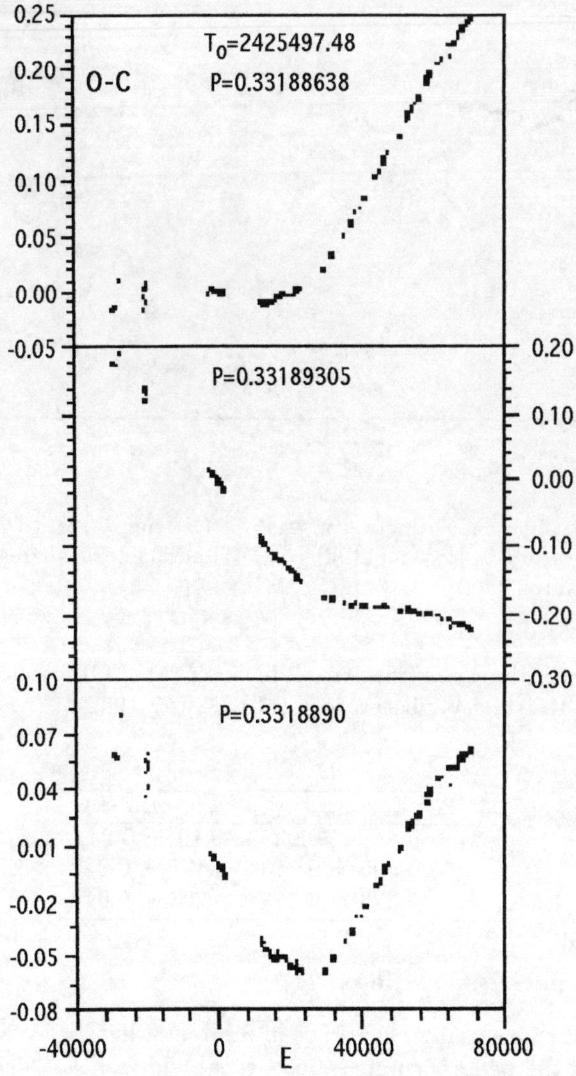

Fig. 2. Three different appearances of the $(O-C)$ diagram of AB And (from Demircan et al. 1993)

The interpretation will be different in each case: respectively, constant periods before and after a period jump around 1950, continuous period increase, and a sinusoidal variation of the period. A detailed analysis (Demircan et al. 1993) proved that the $(O-C)$ diagram is part of a sine curve indicating a sinusoidal variation of the orbital period. The majority of residuals from a sine or parabolic variation are found to be non-random with an amplitude not explainable by the error in our photoelectric data. Such large residuals can be taken as a sign of another mechanism responsible for the period variations. The scatter in the $(O-C)$

diagrams is also too large to be explained in terms of the errors in the photoelectric data. Thus, the $(O-C)$ diagram of W UMa type binaries are formed, in general, by the beat phenomena of two or three variations caused by different mechanisms.

4. Long-term light variations

Apart from the phase-dependent continuous light variation, cycle-to-cycle light-curve variation of up to a few per cent of a magnitude is well known for W UMa systems (cf. Rucinski 1985). The light levels at maxima and minima as well as the times of minima (and thus orbital period) change as a result of cycle-to-cycle variations. To find the expected correlation between the light and period changes we collected all available light curves from the literature for about a dozen W UMa systems on our programme. As an example the master light curve in V of AB And formed by all collected observations is shown in Fig. 3 for which the phase correction according to the sinusoidal ephemeris from Demircan et al. (1983) was taken into account. We read the light levels at maxima and minima on each collected light curve and formed the depths of minima D1 and D2, and magnitude difference between maxima $\Delta Max = max_1 - max_2$ and between minima $\Delta Min = min_1 - min_2$. The variation of these parameters in the V filter is also shown for AB And in Fig. 4, together with the $(O-C')$ residuals from the sinusoidal variation (as shown in Fig. 2c). Due to insufficient and unevenly distributed data, no general trend was found in the variations of D1, D2, ΔMax and ΔMin (see Fig. 4). We noted in our observations that a large fraction of the scatter displayed by D1, D2, ΔMax and ΔMin variations in time are caused in just a few cycles of orbital revolution. No significant correlation was found between the cycle-to-cycle light variation and the orbital-period changes.

5. Conclusion

There are two plausible alternatives for the large-amplitude sinusoidal variation of the orbital period. The first one is the well-known light-time effect due to a third body in the system, and the second is the period modulation due to a magnetic-activity cycle of a component star (cf. Applegate 1992). With the present data we showed that it is not possible to confirm or refute either of these mechanisms. The long-term observational data are insufficient and, more important, unevenly distributed with large gaps in time. Only future systematic and continuous long-term monitoring of contact binaries can help towards a better understanding of their long-term light and period variations and the probable connection between these.

Acknowledgements

I thank my colleagues and co-authors E.Derman, S.Selam, Z.Müyesseroğlu, A.Akalin and B.Gürol for the permission to quote from joint research, and for the pleasant cooperation.

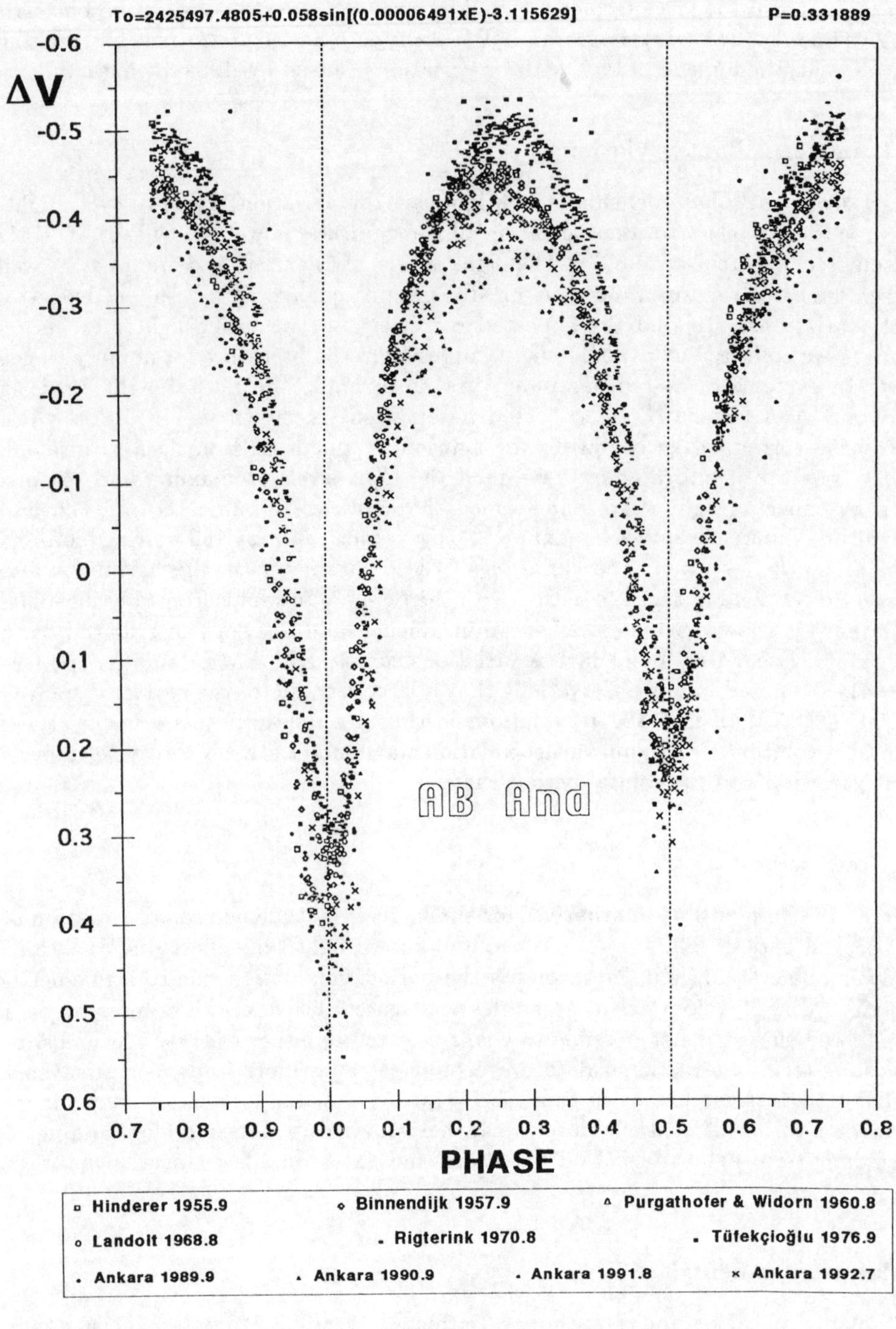

Fig. 3. The master light curve of AB And formed by all available photoelectric light curves from the literature

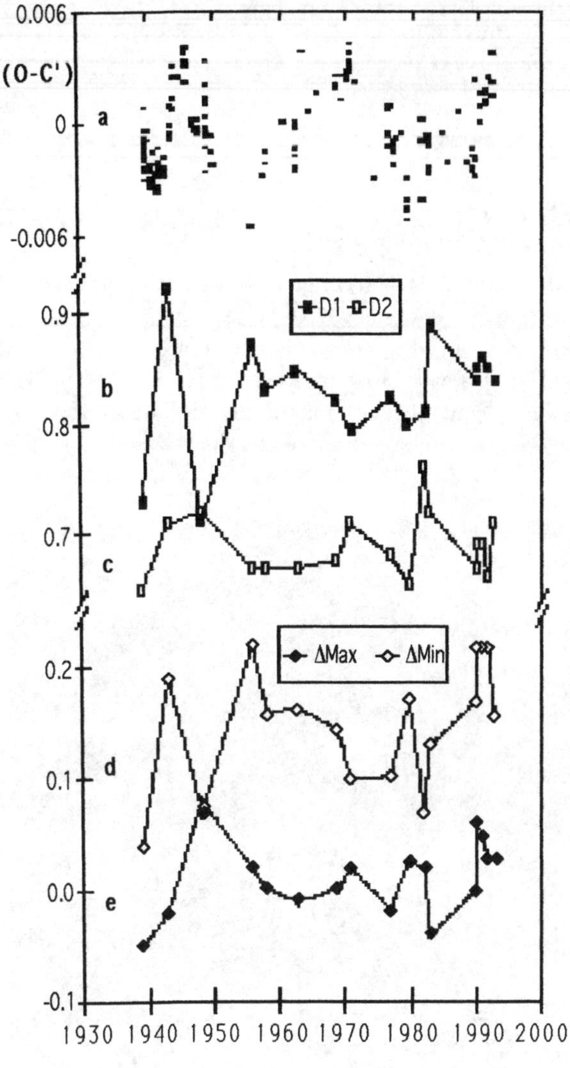

Fig. 4. The variations of a) the $(O - C')$ residuals from the sinusoidal best fit to the $(O - C)$ diagram, b) the depths D1 of the primary minima, c) the depths D2 of the secondary minima, d) the difference between two maxima ΔMax, and e) the difference between two minima ΔMin for the W UMa system AB And (from Demircan et al. 1993).

References

Applegate J.H., 1992, *Astrophysical Journal*, 385, 621

Demircan O., Derman E., Akalin A., Selam S., Müyesseroğlu Z., 1993, *Monthly Notices of the RAS*, in press.

Gürol B., 1992, M.Sc. thesis, Ankara University

Kwee K.K., von Woerden H., 1956, *Bulletin of the Astronomical Institutes of the Netherlands*, 12, 327

Müyesseroğlu Z., 1992, unpublished, Ankara Univ. Obs.
Rucinski S.M., 1985, in "Interacting Binaries", eds. P.P. Eggleton, J.E. Pringle, Reidel Publishing Company, 13

DISCUSSION

Chochol: Can the mass transfer explain the light variations of the binary AW UMa? Do you have an indication of period change?

Demircan: The $O - C$ diagram of this strange system with the lowest known mass ratio indicates a secular period decrease with a rate of 1 to 2 $10^{-7}d\ yr^{-1}$. We think the period decrease may be due to an interaction with the envelope around the system. The system shows very low magnetic activity as the UV and X-ray observations indicate. The recent observations show night-to-night light variations of up to $0^{m}.04$ which are largest at longer wavelengths.

Rons, van Loon and Vyverman contemplating

LONG-TERM MONITORING OF CYCLIC FREQUENCY VARIABILITY IN RAPIDLY OSCILLATING Ap STARS AT THE SUTHERLAND STATION OF THE SOUTH AFRICAN ASTRONOMICAL OBSERVATORY

D. W. KURTZ, and Peter MARTINEZ
Department of Astronomy
University of Cape Town
Rondebosch 7700
South Africa

ABSTRACT. Cyclic frequency variations have been detected in several rapidly-oscillating Ap stars. We describe our long-term monitoring projects, entailing nightly, or weekly, runs of 1 hour or more of high-speed photometry on each star, which have now been running for two years, and which will run for years to come. We place special emphasis on several advantages that the Sutherland site of the South African Astronomical Observatory has over all other observatories for projects of this sort. We give a detailed discussion of many pitfalls in high-speed photometry which, if unheeded, preclude many observatories from producing observations of useful quality for studies of sub-mmag amplitude variations.

Keywords: long-term monitoring – period changes – rapidly-oscillating Ap stars

1. Introduction

Among the A stars there is a subclass of peculiar stars, the Ap stars, which show strongly-enhanced spectral lines of the Fe-peak, rare-earth and lanthanide elements. These stars have global surface magnetic fields several orders of magnitude larger than that of the Sun, 0.3 to 30 kG is the measured range. For stars with the strongest magnetic fields, the spectral lines are split by the Zeeman effect and the surface magnetic-field strength can be measured. Generally, though, the magnetic fields are not strong enough for the magnetic splitting to exceed other sources of line broadening. In these cases the Zeeman effect gives rise to residual polarisation differences between the red and blue wings of the spectral lines, thus giving a measure of the effective magnetic-field strength—the integral of the longitudinal component of the global magnetic field over the visible hemisphere, weighted by limb-darkening (see Mathys 1989 and Landstreet 1992 for reviews of magnetic measurements). In the Ap stars the effective magnetic-field strengths vary with rotation. This is well understood in terms of the oblique-rotator model in which the magnetic axis is oblique to the rotation axis, so that the magnetic field is seen from varying aspect with rotation. See Wolff (1983) for a general review of the A

C. Sterken and M. de Groot (eds.), The Impact of Long-Term Monitoring on Variable Star Research, 185–196.

stars, Ap stars and the oblique-rotator model. See Landstreet (1992) for a general discussion of the magnetic fields in all stars.

Much has been learned about the oblique rotation of Ap stars by long-term monitoring of their magnetic-field strengths, e.g., century-long magnetic variations are inferred in γ Equ (see Mathys 1991, Fig. 37) from nearly continuous magnetic observations that have been obtained since the late 1940s. Much has also been learned from the study of rotational light variations which are caused by spots, or rings, of enhanced abundances of the rare earths associated with the magnetic poles. These rotational light variations are the topic of a review by Hensberge elsewhere in these proceedings, hence they will not be discussed further here.

The rapidly-oscillating Ap (roAp) stars are cool magnetic Ap SrCrEu stars which pulsate in high-overtone, low-degree ($n >> l$) p-modes with periods which range from just under 6 minutes to about 16 minutes (the upper limit is poorly defined). The amplitudes of the light variations are observed to be a tenth of a mmag to a few mmag when the stars are observed photometrically through a Johnson B filter. (The lower limit is observational; it is likely that more precise photometry will detect lower-amplitude variations.) Thus, these stars have pulsation modes similar to the low-degree modes seen in the Sun, but with amplitudes three to four orders of magnitude larger. The roAp stars are currently the only non-degenerate stars known unequivocally to pulsate in p-modes similar to the Sun.

A major difference from the Sun, however, is that the pulsations in the roAp stars are dominated by the strong global magnetic fields of the Ap stars. The pulsation amplitude is modulated with rotation in phase with the magnetic-field modulation. This amplitude modulation of the light curves of the roAp stars can be described by two models: the oblique-pulsator model and the spotted-pulsator model. The oblique-pulsator model assumes that the pulsation and magnetic axes are aligned and are oblique to the rotation axis; the rotational modulation of the light variations is caused by the varying aspect of non-radial pulsation modes (see Kurtz 1990; Shibahashi & Takata 1993). The spotted-pulsator model assumes that the pulsation axis coincides with the rotation axis of the star so that the pulsation modes are always seen from the same viewing angle, but that the ratio of flux to radius variations, f, and the phase lag between the flux and radius variations, ψ, are variable over the surface as a function of the magnetic-field strength (Mathys 1985) leading to the observed amplitude modulation. There are several recent reviews of the roAp stars: Shibahashi (1987), Kurtz (1990), Matthews (1991) and Martinez (1993). The last gives a list of 25 of the 26 roAp stars known as of this writing; 23 of those stars, and references to the literature on them, can also be found in Kurtz & Martinez (1993).

A very interesting newly-discovered characteristic of the roAp stars is that they have intrinsic, cyclic frequency variability on a time-scale of hundreds of days to years (Kurtz et al. 1994a,b; Kreidl et al. 1994). This suggests that a magnetic cycle may be operating in these stars. Theory, on the other hand, argues that such cycles are not possible in Ap stars, in which case there is no current explanation for the phenomenon. It is the long-term monitoring of roAp stars using high-speed photometry that has allowed the discovery of this cyclic frequency variability. The style of operation of the Sutherland station of the South African Astronomical Observatory (SAAO), the high photometric quality of the sky there, and the rigorous maintenance of the equipment have made this long-term monitoring possible.

In Section 2 we will discuss the operation at Sutherland, emphasising how it has allowed us and our colleagues to obtain data that are, at present (we hope that will change), difficult or impossible to obtain elsewhere. In Section 3 we will return to the frequency variability in the roAp stars and explain the interesting astrophysical results we are getting out of our long-term monitoring.

2. Long-term high-speed photometric monitoring at Sutherland

The Sutherland station of the SAAO lies on a plateau 1768 m above sea level several hundred kilometres from Cape Town. The climate is semi-arid and 50% of the nights are photometric with no seasonal variations through the year. Particularly important is that the Atlantic Ocean lies about 250 km to the west, and the Indian Ocean lies about 250 km to the south. Prevailing winds, which commonly blow at 20-40 km h^{-1}, only rarely blow from the Northeast–the only direction which is from the interior of the continent. All the other winds sweep in pollution-free, dust-free air of high transparency for most of the 50% of the nights we call photometric. This gives exceptionally good transparency: We often watch the stars blink out when they set in the west; on many nights we have photometric zero-point shifts in the visible of no more than a few mmag throughout the entire night. Point (1) is that Sutherland is 50% photometric with exceptionally transparency for a large fraction of those nights.

The SAAO welcomes applications for observing time from astronomers throughout the world, and about 20% of the time is awarded to overseas observers. The rest of the time on our 1.9-m, 1.0-m, 0.75-m, and 0.5-m telescopes goes to about 20 astronomers in South Africa, most of whom are resident in Cape Town–either at the SAAO, or in the Department of Astronomy at the University of Cape Town. Many of us regularly are assigned 8 to 12 weeks of observing time per year. Those who work in groups have the potential to devote 2 to 4 times that amount of observing time to a project, should they so wish. This, in and of itself, allows many long-term projects to be accomplished without the administrative complexities of further collaboration of astronomers elsewhere. Point (2) is that the resident astronomers in South Africa get large amounts of telescope time year after year.

Few, or no, other observatories are run like this. This is principally because most major observatories have concentrated on building and operating large telescopes to investigate problems which require more photons, or better spatial resolution. Because of the sociology of astronomy, a large fraction of the world's astronomers choose to do work that requires these large telescopes. Competition for observing time then excludes the use of these large telescopes for long-term projects. The fashion for being a large-telescope observer, which carries substantial prestige in the pecking order of observing astronomers (as we all know, astronomers do peck), has led to an unfortunate neglect of small telescopes. When one of us (DWK) was observing some years ago at CTIO, a young astronomer (observing with the 4-m telescope, of course) over dinner exclaimed that "there is little left to do brighter than 24th magnitude!". A large, prolific, substantially older resident astronomer, who was eating by himself, overheard this, snorted loudly, and said "hmmppff! Anyone can do research with a 4-m telescope. It takes some brains to work with a small one." Well, we have belaboured this point because so few observatories maintain their small telescopes which are essential for long-term work which is the topic of this workshop. We hope that the many papers presented at this meeting will provide support for those who would like to change this.

Another advantage of observing at Sutherland is that the SAAO has a programme of service observing. The SAAO grew out of the Royal Observatory at the Cape, where it was a tradition that the director was the Director, and many, or all, of the astronomers worked for him (that is an historical "him"; no implication about the sex of future directors is implied). There are currently astronomers at SAAO who began their careers as hired observers, and some of them still perform long-term service observing in collaboration with other staff astronomers. Also, at Sutherland, there are two night assistants, Fred Marang and Francois van Wyk, who have learned to observe on the job. They now act as night assistants only to overseas observers, and are otherwise the principal observers on the 0.5-m telescope. Both have family ties in the Sutherland village, so that they are part of the community and are likely to stay in their jobs for many years to come. This is important. Service observing by recently trained people is not of the quality that experienced people produce. Point (3) is that the service observers remain on the job for years and decades (recently an observer retired from SAAO after more than 40 years of service). Their experience, plus the chance for constant feedback from the astronomers initiating the projects, leads to high-quality data. The constancy of the observers leads to long-term consistency in that quality, which is important for long-term monitoring projects.

Equipment is of paramount importance in obtaining high-quality observations. Many observatories are now building small, automated photoelectric telescopes to take over long-term monitoring projects. This is the way of the future; the SAAO will soon complete an 0.75-m automatic photoelectric telescope, too. Largely, amateur astronomers have led the automatic-telescope work–much to their credit. Professional observatories need to follow in their footsteps. We do not yet know whether the automatic telescopes in existence are capable of the kind of long-term monitoring project using high-speed photometry that is the focus of the results presented in the next section. We will discuss that problem further below.

At Sutherland the 0.5-m telescope is dedicated to photometry. The photometer is used on every clear night all year round; it only comes off the telescope for maintenance (see Kilkenny et al. 1988). Mostly, it is used by the small group of professional observers mentioned above. Standard-star observations are part of the service observing, and an astronomer in Cape Town is in charge of the maintenance of the filters and transformation equations. We, therefore, always know exactly how well the system is working and what the quality of the sky is like. For example, the Mt. Pinatubo eruption caused substantial degradation of photometric conditions over much of the world. We in South Africa know precisely what has happened to the Sutherland photometry, because we have continuous observations with the same equipment and the same observers. In Cape Town, Dr. Cousins, who has 70 years observing experience, observes every photometric night year-round and maintains standard-star systems for use by astronomers throughout the southern hemisphere.

Point (4) is that we know our equipment at Sutherland. When something is wrong we detect it quickly. An analogy is that of driving your own car, or driving a rental car: With your own car new noises and vibrations are immediately evident. With a rental car you often don't notice. Too many astronomers travel to strange (to them) observatories on an occasional basis to use equipment which is seldom on the telescope and which they do not have a feel for. Under these circumstances problems are often not noticed and not corrected. Poor-quality data are the result. In our experience, photometers at many observatories suffer from a vast array of problems which lead to poor-quality photometry. Because of the extremely low amplitudes of the roAp stars it is imperative that roAp-star photometrists be aware of, and overcome, many common pitfalls which are:

1. The use of small apertures which causes light losses from tracking errors and seeing fluctuations. Bright stars do not require the use of small apertures; often small apertures are used ritually without thought to the problem or consideration for the brightness of the star.
2. Careless manual guiding which causes spurious dips in the light curve which have to be excised. Such gaps reduce the duty cycle and introduce unwanted sidelobes in the Fourier spectrum of the time series. This is seldom a problem for differential photometry where stars are observed for only a few minutes. It is very important for high-speed photometry where stars are monitored continuously for hours.
3. In some telescopes, moonlight reflects off internal parts of the telescope causing problems, especially when the Moon is close to the programme star. When a bright Moon is close in the sky to the observed star, this can cause step discontinuities in the light curve when the dome is rotated.
4. Bad monitoring of sky background causes the sky-subtraction process to introduce spurious trends which manifest themselves as low-frequency noise. This is especially true for observations acquired near astronomical twilight or at Moon rise and set.
5. Sensitivity variations across the aperture introduce discontinuities in the light-curve when guiding corrections are made. The flatness of the sensitivity of the photometer should always be tested by positioning the star in the four quadrants of a large aperture and noting the count rates at each position. Such variations also exacerbate the deleterious effects of the periodic drive error of the telescope (see (13) below). This is a common problem with many photometers; often the source of the problem is subtle and a great deal of effort is required to fix it. We have had many collaborations where we had to reject the observations obtained at other observatories which were rendered useless (not just degraded) by this problem.
6. Damp photomultiplier tube bases can be caused by power failures unnoticed by the observer during the day. If the power is down for long enough, the cold-boxes warm up–potentially allowing moisture to condense on the tube base, thus providing an electrical conduction path for inter-electrode leakage. This causes greatly increased scatter in the count rates, higher dark counts and

introduces erratic excursions in the light curve. The only solution is to bake out the moisture from the PMT. The presence of dirt or grease on the base of the PMT can lead to similar problems. Another consequence of daytime power cuts is that the PMT sensitivity may not stabilise by nightfall.

7. Moisture can also condense on the Fabry lens or glass window of the photocathode thus fogging these surfaces. To prevent this, the Fabry lens is often heated slightly.
8. Dirt on filter surfaces exacerbates problems with the drive error of the telescope at best, and produces useless data at worst. If a technician installs the filters, they should be checked by the observer. Sometimes they are not the ones the observer thinks! We know one occasion when an observer spent a week observing through masking tape. Alternatively, the filters might be scratched or otherwise damaged or deteriorated. We, ourselves, recently lost a night's observing at CTIO when a 25-year-old, badly deteriorated Strömgren filter was installed in the photometer; its transmission curve had shifted so far that the observations were rendered useless.
9. Dirt can accumulate on the glass window of the photocathode. Many observers keep their filter sets scrupulously clean only to place them in a photometer whose photocathode is dirty. On one occasion in our experience the aluminium dark slide of the University of Cape Town photometer abraded from years of use and deposited a fine swarf on the surface of the glass envelope of the PMT. When we started observing on that occasion, a 2-minute telescope drive error oscillation was immediately apparent in the real-time display of the data. The severity of the problem had a strong hour-angle effect which was probably caused by the swarf rolling around on the surface of the glass envelope. Interestingly, the previous observers had not noted any faults.
10. Vignetting can be a frustrating source of error to find. Several years ago it was discovered that the baffle on the secondary mirror of one of the SAAO telescopes, which was installed for use with an infrared photometer, was barely vignetting the beam with the UCT photometer. The problem was so subtle that it took two years to find its cause. This is always a possibility on telescopes where a variety of instruments are used. Changes can be made for the benefit of one instrument which can (unknowingly) degrade the use of another.
11. Misalignments in the photometer.
12. The long time series acquired in high-speed photometry often reveal flexure problems in photometers. We have often noted flexure problems in runs lasting several hours. At the start of the run the programme star and guide star were both well centred. Hours later, the guide star was centred, but the programme star had moved off centre. The solution here is to interrupt the observations every 1.5-2 hr or so (depending on the declination) to check the centring of the programme star in the aperture. This is particularly important when using autoguiders where complacency can set in.
13. Spurious periodicities are often introduced into the data by telescope-drive oscillations. The most common telescope tracking errors are those with periods of 2 or 4 sidereal minutes (8.356 or 4.178 mHz, respectively). In the case of HD 60435, a 4 mHz oscillation was "studied" for several years before it was realised

that this was an oscillation in the telescope and not in the star! The drive error is strictly periodic, but frequent guiding corrections introduce irregular phase glitches which produce a broad Lorentzian hump of power centred on the drive frequency rather than a single spike. It should also be remembered that the apparent amplitude of the drive error goes as the cosine of the declination. This problem only occurs when there is also a sensitivity variation across the aperture in the photometer. When the tracking period of the telescope is detected in the Fourier transform of a high-speed photometric time series, then one, or more, of the problems listed here is present. Corrective measures must be taken. We have many experiences of this problem going unnoticed by our collaborators at other observatories.

14. A telescope that is too finely balanced and "floats" on a worn drive gear produces erratic excursions of the star in the aperture. This problem can be cured with the simple expedient of putting a small extra weight on the telescope so that it is no longer so finely balanced.
15. Electronic malfunctions such as a drifting or fluctuating high-tension supply introduce spurious trends in the data. The very large variation in gain produced by a small variation in voltage makes it essential to use a very well regulated voltage supply.
16. Inadequately shielded photomultiplier tubes are affected by variations in the local magnetic field as when, for example, an iron observing ladder is moved close to the instrument.
17. It is not uncommon to have electrical switching noise caused by coldbox coolers, lights, dome-rotation motors, wind-blind motors, printers and radios causing spikes in the data stream.
18. Poor thermal regulation in cold boxes can lead to high dark current as the PMT warms up. This often appears as a periodic oscillation in the count-rates as the cooler unit switches on and off with some characteristic period. One should also remember that temperature variations cause changes in the spectral characteristics of the photocathode with a general increase in red sensitivity with increasing temperature. For U-band photometry this means that the red leak will increase as the PMT temperature increases.
19. Inaccurate time is also an insidious source of error. It is vital for roAp star work that the observer should have reliable means, independent of the data-acquisition system, for setting the time to an accuracy of 0.1 s. Most acquisition computers supply a date and time which get written to the header of the data file. Often, these dates and times are wrong by anything ranging from a fraction of a second to a day (often exactly 1 day), or more. In some acquisition systems, the observer enters the date, time and the RA and Dec of the star at the start of a run and no other record is kept of the observations. The data are logged as Heliocentric Julian Dates and counts. This is a disaster waiting to happen. In the event of a mistake in the entry of any of these 4 quantities, it is impossible to recover the actual times. Heliocentric corrections are also often incorrectly calculated. Computing the correct heliocentric times is vital to producing useful roAp photometry, especially when subtle period changes or frequency shifts are being sought.

The above sources of error arise in the instrumentation and can all be eliminated or minimised by careful maintenance. An on-line data display is absolutely essential to detecting many of these problems at the telescope in time for remedial action to be taken. We emphasise here that we have experience with all of these problems, either in our own photometry, or in that of collaborators at other observatories. Many of these problems are severe at some of the biggest, most prestigious, observatories where almost all effort and attention is given to the large telescopes and small telescopes and photometers are neglected.

We end this section by reiterating the principal reasons why we get excellent data at Sutherland:

1. Sutherland is 50% photometric with exceptionally transparency for a large fraction of those nights.
2. The resident astronomers in South Africa get large amounts of telescope time year after year.
3. The service observers remain on the job for years and decades.
4. We know our equipment at Sutherland.

3. Frequency variability in roAp stars

We have been observing the roAp stars for 15 years now. We have found in many cases that the pulsation is constant in amplitude but variable in phase. It is now clear that this is caused by frequency variability, and we have been able to show in several cases that this variability is intrinsic to the star and is cyclic in nature. This has possibly profound theoretical implications. Fig. 1 shows a phase O-C diagram for HR 3831 (Kurtz et al. 1994a). This diagram has been produced by observing HR 3831 with 10-s time resolution for 1 hr on every possible night throughout its observing season for two years now. Similar, but less extensive results have been obtained for α Cir (Kurtz et al. 1994b) and HD 134214 (Kreidl et al. 1994). We have ongoing, long-term monitoring projects for these roAp stars, and some others, which we expect to continue for years. This has mostly been possible because of the four points of advantage of Sutherland discussed in Section 2. We said there that we are unsure whether automatic photoelectric telescopes now in operation can obtain observations of the quality we demand. We are unsure simply because the operators of such telescopes have not yet tried. Many of the pitfalls in high-speed photometry listed in Section 2 are not apparent when observations are only made for 1 or 2 minutes, when observers are satisfied with 0.01 mag accuracy, or when photon statistics limits accuracy.

It has long been observed that the Ca II H and K emission cores in the Sun are correlated with the 11-year solar magnetic cycle. Observations of periodic variations in the H and K emission from other solar-type stars have reasonably been inferred to be evidence of similar cycles in those stars (see e.g. Baliunas & Vaughan 1985). The periods found range from about two years upwards. Recently, Elsworth et al. (1990), Libbrecht & Woodard (1990), Bachmann & Brown (1993) and others have found that the frequency of the solar p-modes is correlated with the solar cycle. The range of that frequency variation is about 0.5 μHz. The time-scale of the frequency variations in HR 3831 and other roAp stars is of the order of a

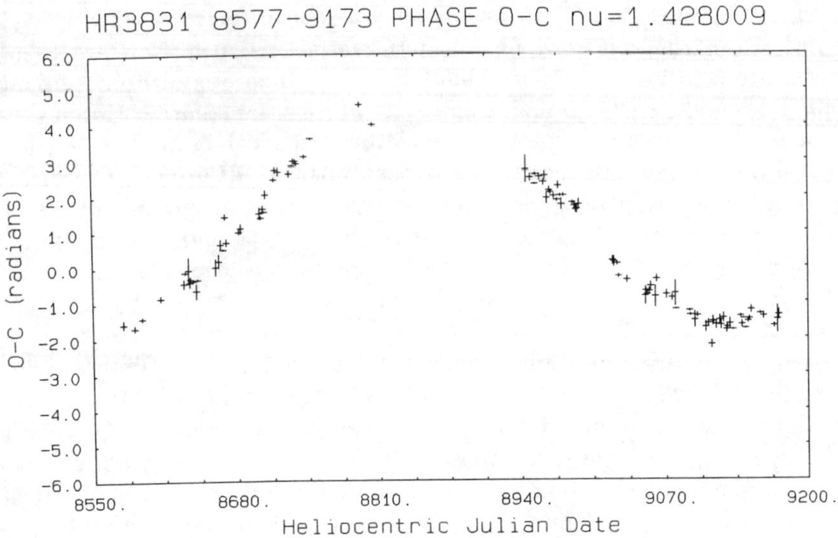

Fig. 1. A phase $O - C$ diagram for HR 3831. See Kurtz et al. (1994a) for a complete discussion of this diagram, which gives the strong impression that the phase, and hence the frequency, of HR 3831 varies cyclically with a period near 1.6 years.

few hundred days to two years —shorter than, but overlapping with, the periods of the magnetic cycles in cool stars. The range of the frequency variation in HR 3831 is 0.12 μHz–similar to that seen in the Sun over the solar cycle. We therefore suggest that changes in the magnetic-field strength, either directly, or indirectly by acting on the stellar structure as in the Sun, cause the frequency variations in HR 3831. There are objections to thinking of these frequency variations as originating in a magnetic cycle analogous to that of the Sun. It is thought that the magnetic fields of hot stars and cool stars are fundamentally, physically different from each other. The cool stars, like the Sun, are thought to have magnetic fields generated by internal dynamos. The correlation of magnetic activity with rotation period in the cool stars supports this. On the other hand, the magnetic stars of the upper main-sequence show no correlation of rotation and magnetic-field strength. Their fields are thought to be fossil in origin. The Ap stars have only a very thin (a few thousand km) surface convection zone (very unlike the Sun's 200,000 km surface convection zone), and since their strong magnetic fields give rise to rigid rotation, as evidenced from the success of the oblique-rotator model, it is predicted that they should not have magnetic cycles like that of the Sun, which depends on both differential rotation and convection to drive it (in the standard interpretation). The possibility of magnetic cycles in Ap stars has been discussed since their discovery by Babcock in the late 1940s. No unequivocal evidence for cycles has previously been found. Even the century-long magnetic variations in γ Equ are taken by most investigators to be caused by oblique rotation. There is also no evidence of chromospheres, and hence Ca H and K emission, in Ap stars (Lanz & Hubeny 1993), although such effects are much harder to detect in A stars than in G stars and thus are not ruled out by the null results of the observational attempts. The discovery and long-term monitoring of some chromospheric spectroscopic signature in Ap stars would be a most welcome test of our magnetic-cycle suggestion. Inference from the Sun suggests that the purported magnetic variations of HR 3831 are orders of magnitude less than those which could be detected directly in magnetic measurements.

Interestingly, similar pulsation-frequency variation has been detected in the ZZ Ceti (pulsating white dwarf) star G29-38 (Winget et al. 1990). There the time-lag (light travel time in a binary model) range is a little over 200 s with a period of about 110 days. Thus, the phase of the pulsation varies by about 1/3 of a period with no amplitude modulation–a very similar situation to that of HR 3831. Winget et al. reasonably assumed that the correct explanation for this frequency variability was a Doppler shift from binary motion and they predicted that the range of the radial velocity would exceed 50 kms^{-1}. A search for radial-velocity variations in G29-38 by Graham et al. (1990) ruled out the 50 kms^{-1} range required by the binary hypothesis. There is no other hypothesis for the frequency modulation in G29-38. Since some white dwarfs have megagauss oblique-dipole magnetic fields (see Landstreet 1992), we suggest that magnetic modulation should be considered. Also interestingly, Soon et al. (1993) have suggested that $(P_{cyc}/P_{rot})^2$, the square of the chromospheric oscillatory time-scale to the rotation period, is a good observational measure of the theoretical dynamo number for cool, chromospherically-active stars. The ratio of the period of the frequency variation cycle to rotation period in HR 3831 is $log(P_{cyc}/P_{rot})^2 = 4.6$, which fits an extrapolation of the Soon et al. relationship to higher temperatures very well.

Here are four points to consider:

1. The Sun pulsates in high-overtone p-modes which are thought to be excited by resonance with stochastic convection. The driving mechanism thus selects the high overtones with periods near 5 minutes. The frequencies of those modes are modulated in phase with the solar cycle on an 11-yr time-scale. From the evidence of the Ca emission, other G stars are thought to behave similarly with cycle times as short as 2 yr. Magnetic fields are thought to be dynamo generated in these stars.
2. The ZZ Ceti stars pulsate in high-overtone g-modes excited by the κ-mechanism operating in the H I ionisation zone near the surface. Their high-overtone pulsation modes, with periods near 10 minutes, are selected by mode trapping in the highly-stratified white-dwarf atmosphere. In G29-38 one pulsation frequency varies cyclically with a period of 110 days. A small percentage of white dwarfs have oblique-dipole magnetic fields with strengths of 1 to 500 MG.
3. The roAp stars pulsate in high-overtone p-modes with an unknown excitation mechanism. It is not known what selects the high overtones with periods of 6 to 15 minutes, but stratification, as in the white dwarfs, does not seem possible.

The frequencies vary cyclically on time-scales of hundreds of days to years. Ap stars have oblique-dipole magnetic fields of 0.3 to 30 kG which are thought to be fossil in origin.

4. The ratio $(P_{cyc}/P_{rot})^2$ is similar in the roAp stars and the cool, chromospherically-active stars.

Here we have three different kinds of pulsating star showing some remarkably similar behaviour. Yet the three different sets of astronomers working on these three different sets of stars have three different sets of explanations for the similar behaviour. That bears some thought. Long-term monitoring was essential to bring these interesting problems to our attention.

Acknowledgements

The long-term projects described in this paper can only be done with the indispensable support of the resident astronomers and technicians at SAAO. We particularly thank Dr. Dave Kilkenny and Messrs. Fred Marang, Greg Roberts and Francois van Wyk.

References

Bachmann, K. T., Brown, T. M., 1993, *Astrophysical Journal*, 411, L45

Baliunas, S. L. & Vaughan, A. H., 1985, *Annual Review of Astronomy and Astrophysics*, 23, 379

Elsworth, Y., Howe, R., Isaak, G. R., McLeod, C. P. & New, R. ,1990, *Nature*, 345, 322

Graham, J. R., McCarthy, J. K., Reid, Niell, Rich, R. M., 1990, *Astrophysical Journal*, 357, L21

Kilkenny, D., Balona, L. A., Carter, D. B., Ellis, D. T., Woodhouse, G. F. W., 1988, *Monthly Notices of the Astronomical Society of South Africa*, 47, 69

Kreidl, T. J., Kurtz, D. W., Schneider, H., van Wyk, F., Roberts, G., Marang, F., Birch, P. V., 1994, *Monthly Notices of the RAS*, in preparation.

Kurtz, D. W., 1990, *Annual Review of Astronomy and Astrophysics*, 28, 607

Kurtz, D. W., Martinez, P. 1993, *Astronomical Society of the Pacific Conf. Ser.*, 44, 561

Kurtz, D. W., Martinez, P., van Wyk, F., Marang, F., Roberts, G., 1994a, *Monthly Notices of the RAS*, in press

Kurtz, D. W., Sullivan, D. J., Martinez, P., Tripe, P., 1994b, *Monthly Notices of the RAS*, in preparation.

Landstreet, J. D., 1992, *Astronomy and Astrophysics Review*, 4, 35

Lanz, T., Hubeny, I., 1993, *Astronomical Society of the Pacific Conf. Ser.*, 44, 517

Libbrecht, K. G. & Woodard, M. F., 1990, *Nature*, 345, 779

Martinez, P., 1993, Ph. D. Thesis, University of Cape Town

Mathys, G., 1985, *Astronomy and Astrophysics*, 151, 315

Mathys, G., 1989, *Fund. Cosmic Phys.*, 13, 143

Mathys, G., 1991, *Astronomy and Astrophysics, Supplement Series*, 89, 121

Matthews, J., 1991, *Publications of the ASP*, 103, 5

Shibahashi, H., 1987, *Lect. Notes Phys.*, 274, 112

Shibahashi, H., Takata, M., 1993, *Publications of the ASJ*, 45, 617

Soon, W. H., Baliunas, S. L., Zhang, Q., 1993, *Astrophysical Journal*, 414, L33

Winget, D. E., et al., 1990, *Astrophysical Journal*, 357, 630

Wolff, S. C., 1983, *The A-type Stars: Problems and Perspective*, Monograph series on non-thermal phenomena in stellar atmospheres, NASA SP-463

DISCUSSION

Sterken: Two comments about the differences between SAAO and ESO:

1. Service observers at ESO are young astronomers with short-term contracts whose job situation depends on scientific production. Your service observers are high-level technicians who stay in their jobs for many years to come and whose job security does *not* depend on scientific production. That dissimilarity not only entails a large difference in *attitude* of service observers, but also in *accumulated experience.* Both these elements are very often misapprehended by advocates of service observing.
2. At the ESO telescope that we used for the last decade, it took typically four to six weeks to bring the data to my home institute. Hence, instrumental deficiencies cannot be discovered promptly. At SAAO you have a plot of the raw data on a screen during the night and the data in the appropriate format on your own computer system the same morning. That is a big difference.

Kurtz: I agree. At SAAO I reduce my data before leaving the telescope at the end of the night. They are then transferred to the mainframe computer in Cape Town where they are analysed by a colleague, or by myself, before the next night of observation. So, we always know exactly what we have already obtained on a night-by-night basis.

Breger: The difficulty with imperfect data from maladjusted photometers and/or telescopes seems to be very common. The δ Scuti campaigns have shown that the reliable data come from experienced observers working with photometers they know, giving precision of $\pm 0\overset{\mathrm{m}}{.}002$ per single measurement. This compares with a best of $\pm 0\overset{\mathrm{m}}{.}004$ for less than ideal observing conditions.

Kurtz: I agree. I singled out problems at ESO only because so many users of ESO are present at this meeting. The problems are widespread.

Dravins: The secular change of oscillation frequency need not have a magnetic origin: it might be some other instability that is modulating the speed of sound and/or its pathlength inside the star.

Kurtz: That is true. The size of the sound cavity in the Sun is about 45 minutes; the changes in the solar-pulsation frequencies over the solar cycle indicate that it changes by about 1 second. Although those changes are in phase with the magnetic changes, the effect on the pulsation may be indirect. In the case of the Ap stars, it may not even be related to the magnetic field, as you say. In that case, we don't know what modulates the frequencies.

Moffat: Don't you get tired of doing only photometry all the time? How about spectroscopic or polarimetric monitoring on your short time-scales: would that not help constrain and understand the physics better?

Kurtz: No, I don't get tired of it. It is relaxing, enjoyable work. We have tried continuum polarimetry with null results. Spectroscopic changes are so small that they have only been barely detected using the CFHT and Palomar 5-m telescopes.

LONG-TERM PHOTOMETRY OF Ap STARS

H. HENSBERGE
Koninklijke Sterrenwacht van België
Ringlaan 3
1180 Brussels
Belgium

ABSTRACT. The importance of long-term monitoring for the understanding of the physics of magnetic chemically-peculiar stars of the upper main-sequence is outlined. The resulting gain in efficiency for complementary high-dispersion work at larger telescopes is discussed. Using the search for long-term variability in CP2 stars within the framework of the LTPV project as an example, the power of long-term organized efforts is compared with that of *a posteriori* data combination.

Keywords: Ap stars – CP2 stars

1. Scientific motivation for long-term observations

The chemically-peculiar stars of B2 - F spectral type offer the possibility to study a number of physical processes which have very different characteristic time-scales. Most accessible from the point of view of the collection of empirical data, is the subgroup forming the magnetic sequence, because here the variability is more pronounced and so are, in general, the observed peculiarities themselves. The terminology "magnetic sequence" should be understood as referring to those stars that show a magnetic field that is strong and global (a large dipolar contribution to the field), so that it is detectable with the present techniques. It does not imply that HgMn stars, or metallic-line (Am) stars, etc. would have no magnetic field at all. Stars in the non-magnetic sequence may be either without field, with a significantly weaker global field, or with a strong field of complicated structure, such that the measurable effect, averaged out over the visible disk, is insignificant.

Table 1 summarizes the time-scales of interest in those stars with reference to the information content on physical processes and the immediate reason for the observed variability.

The rapid oscillations have amplitudes typically of the order of millimagnitudes and are restricted to the cooler magnetic Ap stars. Their information content, and the need for long-term monitoring in this field, are discussed in this volume in the contribution by D.W. Kurtz. For all aspects related to rapid oscillations, I refer to that paper.

C. Sterken and M. de Groot (eds.), The Impact of Long-Term Monitoring on Variable Star Research, 197–208.

Ap stars are intrinsically slow rotators, which offer the possibility to measure straightforwardly the length of the rotation period as the aspect of their spotted surface varies periodically with time. Most periods are of the order of one day to one week, with a tail towards longer periods. Statistically, the hotter stars rotate faster than the cooler ones. Signal-to-noise and time-efficiency considerations have made differential intermediate-band multicolour photometry the preferred technique for detecting the periodicity. The length of a common observing run on a small photometric telescope (which exceeds the period length by a factor of 3 or more for the majority of stars), multiple observations per night (in order to reduce aliasing ambiguity) and, especially in the cooler stars, the change in the shape of the light curves obtained in different filters, conveniently provide additional constraints when applying period-search algorithms.

TABLE I
Time-scales of variability and related physical processes

log time scale (sec)	(years)	obs. technique	cause	physics info
2-3	-6	phot.,(v_{rad}).	rapid oscillations	astroseismology (star interior)
5-6	-2	phot.,spec.,magn.	stellar rotation (surface inhomog.)	chem. diffusion magnetism
7-8	0	spec.,phot.,magn.	very slow rotators (surface inhomog.)	rot. braking + see above
9	1-2	spec.,phot.,magn.	still slower rot.? precession?	see above
			stability of surface structures	MHD - migration, life time spots

Today, it is well established that rotational modulation is common to all variable magnetic Ap stars. In specific cases, other sources of variability may be superposed on it. The best-documented examples are the rapid oscillations in the cooler magnetic Ap stars. Some other stars show multiple periods, but these cases are quite rare. Among them, HD 25267 (τ^9 Eridani) is a binary that shows, in addition to the binary period of $5\overset{d}{.}95$ in the radial velocities and the rotational period of $1\overset{d}{.}21$ in the light and magnetic-field variations, also a third, still unexplained, period (Manfroid et al. 1985) of $3\overset{d}{.}84$ and, as judged from the photometry with Maitzen's Δa system, probably still a further (poorly documented, see Hensberge et al. 1981) longer-term variability. HD 152107 (52 Her) shows, in addition to the rotational modulation over $3\overset{d}{.}86$, a near to 10 year periodic variability interpreted as due to forced precession under the gravitational influence of its companion (Gerth, 1990 and references therein). Evidence for non-periodic phenomena, such as flares or magnetic instabilities, is very scant.

Observational quantities which are modulated by rotation put strong constraints on theoretical models. The review paper of Landstreet (1992) and the recent report by Mathys (1993) show how the combination of different observing techniques lead, on the one hand, to the disapproval of simple dipole or axisymmetrical field structures for these stars and, on the other hand, to the confirmation of the global character of the field; there are essentially no non-magnetic regions on the stellar surface, in contrast to the situation encountered in solar-type and cooler stars. The mapping of surface line strengths and chemical abundances indicate different surface distributions for different ions. It provides basic constraints for theories of chemical diffusion—a confrontation which still lies in the future given the complicated character of the diffusion theory in the presence of a strong magnetic field—or for any competing theory that aims to explain the abundance patterns and their surface and depth distribution.

Some stars show periodicities up to a few orders of magnitude longer than the bulk of the Ap stars. While periods of the order of 100 days in sharp-lined stars were commonly identified with extremely slow rotation, more doubt exists about the periods of years to tens of years. If these are rotation periods, then the fact that several bright Ap stars are almost not rotating, suggests a braking mechanism that works very efficiently only in a small subsample. At present there exists no theoretical prediction of such mechanism. If these periods, on the contrary, are related to another physical process, like e.g. free or forced precession, then these stars may offer new insights in hitherto disregarded physics.

Available evidence suggests that the surface inhomogeneities and the magnetic structures on Ap stars remained quite stable over the time interval wherein these stars were monitored. A (small) number of magnetic stars have been observed, magnetically and/or photometrically throughout their rotational cycle, at epochs differing by decades up to more than half a century. The same period is found, without need for introducing any phase shift. The position of the spots must therefore be stable over the time interval concerned. Adelman and Pyper (1993), however, note a change in the shape of the U light curve of HD 19832 (56 Ari) over a time span of decades, and suggest precession as the cause.

2. Impact of and need for long-term efforts

2.1. REVIEW OF LONG-TERM EFFORTS

Although the oblique-rotator model was suggested by Stibbs (1950) soon after the detection of magnetic fields in these stars, it was not readily accepted. Babcock's long-term efforts on the measurement of magnetic fields, collected in his catalogue (Babcock, 1958), initially gave rise to his definition of a subclass of irregular variables, and to his interpretation in terms of real magnetic variability rather than in terms of aspect effects. With additional photometry as a more sensitive period tracer, the huge collection of Babcock's magnetic field data is thereafter interpreted as supporting evidence for periodic variability. This would have been much more cumbersome with a more restricted or less homogeneous set of magnetic data. Once the expectation to find periodicity was high, more periods were searched for, the inverse relation between period length and line width was established, and efforts were directed to theories involving the oblique-rotator model.

It was also Babcock who found, as a consequence of the long-term character of his work, that some stars had extremely long magnetic periods: 224 days in the case of HD 188041 and almost 6.5 years for HD 187474. More long periods were detected for stars exhibiting extreme variation amplitudes. Osawa et al. (1965) used the unusually large photometric variability over a range of $0\overset{m}{.}1$ in V to reveal the 159-day period of HD 221568. Preston (1970) exploited the appearance of Zeeman-split lines in the strong field of HD 126515 to discover its 130-day period from spectroscopic observations spread over 12 years. Preston and Wolff (1970) detected the 23-year period of HD 9996 from the drastic intensity changes in its rare-earth spectrum. Wolff (1975) used her own and published absolute photometry to propose a 13.5-year period for HD 110066 which has an amplitude in V of almost $0\overset{m}{.}2$. In all these latter cases, the success of detection may be considered an immediate consequence of the uncharacteristically high level of variability presented by the star.

Much less effort has been put into the detection of more slowly-variable magnetic stars with lower amplitudes as commonly observed in the bulk of Ap stars. Between 1963 and 1973 Bonsack and Pilachowski (1974) extended Babcock's magnetic measurements for HD 201601 (γ Equ) and found further evidence for a monotonous change in the strength of the averaged field component along the line of sight, suggesting that the period might be as large as 72 years. The slow magnetic variation was later confirmed by Scholz (1979) and Mathys (1990). Photometric studies suggested shorter periods (from 9 to 1786 days, see references in Catalano and Renson 1984), with low amplitude and without agreement between different observers. Wolff and Morrison (1973) made differential photometry for 11 very slow rotators from Preston's (1970) list. Four of these stars showed periods in the range 10 – 100 days, the others remained constant over an observing season. Comparison of photometric results obtained in observing runs separated by 8 years gave evidence for long-term variability in two of the seven remaining stars. Despite the fact that the comparison had to be made on an absolute level, HD 2453 was selected as a good candidate and Wolff (1975) presented the 525 day period from photometry and magnetic data obtained over 3 subsequent years. The period of the second candidate, the northern-hemisphere 8th magnitude star HD 18078 is still unknown.

More recently, a group of European astronomers started to monitor candidate long-term magnetic Ap stars at ESO (Hensberge et al., 1984) and this initiative was then incorporated in the Long-Term Photometry of Variables project. This project has given first-time evidence for long-term photometric variability in HD 201601, in agreement with the magnetic variations (Hensberge et al. 1992). Very recently, Leroy et al. (1994) have observed the rotation of the magnetic axis over nearly 15°in the period 1990-93, through linear polarimetry. The inferred rotation period, in accordance with other observing techniques, is close to 77 years! The LTPV project has also detected new long-period variables. The cobalt-rich binary star HD 116458 varies periodically in $147\overset{d}{.}9$, while its orbital period is $126\overset{d}{.}18$ (Dworetsky, 1982). HD 94660 has a period which is at least 2700 days. Moreover, the light curve of Babcock's long-period variable HD 187474 was measured for the first time over the whole cycle of 2345 days, and the period for HD 188041 is refined to $223\overset{d}{.}9 \pm 0\overset{d}{.}2$. All these results are displayed in Hensberge (1993). The 5 stars with known rotationally-broadened lines, originally included through a lack of line-width information, did not show any long-term variability. They are intrinsic low-amplitude variables with a shorter period; at least two have a range of variability below 3 mmag, and can be suitable for check purposes on instrumental systems. The absence of long-period stars with rotationally-broadened lines is an additional argument in favour of the applicability of the oblique-rotator hypothesis to the case of the longest known periodicities.

With a totally different astrophysical motivation, Manfroid and Renson (1989) and Renson and Manfroid (1992) searched, partly in the LTPV project, confirmation for the appearance in rotational light curves of structure that might be difficult to explain within the oblique-rotator model. Especially in the case of HD 41089, they suggest that the narrow dip at light maximum might be related to the transit of a darker orbiting body.

Recently, Adelman and Pyper (1993) have initiated, as part of the *Four College Automated Telescope Project* at Mount Hopkins, a long-term effort in order to find and cover long periods and secular changes in the shape of the light curve of shorter period Ap stars, and to refine rotational periods.

Further indications for long-term variability are undoubtedly hidden in remarks about spectral variability and in discordant spectral classifications. Without the original material at hand, however, it is difficult to evaluate the qualitative remarks found in the literature. Additional interesting information might be found in the photometric catalogues based on so called "closed" photometric systems, in particular in the catalogue(s) of Geneva photometry (Rufener, 1988).

In conclusion, up to now little effort has been directed specifically at the detection of long-term phenomena at the level of variability seen in shorter time scales. It appears to me that the main reason for this lies in the limitations imposed by data inhomogeneity, partly through a lack of interest in investing observing time in instrument calibration and in obtaining reference data. As a consequence, information has been useful only for long-term purposes when either the same researcher(s) could continue their efforts over a long time interval on the same instrument, or when the long-term character of the project had been a priori defined.

2.2. LONG-TERM EFFORTS AND OBSERVING EFFICIENCY

Coverage of the rotation cycle is very telescope-time consuming, especially in the high-resolution spectroscopic mode needed in studies of magnetic field effects and of spectral lines of trace elements. Information on period, rotation phase and line width is essential in order to avoid waste of telescope time. Zeeman-split lines in integrated light can be observed only when the magnetic broadening exceeds the rotational line broadening. Also, abundance analyses are most informative at minimum rotational broadening (still, abundance analyses on faster rotators are physically interesting, since abundance patterns may depend on rotation). Mapping, on the contrary, benefits most from an intermediate situation: rotation should be slow, in order to restrict spectral-line blending, but not absent, since the Doppler shifts of the local line profile contain essentially the latitude information, while different rotational phases refer to different longitudes. An apparent rotational velocity in the vicinity of 20 km s^{-1} seems the most appropriate compromise. The uniqueness of the maps is most compromised when the rotation equator crosses the centre of the disk.

Thus, line width is important as a first selection criterion for spectroscopic studies. Next, the length of the rotation period determines the distribution of the repeated observations. Once the period is known, the more costly spectroscopic and magnetic observations over the rotation cycle can be planned with increased efficiency. Usually, phase coverage requires multiple runs as guest-observer runs at larger telescopes are commonly short. Therefore, sufficient precision in the rotation period is mandatory for predicting the ephemeris in the next observing runs.

These needs of spectroscopists differ from the practical interests of photometrists. The characteristic length of the observing runs of the latter, two weeks or somewhat less, implies that results are easiest obtained in one single run for faster rotating stars. Since it is difficult to obtain additional observing time for refinement purposes, and also because that work is considered less exciting, the majority of the published periods lack the accuracy required by spectroscopists (see also Adelman and Pyper, 1993). In addition to a loss in efficiency at larger telescopes, this leads finally to a loss in interpretation: phase relations between spectroscopic and photometric quantities, or between light variations in different wavelength regions, become unclear.

In conclusion, efficiency considerations already suggest two kinds of long-term projects:
a. many periods should be refined, and this can be done most efficiently in an optimized long-term effort.
b. many extremely sharp-lined stars have still unknown or uncertain periods, while their spectroscopic potential is large but for that reason often largely unexploited. Such a photometric study should be based ideally on the results of a line-width survey study for many stars. The work of Abt and Morrell (1994) will be useful in this respect.

2.3. LONG-TERM EFFORTS AND SAMPLE BIAS

Practical limitations imposed on observational efforts often lead to data acquisition in a biased sample. Spectroscopy is obviously biased with respect to line width. Variability searches are biased with respect to amplitude. Period searches are additionally biased with respect to period length. These biases often lead to a lack of discrimination with regard to theoretically important questions.

As an example, consider the data obtained for stars with apparent periods of the order of several years. What empirical evidence could favour the interpretation in terms of rotation for both groups or, on the contrary, could support an alternative theory? The common presence of long-term variations superposed on the rotational modulation for faster rotators would immediately prove that new physical processes have to be considered. The general lack of data homogeneity, however, reduces heavily the utility of *a posteriori* searches for such effects. The variations seen in different realisations of the Strömgren u filter and the influence of standardisation transformations on the shape of the light curve of HD 188041 (or of any other star with strongly wavelength-dependent variability) show that instrumental factors can produce systematic differences. Our experience with several ESO telescopes and filter systems has taught us that a restricted number of Ap stars is very sensitive to instrumental factors. While systematic instrumental effects, differential to a comparison star, remain below the 1% level for the large majority of Ap stars, for some of them we could measure, in a repeatable manner, instrumental differential changes of up to $0^{m}_{.}07$ in u. One of the largest effects was seen in the Am star HD 76270, successfully used as (constant) comparison star in any fixed instrumental system. Such differences do not need to remain constant over the rotation cycle and, thus, may produce changes in the shape of the light curve. According to our experience, however, the effect is much more dependent on the instrumentation than on the stellar variability during the rotation cycle, and may be particularily pronounced in the Strömgren u and Johnson U passbands. Therefore, the fact that Adelman and Pyper (1993) suggest that the light curve of HD 19832 changed beyond doubt in U and only possibly in B and V, awaits confirmation from a set of homogeneous data, or from measurements with different U filters, before it can be taken as strong evidence for secular stellar variability.

In the case of periods differing by orders of magnitude supposedly having different origins, it seems probable that the combined distribution of periods would be bimodal. It seems also probable that, at least for some quantities, the amplitude of the variation would (statistically) depend on the process involved. When different observing biases are present in both samples, like the large bias towards extreme amplitudes in the 'accidental' detection of long-period variables, such comparisons become meaningless. The search for long-period stars must then first be organized in a more systematic manner, so that e.g. all members in a magnitude-limited sample and above a certain detection level of variability are included. Since the required frequency of observation is low, but the needed time-base long, such projects are likely to be performed only in a more globally-organized environment with prospects of long-term stability.

In conclusion, the theoretical requirement for unbiased data over properly defined samples, suggests that organized long-term efforts may give much bigger returns than the accidental a posteriori combination of available data. The search for secular variability in the shape of the light curves is one of the programmes that might benefit most from an organized long-term effort.

2.4. LONG-TERM EFFORTS AND OBSERVING ACCURACY

I will use the data obtained in the LTPV project, in the search for long-period Ap stars, as a basis to discuss what accuracy is obtainable in a long-term effort. I will also evaluate how the quality has been influenced by such factors as the use of simultaneous or sequential photometry, the use of different instruments and the contribution of many observers.

In order to have a minimum check on quality, I required every observation to be composed as a series of measurements of one programme star, P, and two comparison stars, C1 and C2, in the order C1-P-C2-P-C2-P-C1. Minimum integration times of 40 seconds were specified for bright stars. In this way, an internal consistency estimate can be made. In the case of constant comparison stars C1 and C2 this estimate includes the use of system-defined constant differential magnitudes and colours between C1 and C2. For most stars the requested frequency of observation

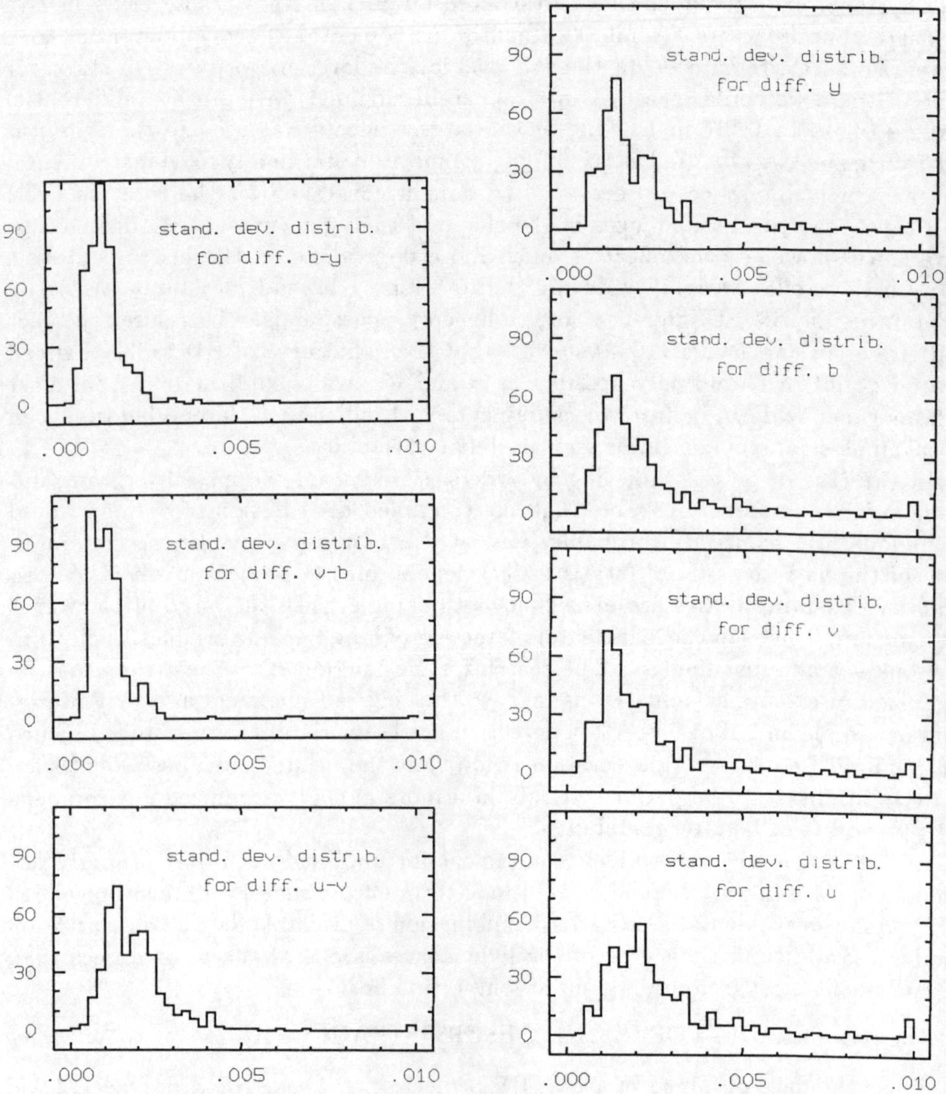

Fig. 1. Histograms of internal accuracy of the final differential magnitudes and colours as derived from the combination of the seven measurements in each series.

lies between 2 and 4 times per month.

The relevant histograms for observations distributed to the investigators are shown in Figure 1 for the 50-cm SAT (Danish Strömgren Automatic Telescope) at ESO. The error estimates in y, b and v are well correlated (with the exception that y was systematically slightly less-precise for fainter stars because of the lower count-rate) so that a more robust estimate can be obtained from the combination of the error information in the three passbands. The fourth passband, u, generally has a significantly lower count-rate and is also more susceptible to atmospheric effects. The dependence on atmospheric effects and/or lower count-rate still survives in the $u - v$ colour. Notice, however, the high precision in the $y - b$ and $b - v$ colours which is a direct consequence of the simultaneity of the photometry (also their information can be combined in a final error estimation). It follows that the errors in y, b, and v are dominated by systematic effects (presumably atmospheric), and not by photon noise. In the case of the non-simultaneous photometry at the ESO 50 cm and Bochum 60 cm telescopes (histograms not shown here), the errors in the colours are as large as in the magnitudes, indicating that atmospheric variations now contribute significantly in the error balance. Also, the absolute internal errors in the single passbands are statistically worse, as a consequence of the much longer time interval needed to perform the whole sequence of observations.

About 10% of the observations have a quality that is at least a factor 3 worse than most data, and could compromise the data interpretation when no error estimates would have been available. Screening of data by the different observers and during the reduction process is thus reasonable, but the final investigator should not accept all data as being of comparable quality. Lower-quality data may enter the analysis stage because different observers have applied a personal cut-off level for observability. (In fact, the cut-off level in a hybrid programme like the LTPV should depend on the sub-programme accuracy requirements for different purposes.) Very few data in few runs have such a low quality that personal errors or a lack of observing experience are responsible. Some runs have achieved low weight because the observing sequence had been changed subjectively to C1-P-C2. The accuracy obtained in different runs is more closely correlated with weather conditions than with observational experience. Rather, observational experience (or, maybe more correctly, dedication) enters into the distribution of the observations over the objects and over the run, and in taking into account individual requirements for specific objects. In a difficult case, like observing an object with a companion at six arcseconds with a prescribed centring, many runs proved to be unusable. The frequency of these cases increased drastically when, because of other programme requirements, an automated centring mode was applied. Consequently, the object was then removed from the observing list. No doubt the weakest point in the involvement of a large number of observers lies in the application of special requirements, and in ensuring that these special requirements are brought to the observer's attention.

An estimate of the external accuracy in a given instrumental system was made from a comparison between the scatter of the values measured for each couple of (constant) comparison stars and the prediction from the internal errors. After inclusion of a time-dependent gradient in the dead-time correction for the v photomultiplier (Hensberge et al. 1992), the ratio of external to internal errors is, on the average, only slightly higher than 1.0 (Fig. 2). The largest excess relates to the u channel (1.25) and suggests a contribution of atmospheric origin. This error analysis indicates low-amplitude variability in a number of stars (<5 mmag for HD 59256, <8 mmag for HD 71066) and allows to specify stronger upper limits to the total amplitude in those cases where no significant variability is indicated (<3

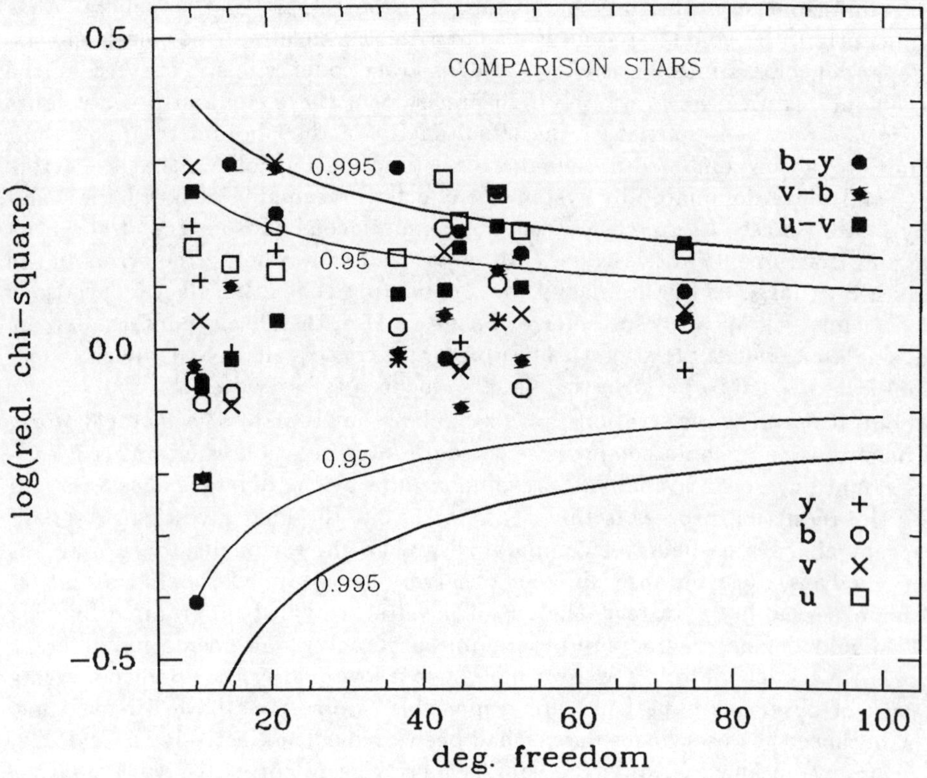

Fig. 2. Comparison of external and internal errors for 50-cm SAT data on a logarithmic scale. Symbols represent different passbands and colours. Curves refer to the 0.05 resp. 0.005 significance level at which the hypothesis "external error = internal error" can be rejected.

mmag, HD 3326, HD 107696, HD 221760).

In conclusion, with proper precautions in the observing procedure, the accuracy obtained in single observing runs using differential photometry can be extended over very long periods. The homogeneity in instrumentation, observing and data reduction is essential in low-amplitude applications. The availability of homogeneous data over long time intervals opens new research possibilities related to long-term stability and extremely slow rotation.

3. Conclusions

Homogeneous data over long time intervals give access to empirical checks on otherwise unaccessible physical processes. In the specific case of magnetic stars, long-term homogeneity will define better constraints on magneto-hydrodynamical stability and can provide clues with regard to the mechanism that produces the rotational braking.

The final accuracy in long-term investigations based on differential photometry is comparable to that obtained in short runs and depends more on advantageous observing conditions (simultaneous photometry, stable instrumentation, good atmospheric conditions) and on the internal group organisation (instructions for observing run and consistent data reduction), than on the observer.

Long-term efforts form a sound basis for the detection of systematic effects, which can then also be removed from data sets of other observers. They also provide homogeneous information on atmospheric parameters, that may be of value to other users which cannot afford to measure them.

Acknowledgements

I am most grateful to all the observers in the framework of the LTPV programme, to Jean Manfroid for his huge and continuously up-to-date reduction work, and to Chris Sterken for his persistent management of the LTPV project.

References

Abt, H.A., Morrell N.I. 1994, in *The MK process at 50 years*, ed. C. Corbally (Vatican Obs. Workshop, Tucson), in press

Adelman, S.J., Pyper, D.M. 1993, in *Peculiar versus normal phenomena in A-type and related stars* (IAU Coll. 138), ed. M.M. Dworetsky, F. Castelli, R. Faraggiana, p.644

Babcock, H.W. 1958, *Astrophysical Journal, Supplement Series*3, 141.

Bonsack, W.K., Pilachowski, C.A. 1974, *Astrophysical Journal* 190, 327.

Catalano, F.A., Renson P. 1984, *Astronomy and Astrophysics, Supplement Series*55, 371.

Dworetsky M.M. 1982, *The Observatory*102, 138.

Gerth E. 1990, in *Hot Chemically Peculiar and Magnetic Stars*, ed. G. Scholz Mitteilungen des Karl-Schwarzschild-Observatoriums Tautenburg nr. 125, 33.

Hensberge H., Maitzen, H.M., Deridder G., Gerbaldi M., Delmas F., Renson P., Doom C., Weiss W.W., Morguleff N. 1981, *Astronomy and Astrophysics, Supplement Series*46, 151.

Hensberge H., Manfroid J., Schneider H., Maitzen H.M., Catalano F.A., Renson P., Weiss W.W., Floquet M. 1984, *Astronomy and Astrophysics*132, 291.

Hensberge H., Manfroid J., Sterken C. 1992, The Messenger 70, 35.

Hensberge H. 1993, in *Peculiar versus normal phenomena in A-type and related stars* (IAU Coll. 138), ed. M.M. Dworetsky, F. Castelli, R. Faraggiana, p.547.

Kurtz, D.W. 1994 (this volume)

Landstreet, J.D. 1992, *Astronomy and Astrophysics Review*4, 35.

Leroy J.L., Bagnulo S., Landolfi M., Landi Degl'Innocenti E. 1993, preprint (submitted to *Astronomy and Astrophysics*)

Manfroid J., Mathys G., Heck A. 1985, *Astronomy and Astrophysics*144, 251.

Mathys, G. 1990, *Astronomy and Astrophysics*232, 151.

Mathys, G. 1993, in *Peculiar versus normal phenomena in A-type and related stars* (IAU Coll. 138), ed. M.M. Dworetsky, F. Castelli, R. Faraggiana, p.232.

Osawa, K., Nishimura, S., Ichimura, J. 1965, *Publications of the ASJ*17, 199.

Preston, G.W. 1970, *Astrophysical Journal*160, 1059.

Preston, G.W., Wolff, S.C. 1970, *Astrophysical Journal*160, 1071.

Rufener F. 1988, *Catalogue of Stars Measured in the Geneva Observatory Photometric System*, 4th edition, Geneva Obs., Switzerland.

Scholz, G. 1979, *Astronomische Nachrichten*300, 213.

Wolff, S.C., Morrison, N.D. 1973, *Publications of the ASP*85, 141.

Wolff, S.C. 1975, *Astrophysical Journal*202, 127.

DISCUSSION

Kurtz: The geometry of the rotational inclination and magnetic obliquity of TW Col may be such that we just get a glimpse of the other pole with its spot at the peak of the rotational light curve. That may explain the small dip you observe there.

Hensberge: If the shape of the dip is sufficiently smooth, this is certainly a possibility. If the dip is steep and narrow, an explanation of this kind becomes more difficult. It might still be worthwhile to perform continuous monitoring during one night, at the right rotation phase, in order to measure the profile of the dip with precision, so that a quantitative evaluation is possible. This continuous monitoring is not supported in the LTPV observing procedures.

Hall and Szabados enjoy a moment of rest

DISCUSSION No 3: BINARIES AND Ap STARS

Bruch: We have recently submitted to Astronomy and Astrophysics a paper on the variable light curve of the eclipsing symbiotic star AR Pavonis (in collaboration with Michael Niehues and Albert F. Jones). Photometric measurements in four colours and visual observations of the eclipsing symbiotic star AR Pav covering almost 7 cycles of the $604^{d}_{.}5$ orbit between 1982 and 1993 were presented. These data allow the determination of improved orbital ephemeries. The eclipse width is found to be smaller than previously estimated. Out-of-eclipse, AR Pav is found to be highly variable. In some cycles a marked peak appears which is absent in others. Several explanations for the out-of-eclipse variability were suggested, the most satisfying of which is a brightness modulation due to variations in the mass-transfer rate from the red giant to the hot component.

Harmanec: Would it be possible to explain the single-wave colour variations for the symbiotic star BD-21 3873 by assuming that the star is filling its Roche lobe (the Roche lobe is cooler on the side facing the companion—or better if there is a significant reflection effect)?

Bruch: Colour variations of the observed kind may, of course, be explained in principle by the effect which you mentioned. However, I suppose that it is difficult to explain the extremely large amplitude of $1^{m}_{.}2$ of the c_1 index in this way because it means that, depending on orbital phase, we see a strongly-variable Balmer-continuum radiation. While, without a more detailed investigation, I would not completely exclude such a scenario, I think that we should apply Occam's razor in view of the fact that we expect to have an emission nebula around the hot component of the symbiotic system which could well be responsible for the observed light curve. If my current efforts to model this scenario fail, I can still consider the alternative you propose.

Harmanec: Have you investigated the orbital light curves for possible cyclic long-term variability?

Bruch: So far only about 45 observations of BD -21 3873, scattered over 10 years, are available, and the fase-folded light curve does not contain much which might have been introduced by variations on other time scales than the orbital one. Therefore, any other long-term variations can at most have a low amplitude and will, thus, be difficult to detect in our limited sample of data. But, in fact, I have not yet looked

C. Sterken and M. de Groot (eds.), The Impact of Long-Term Monitoring on Variable Star Research, 209–212.

for them.

Bruch: We also carried out a photometric study (*uvby*) of the β Lyrae-type binary FY Vel (with Michael Niehues) over a time base of 6 years. Our data confirm the picture resulting from the observations of Thackeray et al. (1970, *Information Bulletin on Variable Stars*, No. 485) and Thackeray (1971, *Monthly Notices of the RAS*, 154, 103).The orbital period is slightly improved and the shape of the mean light and colour curves is discussed. The scatter in the mean light curve is caused by irregular variations of FY Vel which appear not to contain any further periodic contribution. Densely spread observations around two orbits give an impression of the scale of light and colour variations at different epochs. A Wilson-Devinney type fit to the light curves simultaneously in the v, b and y bands yields results which, albeit uncertain in some cases, confine many important system parameters. In agreement with Thackeray (1971) the secondary star of FY Vel is found to be highly underluminous for its mass.

Niarchos: In deriving the photometric solution using the Wilson-Devinney code, have you analysed all the light curves (in different wave lengths) simultaneously? Some of the light curves show different variability outside the eclipses. It would be very interesting if you could solve each light curve separately, and compare the results with those of a simultaneous solution.

Bruch: We started with fitting light curves in different bands separately. The wavelength-independent parameters turned out to be practically identical in all cases. Therefore, we finally performed a simultaneous fit to the v, b and y light curves to give us the final results. The u band was excluded because irregular variations, mainly seen in u, deteriorate considerably the quality of the results.

Hall: (to Hensberge) In your talk you described that photometry from different groups with different equipment, when combined, does not agree. Then you said the remedy is to use only photometry from one group with unchanging equipment. But I believe this is not the proper strategy. First, without the help from the other groups, you cannot obtain the necessary complete phase coverage in a long-term campaign. In other words, you *must* use the other photometry. Of course this requires that all photometry be transformed, etc., so that it will be consistent. This can be done if all groups perform their photometry properly and if the transformations are done properly. Second, There is no instrumental system that can remain constant forever; it also requires internal transformations, etc. to produce consistent photometry.

Hensberge: In my opinion, one important aspect of the difficulties encountered in comparing photometric results available in the literature and made with different instrumentation, is that commonly there is not enough information in the paper to judge the systematic differences. When the characteristics of the instrumentation are well specified, and the transformation equations given, together with either instrumental or transformed differential photometry, then it could be possible to judge the relevance of systematic differences. Unfortunately, referees do not seem to encourage such practice.

Poretti: In the framework of our differential photometry of δ Scuti stars (Poretti and Mantegazza 1992, *The Messenger*, 68, 33) we were able monitor the short–term variability of the extinction coefficient k'_λ at La Silla. The method to determinate instantaneous values of k'_λ is described by Poretti and Zerbi (1993, *Astronomy and Astrophysics*, 268, 369) and is based on the knowledge of the value of the extra-atmospheric instrumental magnitude m_o for a comparison star. This quantity can be determined assuming a trial value and then applying a least–squares linear fit to the magnitude differences between two comparison (constant) stars as a function of $\Delta X/X$, where X is the airmass and ΔX the airmass differences between the two stars. In this way the error in the evaluation of m_o can be derived (it is the slope of the fit) and the true value of m_o calculated.

We observed slow drifts and/or small-amplitude oscillations of $k'_\lambda(t)$ values; Fig. 1 shows some examples. The non–linear behaviour of the extinction law, even at good photometric sites, gives us the warning that instantaneous values should be considered in high-accuracy photometry and spurious periodicities can be enhanced if constant values are used.

We are planning the preparation of an atlas of $k'_\lambda(t)$ curves and a power-spectrum analysis.

Breger: (to Poretti) If you observe two stars separated by a few degrees all night, your technique should allow the separation of the variable extinction and equipment-sensitivity drifts. How large is the remaining equipment drift after removal of variable extinction?

Poretti: The published drifts are surely extinction-dependent, since they are related to the quantity $\frac{\Delta \sec z}{\sec z}$. The instrumental drift is expected to be small if the photomultiplier is cooled, the high-voltage supply is stabilized and the filters are not changed. Indeed, observed drifts are smaller than $0\overset{m}{.}01$.

Hall: (to Poretti) Do you realize that your description of extinction versus time for a given night will pertain to just the part of the sky traversed by your comparison star on that night, and is not necessarily a faithful description of the extinction all over the sky during the entire night? Therefore, your proposed atlas may not be very useful in a quantitative way to other photometric observers at ESO on that same night.

Poretti: The knowledge of the extinction behaviour will be useful in a statistical sense to understand better the "transient" and/or pseudo-periodic phenomena observed in light curves. If short time-scale variations cannot be applied to the entire sky, the slow drifts observed on some nights can be regarded as large spatial-scale events.

Schwarzenberg-Czerny: On several occasions, analysis is concerned with comparison of variances of programme-star observations and of comparison-star observations, or whether variances depend on instruments. Statisticians use a simple method, called Analysis of Variance to deal with that sort of questions. I think the method could be used successfully in astronomy for the above-stated purpose.

Sterken: (to Chochol) The $0\overset{m}{.}002$ precision on Nova Cyg data from Skalnaté Pleso is certainly an indication that the site is a place with possibilities of good skies. My experience from Jungfraujoch High-Alpine Observatory (Switzerland, 3600 m) is that there are—from the point of view of extinction—three kinds of "photometri-

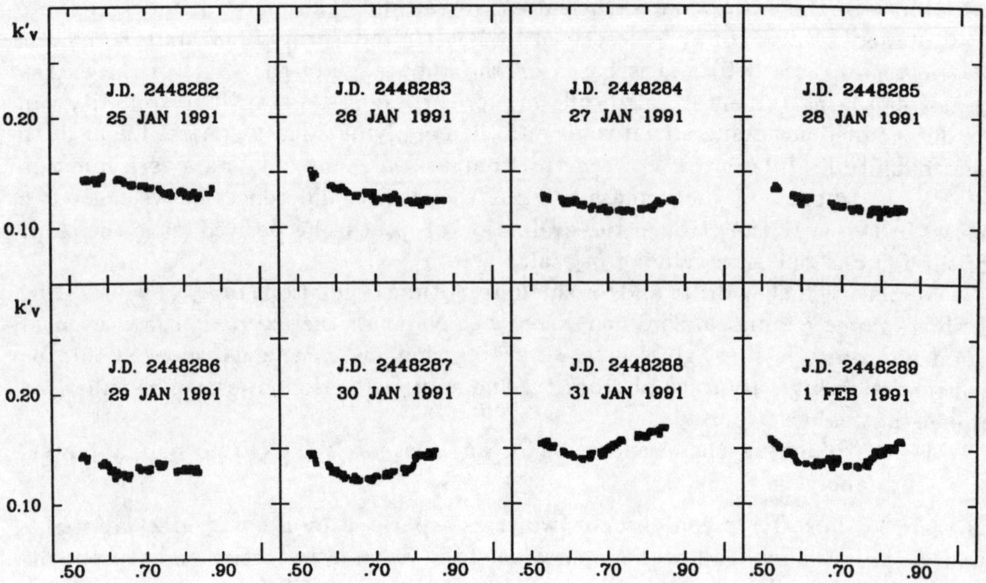

Fig. 1. Evolution of k'_V on 8 consecutive nights

cally useful" nights: there are excellent nights, there are good nights, and there are rather poor (but acceptable) nights. But in LTM campaigns, involving other European sites (in summer), I noted that the best nights at other sites do not even match in quality with what we consider the poorest nights at Jungfraujoch (see Sterken et al. 1993, *Astronomy and Astrophysics, Supplement Series*, 98, 383, Table 2). How do you cope with such situations in your campaigns with other European partners?

Chochol: We have reached $0\overset{m}{.}001$ precision at Skalnaté Pleso Observatory only during excellent nights. Although the precision reached at other observatories contributing to our campaign is worse, it is sufficient for our purpose, because we rarely discuss light variations with amplitudes smaller than $0\overset{m}{.}5$.

PROBLEMS AND PERSPECTIVES OF LONG-TERM OBSERVATIONS OF POPULATION I AND II CEPHEIDS, RR LYRAE-TYPE AND OTHER LUMINOUS PULSATING STARS

L. SZABADOS
Konkoly Observatory
H-1525 Budapest XII
P.O. Box 67
Hungary

ABSTRACT. Phenomena appearing on long time-scales are reviewed for Cepheids of both Populations, RR Lyrae-type variables, RV Tauri stars, semi-regular variables and other luminous pulsators. Stellar evolution, binary nature, stellar magnetism and other effects cause observable changes in the pulsation period and variations in the light curve.

Keywords: period changes – variability types: Cepheid, RR Lyrae, BL Herculis, W Virginis, RV Tauri, SRd, UU Herculis, hybrid, Mira, R Coronae Borealis – velocity curve – light curve

1. Introduction

Recently, with dozens of types of variable stars recognized, the observation of classical pulsating variables is a project out-of-fashion in some astrophysicists' views. The aim of this paper is to demonstrate that the amazingly regular pulsators are subjected to various changes on a longer time-scale which hints at their unique astrophysical importance.

The long-term monitoring of these stars is an especially time-consuming observational task because the shortest useful unit of a data series —a complete light curve—is commensurable with the pulsation period of the given star, i.e. as long as several days, weeks or months. In addition to the Cepheids of both populations and RR Lyrae-type variables, this group also includes less regular variables: RV Tau, SRd, UU Her, hybrid and Mira stars and a small subgroup of R CrB-type variables.

The general problem is that the physical effects intrinsic to the variable star should be separated from the effects due to the inaccuracy and inhomogeneity of the observations. In this respect, not only is the transformation of various observers' data into a common system a difficult task, the same observer can also produce inhomogeneous data if seasonal or ageing drift appears in the equipment used, not to speak of the obvious case when any element of the instrument is changed. A detailed discussion of the problems that may cause spurious variations in the stellar

C. Sterken and M. de Groot (eds.), The Impact of Long-Term Monitoring on Variable Star Research, 213–228.

brightness is not given here because this topic has been treated in other reviews of this conference and in Sterken's (1993) recent summary.

Instead, we concentrate on the physical effects which can be pointed out only if long series of observations are available. These effects have been caused by various phenomena including stellar evolution, stellar magnetism and binarity. In the case of several types of pulsating variables stellar structure can be tested (astroseismology). Generally speaking, a wealth of valuable astrophysical information (e.g. stellar mass) can be revealed with the help of long series of observations of luminous pulsating stars.

In this paper some results relating to the long-term behaviour of Cepheids, RR Lyrae stars and other luminous variables in or near the instability strip are summarized with an emphasis on the perspectives of observing such variables. For practical purposes the discussion is arranged according to the type of variability. In addition to the long-term photometric monitoring, results of radial-velocity studies are mentioned in some cases. Percy's (1992) recent review contains important additional information about the long-time-scale nonlinear behaviour of the supergiant variables.

2. Cepheids

Classical, i.e. Population I, Cepheids belong to the most regular pulsators. It is this fact that facilitates discovery of any significant change in their observed parameters. The following long-time-scale phenomena have been discovered or suspected in Cepheids:

- evolutionary period changes
- phase jumps in the pulsation period
- period changes and/or γ-velocity variation caused by companion star(s)
- changes in the photometric amplitude
- mode energy transfer in beat Cepheids
- phenomena related to solar-type activity.

A short description of these phenomena (including relevant examples) is given below.

The period of a pulsating variable star can be determined very easily and with remarkable accuracy. The long time-base especially favours the discovery of cumulative effects, such as evolutionary period changes in the $O - C$ diagram. Several bright Cepheids have been observed for more than two centuries. Although the photometric accuracy has become satisfactory only since the advent of photoelectric techniques, earlier observations are also useful for tracing changes in the pulsation period.

When crossing the instability strip, the pulsation period of a Cepheid is subjected to secular variations. The first, third and fifth crossings result in a period increase, while during the second and fourth crossings the pulsation period slowly decreases. A monotonous period change manifests itself as a parabolic graph in the $O - C$ diagram (see Figure 1). The pulsation period of δ Cephei has been decreasing. The amount of variation is tiny but still noticeable: the period is shortening with a rate of 9 sec/century. In Figure 1 other interesting facts are also seen. When

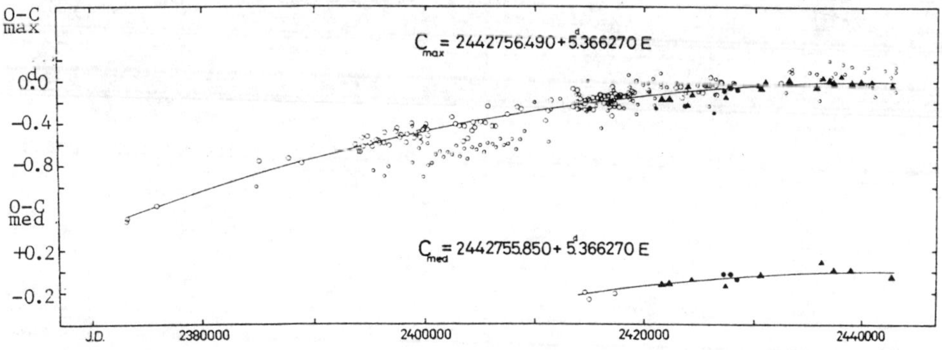

Fig. 1. $O - C$ diagram of δ Cephei, an example of continuous period decrease (Szabados 1983). Triangles indicate residuals determined from photoelectric data, filled circles those from photographic, and open circles those from visual observations. The sizes of the symbols refer to the weight assigned to the given residual. A parabolic least-squares fit is also shown.

the repeatability of various light-curve features is studied, one can also discern subtle changes in the light curve with the help of the $O - C$ diagram. In the case of Cepheids it is the moment of the median brightness that can be readily determined in addition to that of the light maximum. Figure 1 indicates that the phase difference between the maximum and median brightnesses has been constant during this century, i.e. the asymmetry of the light curve is stable. Another important lesson suggested by this Figure is that the observational data have to be published. The systematic deviation of the visual $O - C$ residuals determined from the data obtained in the middle of last century is due to the fact that these residuals are based on the published moments of maxima because the individual brightness data are not available in the literature. Such examples of inhomogeneity often lead to false interpretations of the $O - C$ diagram. When studying period changes, the original observations have to be analysed in order to minimize the effect of inhomogeneity.

Deviations may also arise because of a lacking or improper transformation of the moments of observation into a common system. For the majority of stars (and phenomena) Heliocentric Julian Date is the best choice but for the rapidly oscillating or orbiting stars, Barycentric Julian Dynamic Date (BJDD) is more adequate. In this latter system, allowance is made for the leap seconds cumulated because of the uneven rotation of the Earth.

The large number of parabolic $O - C$ graphs and their overwhelming occurrence toward longer pulsation periods indicate that we are really witnessing stellar evolution (Szabados 1984). Fernie (1984) also compared the observed period changes with the values expected from the theoretical calculations made by Becker et al. (1977). The agreement is promising but not perfect as is seen in Figure 2.

The study of period changes is a very fruitful area but requires much patience and long observational series. One aim is the separation of period changes due to various effects (e.g. duplicity – see below), another goal is a more thorough analysis of period changes related to stellar evolution. In this respect Fernie's (1990) recent

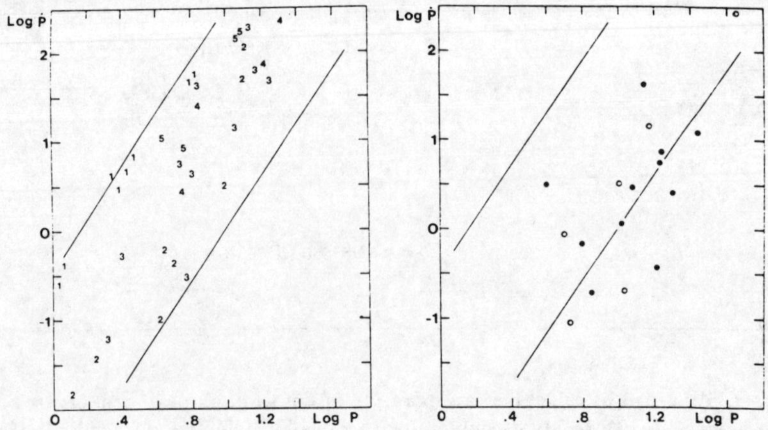

Fig. 2. Theoretically predicted (left) and observed (right) rates of period changes for Cepheids (Fernie 1984) in sec yr^{-1}. The small numbers refer to the crossing, open circles represent negative changes, while filled circles denote positive changes. The probability of observing the very fast first crossing through the instability strip is extremely low.

work has to be mentioned. He put forward arguments that the pulsation period of Cepheids does not necessarily vary linearly with time. The actual direction and speed of period variation depends on the changes in luminosity and temperature which are nonlinear with time. In addition, the motion along the evolutionary track has a variable rate. In Fernie's example (Y Oph), the parabolic fit is satisfactory for a fifty-year-long interval while the whole $O-C$ diagram can be better approximated by a cubic polynomial fit. As a matter of fact, there are very few similar examples (S Sge, SV Vul – see references in Szabados, 1983) where secular deviations from the parabolic $O-C$ graph are definitely present. This hints at the need of continuous monitoring of long-period Cepheids because these stars evolve more rapidly than their short-period counterparts.

Another interesting feature in the $O-C$ diagram of Cepheids is a subtle phase jump pointed out in more than 20 cases (Szabados 1992a). In order to reveal this effect, regular observations spread over a sufficiently long time interval are necessary, because the phase jump is usually not larger than several hundredths of the pulsation period. A typical example for a recent phase shift is seen in Figure 3. An independent confirmation of this phase jump in the pulsation of FF Aql comes from radial-velocity measurements (Evans et al. 1990). The phase jump is not always sudden. In several cases the Cepheid variable oscillates with an intermediate period before returning to the original pulsation period. The phase jump might be accompanied with changes in the shape of the light curve as shown in Figure 4.

The cause of the phase jump has not been clarified yet. Cepheids showing this phenomenon belong to binary or multiple systems. Therefore, it is plausible to assume that the effect has to do with duplicity. Kolláth and Nuspl (Kolláth 1993) performed calculations of pulsation models perturbed by the presence of a companion star and, in fact, some of their models show sudden changes in the phase of the pulsation.

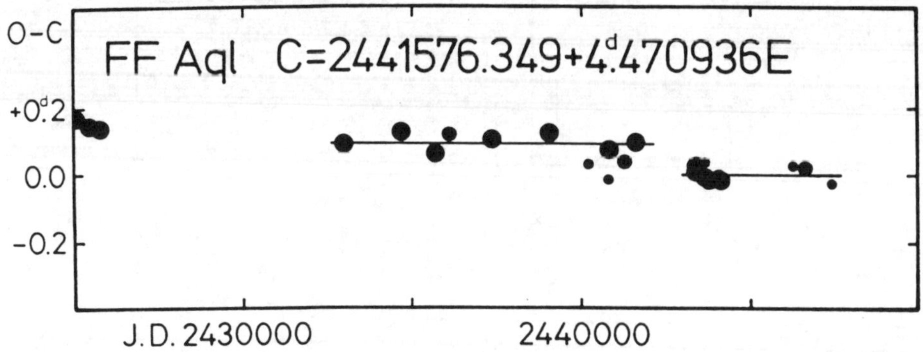

Fig. 3. Phase jump in the $O-C$ diagram of FF Aql. All $O-C$ residuals are based on photoelectric observations.

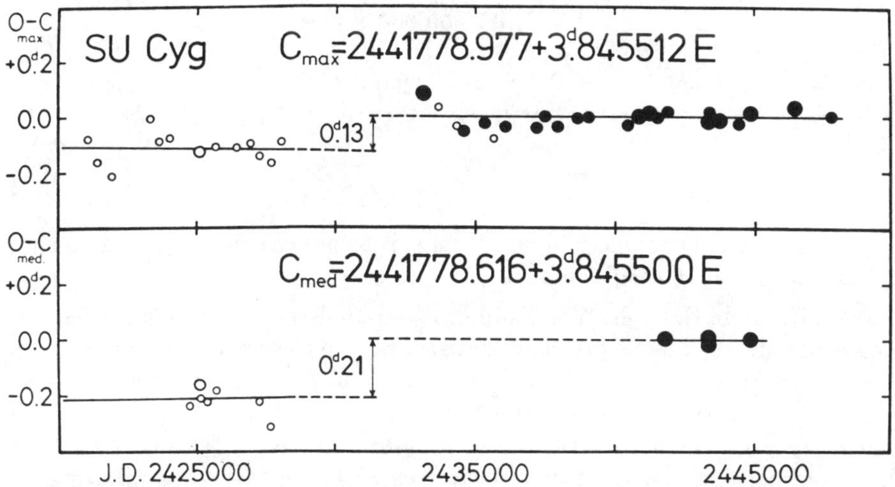

Fig. 4. A phase jump in the $O-C$ diagram of SU Cyg is accompanied by a change in the shape of the light curve. The different values of the jump for the maximum and median brightnesses indicate that the upper part of the ascending branch has become steeper. Visual observations made before J.D. 2 420 000 have been omitted from the figure but not from the fit.

Duplicity gives rise to another kind of period change. In this case the period variation is apparent and is caused by the light-travel-time effect due to the finite speed of light. This effect can be pointed out for systems having suitably large orbital inclination and long orbital period. Although the frequency of binaries (and multiple stars) among Cepheids exceeds 50% (Szabados 1992b), there is only one case (AW Per – Vinkó 1993) where the light-time-effect interpretation of the $O-C$ diagram is indisputable thanks to the ample radial-velocity data. The large value of the orbital period (about 35 years) exemplifies that this effect can only be pointed out if an extremely long data series is available. Binary Cepheids, however, play an important rôle in the calibration of the period – luminosity relationship. Therefore,

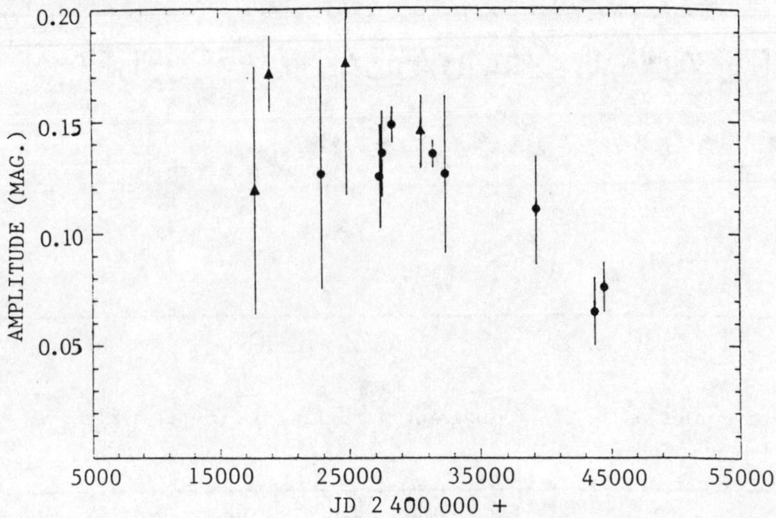

Fig. 5. Photometric B amplitude of Polaris as a function of time. Dots indicate values derived from photoelectric observations, triangles are based on photographic data (Arellano Ferro 1983).

it is important to study as many binary Cepheids as possible in order to determine their physical parameters individually.

The separation of the pulsational and orbital radial-velocity variations can only be performed with the help of extensive data series. Up to now a spectroscopic orbit has been determined for twelve Cepheids having a physical companion and variations in the γ-velocity have been pointed out for more than fifty other Cepheids (Szabados 1992b). Further spectroscopic monitoring of these stars is highly desirable. Long-term radial-velocity measurements are especially useful when the amplitude of the orbital effect is very small or the orbital period is very long. In the case of T Mon (Gieren 1989) the orbital effect became apparent in the eighties because of the almost two-century-long period of revolution.

Returning to the photometric observations, an important result of the long time-scale monitoring is the discovery of secular changes in the amplitudes. There are two Cepheids for which this effect has been revealed beyond doubt: Polaris and Y Oph. Curiously enough Polaris has been the lowest-amplitude Cepheid for decades. The decrease in its photometric amplitude was discovered by Arellano Ferro (1983). Figure 5 shows the amplitude in the B-band against Julian Date. The same behaviour can be detected in the pulsational radial-velocity amplitude of this Cepheid (Dinshaw et al. 1989). If the observed tendency continues, the pulsation of Polaris will cease as early as 1995. The interpretation of this phenomenon is, however, not trivial because Polaris is situated in the middle of the instability strip, similar to the other Cepheid, Y Oph. The latter shows a secular decrease in the amplitude of brightness variation (Fernie 1990). Thus, neither star is leaving the instability strip.

In addition to the secular amplitude variation, changes in the pulsational amplitude are also observed on a shorter time-scale. The peculiar Cepheid V473 Lyrae (HR 7308) is the shortest-period pulsator ($P=1\overset{d}{.}49$) among the known galactic Cepheids. The amplitude of pulsation is modulated by at least a factor of 15 with a cycle length of about 1400 days (Burki et al. 1986). Several hypotheses have been put forward to explain this strange behaviour (a star entering or leaving the instability strip; interaction between pulsation and convection; instability of the limit cycle). The available observational material is not sufficient for a complete phenomenological description; even the simple question about how regular the amplitude modulation is (periodic, cyclic, none) remains unanswered.

Beat Cepheids pulsate with more than one period. In most cases the observed two periodicities correspond to the fundamental mode and first-overtone oscillations. The dozen or so galactic Cepheids of this kind also deserve regular photometric monitoring. Long observational series of these stars have two advantages:

1. Easy identification of the component frequencies. Short data series result in a wide spectral-window function sometimes exhibiting complicated alias structure, so that Fourier-analysis, the most sensitive conventional method for pointing out periodicities, cannot be applied successfully. In order to obtain very narrow peaks in the power spectrum, long data series with a minimum of interruptions are necessary. A data series of 150 observations obtained during the seventies made it possible to identify three modes in the pulsation of BQ Serpentis: in addition to the known frequencies of the fundamental mode and first-overtone oscillations, the harmonics and linear combinations of these frequencies also appear and, surprisingly the second overtone and its coupling terms with f_0 and f_1 are also present (Szabados 1993). A more detailed analysis of TU Cas, one of the brightest beat Cepheids, also shows triple-mode pulsation. The amplitude of the third excited mode is very low. Accurate, homogeneous and extensive photometry is, therefore, necessary to reveal the second overtone. As a matter of fact, Faulkner (1977) already claimed the presence of the second overtone in the light curve of TU Cas but Hodson et al. (1979) attributed this discovery to inhomogeneity of data analysed together. Faulkner's claimed third period, however, does not coincide with the recently discovered one.

From a theoretical point of view, even a stable double-mode pulsation cannot be modelled for Cepheids (Moskalik et al. 1992). Therefore, triple-mode oscillation is a real challenge for theoreticians.

2. The investigation of the stability of the modes is also impossible because of a lack of long observational series. Stability of the modal amplitudes has been studied in the case of TU Cas and U TrA, the most thoroughly observed beat Cepheids. In spite of several weak indications concerning long-time-scale amplitude variations the problem has not been solved yet. Observation of mode-switching would be of crucial importance from points of view of both pulsation theory and stellar evolution.

In addition to the phenomena observed in the light curve, variations of the radial-velocity curve can also appear on a long time-scale. W Sgr, a bright Cepheid belonging to at least a triple system, is a good example. The variations occur at a definite phase of the radial-velocity curve where a bump of changing strength appears (see Babel et al. (1989) and references therein).

The first indication of starspot(s) on a Cepheid, a really unexpected phenomenon, is also the result of an extensive radial-velocity measurement programme. Dinshaw et al. (1989) pointed out that if the pulsational radial-velocity variations ($3^{\mathrm{d}}_{\cdot}97$ period) and the radial-velocity component due to the spectroscopic orbit (about 30-year period) are removed, the residuals still contain a periodic component. The only reasonable explanation for this 45.3 day periodicity (semiamplitude 1.0 km s^{-1}) is the rotational modulation caused by surface feature(s). Long-term monitoring is necessary to have a more detailed picture on stellar activity on Cepheids. At first, stability (lifetime) of the spots can be studied, then analogy to the solar activity cycle can be searched for (see Hall's (1990) suggestion about explaining the long cycles in the $O-C$ diagrams as due to magnetic cycles).

3. RR Lyrae type variables

The RR Lyrae variables are less massive but older than the classical Cepheids and they are found on the horizontal branch of the HRD within the instability strip. RR Lyrae variables may show two kinds of double periodicity: both the Blazhko-effect and double-mode pulsation give rise to long-term photometric behaviour worth following.

The Blazhko-effect is a long-period (20–100 days) modulation of the light curve. The characteristic variations, both in the amplitude and the shape of the light curve, typically occur in RRab stars (fundamental-mode pulsators) with a pulsation period shorter than $0^{\mathrm{d}}_{\cdot}7$ (Szeidl 1988). Although the Blazhko-effect has been pointed out in almost 30% of RRab variables brighter than 12th magnitude, there are only few thoroughly-studied cases, especially as regards the long-time-scale variations. The variety of interesting phenomena is visualized in Figure 6 showing the behaviour of RR Lyrae itself during a five-year interval (Szeidl 1975). The lower panel shows the cyclic variations of the maximum brightness while on the middle panel the oscillation of the phase of the median brightness (measured on the ascending branch) is seen superimposed on the erratic period change in the $O-C$ diagram of the $0^{\mathrm{d}}_{\cdot}567$ pulsation period. It is worthy of note that whenever the amplitude of these oscillations becomes significantly smaller than its normal value, a phase shift occurs in the $O-C$ diagram of the $40^{\mathrm{d}}_{\cdot}8$ Blazhko-period (top panel). This phase discontinuity in either direction seems to be a cyclic phenomenon, the cycle length being about four years which can be traced back to J.D. 2 431 000 with the help of an extremely long data series (Detre and Szeidl 1973).

This strange behaviour can be explained in terms of a magnetic-activity cycle where the observed 4-year cycle is the magnetic one and the Blazhko-period is the rotation period of the star (oblique rotator). The analogy with the solar activity cycle is supported by direct magnetic measurements (see references in Detre and Szeidl 1973).

Unfortunately, other RR Lyrae stars have not been studied so thoroughly, although other RRab variables may also possess this kind of long cycle: XZ Cyg (9.2 years), RW Dra (7.4 years), Y LMi (7.7 years) (Szeidl 1988). Because of the unique opportunity to study stellar activity in pulsating variables, RR Lyrae stars with Blazhko-effect are important targets for long-term monitoring.

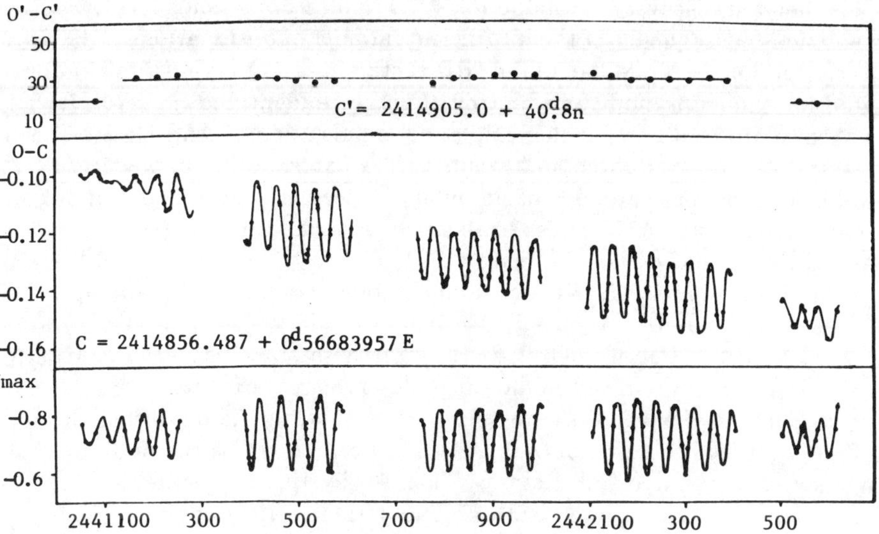

Fig. 6. The behaviour of RR Lyrae on long time-scale. Top panel: $O-C$ diagram of the $40^{\rm d}_{\cdot}8$ Blazhko-period. Middle panel: $O-C$ diagram of the $0^{\rm d}_{\cdot}567$ pulsation period. Lower panel: oscillation of the height of the brightness maximum in V (Szeidl 1976).

It has to be noted that the RR Lyrae variables are usually observed during a certain interval of the pulsational cycle because, in order to study the period changes, enough information can be obtained from a section of the light curve (e.g. covering phases between the light minimum and the subsequent maximum). For performing a conventional period analysis, however, the whole light curve has to be covered which is an even more time-consuming task. In fact, some RR Lyrae variables are reported to have additional periodicities. According to Szeidl (1988) each of these third periods may be spurious caused by the yearly gap in the data sample (except RS Boo). Similarly, the Blazhko-effect reported for RRc stars may be somewhat uncertain. In these shorter-period pulsators the first overtone is excited and some of them really have a modulated light curve caused by double-mode pulsation (see below).

While the contradictory results concerning the Blazhko-effect in RRc stars cannot be explained with the temporal evolution of the Blazhko-effect itself (it would be premature to do so), there are pieces of evidence that this phenomenon ceased to exist in RRab stars. For example, in RR Gem cessation of the Blazhko-effect occurred around 1940, while the same happened to SW And in the mid-fifties (Szeidl 1988 and references therein). Commencement of the Blazhko-modulation in a "normal" RR Lyrae star still awaits discovery. One aim of the long-term photometry of RR Lyrae stars is certainly to detect such substantial changes. The transition between the Blazhko- and non-Blazhko-state (and vice versa) is an additional evidence that there is no sharp distinction between the RR Lyrae stars with and without Blazhko-effect. Because this effect is a manifestation of the stellar magnetic activity, its absence may be a kind of stellar Maunder minimum.

The belated discovery of the double-mode pulsators among RR Lyrae variables is an obvious consequence of the insufficient number of observations. The first representative of double-mode RR Lyrae (RRd) stars was AQ Leonis; its simultaneously excited fundamental mode and first overtone were identified in 1976 (Jerzykiewicz and Wenzel 1977). (The peculiar triple-mode pulsator AC And has not been taken into account here – see Fitch and Szeidl 1976.) Later, double-mode RR Lyrae stars have been found in a number of globular clusters and dwarf spheroidal galaxies. The globular cluster M 15 is especially rich in such variables (see e.g. Cox et al. 1983). Quite recently Clement et al. (1991) discovered two more RRd variables in the field. Promising candidates for double-mode variability are, among others, the RRc stars BT Aqr, BV Aqr, AH Cam and AS Vir whose light-curve modulation has been reported but no detailed study of these stars has yet been performed. Stars pulsating in the mixed-mode range deserve a closer attention from an observational point of view. It is interesting to note that, unlike double-mode Cepheids, the first overtone has an amplitude at least twice as large as the fundamental-mode frequency component in RRd stars. This is why these variables are expected to occur among the RRc-type pulsators. The period ratio is about 0.75. Long series of observations or repeated measurements at considerably different epochs are necessary in order to study the stability of the amplitude of the excited modes. Since both Oosterhoff I and II type globular clusters contain RRd stars, the mixed-mode pulsation is not confined to a certain age or metal content.

Monoperiodic RR Lyrae variables are also important on a longer time-scale. Stellar evolution is expected to have an impact on the change in the period of pulsation (see Lee 1991). RR Lyrae stars with Blazhko-effect usually exhibit very complicated period changes. Therefore, the singly periodic RR Lyrae stars are the suitable targets for revealing period changes of evolutionary origin. Not only good coverage on the $O - C$ diagram is necessary, but numerous stars have to be studied in order to obtain statistically-significant results. In a recent paper e.g., Wehlau et al. (1992) pointed out the existence of a significant radial gradient in the rate of period change for the RR Lyrae stars in the globular cluster NGC 7006 (their sample consisted of 46 variables). Sweigart and Renzini (1979) suggested that random mixing events in the semi-convective layer may cause sudden period changes which show up as spikes in the $O-C$ diagram. Such episodic period change was then found e.g. in the $O - C$ diagram of V 15 in M 15 (Barlai 1982).

Binary stars among RR Lyrae variables are less frequent than among Cepheids. Nevertheless, wave-like pattern in the $O - C$ diagram may indicate the presence of a companion (see e.g. Coutts 1971). Radial-velocity observations spread over a long time interval are particularly useful for detecting binary companions. For example, TU UMa has a variable γ-velocity. Therefore, its cyclic $O - C$ graph is correctly interpreted in terms of the light-time effect (Saha and White 1990).

4. Population II Cepheids

The low-mass counterparts of the classical Cepheids are separated into two groups, both at a later stage of evolution. The BL Her group contains horizontal-branch variables (pulsation period: 1–3 days), while the stars belonging to the W Vir group are pulsating with longer periods because they enter the instability strip while executing blueward excursions from the AGB due to shell flashes in the stellar interior.

The evolutionary time-scale of these Population II stars is much shorter than that of the much younger classical Cepheids. Monitoring of their period change is especially important. For example, pulsations of AU Pegasi are subjected to unusually large period variations (1% increase during 15 000 days – Vinkó et al. 1993).

In addition to the change in the pulsation period, variations in the shape of the light curve may also occur. Various model calculations indicate RV Tauri characteristics, i.e. alternating larger and smaller amplitude oscillations where it is the minimum brightness level that alternates. Regular observations of some bright Population II Cepheids with APTs are therefore desirable to point out this behaviour observationally.

RU Cam, an atypical Population II Cepheid, belongs to the best observed variables owing to its peculiar behaviour. The normal pulsation ($P = 22$ days) of this carbon-rich star ceased to exist in the sixties and since that time low-amplitude oscillations of a complicated pattern have been observed. An impressive light curve of RU Cam covering more than a decade (consisting of more than 2000 individual observations) was published by Kolláth and Szeidl (1993). The conclusion of their thorough analysis is that multiperiodic variation and low-dimensional chaos can both be excluded.

5. Other pulsating supergiants

Pulsational model calculations have shown that, keeping the mass and luminosity unchanged, a decrease in effective temperature leads to a systematic progression from regular pulsation (Population II Cepheids) through period doubling (RV Tauri variables) to chaotic behaviour (semi-regular variable stars) in the stellar oscillation (Buchler 1990). This physical sequence justifies that this review cannot be finished with Population II Cepheids without mentioning their less regular counterparts. As a rule, variables cooler than the pulsators in the instability strip lose mass to a larger extent. Mass loss and circumstellar matter enhances variability on both short and long time-scales. For this reason, luminous pulsating variables outside the instability strip are promising targets for long-term photometric and spectroscopic projects.

The region of the HRD adjacent to the red side of the instability strip contains numerous types of pulsating variables. It is not intended to sum up the aspects of long-term monitoring of all these variables here. Instead, some interesting phenomena (which are thought to be the most important) are selected in order to direct attention to importance of these observationally neglected variables.

RV Tauri stars are yellow supergiants in the post-AGB phase of evolution. Secular period changes of evolutionary origin have been detected (Percy et al. 1991) in accordance with theoretical predictions. In addition, cyclic period changes have been found in some RV Tauri stars (e.g. in AC Her, see Zsoldos 1988). The amplitude of these several decades long cycles is incompatible with the orbital light-time effect interpretation. Hall (1990) proposed that this phenomenon is caused by stellar magnetism. As a matter of fact, such cycles have also been found in the $O-C$ diagrams of other variables (e.g. Cepheids).

As to variability of light curve, the long-standing puzzle of the RVb phenomenon has not been solved yet. The subgroup of RVb stars exhibit long-term changes in mean magnitude (the long period is about 15 times longer than the pulsation period). A more complete review on long-term phenomena in RV Tauri stars can be found in Percy's (1993) paper.

Semi-regular yellow supergiants (SRd stars) form a very heterogeneous group. Their pulsation is even less regular than the RV Tauri-type variation. The distinction between RV Tauri and SRd stars is, however, not sharp (Wahlgren 1993) because the temperature which controls regularity of the pulsation is a continuous variable. Inhomogeneity of the SRd type is even more pronounced by the fact that it includes giant, supergiant and hypergiant variables.

The massive Population I hypergiant ρ Cas has a very complicated light curve (Zsoldos and Percy 1991). In addition to the pulsational light variation whose 300-day period is discernible, the other physical process having a strong impact on the light curve is the shell ejection. The latest such episode occurred in 1985–1986 (its manifestation was fading in V and reddening in the colour indices). V509 Cas, the "spectroscopic twin" of ρ Cas, underwent a much slower shell episode, and a secular increase in its brightness has been detected (Zsoldos 1986).

A unique group among the semi-regular stars are the so-called *UU Herculis -type variables.* These stars are supergiants at high galactic latitude. The low-amplitude pulsation of UU Her-type stars has a period of several weeks. They are probably low-mass stars in the post-AGB stage of evolution.

The photometric behaviour of the prototype star, UU Her, is very interesting. As pointed out by Zsoldos and Sasselov (1992), the light curve of UU Her can be interpreted in terms of double-mode pulsation. The ratio of the fundamental mode and first-overtone frequencies is 0.60, a value differing significantly from the ratio valid for classical Cepheids. The other important difference is the interchange of the modal energy which results in an amplitude modulation with a period of 4.6 years superimposed on the light curve composed of the mixture of the 72 and 45-day periods. The mode energy change in Cepheids appears on a much longer time-scale, if the interpretation of the inhomogeneous data series is correct.

Detection of binary companions to UU Her stars also requires observations spread over a long time interval. A promising radial-velocity project and its initial results (binary nature of V441 Her) have been described by Waelkens and Waters (1993). Being supergiants, UU Her stars cannot belong to binary systems with orbital periods shorter than several hundred days.

Long-term spectroscopic monitoring of the so-called *hybrid stars* is important for understanding this peculiar group of stars. The bright giant and supergiant (G0Ib – K2II) hybrid stars share their position on the HRD with V441 Her, but they show substantial winds and high-temperature UV emission lines. As Rao et al. (1993) pointed out, time series studies of chromospheric variations can reveal important information about rotational modulation of atmospheric inhomogeneities, evolutionary time-scales of such features, and even pulsation can be detected by studying the Ca II H and K emission cores.

The reddest pulsating stars, *Miras*, also deserve long-term monitoring. Their long pulsation period and large photometric amplitude, however, are unfavourable for long-term projects. Nevertheless, exciting events may happen to Mira variables. A good example for supporting this statement is the development of long-period oscillations in the red component of RR Tel (a symbiotic nova) whose enhanced mass loss eventually led to an outburst of the more compact component (Robinson 1975). Moreover, detailed maps of water masers around Miras and variable supergiants show marked changes in structure over several years (Bowers et al. 1993). Other aspects of long-term changes in Mira variables have been summarized by Percy et al. (1990).

The characteristic feature of the carbon-rich, hydrogen-deficient yellow supergiant *R CrB variables* is the occasional decrease in brightness caused by dust clouds getting into the line of sight. A less spectacular variation, but relevant to the topic of this paper, is the pulsation in those R CrB stars whose effective temperature places the star into the instability strip. Regarding only pulsation, a number of interesting phenomena have been observed which calls for long-term monitoring of the handful known stars of this type. There is evidence that the 38-day pulsation period of RY Sgr is modulated on a 1000 to 2000-day time-scale (Marraco and Milesi 1982). Another R CrB star, S Apodis, may have switched from its 120-day pulsation to an overtone of nearly 40-day period (Kilkenny 1983). As to R CrB itself, the pulsations are somewhat irregular: from various subsets different periods can be determined (for references see Percy (1992)). Further long series of observations are necessary in order to answer the questions whether their pulsation is multiperiodic, chaotic or irregular and whether the ejection of the dust cloud is triggered by the pulsation.

Finally, we have to refer to FG Sge which, in various aspects, is similar to many types of variable stars, including pulsating supergiants. On account of its very rapid evolution, this key-object really deserves continuous monitoring. FG Sge, the central star of the planetary nebula He 1-5, has been pulsating since the sixties when it entered the instability strip from the blue side. While crossing the strip, its pulsation period increased in an unprecedented manner: from 15 days in 1962 to 107 days in 1979 (Jurcsik and Szabados 1979). In addition, this star shows a wealth of photometric and spectroscopic peculiarities. The latest spectacular event on FG Sge happened in the second half of 1992 when an R CrB-type dimming and recovery were observed (Jurcsik 1993). This is an example of the most direct observation of stellar evolution.

Such outstanding objects are really very rare but, fortunately, there are hundreds of pulsating variables whose regular, persistent monitoring helps understand stellar structure, evolution and other astrophysical problems.

Acknowledgements

I would like to express my appreciation to the organizers of this Advanced Research Workshop, especially to Dr. Christiaan Sterken for the invitation to present this paper and for the financial support. Additional travel support from Hungarian

OTKA grant No. T7438 is acknowledged. Thanks are also due to staff members at Konkoly engaged in variable star research for lots of discussions: the shortest one took just a minute, the longest has been continuing for years. Long time-scale phenomena appear and are important everywhere ...

References

Arellano Ferro, A. 1983, *Astrophysical Journal*, 274, 755

Babel, J., Burki, G., Mayor, M., Waelkens, C. and Chmielewski, Y. 1989, *Astronomy and Astrophysics*, 216, 125

Barlai, K. 1982, *Comm. Konkoly Obs. Hung. Acad. Sci.*, Budapest, No. 83, 223

Becker, S.A., Iben, I. and Tuggle, R.S. 1977, *Astrophysical Journal*, 218, 633

Bowers, P.F., Claussen, M.J. and Johnston, K.J. 1993, *Astronomical Journal*, 105, 284

Buchler, J.R. 1990, in *"Non-Linear Astrophysical Fluid Dynamics"*, eds. J.R. Buchler and S.T. Gottesman, Ann. New York Acad. Sci., 617, 17

Burki, G., Schmidt, E.G., Arellano Ferro, A., Fernie, J.D., Sasselov, D., Simon, N.R., Percy, J.R. and Szabados, L. 1986, *Astronomy and Astrophysics*, 168, 139

Clement, C.M., Kinman, T.D. and Suntzeff, N.B. 1991, *Astrophysical Journal*, 372, 273

Coutts, C.M. 1971, in Proc. IAU Coll. No. 15, Bamberg Veröff., IX, Nr. 100, 238

Cox, A.N., Hodson, S.W. and Clancy, S.P. 1983, *Astrophysical Journal*, 266, 94

Detre, L. and Szeidl, B. 1973, in *"Variable Stars in Globular Clusters and in Related Systems"*; ed.: J.D. Fernie, Reidel, Dordrecht, p. 31

Dinshaw, N., Matthews, J.M., Walker, G.A.H. and Hill, G.M. 1989, *Astronomical Journal*, 98, 2249

Evans, N.R., Welch, D.L., Scarfe, C.D. and Teays, T.J. 1990, *Astronomical Journal*, 99, 1598

Faulkner, D.J. 1977, *Astrophysical Journal*, 218, 209

Fernie, J.D. 1984, in *"Observational Tests of the Stellar Evolution Theory"*, Proc. IAU Symp. No. 105; eds.: A. Maeder and A. Renzini, Reidel, Dordrecht, p. 441

Fernie, J.D. 1990, *Publications of the ASP*, 102, 905

Fitch, W.S. and Szeidl, B. 1976, *Astrophysical Journal*, 203, 616

Gieren, W.P. 1989, *Astronomy and Astrophysics*, 216, 135

Hall, D.S. 1990, in *"Active Close Binaries"*, NATO ASI C319; eds.: C. Ibanoglu and I. Yavus, Kluwer, Dordrecht, p. 95

Hodson, S.W., Stellingwerf, R.F. and Cox, A.N. 1979, *Astrophysical Journal*, 229, 642

Jerzykiewicz, M. and Wenzel, W. 1977, *Acta Astronomica*, 27, 35

Jurcsik, J. 1993, *Acta Astronomica*, (in press)

Jurcsik, J. and Szabados, L. 1979, *Information Bulletin on Variable Stars*, No. 1722

Kilkenny, D. 1983, *Monthly Notices of the RAS*, 205, 907

Kolláth, Z. 1993, personal communication

Kolláth, Z. and Szeidl, B. 1993, *Astronomy and Astrophysics*, 277, 62

Lee, Y.-W. 1991, *Astrophysical Journal*, 367, 524

Marraco, H.G. and Milesi, G.E. 1982, *Astronomical Journal*, 87, 1775

Moskalik, P., Buchler, J.R. and Marom, A. 1992, *Astrophysical Journal*, 385, 685

Percy, J.R. 1992, in *"Nonlinear Phenomena in Stellar Variability"*; Proc. IAU Coll. No. 134; eds.: M. Takeuti and J.R. Buchler, Kluwer, Dordrecht (in press)

Percy, J.R. 1993, in *"Luminous High-Latitude Stars"*; ed.: D.D. Sasselov, *Astronomical Society of the Pacific Conf. Ser.*, 45, 295

Percy, J.R., Colivas, T., Sloan, W.B. and Mattei, J.A. 1990, in *"Confrontation between Stellar Pulsation and Evolution"*; eds. C. Cacciari and G. Clementini, *Astronomical Society of the Pacific Conf. Ser.*, 11, 446

Percy, J.R., Sasselov, D.D., Alfred, A. and Scott, G. 1991, *Astrophysical Journal*, 375, 691

Rao, L.M., Baliunas, S.L., Robinson, C.R., Frazer, J., Woodard, L. and Donahue, R.A. 1993, in *"Luminous High-Latitude Stars"*; ed.: D.D. Sasselov, *Astronomical Society of the Pacific Conf. Ser.*, 45, 301

Robinson, E.L. 1975, *Astronomical Journal*, 80, 515

Saha, A. and White, R.E. 1990, *Astronomical Journal*, 102, 148

Sterken, C. 1993, in *"Precision Photometry"*, eds.: D. Kilkenny, E. Lastovica and J. Menzies, SAAO, 57
Sweigart, A.V. and Renzini, A. 1979, *Astronomy and Astrophysics*, 71, 66
Szabados, L. 1983, *Astrophysics and Space Science*, 96, 185
Szabados, L. 1984, in *"Observational Tests of the Stellar Evolution Theory"*, Proc. IAU Symp. No. 105; eds.: A. Maeder and A. Renzini, Reidel, Dordrecht, p. 445
Szabados, L. 1992a, in *"Complementary Approaches to Double and Multiple Star Research"*, Proc. IAU Coll. No. 135; eds.: H.A. McAlister and W.I. Hartkopf, ASP Conf. Ser., 32, 255
Szabados, L. 1992b, ibid. p. 358
Szabados, L. 1993, in *"Applications in Time Series Analysis in Astronomy and Meteorology"*, Padova, p. 247
Szeidl, B. 1976, in *"Multiple Periodic Variable Stars"*, Proc. IAU Coll. No. 29; ed.: W.S. Fitch, Akadémiai, Budapest, Vol. 1, 133
Szeidl, B. 1988, in *"Multimode Stellar Pulsations"*; eds.: G. Kovács, L. Szabados and B. Szeidl, Konkoly Obs., Kultura, Budapest, p. 45
Vinkó, J. 1993, *Monthly Notices of the RAS*, 260, 273
Vinkó, J., Szabados, L. and Szatmáry, K., 1993, *Astronomy and Astrophysics*, (in press)
Waelkens, C. and Waters, L.B.F.M. 1993, in *"Luminous High-Latitude Stars"*; ed.: D.D. Sasselov, *Astronomical Society of the Pacific Conf. Ser.*, 45, 219
Wahlgren, G.M. 1993, ibid. p. 270
Wehlau, A., Nemec, J.M., Hanlan, P. and Rich, R.M. 1992, *Astronomical Journal*, 103, 1583
Zsoldos, E. 1986, *The Observatory*, 106, 156
Zsoldos, E. 1988, *Information Bulletin on Variable Stars*, No. 3192
Zsoldos, E. and Percy, J.R. 1991, *Astronomy and Astrophysics*, 246, 441
Zsoldos, E. and Sasselov, D.D. 1992, *Astronomy and Astrophysics*, 256, 107

DISCUSSION

Sterken: The phase shifts shown in the $O-C$ diagram of SU Cygni look very attractive, but if I remove the lines you have drawn, I would as well like to try fitting a parabola, and explain the deviations from it by invoking another phenomenon. The paucity of data in some LTM studies may give rise to speculative interpretations of the $O-C$ diagram.

Szabados: In the case of SU Cyg, the earliest observations (confirming the approximation by linear sections) have been omitted. This $O-C$ diagram is shown as an example of the variation in the shape of the light curve.

Moffat: What are the error bars in the case of SU Cyg?

Szabados: The error bars are not larger than the size of the dots.

Kurtz: The constant phase with jumps in your $O-C$ diagram is indicative of a constant frequency with the phase being reset occasionally—probably by the driving mechanism. I don't understand your suggestion that the solution to this may lay in the binary nature of these stars.

Szabados: It is an observational fact that such jumps exclusively occur in binary Cepheids. The physical mechanism, of course, needs theoretical explanation.

Poretti: There is another relevant aspect in the long–term study of Cepheid variables, related to the possibility to obtain a large number of well–covered light curves. The Fourier decomposition is a powerful method to describe them as a function of the period value (i.e., to describe quantitatively the Hertzsprung progression). By using this method, Antonello, Poretti and Reduzzi (1990, *Astronomy and Astrophysics*, 236, 138) found evidence for a much clearer separation between Classical and s–Cepheids. A different pulsation mode (fundamental for the Classical Cepheids, first overtone for the s–Cepheids) was proposed as the physical cause for the separation. This hypothesis was confirmed by further analyses (Mantegazza and Poretti 1992, *Astronomy and Astrophysics*, 261, 137; Poretti 1994, *Astronomy and Astrophysics*, in press). A sample of 225 Cepheids, 30 of which can be reliably identified as first overtone pulsators, was selected subsequently; there is a strong interest in increasing this number. The presence of a discontinuity in the Fourier parameter space at $P = 3\overset{d}{.}2$ was also shown. A possible physical explanation for such a discontinuity is the resonance between the fourth and the first overtones.

Walraven – on the anti-smoking path?

THE STORY OF AI VELORUM: A REVIEW

Th. WALRAVEN, and J.H. WALRAVEN
920 Keyter Street
Daspoort 0082
South Africa

ABSTRACT. We give an historical account of our photometric observations since 1951 of the δ Sct type double-mode pulsating star AI Velorum. Our method using the *mean light surface* is explained and the derivation of pulsation periods, amplitudes and phase-shifts illustrated. More recent results obtained through Fourier analysis of the same data set confirm our earlier results.

Keywords: AI Velorum – pulsating stars

1. Introduction

The deepest root of our interest in pulsating stars goes down to 1938 when the first author, as a student of Professor Antonie Pannekoek, investigated high-dispersion spectra of δ Cephei at the University of Amsterdam. The results showed that the motions of the atmosphere resembled a running wave or even a shock wave. Pannekoek liked to compare this with ocean waves which are sinusoidal when still far away, but approaching the beach, get more and more steep fronts and finally topple over as surf. While this work was in progress, the first author married the second author who, from the start, assisted her husband in his work. This was the beginning of a lifetime of very close cooperation and of shared interest especially in pulsating stars.

After the war we obtained a position at the Leiden Observatory. Professor J.H. Oort, having just received from Gerald Kron a box of 931A photomultipliers as war surplus, asked us to try these for photoelectric observations. We started observing RR Lyrae and, favoured by—for Holland—unusually clear, weather, we worked night after night and collected a remarkable series of light curves which showed, in sharp detail, the strong variations in amplitude and phase already known as the Blazhko effect.

Jan Oort was enthusiastic about these results and asked us to start photoelectric photometry also at the Leiden Southern Observatory in Johannesburg. Professor Ejnar Hertzsprung, hearing about this plan, asked us to have a close look at AI Velorum. He had discovered the variability of this star on plates taken by H. van Gent with the Franklin Adams camera. By just looking at these plates he estimated the brightness and was able to determine, from several high maxima, a good fitting

C. Sterken and M. de Groot (eds.), The Impact of Long-Term Monitoring on Variable Star Research, 229–246.

period of $0\overset{d}{.}1115741 \pm 0\overset{d}{.}000002$. However, he was puzzled by the strong irregularity in the light curves. We promised to observe AI Vel, not knowing then that we were about to study the most fascinating pulsating star in the sky.

2. The Observations of AI Velorum

After our arrival in South Africa we mounted a photometer on the Rockefeller astrograph and started observing in January 1951. We soon noticed that AI Velorum was not irregular at all but showed, like RR Lyrae, a modulation of amplitude and phase. Only, in this case, the modulation was much faster, taking only $3\frac{1}{2}$ primary cycles, whereas RR Lyrae shows a beat of 72 cycles, and the modulation in amplitude was much deeper in AI Velorum.

It was clear that we were seeing an exceptional case of interference of two pulsations of nearly equal strength, so that strong light variations alternated with nearly standstills. Moreover, while in Holland we had been excited by a hump which occasionally appeared on the rising branch of the light curves of RR Lyrae, AI Velorum showed many humps and every time at other places and with other shapes. We realised that it was urgently necessary to make a complete study of this remarkable star. But there was a serious problem. The astrograph constructed by Grubb and Parsons, a ship-building company, was very strong and heavy and was intended for photography, not for rapidly moving to and fro. Our experience from the nights during which we had observed had taught us it was humanly impossible to observe the whole night every night and as quickly as was necessary. Considering that we were still in the cloudy season and that better skies were to come a few months later, we decided to construct rapidly an automatic photometer.

So the second author ground and polished a precision wedge of neutrally-tinted glass while the first author made a servo system in which the wedge is moved in the beam of the brighter comparison star and thereby weakens it until it has the same brightness as the variable star. Now observing became really easy. Once the stars are adjusted in their respective diaphragms the observer can sit down and watch the light curves being registered. Only an occasional look in the finder was required to check the guiding of the telescope and, because the polar-axis wormgear is very precise, a check every quarter of an hour was sufficient.

An unexpected advantage of the instrument soon became apparent. It enabled us to observe on partially clouded night. As every photometrist knows, observing in such nights is very frustrating. Not so with the wedge photometer. With clouds in front of the stars the servomotor runs wildly up and down, full scale. But as soon as both stars are clear the registering pen instantaneously resumes the scribbling of a piece of useful registration. So we used to observe every night even when we saw only a few stars between clouds and this considerably increased the amount of data per season. We resumed observations in April 1951 and continued in 1952 and 1953.

In 1952, O. Eggen discovered the variability of the star HD 223065, later renamed SX Phoenicis. We observed this star in 1952 and it turned out to be another case of double-period pulsations, although the modulation is not as deep as for AI Velorum. In Fig. 1 some registrations, as obtained with the wedge-photometer,

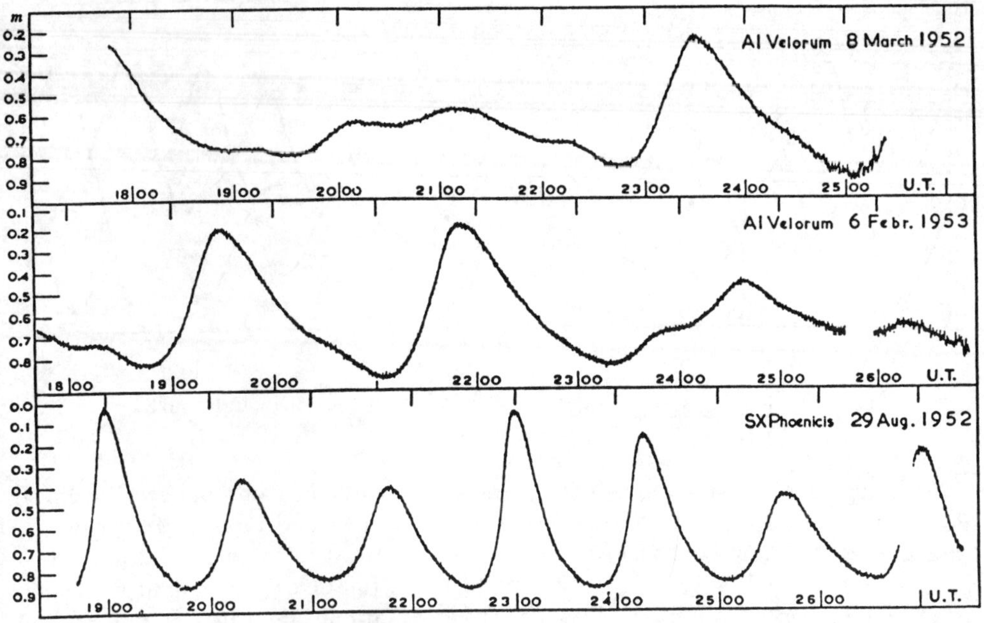

Fig. 1. Sample records of light curves. The noise is caused by seeing, it increases when the star approaches the horizon

are shown.

After collecting the 1951 observations we studied the results and determined first the periods P_0 and P_1 of the two interfering pulsations. Hertzsprung's period $P_0 = 0^{\rm d}\!.111574$ was confirmed. The beat period P_b was found from all observed maxima and minima and we got $P_b = 0^{\rm d}\!.379182 \pm 0^{\rm d}\!.000010$. Then, using $1/P_b = 1/P_1 - 1/P_0$ we obtained $P_b = 0^{\rm d}\!.0862075$.

With the two phases, $\phi_0 = t/P_0$ and $\phi_1 = t/P_1$ known, all observations were grouped together in a two-dimensional array. After smoothing, this gives the mean magnitude in the form of a table with twenty rows for ϕ_0 in steps of 0.05 and twenty columns for ϕ_1, also in steps of 0.05. The 400 data points give what must be named the "mean light surface" in analogy to the "mean light curve" of a single-period variable. With the mean light surface known we could determine residual light curves where the mean value is subtracted from the observed magnitudes. The residual curves showed the presence of small waves, with amplitudes of a few hundredth of a magnitude, which caused the humps seen in the light curves. We could determine two periods, i.e. $P_2 = 0^{\rm d}\!.0464600$ and $P_3 = 0^{\rm d}\!.0317584$.

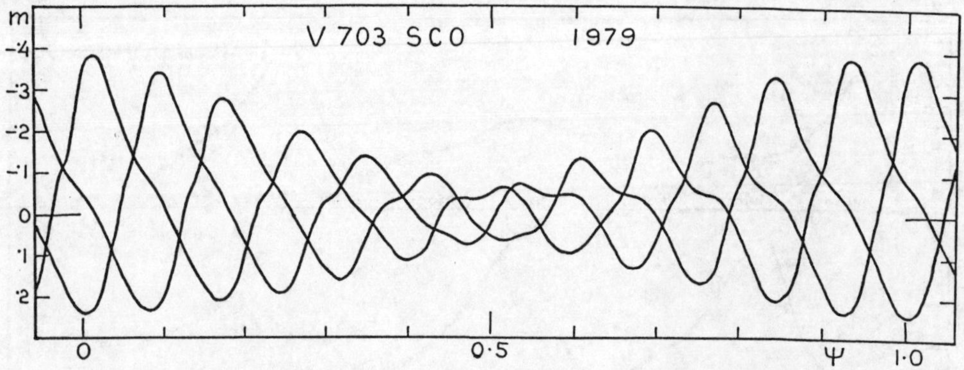

Fig. 2. V703 Sco, a set of light curves read from the mean light surface

Although 2317 observations were used, we concluded in our paper (Walraven 1952) that more observations were necessary. We obtained more observations in 1952 and 1953. Then we returned to Leiden where we studied them and published a paper (Walraven 1955). In 1952 the amplitude of the pulsation P_1 had weakened considerably, in 1953 still more. Likewise the small-amplitude waves P_2 and P_3 had diminished and were more difficult to see. The overall picture we had of AI Velorum was one of extreme complexity. There are three, perhaps four, different oscillations and moreover their strengths vary over the years.

We were busy for several years with designing a new telescope, equipped with a five-channel photometer, the *VBLUW* system. This telescope was so good that it was overloaded with programmes of Leiden observers so that we could not observe AI Velorum with it. The next observations were obtained with the wedge photometer which had been severely damaged during the transport of the astrograph from Johannesburg to a new site but had been reconstructed again in 1978. In 1979 we collected with it the largest amount of data of all seasons. We also observed V703 Sco in August 1979. A year later the Leiden Southern Station was closed down, and at the same time the first author reached the age of retirement.

According to the 1979 observations AI Velorum had become slightly more active than in 1953 but we could not detect the waves P_2 and P_3. We were curious as to whether the star would return again to the strong modulation of 1951. For observing such an event we built our own observatory. The telescope, with a 40-cm primary mirror, carried a six-channel photometer, on the *RVBLUW* system, based on the same principles as the *VBLUW* photometer but now including a red channel *R*. All optical parts, for both the 5-channel and the 6-channel photometer, were made by the second author.

In 1987 we observed AI Velorum again and it appeared that the modulation in amplitude had become nearly as strong as in 1951. We then experienced problems with the reduction of the observations, now in 6 channels, whereas all previous observations had been in blue light only. Nevertheless, some results have been obtained because the variations in radius and gravity could be found using the colour indices $B - R$ and $U - B$. We submitted a short paper on this subject at the Conference in Bologna (1990).

At this conference Dr. L.A. Balona was present and he kindly offered to help with the reductions. And he did so effectively. He analysed 13394 observations and firmly established the presence of the wave P_2 in all seasons, including those where we could not detect it with our method of reduction. Moreover, he could identify this wave with the fifth radial overtone, which is a very important result because large growth rates for high overtones had been predicted by Cox and by Stellingwerf but they were hitherto not observed in stars. This and other results have been published in a paper (Walraven & Balona 1992).

3. The mean light surface

A graphic representation of a light surface can only be given as a systematic set of cross sections. We actually made such graphs, while the mean light surface was constructed from the observations. From these graphs the table with twenty rows (for ϕ_0) and twenty columns (for ϕ_1) was obtained.

But for the study of the mean light surface such ϕ_0, ϕ_1 cross sections are not suitable because the curves are totally unlike the observed light variations. It is much better to study cross sections read diagonally from the table—that is, making graphs for fixed values of $\psi = \phi_1 - \phi_0$ and as a function of $\chi = \frac{1}{2}(\phi_1 + \phi_0)$. From the table twenty such graphs can be read.

In the upper part of Fig. 3 we show a set for ten values of ψ only, otherwise the picture would be overcrowded with lines. Now the curves are systematic and periodic and yet they resemble the actual light variations such as shown in Fig. 1. The shape of the high light curves, $\psi = 0$, and the flat light curves, $\psi = 0.5$, is now clearly demonstrated. It may be noted that the curve $\psi = 1$ is the same as for $\psi = 0$ but is shifted in χ by half a cycle. The set shown in Fig. 3 is derived from the 1979 observations.

In the lower part of Fig. 3 a set of synthetic light curves is also shown, computed from a model. The model used is a transformation of the sum of two sine curves; acronym $TS2S$. As is fully described in our paper (Walraven 1952) a decomposition of flat light curves produced two components which both had a shape of pure sine curves. Typical values for the semi-amplitude of these components is $a = 0\overset{m}{.}17$ and $b = 0\overset{m}{.}12$. The sum of the two components, $S2S$, we write as

$$u = a sin 2\pi\phi_0 + b sin 2\pi\phi_1 \tag{1}$$

where u stands for undistorted. The $S2S$ reproduces the observed light-curves, i.e. $-m = u$, as long as u is zero or near to zero. We consider the $S2S$ as the basic, underlying pulsation of the star. In our paper (Walraven 1955) we described how a systematic transformation is sufficient to convert the $S2S$ into the observed light curves $-m$, also when u is large.

We consider the transformation as the anharmonic distortion, a well-known

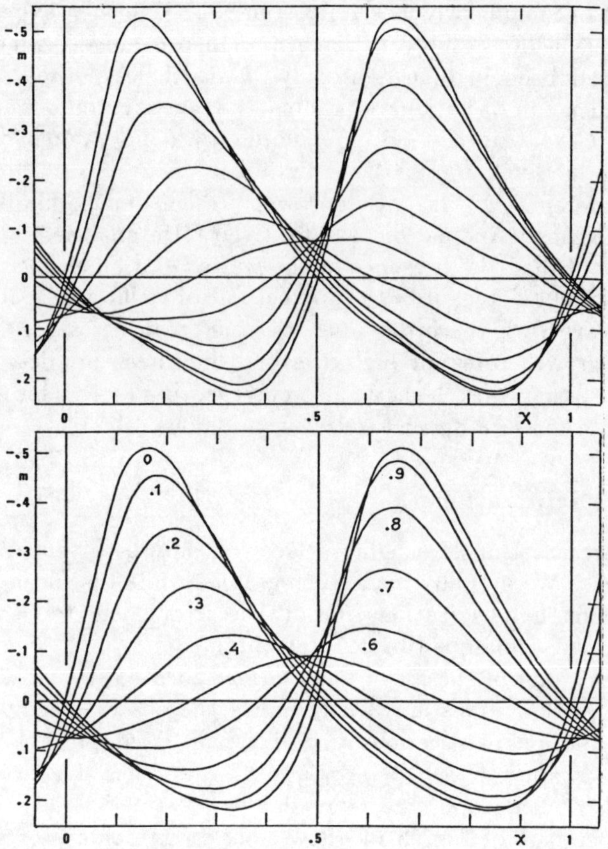

Fig. 3. Cross sections through light surface for ten values of ψ as a function of χ. Top: observed; bottom: model

name in acoustics, which grows when $S2S$ deviates more from zero. How the transformation works is illustrated in Fig. 4. A shows the two basic pulsations; in B they are added to $S2S$. In C each point is shifted in time, proportional to u. Since this makes the curves skewed we named this part of the transformation the S-distortion. The curves now have the same shape as the radial-velocity variations of the star (Gratton 1953). Finally, in D, a shift in magnitude is made, called the M-distortion, and this gives the model light surface, $TS2S$.

At first sight there is not much difference between the two parts of Fig. 3. The observed light surface is somewhat buckled at some places although less than the light curves shown in Fig. 1. Most of the humps in Fig. 1 are caused by small-amplitude waves which run independently of ϕ_0 and ϕ_1 and which are averaged out in the mean light surface. The remaining humps have a fixed position on the light surface.

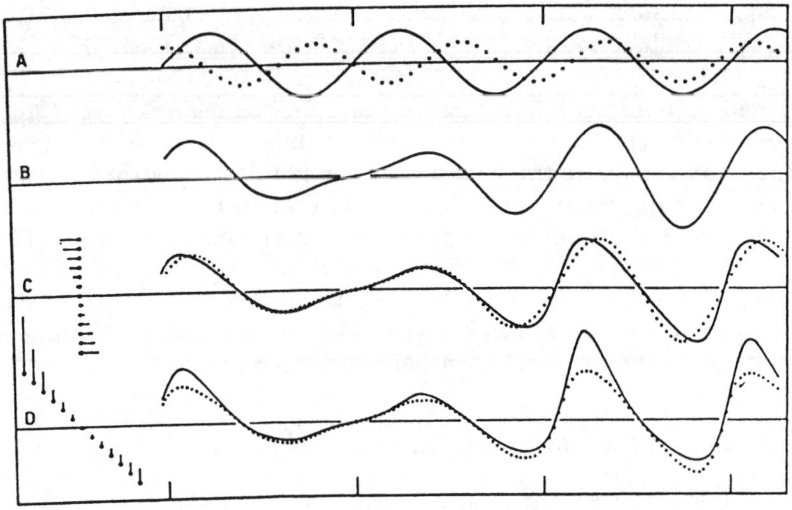

Fig. 4. Construction of radial-velocity curve and light curve from two harmonic pulsations. A: harmonic curves; solid line = $a sin 2\pi\phi_0$; dotted line = $b sin 2\pi\phi_1$; B: sum of curves in A; C: curve B shown as dotted line is transformed into radial-velocity curve by displacement in time proportional to height (S-distortion); the way in which the displacement varies with height is shown on the left; D: curve C shown as dotted line transformed into light curve by M-distortion; the vertical lines in the left-hand corner show how the M-displacement varies with height

The situation shown in Fig. 3 can be summarized as follows: 95% of the light surface is accounted for by the $TS2S$ model. The remaining 5% is a not yet solved problem. We could dismiss these residuals as observing errors. But we had a similar situation before with the residuals of observed light-curves from the mean light surface. We could see small-amplitude waves running independently of ϕ_0 and ϕ_1 but with much variation in certainty, both from one night to another and from one year to another. In 1979 we did not see these waves at all: the residuals were entirely irregular. But later, Balona made an exhaustive investigation on a powerful computer and established the reality of these waves for all seasons and discovered even more such waves. His results show that the amplitude has dropped from 16 millimagnitude in 1951 to only 3 millimagnitude in 1979. It follows that our method of study has a limit of detection somewhere near 10 millimagnitude.

Now, studying the residuals of the mean light surfaces from the $TS2S$ model we have the same situation. In nearly every season we see three humps in one cycle χ. They are more pronounced for $\psi = 0.7$ or 0.8 and very uncertain for $\psi = 0.2$. But the irregularities are such that we cannot describe the humps by a clearly defined set of periods. The problem is aggravated by the fact that small changes in the parameters of the $S-M$ distortion give sharp changes of the residuals, especially on the rising branch and maximum of the high light-curves. We shall go on studying the residuals, but we shall not discuss this matter further in this paper.

With regard to the 95% agreement between the observations and the $TS2S$ model it must be stated that the concept of the $TS2S$ model was forced upon us by AI Velorum itself. This star is unique amongst pulsating stars in showing the complete range from small, undistorted to large-amplitude and strongly distorted pulsations. The most important conclusion we can draw is that the amount of anharmonic distortion depends only on the amount of excursion from the equilibrium state, or, it is irrelevant by which combination of phases ϕ_0 and ϕ_1, the excursion is produced. Another way of saying this is: there is no interaction between the two pulsations or, at least, the interaction is very weak. The two basic pulsations make each other anharmonic by not more than ten millimagnitude.

4. Fourier Analysis of Mean Light Surface

In the past few years we used the Fourier method to study the light surface. Although we cannot use the thousands of observations per season, our 64K computer can handle the two-dimensional analysis of our 20 by 20 point data grids. For each of the twenty values of ϕ_0 we have twenty data points as a function of ϕ_1. We fit through these an expression:

$$A_0 + \sum(A_k cos 2\pi k\phi_1 + B_k sin 2\pi k\phi_1)$$

where k runs from 1 to n, the order of the analysis. This gives $2n + 1$ components A_k or B_k or A_0. We then apply the same program to each component considered as a function of ϕ_0. This gives $(2n+1)^2$ components which, after suitable arrangement and combination, give the two-dimensional form:

$$A_{(0,0)} + \sum A_{(i,j)} sin 2\pi(i\phi_0 + j\phi_1 + \Phi_{(i,j)})$$

where the indices i and j run over all positive and negative integers, including zero, subject to the restrictions:

$$0 <\mid i \mid + \mid j \mid \leq n; i + j \geq 0 \text{ and } j > 0 \text{ when } i + j = 0.$$

This gives us a constant $A_{(0,0)}$, $n(n+1)$ amplitudes $A_{(i,j)}$ and $n(n+1)$ phases $\Phi_{(i,j)}$.

It may be noted that in the expression we use a ***sin*** instead of the customary ***cos***. This gives less confusing phases in connection with formula (1). We repeated the procedure, with ϕ_0 and ϕ_1 interchanged and this produced identical results. Again the same results were obtained by analysing first according to χ and then according to ψ.

We performed the analysis with $n = 4$ as Balona had shown that there are no significant components of higher order. So we get, besides the constant, twenty amplitudes and twenty phases. By combining the components again and subtracting the result from the original data, residuals are obtained. The RMS value of the residuals over the entire light surface is found to be of the order of 0.007 magnitude but the residuals are rather systematic. When plotted as a function of χ for fixed values of ψ, the residuals for the high light-curves show five waves which are narrow and high on the ascending branch. This wave train has much the same shape as the illustrations found in handbooks where the Gibbs phenomenon is discussed.

Peak oscillations on the steep rising branches of $0^{m}_{.}03$ or $0^{m}_{.}04$ semi-amplitude are common. So the Fourier analysis cannot handle the steep rising branches.

It should be remarked that the steep rising branches are also difficult to handle as the residuals of observed light-curves from $TS2S$ models and these often show the same oscillations. This shows that an adequate study of steep rising branches requires a special technique if one wants to study such things as humps there. This localized problem is not important if one studies the amplitude and phase of the strong low-order components. Even so, care must be exercised in the use of these quantities.

By studying Fourier decompositions of $TS2S$ models we discovered that the first-order components have a larger amplitude than the basic $S2S$ components and, also, that they are displaced to an earlier time. On the other hand, the second-order components arrive later in time.

One can easily understand these phase shifts by considering the sine wave which is subject to S and M distortion. The application of S distortion to a single sine wave requires the addition of a wave with double frequency but with the same zero point. The Fourier decomposition then finds as first harmonic the original sine wave and as second harmonic the added wave, both with phase zero. If S distortion increases, a third harmonic is added, again with the same zero point. But if M distortion is applied to the $S-$distorted wave, both maximum and minimum are raised. Since these two places are both close to the zero point of time, the Fourier decomposition adds a cosine component to the first harmonic; that means it shifts to an earlier time. However, the curve must remain zero at the zero point of time and, therefore, to compensate for the increase caused by the cosine component in the first harmonic, a negative cosine component is added to the second harmonic which, therefore, shifts to a later time. If all harmonics are plotted in a cosine-sine diagram the points are seen to form a kind of spiral. However, this spiral is a prerogative of the light curve only. The radial velocity and gravity show in a Fourier decomposition the same phase for all components.

The first harmonic of the light curves for constant ψ —such as shown in Fig. 3— is shown as a vector in a phase diagram in Fig. 5. The abscissa is the *cos* component and the ordinate is the *sin* component. The origin is denoted by a circle. The little dots indicate the vectors for the various values of ψ, which are indicated for SX Phe only. For other stars the ψ values are in corresponding positions.

For increasing values of ψ the vectors rotate in anticlockwise direction, except for V703 Sco where the vectors rotate in opposite sense because in that star the amplitude of the first overtone is stronger than that of the fundamental mode. Starting with the largest light-curve for $\psi = 0$, represented by a dot on the top, for successive values of ψ the amplitude decreases as the phase decreases. The smallest amplitude is at $\psi = 0.5$ where the phase is -0.25. After that the amplitude increases again while the phase continues to decreas until full amplitude is reached again at the bottom.

The S-shaped median line is more or less the same in all diagrams and indicates the increased phase for the high light curves, which is an artifact caused by $M-$distortion and has no physical meaning. The variations in relative strength of the amplitudes of the two modes is clearly seen in different seasons for AI Velorum

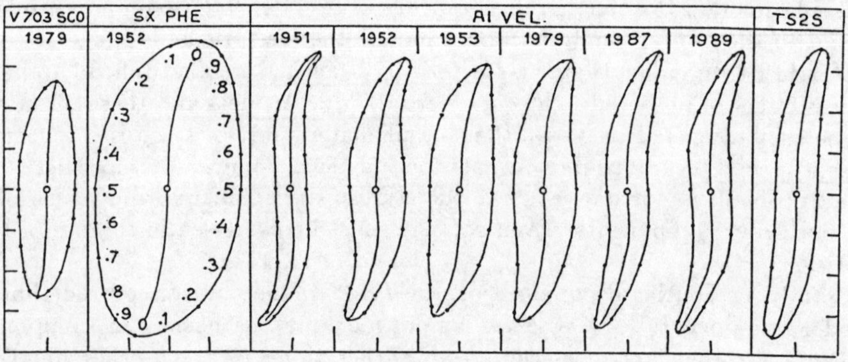

Fig. 5. Phase diagram of first harmonic of light curves for constant ψ. Horizontal: the *cos* component; vertical: the *sin* component. Scale divisions are 0.1 magnitude. ψ is indicated in loop of SX Phe only and is the same in other loops, except for V703 Sco which is described in clockwise direction

and in different stars.

A phase diagram of a *TS2S* model is included to show a small difference in phase relative to the observations which is a true deviation from the *TS2S* model, described in our paper (Walraven 1955) as a small shift in phase, periodic in ψ and with amplitude 0.015 in χ. In an attempt to find residuals between observed light surfaces and *TS2S* models, we computed for each season, but always with the same *S* and M distortion, the best fitting *TS2S* model by varying the amplitude and phase of the *S2S* components until the Fourier decomposition of the model produces the same $A_{(1,0)}, A_{(0,1)}, \Phi_{(1,0)}$ and $\Phi_{(0,1)}$ as found for the observed light surfaces. In this way we can eliminate the phase errors which are present in the mean light surfaces. Such errors are present because, when the tables of the mean light surface were constructed, we knew the phases, from maxima and minima, with limited accuracy. We found, as the mean of all seven seasons

1. the phase of $A_{(1,0)}$ is $+0.0150 \pm 0.0009$ cycle or $+5.40°$ and the phase of $A_{(0,1)}$ is $+0.0238 \pm 0.0007$ cycle or $+8.57°$ at the zero point of the *S2S* components;
2. the amplitude $A_{(1,0)}$ is 1.117 times a of *S2S*; the amplitude $A_{(0,1)}$ is 1.121 times b of *S2S*.

The ratios are not constants but increase slightly when $a + b$ is larger.

Let us now consider the second-order components which are $A_{(2,0)}, A_{(1,1)}, A_{(0,2)}$ and $A_{(-1,1)}$. The numbers in the table are the mean of all seasons. The amplitudes are in magnitude and the phases in cycles referred to the zero point of *S2S*. It may be noted that the three components $A_{(2,0)}, A_{(1,1)}$ and $A_{(0,2)}$ are all shifted to a later time, in accordance to the shift mentioned above of the second harmonic of a single distorted wave. The difference between the phase shifts are sufficiently small to allow the following simple reasoning. At $\psi = 0$ the three amplitudes can be added as $A_{(2,0)} + A_{(1,1)} + A_{(0,2)} = 0^{m}_{.}1276$. A one-dimensional analysis of the light curve $\psi = 0$ gives us the amplitude of the second harmonic $0^{m}_{.}1310$.

At $\psi = 0.5$ both $A_{(2,0)}$ and $A_{(0,2)}$ have opposite phases relative to $A_{(1,1)}$; so now the three components are added as $A_{(1,1)} - A_{(2,0)} - A_{(0,2)} = -0^m_.0068$. A one-dimensional analysis of the light curve for $\psi = 0.5$ gives as the amplitude of the second harmonic $+0^m_.0057$.

The sign seems incorrect but, at $\psi = 0.5$ the first-order components $A_{(1,0)}$ and $A_{(0,1)}$ combine so as to give a low-amplitude sine wave which is shifted by $\frac{1}{4}$ cycle in ψ (see Fig. 3) which is half a cycle for the second harmonic; so the sign is correct.

Component		$A_{(1,0)}$	$A_{(0,1)}$	$A_{(2,0)}$	$A_{(1,1)}$	$A_{(0,2)}$	$A_{(-1,1)}$
Combination		ϕ_0	ϕ_1	$2\phi_0$	$\phi_0 + \phi_1$	$2\phi_1$	$\phi_1 - \phi_0$
		$\chi - \frac{1}{2}\psi$	$\chi + \frac{1}{2}\psi$	$2\chi - \psi$	2χ	$2\chi + \psi$	ψ
Observed	ampl.	0.1906	0.1360	0.0413	0.0604	0.0259	0.0400
	phase	+0.0150	+0.0238	-0.0796	-0.0674	-0.0303	+0.2893
TS2S	ampl.	0.1906	0.1360	0.0430	0.0635	0.0264	0.0385
	phase	+0.0150	+0.0238	-0.0735	-0.0647	-0.0310	+0.2854

So for both the highest and the flattest light curves the three second-order components cooperate to produce the right amount of anharmonic distortion which is necessary to reproduce the observed shapes. Here only the sum $A_{(2,0)} + A_{(0,2)}$ was used. It is probable that the specific values of these two components are required to account for the shapes of light curves other than for $\psi = 0$ and $\psi = 0.5$. Evidence for this is the agreement with the three amplitudes found for the *TS2S* model.

So it seems that the task of the three second-order components is only to produce correct shapes for all light curves. Further finishing touches are provided by the third-order and higher components. But by adding sine waves together, the average value of the curve, over the full range of χ, remains zero, so the shape may be all right but the mean level is not. As is seen in Fig. 3, the mean level of the high light-curves is raised due to the M–distortion. The Fourier component $A_{(-1,1)}$ takes care of this raising of the level. As seen in Table 1, the phase of this component is near to 0.25; that means it is a cosine component, high positive at $\psi = 0$.

Considering that the four second-order components are similar for observations and models, we may conclude that it is certain that the task of the four second-order components is only to reproduce correct shape and level for all the light curves. The same can be said for the six third-order components. They all show differences "observed minus model" of some millimagnitudes only and these vary erratically in different seasons. There is one exception. The component $2\phi_0 - \phi_1$ has a difference of ten millimagnitude and with the same phase in all seven seasons. This may be a real interaction. So the Fourier components are mainly ghosts of the two real spectral lines $A_{(1,0)}$ and $A_{(0,1)}$ introduced in the spectrum by anharmonic distortion.

5. The Radius and Gravity

Multicolour observations can be used to study the variations of radius and effective gravity. In Fig. 6 some *VBLUW* observations of short-period variables are shown

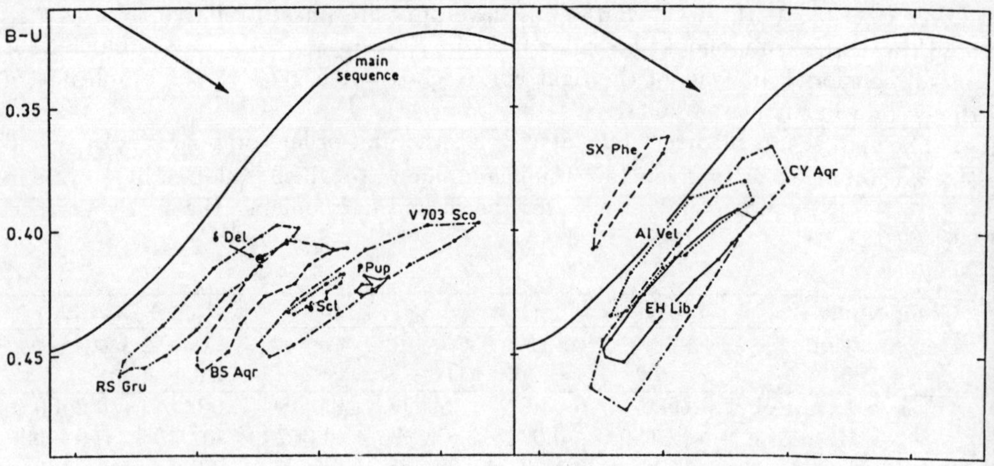

Fig. 6. Loops of short-period variables in $V - B$, $B - U$ diagram. Periods between $0\overset{d}{.}12$ and $0\overset{d}{.}20$ in the left part, periods between $0\overset{d}{.}05$ and $0\overset{d}{.}12$ in the right-hand part.

on a two-colour diagram. All variables, either single-period or double-period, describe loops in counter-clockwise directions. The arrows indicate in which direction the loops are displaced when the star is reddened by interstellar dust. AI Velorum and SX Phoenicis are not reddened. The peculiar position and slope of the loop of SX Phoenicis indicate that it is a Pop II object.

Note that the loops all have roughly the same ratio of width to length and that the slope is the same for a given period. Fig. 6 shows what kind of information is obtained by observing many stars each for a short time. In Fig. 7 it is shown what you get by observing one star for a long period. This figure shows where AI Velorum moves in the two-colour diagram $U - B$ versus $B - R$ according to observations with our *RVBLUW* photometer in 1987, 1989 and 1992. Here again we do not follow time because that would produce an inextricable tangle. But by using constant values of ψ we obtained a closed loop for every value of ψ. The largest loop is for $\psi = 0$, or $\psi = 1$. At the top, loops for $\psi = 0$ through $\psi = 0.5$ are shown and at the bottom the loops for $\psi = 0.6$ through $\psi = 1.0$.

All loops are described in anticlockwise direction. The right-hand tip of a loop corresponds to minimum light and temperature. From there the upper part is quickly described while the light increases until, at maximum light and temperature, the left-hand tip of the loop is reached. Then, much more slowly, the lower part of the loop is described, from left to right again. The loop for $\psi = 0.1$ is only slightly smaller and for a large part coincides with the loop $\psi = 0$. For larger ψ values the loops become smaller and for $\psi = 0.5$ it is reduced to a crinckle. It may be noted

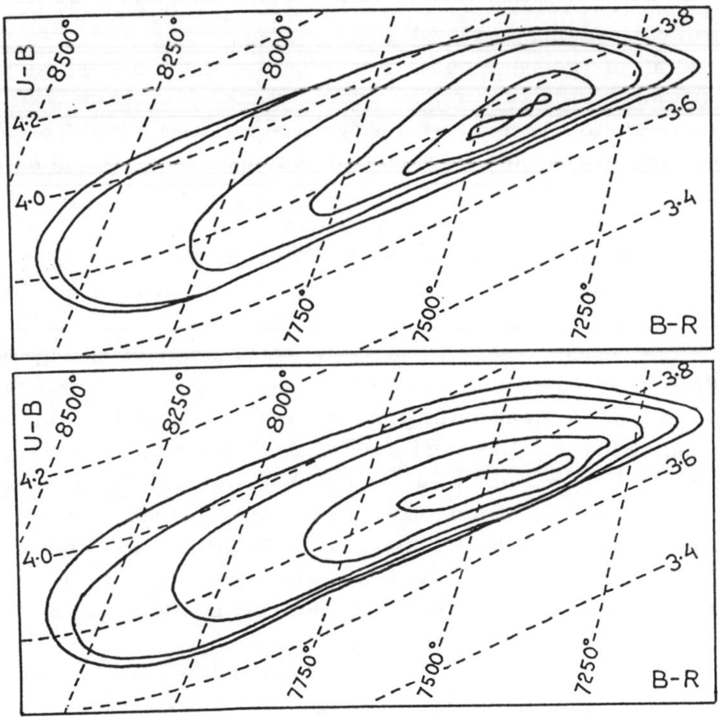

Fig. 7. AI Velorum describes loops in anti-clockwise direction as a function of χ, for fixed values of ψ. Top diagram shows loops for $\psi = 0.0, 0.1, \ldots$ to $\psi = 0.5$, bottom shows loops from $\psi = 0.6, 0.7, \ldots$ to $\psi = 1.0$. Theoretical colour temperatures computed by Kurucz give lines of constant temperature, T_{eff}, and of constant $\log g_{\text{eff}}$. The length and height of the frames are 0.4 and 0.2 magnitude

that for the medium-sized loops there is a systematic difference in shape between the two figures.

Also shown in Fig. 7 is a grid of lines for constant values of T_{eff} and $\log_{\text{eff}}$. The grid is based on computed fluxes from model atmospheres by Kurucz (1979). The Kurucz grid has been carefully matched to the Walraven photometic system by Pel (1978). Also a few observations of AI Velorum have been made in that system (Oosterhoff and Walraven 1966). From these observations, assuming no interstellar reddening, we estimated the position of the equilibrium point in the Kurucz grid. This gave $T_{\text{eff}} = 7450°$ and $\log g_{\text{eff}} = 3.75$. We used these values to match the colours in Fig. 7.

The $\log g_{\text{eff}}$ as read in the Kurucz grid is plotted against phase χ in Fig. 8. We see here a dramatic confirmation of our finding from the light curve that the basic pulsations are two sine waves. In this case the waves have equal amplitudes and are cosine waves. At first sight it may seem strange to associate the sharp peaks with cosine waves but just for this reason it is so remarkable that the observed variation can only be interpreted as the result of the addition of two equal cosine waves. This conclusion is firmly based on two aspects:

1. the peaks are all at the same phase $\chi = 0$ for $\psi =< 0.5$ and all at the same phase $\chi = 0.5$ for $\psi > 0.5$;
2. apart from some ripples there is no gravity variation at $\psi = 0.5$. These aspects are both produced by formula (1) where *sin* is replaced by *cos* and b by a with the result
$$u = 2acos2\pi\chi \; cos\pi\chi \tag{2}$$
This formula shows both the behaviour in phase, as a function of ψ, and the complete disappearance of amplitude at $\psi = 0.5$, exactly as in the gravity variations.

The transition from cosine waves to peaks can be understood as follows. The variation in effective gravity on the surface of a pulsating star consists almost entirely of acceleration. A small term $g_0(r_0/r)^2$ must be added to take account of the variation in radius r. As stated earlier, the radial-velocity variations are represented by skewed waves, obtained from sine waves subject to S-distortion. In order to obtain acceleration from velocity we must differentiate in time. The time derivative of the skewed velocity curves gives the peaked acceleration curves. At the same time sine waves are changed into cosine waves and also the ratio of the amplitudes b/a is changed to bP_0/aP_1. By this change the amplitudes become equal.

So the remarkable shape of the variations of gravity, the sharp pulses and the phase jump, can be fully described by a $TS2S$ model. Of course, the question arises, If the velocity and acceleration agree so well is there also a close relation between velocity and radius? And indeed there is. We can study the radius variations in AI Velorum using the Kurucz grid. This given the effective temperature. The difference between observed brightness and the theoretical brightness that is given by the temperature indicates the change in area of the star.

The radius effect is strongly disturbed by photometric errors but nevertheless the least-squares Fourier analysis brings out the amplitude and phase of the radius variations of the two pulsations. The three seasons 1987, 1989 and 1992 give consistent results. It appears that the radius variations have the shape of negative cosine waves, that they have the same phase as the gravity variations, and that the ratio of the amplitudes for the two pulsations is equal to 0.60 ± 0.04. This shows that they can be considered as the result of integrating the velocity variations. We have now arrived at a remarkably simple, yet consistent, picture of the pulsations in radius of AI Velorum. The radius varies like negative cosine waves with minima sharper than maxima. The ratio of the amplitudes is 0.60. Differentiating the radius gives the velocity as skew sine waves, with ratio of amplitudes $0.60 \times P_0/P_1 = 0.77$. Differentiating again gives the acceleration which are strongly distorted cosine waves and the ratio of the amplitudes is $0.77 \times P_0/P_1 = 1.0$.

Furthermore, the anharmonic distortion is apparently quite different in each case. But the changes are automatically made by the differentiation. So, basically, there is only one kind of anharmonic distortion.

We presented a short paper about this result at the conference at Bologna on the confrontation between stellar pulsation and evolution (1990). In this we commented

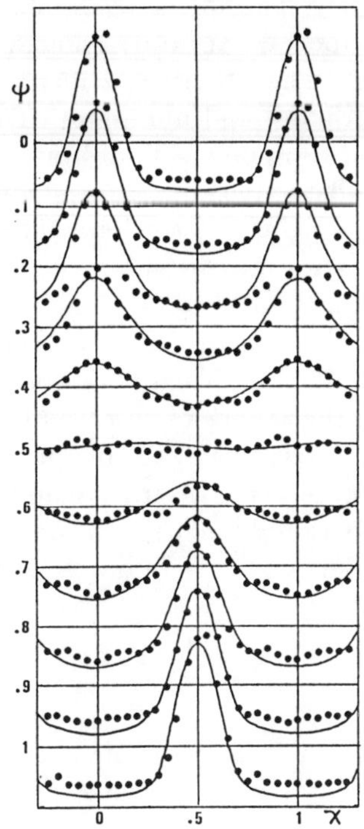

Fig. 8. $log g_{\rm eff}$ read in Kurucz grid for fixed values of ψ as a function of χ. Solid lines: $TS2S$ model. The dots are shifted by 0.06 cycle χ to the left. The horizontal lines correspond to $\log g_0 = 3.75$, i.e. the equilibrium state. The separation between horizontal lines is 0.15 in $\log g$

on the fact that the gravity pulses are about 0.06 cycle late with respect to the light curves. At that time we still believed that light curves have the same phase as radial-velocity curves and were puzzled by this shift. Since then we read in a paper by Balona and Stobie (1980) about the phase shift between radial velocity and light. This shift was even predicted theoretically by Stellingwerf (1979). So it turns out that it is not the gravity pulses which are late, but the light curves which are too early. The shift found by Balona and Stobie (1980) is 0.05 cycle.

The light curves must be considered as deviating from the radius variations by arriving too early and by showing $M-$distortion besides $S-$distortion. Obviously, the process which produces the light variations is much more complicated involving periodic ionization and recombination of hydrogen. However, the ratio of the amplitudes of the light variation is close to that of the velocity variations.

Our conclusion about the close relation between radius, velocity and acceleration is based, with regard to the velocity, on very few observations and on indirect evidence from light curves. It seems that in the near future this shall be remedied considering the truly amazing radial-velocity curves published by Matthews in the Delta Scuti Newsletter (1991).

Another aspect following from the Balona and Stobie (1990) paper is that the radial velocity cannot be precisely reproduced by $S-$distortion alone. The radial-velocity curves shown in this paper are accurate enough to show a clear deviation from an S-distorted sine wave. The observed velocity curves have a more pronounced and more pointed maximum and minimum. This we could reproduce by introducing a vertical distortion component in addition to the $S-$distortion. This component we designate as $V-$distortion since it is an effect present in the velocity curves and has nothing to do with $M-$distortion. It consists of adding a third-degree term to u after $S-$distortion has been applied. We have not noticed this $V-$component in the old, inaccurate observations of radial velocity.

The $TS2S$ curves shown as full lines in Fig. 8 are computed using $S2S$, applying $S-$ and $V-$distortion and differentiating. The correction term $g_0(r_0/r)^2$ was added. The agreement between the shapes of observed $\log g$ and the model is much improved with respect to $TS2S$ curves using $S-$distortion only.

It should be noted that a comparison of gravity and integrated radial velocity such as shown in Fig. 8 leads to a fully independent and accurate determination of equilibrium gravity. Future observations of radial velocity with an instrument as used by Matthews are urgently required.

6. Conclusions

AI Velorum shows a beat-effect which can be explained by simultaneous pulsations in two different periods. The periods have a ratio $P_1/P_0 = 0.7726$.

It is now generally accepted that P_0 represents radial pulsation in the fundamental mode of P_1 radial pulsation in the first overtone. Several other such double-period stars are known but AI Velorum is the only one which, in its light variations, exhibits nearly equal amplitudes of pulsation. At the phase of the beat period where the two pulsations have opposite phase they cancel each other to such an extent that both pulsations must have a sinusoidal shape.

A detailed study of many years observation revealed that the large-scale features of all the light variations can be described as the anharmonic distortion of the sum of two sinusoidal pulsations. From this it is concluded that real interaction between the two pulsations does not exist or is so small that it cannot be detected. By subjecting the observation in the B band of six seasons to a Fourier analysis, Balona firmly established the presence of two independent radial pulsations with small amplitude, which varied by a factor of 5 in the course of the years. One of these periodicities could be identified by Balona as the fifth radial overtone, thereby confirming predictions by Cox (1974) and Stellingwerf (1979).

Observations in colours made in 1987, 1989 and 1992 enabled a study of radius variations and gravity variations. From this a consistent system could be established involving the three kinematical quantities radius, velocity and acceleration, each with their own anharmonic distortion. These three quantities are connected to each other, including distortion, by the process of differentiation or integration. It follows that there is one basic anharmonic distortion.

It turned out that the quantity which has been most extensively observed, i.e. light variation, is only indirectly connected to the variation of velocity, with the result that it is subjected to a different type of anharmonic distortion called

$M-$distortion and that the pulsations in light arrive slightly earlier in time than the velocity variation. This phase shift is present in the theoretical calculations by Stellingwerf (1979).

Attempts to resolve humps or wavelets on the light curves have been unsuccessful hitherto. Although some humps are persistently seen in all seasons, the irregularity of their shape and the variability over the seasons prevent a clear description in terms of periodic components.

7. Table of Fourier Components

The Fourier components of light curves, in the photometric band B, or blue light, are given for AI Velorum (1987/89/92), for SX Phe (1952) and for V703 Sco (1979), in millimagnitude. The phases $\Phi_{(i,j)}$ are reduced to the first-order components $A_{(1,0)}$ and $A_{(0,1)}$ to show more clearly the relative phases.

		AI Velorum 1987/89/92		SX Phe 1952		V703 Sco 1979		AI Velorum 1987/89/92			
i	j	$A_{(i,j)}$ $-m_B$	$\Phi_{(i,j)}$ (sin)	$A_{(i,j)}$ $-m_B$	$\Phi_{(i,j)}$ (sin)	$A_{(i,j)}$ $-m_B$	$\Phi_{(i,j)}$ (sin)	$A_{(i,j)}$ g/g_0	$\Phi_{(i,j)}$ (cos)	$A_{(i,j)}$ $5lr$	$\Phi_{(i,j)}$ (cos)
1	0	193.1	0.000	267.7	0.000	92.3	0.000	0.174	0.000	-25	0.00
0	1	146.3	0.000	82.4	0.000	161.3	0.000	0.174	0.000	-15	0.00
2	0	42.1	-0.098	81.4	-0.163	11.9	-0.138	0.062	0.007	-4	0.08
1	1	68.1	-0.107	45.1	-0.162	36.0	-0.036	0.136	0.018	-7	0.07
0	2	31.4	-0.077	11.1	-0.140	32.4	0.036	0.065	0.026	-2	0.08
-1	1	43.7	0.280	43.2	0.224	24.5	0.332	0.006	-0.08	3	0.03
3	0	10.6	-0.207	27.2	-0.296	1.9	-0.21	0.021	0.016		
2	1	22.9	-0.194	21.0	-0.305	5.0	-0.094	0.071	0.014		
1	2	16.2	-0.128	7.4	-0.29	13.2	-0.424	0.077	0.009		
0	3	5.0	-0.285	1.7	-0.30	6.4	-0.126	0.010	0.043		
2	-1	13.1	0.175	20.3	0.140	5.1	0.057	0.012	0.16		
-1	2	11.8	0.13	8.3	0.024	6.8	0.056	0.007	0.30		
4	0	2.2	-0.26	11.9	-0.389	0.9	-0.18	0.005	-0.02		
3	1	6.9	-0.279	11.6	-0.363	1.5	-0.32	0.030	-0.012		
2	2	9.3	-0.338	5.9	-0.42	3.3	-0.527	0.034	-0.01		
1	3	4.1	-0.238	1.4	-0.44	7.4	-0.549	0.027	0.002		
0	4	1.5	-0.46			0.4	-0.20	0.003	-0.19		
3	-1	4.1	0.08	9.7	-0.03	1.4	-0.14	0.004	0.04		
-2	2	0.7	-0.47	1.4	0.12	1.1	-0.32	0.011	-0.22		
-1	3	4.2	-0.18	1.3	-0.08	4.2	0.020	0.005	-0.08		

The components of SX Phe agree well with the Fourier analysis made by Rodriguez, Rolland and Lopez de Coca (1991). The comparison shows that the first overtone is more excited in 1988 than in 1952. V703 Sco is remarkable in that the first overtone is stronger than the fundamental mode.

The Fourier components of the gravity and radius variations are shown, for $5\log(r/r_0)$, in millimagnitude [1], and again with the phases reduced to those of the first-order components, now as cosine components.

Acknowledgements

We wish to express our gratitude and admiration to Dr. L.A. Balona for the remarkable feat of retrieving valuable information hidden in our observations. We wish to thank Dr. F. Walraven, our son, geologist and head of the Isotope Laboratory of the Geological Survey, who undertook the typing and preparation of the camera-ready version of this paper.

Dedication

This paper is dedicated to the memory of my wife, Johanna Helena, who died in January 1989.

Th. Walraven.

References

Balona, L.A., Stobie, R.S., 1980, *Monthly Notices of the RAS*, 192, 625
Cox, J.P., 1974, *Rep. Prof. Phys.*, 37, 563
Gratton, L., 1953, *Bulletin of the Astronomical Institutes of the Netherlands*, 12, 31
Pel, J.W., 1978, *Astronomy and Astrophysics*, 62, 75
Kurucz, R.L., 1979, *Astrophysical Journal, Supplement Series*, 40, 1
Matthews, J.M., 1991, Delta Scuti Star Newsletter, 3, 9. ed. Breger, M., Vienna
Oosterhoff, P.Th., Walraven, Th., 1966, *Bulletin of the Astronomical Institutes of the Netherlands*, 18, 387
Rodriguez, E., Rolland, A., Lopez de Coca, P., 1991, Delta Scuti Star Newsletter, 3, 17. ed. Breger, M., Vienna
Stellingwerf, R.F., 1979, *Astrophysical Journal*, 227, 935
Walraven, Th., 1952, *Bulletin of the Astronomical Institutes of the Netherlands*, 11, 421
Walraven, Th., 1955, *Bulletin of the Astronomical Institutes of the Netherlands*, 12, 223
Walraven, Th., J.H., 1990, in *Confrontation between stellar pulsation and evolution*, 50, eds. C. Cacciari, G. Clementi, *Astronomical Society of the Pacific Conf. Ser.*, 11
Walraven, Th., J.H., Balona, L.A., 1992, *Monthly Notices of the RAS*, 254, 59

DISCUSSION

Breger: Prof. Walraven's work on AI Vel is a spectacular example of the value of *LTM*. The discovery of four radial modes and one nonradial mode for a large-amplitude δ Scuti variable still needs to be explained theoretically. Since Prof. Walraven was too modest to mention where the results are published: the discoveries are published in *Monthly Notices of the RAS*, 254, 59, 1992.

Sterken: May I add that *LTM* in Walraven mode—that is, when telecopes and photometers are built by yourself and operated in your own garden—has proven to be a scientifically very rewarding approach.

[1] indicated by $5lr$ in the Table

LONG-TERM MONITORING OF THE EXTREME GALACTIC CEPHEIDS V810 CENTAURI AND V473 LYRAE

G. BURKI
Observatoire de Genève
CH-1290 Sauverny
Switzerland

ABSTRACT. Two peculiar galactic supergiants have been monitored in photometry and radial velocity during more than 15 years, using the telescopes of Geneva Observatory at La Silla (ESO, Chile) and Haute-Provence Observatory (France). These two supergiants are found to be the classical cepheids with the longest (V810 Cen) and shortest (V473 Lyr) periods in our Galaxy, i.e. 153 d and $1^{d}.5$. In addition, V810 Cen is a double-mode cepheid with a period ratio of 0.68, strongly suggesting pulsation in the fundamental mode and first overtone, and V473 Lyr exhibits large variations of amplitude in a time-scale larger than 1000 d. The understanding of the complex variability of these two stars requires **long-term–, continuous– and high-precision monitoring**.

Keywords: Cepheids – Geneva photometry

1. Introduction

Many long-term monitoring campaigns have been organised by astronomers of Geneva Observatory by using our own telescopes. Actually, one of these telescopes, located in the southern hemisphere at La Silla (Chile), is dedicated to Geneva stellar photometry. A second, the telescope, dedicated to stellar radial-velocity measurements (CORAVEL spectrometer), is located in the northern hemisphere at the Observatoire de Haute-Provence (France).

Numerous supergiant stars have been monitored by using these telescopes in photometry or in radial velocity in order to try to describe the complex variability of these objects. Partial results on these works have been published in an earlier phase of this programme, such as the description of a period-luminosity-colour relation (Burki, 1978) and the discovery of many binary supergiants (Burki & Mayor, 1983). The publication of the complete results on this survey of more than 15 years is in preparation and the analysis of only two of these stars, V810 Centauri and V473 Lyrae, will be presented here.

These two stars define the actual observational upper and lower limits for the periods of the classical cepheids in our Galaxy. **V810 Cen** is the galactic cepheid with the **longest period** (153 d, see below). Between February 1976 and October 1993, we obtained 444 photometric measurements of this star at the Swiss station

C. Sterken and M. de Groot (eds.), The Impact of Long-Term Monitoring on Variable Star Research, 247–254.

at La Silla Observatory. **V473 Lyr** is the galactic cepheid with **the shortest period** ($1^{\rm d}\!.5$). Between July 1977 and July 1992, we obtained 805 radial-velocity measurements at the Swiss telescope at the Haute-Provence Observatory.

In addition to their extreme periods, these two cepheids exhibit a series of uncommon characteristics with respect to the other members of this group of variable stars. They are ideal examples to illustrate the main purpose of this meeting on the impact of long-term monitoring on variable-star research. Indeed, the understanding of the complex variability of these cepheids requires:

a. Long-term monitoring. The periods or characteristic times of the variations require observations distributed over several thousand days.

b. Continuous monitoring. In such stars, it is especially important to reduce the gaps between different series of observations. Fortunately, it is possible to observe these two stars during the whole year from Chile and France respectively and, therewith, these gaps have been reduced as much as possible.

c. High-precision measurements. Due to the complex and low-amplitude (at least in phases of minimum amplitude in the case of V473 Lyr) variability of these stars, it is not possible to compensate a low precision of measurements by a large number of them, as is sometimes done in the case of strictly periodic and large-amplitude variable stars. For the monitoring presented here, the typical precision of the data is $0^{\rm m}\!.005$ in photometry and 0.35 km s^{-1} in radial velocity.

2. V810 Centauri = HD 101947 = HR 4511

This G0Ia supergiant star is the brightest (probable) member of the open cluster Stock 14. Its long-period variability was discovered by Eichendorf and Reipurth (1979) who gave the first estimate of the period, i.e. about 125 d. In addition, they located the star at the hot border of the cepheid instability strip. On the basis of IUE spectra, Parsons (1981) discovered the existence of a hot companion, a B-type supergiant, fainter than the G-type supergiant by about $3^{\rm m}\!.2$ in the optical V band. A long-term analysis (Burki, 1986), based on the photometric data of Eichendorf and Reipurth (1979) and Dean (1980), together with the first data of the Geneva survey, allowed to show that the variation is multiperiodic. Unfortunately, due to the too short duration of the survey (five years) and to the gaps affecting the data obtained during southern summers only, the proposed periods were not the correct ones, as shown in the following.

Since 1989, continuous (i.e. the whole year round) photometric monitoring has been organised from the Swiss station at La Silla and it is now possible to analyse the data covering a total interval of 16 years (see Figure 1). The power of the Fourier spectrum is given in Figure 2a for the first 9 years of the survey (i.e. $3500 < \mathrm{HJD}\text{-}2440000 < 6800$), and in Figure 2b for the complete survey of 16 years. It is clearly apparent that:

a. The light variation is dominated by two frequencies (Figure 2b), $\nu_0 = 0.006528$ and $\nu_1 = 0.009573$ d^{-1}, corresponding to the periods $P_0 = 153^{\rm d}\!.19$ and $P_1 = 104^{\rm d}\!.46$.

b. These peaks were already present during the first part of the monitoring (Figure 2a), but were masked by the strong aliasing resulting from the non-continuous monitoring (observations done exclusively during summers).

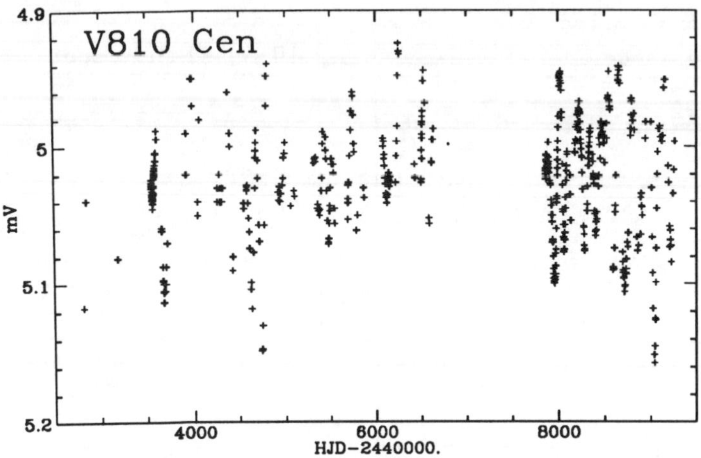

Fig. 1. The V-filter measurements of V810 Centauri

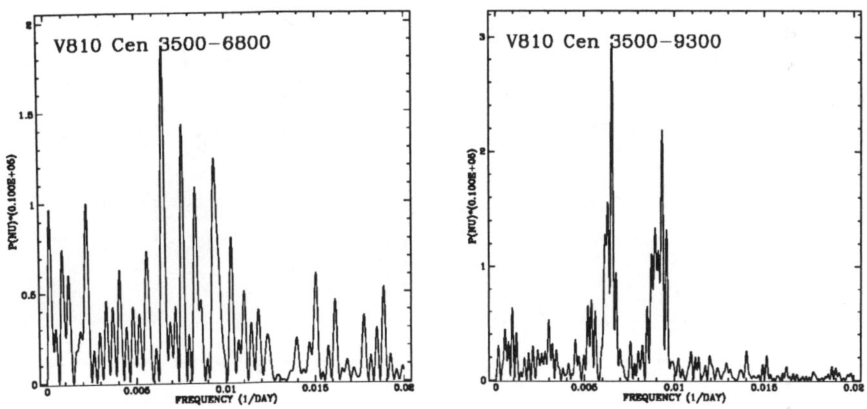

Fig. 2. Power spectrum of the Fourier Transform for the V measurements in two intervals in time (noted in the figures)

c. In addition to the two periods P_0 and P_1, a long-term variation, P $\geq$ 4000 d, can be noted in Figure 1. This variation could be connected to the long-term stellar or/and binary evolution.

The value of the period ratio $P_1/P_0 = 0.68$ is a strong indication that P_0 corresponds to the fundamental radial mode and P_1 to the first overtone. Indeed, in the case of the known double-mode cepheids, the values of the period ratio are in the range 0.697 to 0.711 (Balona, 1985), in agreement with the theoretical calculations for the fundamental mode and first overtone. The only exception in the case of the classical cepheids is the double-mode star CO Aur, for which the period ratio is 0.80, attributed to a pulsation in the first and second overtones (Mantegazza, 1983; Babel & Burki, 1987).

The double-mode-cepheid classification for V810 Cen is confirmed by the phase

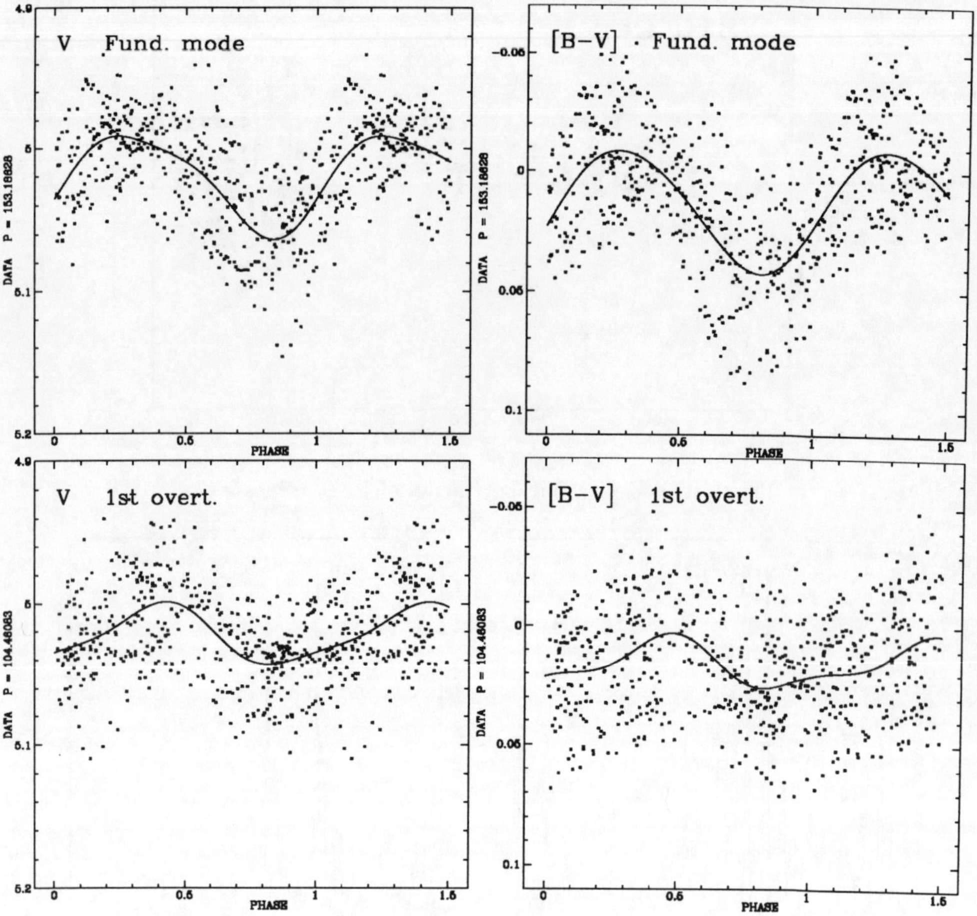

Fig. 3. V and $[B-V]$ curves for the fundamental mode of V810 Cen. Only the measurements in the Geneva photometry have been taken into account. The large residual standard deviations around the fitted curves show that these two modes explain only about half of the total variability

differences between the light and colour curves for each mode. Figure 3 shows that, for each mode, the V and $[B-V]$ curves are roughly in phase, i.e. the star is bluer when brighter. This is the behaviour expected in the case of radial-pulsation modes.

Consequently, the variability of V810 Cen can be described as follows:

a. There are two radial modes of periods $P_0 = 153\overset{d}{.}19$ and $P_1 = 104\overset{d}{.}46$.

b. The period ratio $P_1/P_0 = 0.68$ and the fact that the light and colour curves are roughly in phase, strongly indicates that the two modes are radial, the fundamental one and the first overtone.

c. The fundamental period $P_0 = 153\overset{d}{.}19$ is the longest known period for a classical cepheid in our Galaxy.

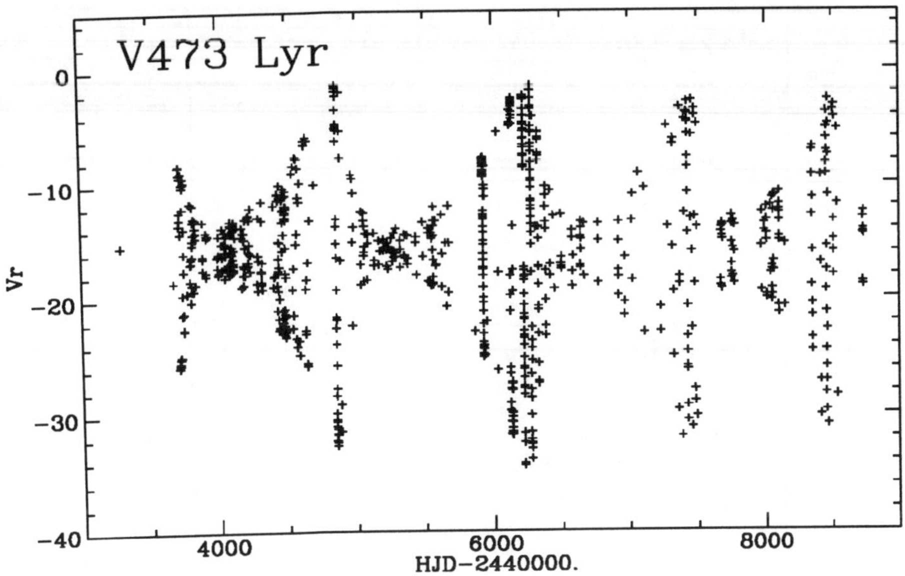

Fig. 4. The radial-velocity measurements of V473 Lyrae

d. The peak-to-peak amplitudes for these two modes are very small, i.e. only $0\overset{d}{.}07$ and $0\overset{m}{.}04$.
e. The residual standard deviation, after removing these two modes, is $0\overset{m}{.}03$, i.e. about 6 times larger than the precision of the measurements. Thus, the star is affected by other kinds of variabilities: a long-term (evolutionary ?) effect and a variation of the amplitudes of the two radial modes are clearly present. The other kinds of variabilities, still to be examined, are other radial (and non-radial ?) modes and/or irregular supergiant-like instabilities.

3. V473 Lyrae = HD 180583 = HR 7308

Semi-regular photometric variations were first detected by Breger (1969) in this F6Ib-II star. Percy et al. (1979) found that this star is a new small-amplitude cepheid. Burki and Mayor (1980) established the main characteristics of this star, i.e. its period is the shortest known for a classical galactic cepheid ($1\overset{d}{.}5$) and its amplitude of pulsation varies by a factor of over 13 in a time of about 1200 d. The Baade-Wesselink method was applied by Burki et al. (1986) to simultaneous photometric and radial-velocity measurements and a pulsation in the second (or higher) overtone was postulated.

Fig. 4 shows the 805 radial-velocity measurements obtained for this star. It can be noted in particular that more than 4 cycles of amplitude variations have been covered and that the characteristics of these cycles (amplitude, duration) vary from one to another.

The data have been divided into 48 successive ranges according to HJD and, in each range, the radial-velocity measurements have been fitted by least squares to a

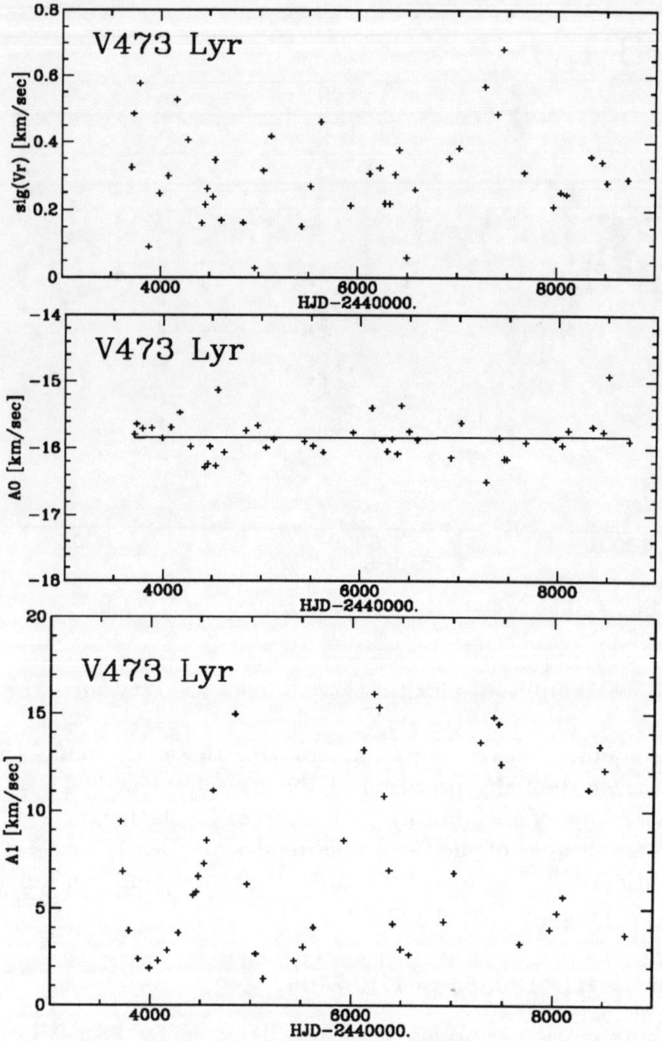

Fig. 5. Variation of residual standard deviation sig(V_r) and of the main parameters A_0 and A_1 of the fitted radial-velocity curve of V473 Lyr.

function of the form:

$$V_r(t) = A_0 + \sum_{i=1}^{n} A_i \cos(2\pi\nu_0(t - t_0) + \phi_i)$$

The adopted value for the frequency is $\nu_0 = 0.670736\ \mathrm{d}^{-1}$, corresponding to a period of $P_0 = 1^{\mathrm{d}}.49090$, and giving a constant phase (or $O - C$) diagram from minimum to maximum phases. The adopted value for t_0 is 2446000.811, giving a phase $\phi_1 = 0$ in the last range of HJD considered. The number of harmonics was chosen between 1 and 5, depending on the shape of the radial-velocity curve, which is asymmetrical at large-amplitude phase (see Burki et al., 1982).

The global results of these fits are given in Fig. 5:

a. Fig. 5a shows that the quality of these fits is very good (i.e. the residual standard deviation is smaller than 0.4 km s^{-1}) in the large majority of the intervals. The values larger than 0.5 km s^{-1} are due to the too small number of measurements in these intervals.

b. Fig. 5b shows that the term A_0 remained constant during the survey. Thus, no evidence for an orbital effect due to a companion is found.

c. Fig. 5c shows that the characteristics of the cycle of the amplitude variation change with time. During the last 3 cycles, the cycle duration decreased and the value of the amplitude maximum decreased while the value of the amplitude minimum increased. Thus, the following question arises: is the star evolving towards a stable pulsation ?

4. Conclusion

V810 Cen is a double-mode cepheid and its fundamental period is the largest among the galactic cepheids. However, these two modes allow to describe only about half of the observed light variability. This star is a powerful test of the evolution and pulsation theories of massive stars because: i) The star is a member of an open cluster. ii) The periods and their ratio depend on the stellar parameters, i.e. age, mass, luminosity, temperature, etc. iii) The G0 supergiant has a blue companion.

V473 Lyr could evolve towards a stable pulsation amplitude. This star is unique for the study of the variation of the non-linear effects related to the variation of the pulsation amplitude in the same star. The evolutionary status of this star pulsating in a high overtone (at least the 2nd one), and the cause of the amplitude variations are to be precised in the future. In particular, a simultaneous monitoring in photometry and radial velocity would be very useful in order to determine more precisely the mean colour and radius variations during the long cycle of amplitude variation.

References

Balona, L.A. 1985, in it Cepheids: theory and observations, Ed. B.F. Madore, Cambridge Univ. Press, Cambridge, 17

Babel, J., Burki, G. 1987, *Astronomy and Astrophysics*, 181, 34

Breger, M. 1969, *Astrophysical Journal, Supplement Series*, 19, 79

Burki, G. 1978, *Astronomy and Astrophysics*, 65, 357

Burki, G. 1986, in *Space research prospects in stellar activity and variability*, Eds. A. Mangeney, F. Praderie, Obs. de Paris-Meudon, Paris, 69

Burki, G., Mayor, M. 1980, *Astronomy and Astrophysics*, 115, 121

Burki, G., Mayor, M., Benz, W. 1982 *Astronomy and Astrophysics*, 109, 258

Burki, G., Mayor, M. 1983, *Astronomy and Astrophysics*, 124, 256

Burki, G., Schmidt, E.G., Ferro, A.A., Fernie, J.D., Sasselov, D., Simon, N.R., Percy, J.R., Szabados, L. 1986, *Astronomy and Astrophysics*, 168, 139

Dean, J.F. 1980, *Information Bulletin on Variable Stars*, 1892

Eichendorf, W., Reipurth, B. 1979, *Astronomy and Astrophysics*, 77, 227

Mantegazza, L. 1983, *Astronomy and Astrophysics*, 118, 321
Parson, S.B. 1981, *Astrophysical Journal*, 245, 201
Percy, J.R., Baskerville, I., Trevorrow, D.W. 1979, *Publications of the ASP*, 91, 368

DISCUSSION

Sterken: There is no real need for two telescopes: one telescope of the 50 or 70-cm class equipped with an echelle spectrograph like the one built by Mandel (see this conference), supplemented with a 30-cm telescope (mounted piggy-back) and a commercial CCD camera will do the job of simultaneous spectrographic and photometric observations at virtually no extra cost.

Szabados: The double-mode pulsation of V810 Cen is especially important because the other known beat Cepheids have much shorter pulsation periods (in the period range 2-6 days).

Waelkens

HOT PULSATING STARS IN THE GENEVA SYSTEM

Christoffel WAELKENS
Instituut voor Sterrenkunde
Celestijnenlaan 200B
B-3001 Leuven (Belgium)

ABSTRACT. We report on our observations of pulsating B stars in the Geneva Photometric System since 1980, emphasizing the importance of long-term monitoring for such objects. A recent theoretical breakthrough has now finally settled the long-standing question of the pulsation mechanism of the β Cephei stars, providing also a unified picture for the β Cephei stars and the slowly-pulsating B stars (SPBs), which are slightly cooler stars that pulsate in g-modes. We emphasize how an ambitious long-term observing programme of SPBs may provide important constraints on interior models of massive stars, eventually solving some important problems.

Keywords: slowly-pulsating B stars – Geneva photometry

1. Introduction

The study of stellar stability and variability is an important diagnostic for our understanding of stellar-interior structure. This is especially true for massive stars, where major uncertainties concerning interior models still persist, and where variability indeed is very common. In many stars, e.g. the Be stars, the cause of the variability is not yet well understood. Recent developments, theoretical as well as observational, have led to important progress in our understanding of classical stellar pulsation among early-B stars (β Cephei stars) and mid-B stars (slowly pulsating B stars). In the subsequent sections we describe and situate the research concerning these stars that has been carried out in the Geneva Photometric System, putting special emphasis on the importance of long-term observations.

2. The β Cephei Stars

The β Cephei stars are a group of early-B stars that pulsate in radial modes or non-radial p-modes. The variability of these stars can most often be described in terms of multiperiodic pulsations with stable amplitudes; in a few stars the pulsation seems to be truly monoperiodic. The β Cephei stars occupy a well-defined instability strip in the upper HR diagram. For a recent review of the observational properties of the β Cephei stars, we refer to the study by Sterken & Jerzykiewicz (1993).

The astrophysical importance of the β Cephei stars resides in the fact that the pulsation mechanism operating in them has been elusive for many decades, thus

C. Sterken and M. de Groot (eds.), The Impact of Long-Term Monitoring on Variable Star Research, 255–266.

causing a vexing challenge for our understanding of the structure of massive stars. It now seems that this question has been solved: in fact, β Cephei stars appear to oscillate because of an envelope mechanism much like the one operating in Cepheids and other well-known pulsating stars. The breakthrough came with the so-called 'new opacities': recent calculations (Rogers & Iglesias 1992, Seaton 1993) show that the metal opacity in the subphotospheric layers of the β Cephei stars had been underestimated by a factor of up to three. The fact that the cause of the pulsations has been finally identified does not imply that the β Cephei stars have become uninteresting, however. Indeed, if the pulsational behavior of these stars is so critically dependent on metal opacity, it is clear that the β Cephei stars may be an interesting probe of metal abundance (Sterken & Jerzykiewicz 1988; Waelkens, Van den Abeele & Van Winckel 1991; Balona 1992).

The importance of long-term monitoring of these short-period variables is twofold. First, the high masses of the β Cephei stars imply that evolutionary period variations can be detected (for a review, see Sterken 1993); in several stars, apparent period variations are due to light-time effects induced by orbital motion (Chapellier 1990; Pigulski & Boratyn 1992). A second effect that requires long-term observational effort, is the amplitude variation observed in some stars. While most β Cephei stars appear to present stable pulsation amplitudes on a time-scale of decades, this is not the case for others. A remarkable case is the star HD 129929 (V386 Centauri). Comparing our 1981 observations with older data obtained with the same equipment, we were unable to explain the latter with the three frequencies and amplitudes found in 1981 (Waelkens & Rufener 1983). Subsequent observations in 1985, 1988, and 1989 have shown that the amplitudes, but not the frequencies, are variable: in all data two frequencies near 6.5 cycles per day and two others near 7.0 cycles/day appear, but with different amplitudes, so that the power spectra are markedly different from year to year (Fig. 1).

3. Variable B Stars in the Geneva System

The characteristics of the Geneva Photometric System are described elsewhere in this volume. The main merits for variable-star research reside in the following:

- High accuracy of the data, so that small-amplitude variations can be detected and studied.
- Nearly continuous operation with the same equipment, so that long-term monitoring can be organized.
- Extreme care for long-term stability, so that small-amplitude phenomena can be studied on a long time-scale and so that statistically meaningful variability criteria can be defined.
- Colour indices that allow meaningful estimates of astrophysical parameters, so that variable stars can be classified accurately.

The two last items above enable one to discuss stellar variability as a function of locus in the HR diagram, as has last been done by Grenon (1993). Waelkens and Rufener (1985) took a closer look at the scatter on the measurements of OB-type stars as a function of position in the XY-diagram, which is the Geneva observational HR diagram for hot stars (Cramer & Maeder 1979). They confirmed the existence of

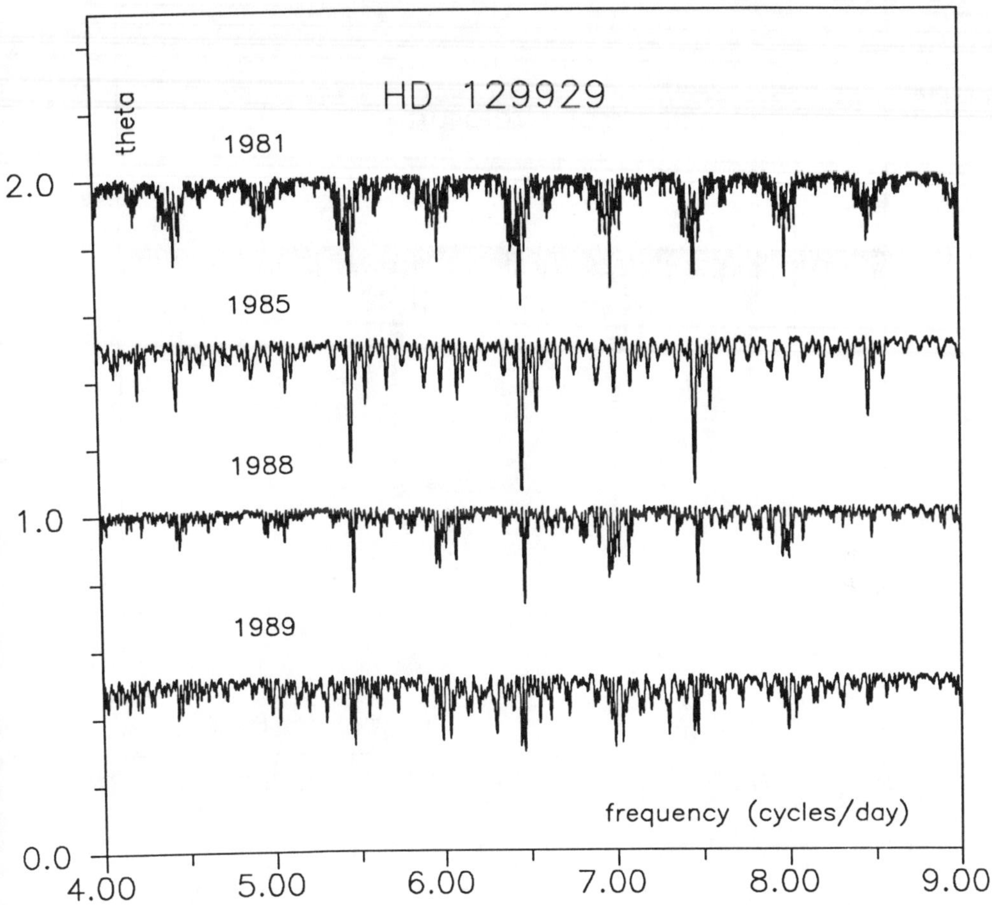

Fig. 1. Stellingwerf θ statistiscs for the visual-magnitude data of HD 129929 during four different seasons. The same frequencies are present in all data, but the amplitudes are markedly different.

well-known higher-than-average variability for OB stars and supergiants and could recover the zone defined by the β Cephei stars. In addition, they detected a zone of variability for mid-B stars on the main-sequence. It turned out that these mid-B stars form an independent new group of photometric variables.

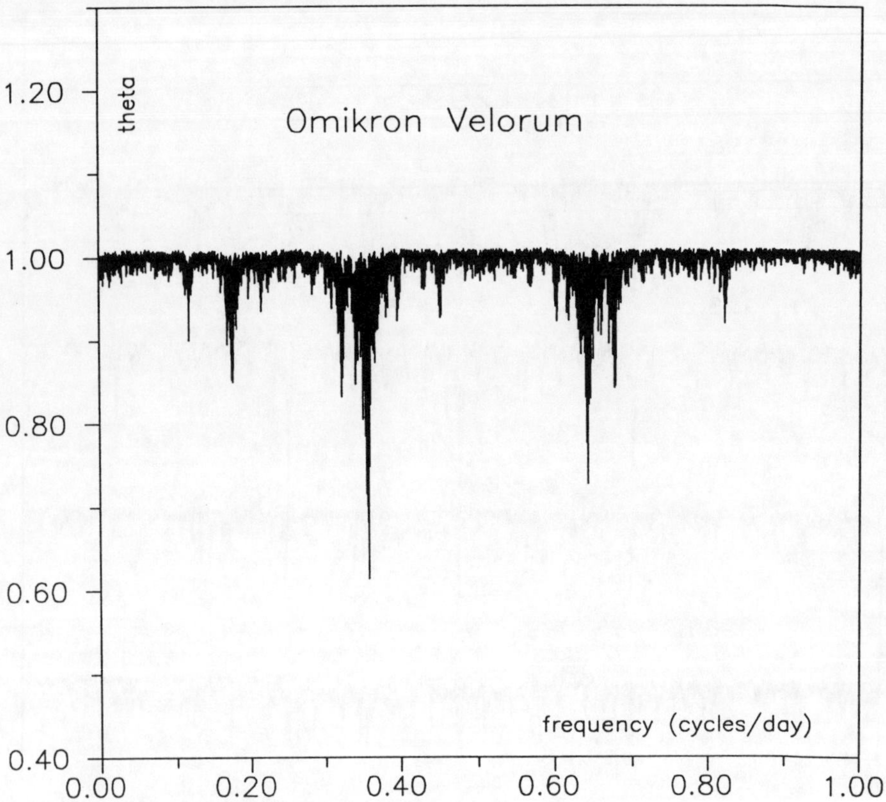

Fig. 2. Stellingwerf θ statistiscs for the visual-magnitude data of o Velorum. This power spectrum can be completely explained by physical frequencies near f = 0.356 cycles/day and aliases and multiples of f. Note that no power occurs at the lowest frequencies.

4. Slowly Pulsating B Stars

We have carried out extensive campaigns on several of the mid-B stars responsible for the higher-than-average scatter since 1981. All observations were obtained with the 'P7' photometer which is permanently attached to the 70-cm Swiss Telescope at La Silla Observatory; the data are absolute photometric measurements obtained with the procedures described by Cramer elsewhere in these Proceedings. The precision of the observations was optimized by increasing the frequency of the measurement of standard stars. We preferred absolute photometry over differential photometry, in order to avoid long-term instabilities in the data and the corresponding low-frequency noise in the power spectra. The absence of low-frequency noise is illustrated in Fig. 2, where we show the θ statistics for our 1980-87 data for the variable star o Velorum (Waelkens 1991).

It turned out that all these stars are multiperiodic variables with periods in the range between 1 and 4 days. A second common characteristic of all stars is the ratio of some 50% of the $[U-B]$ and m_v amplitudes. The multiperiodicity unambiguously points to non-radial g-mode pulsation as the explanation for the variability of these stars. The length of the periods points to modes with high radial order k. The amplitude ratios are well explained in terms of pulsation with modes of degree ℓ = 1 or 2 (Watson 1987, Heyndericks 1991). Accordingly, we have renamed these stars the "slowly pulsating B stars" (SPBs).

The critical issue concerning the interpretation of the variability of the SPBs clearly is the multiperiodicity. The beat periods often are rather long, so that data spread over several years are necessary to unravel the power spectra; for such data the long-term stability of the instrumental system is essential.

We illustrate the period analysis of SPB stars in Fig. 3 for the star HD 45284, where we show the Stellingwerf θ statistics for all data gathered between 1982 and 1989, for frequencies between 1.0 and 1.6 cycles/day. The upper figure shows the θ statistics for all data, the second figure for all data prewhitened with the first frequency, and so on. For pictural reasons we have added an arbitrary number to the θ statistics; after subsequent reductions the data have been rescaled according to the variance already accounted for. It is clearly seen that several peaks occur, and that they remain present with the same power unless they are removed from the data by prewhitening. Three physical frequencies are easily identified, at 1.2385 cycles/day, 1.1275 cycles/day, and 1.5061 cycles/day, respectively. It is quite likely that other physical frequencies occur in the residuals (lower figure), but the amount of data and the aliasing pattern did not allow us to determine them with certainty. We note that, so far, HD 45284 is the SPB with the shortest periods (it also is the least evolved one), and with the shortest beat periods.

The interpretation of SPBs in terms of non-radial pulsations is strenghtened by the behaviour of their spectral-line profiles. In Fig. 4 we show the Si II 4128-30 lines for HD 123515 during three subsequent nights: the line-profile variations are consistent with low-degree modes. It is interesting to note that all stars for which we obtained spectra turned out to be slow rotators, though the observational selection did not contain any criterion concerning $v \sin i$. Probably then, the SPB phenomenon attains largest amplitudes in slow rotators.

In Table 1 we summarize the data for the SPB stars that have been studied in the Geneva system. The first such star to be studied fairly intensively was HR 3462, for which Burki (1983) first detected multiperiodic behaviour; another interesting object is HD 37151 (North & Paltani 1994). Other stars that show similar photometric behavior and that occur in the same part of the HR diagram are 53 Persei (Buta & Smith 1979), HD 27563 (Manfroid & Renson 1989) and HD 80859 (van Genderen et al. 1992).

5. The β Cephei Instability Strip

We have mentioned in Section 2 that the pulsations of the β Cephei stars have now finally been explained in terms of an envelope-opacity mechanism, to which the revised contribution of metals is essential (Cox et al. 1992; Moskalik & Dziembowski 1992; Dziembowski & Pamyathnyk 1993). In the region in the HR diagram defined by the β Cephei stars, the first few radial overtones and several non-radial p-modes are found to be unstable.

In addition, Gautschy & Saio (1993) and Dziembowski et al. (1993) have shown

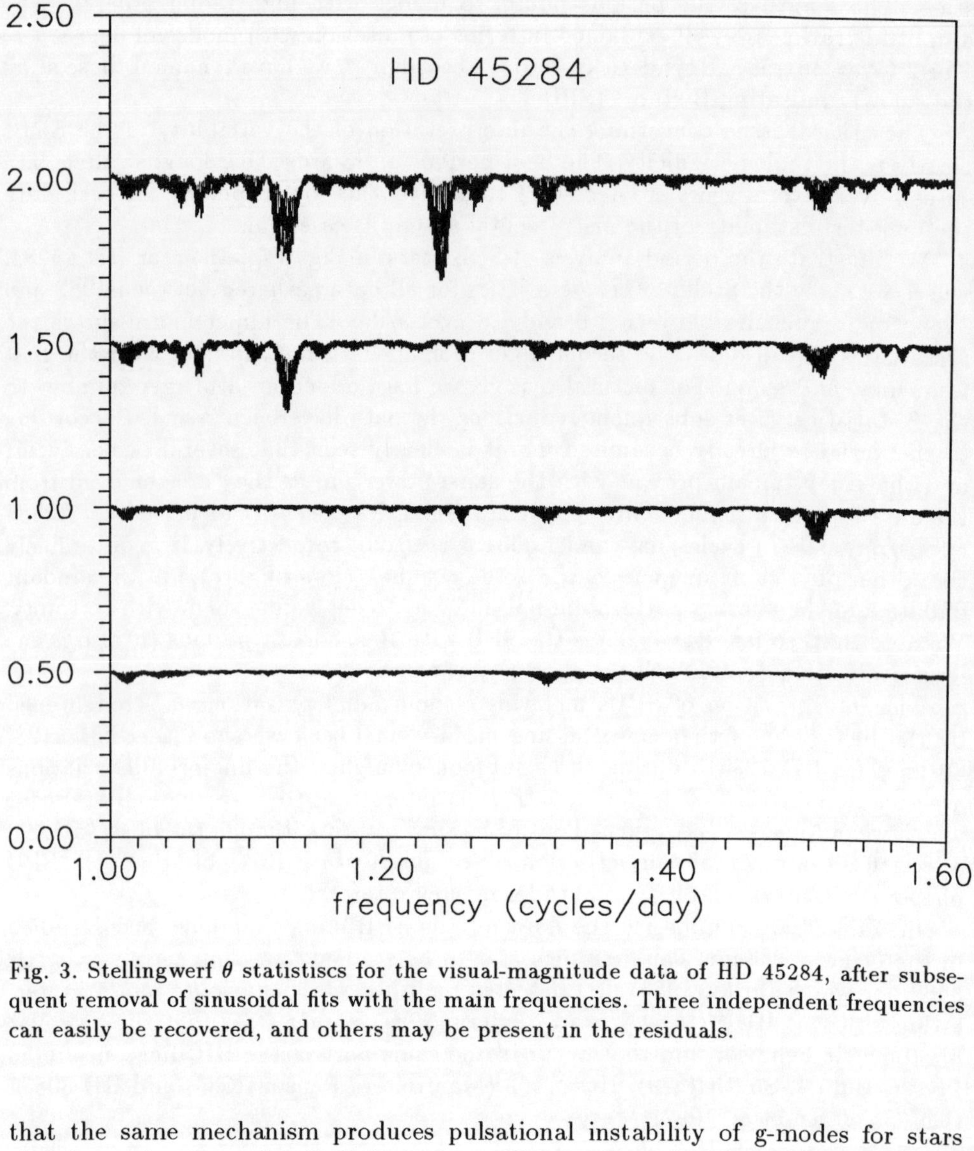

Fig. 3. Stellingwerf θ statistiscs for the visual-magnitude data of HD 45284, after subsequent removal of sinusoidal fits with the main frequencies. Three independent frequencies can easily be recovered, and others may be present in the residuals.

that the same mechanism produces pulsational instability of g-modes for stars slightly cooler than the β Cephei stars. We reproduce in Fig. 5 the theoretical limits of this g-mode instability region, as computed by Dziembowski et al. (1993), and represent the known SPB stars by asterisks. Clearly, the SPB stars fall nicely within the region in which theory predicts g-modes to occur.

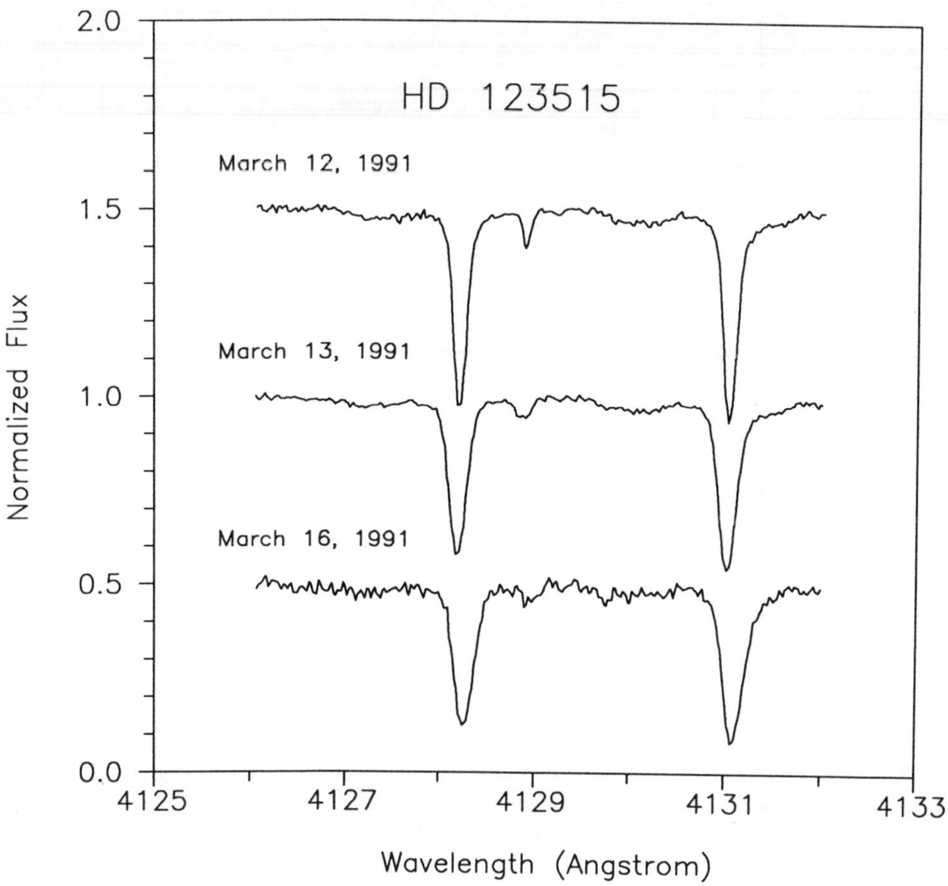

Fig. 4. Line-profile variations of HD 123515.

TABLE I

Photometric and physical parameters and main pulsation period for the SPB stars that were studied in the Geneva system. The physical parameters have been derived by North and Paltani (1994).

Star	X	Y	log T_{eff}	log L/L_{sun}	Main Period (days)
HD 34798	0.701	-0.004	4.202	2.87	1.276
HD 37151	1.029	0.007	4.110	2.36	0.804
HD 45284	0.784	-0.018	4.174	2.71	0.807
HD 74195	0.670	0.030	4.217	3.20	2.797
HD 74560	0.656	0.004	4.220	2.99	1.551
HD 76566	0.646	0.018	4.225	3.12	1.97
HD 123515	1.188	0.024	4.078	2.15	1.459
HD 143309	0.875	0.026	4.152	2.74	1.671
HD 160124	0.756	0.013	4.186	2.86	1.923
HD 177863	0.958	0.027	4.130	2.58	1.190
HD 181558	0.798	0.013	4.173	2.78	1.238

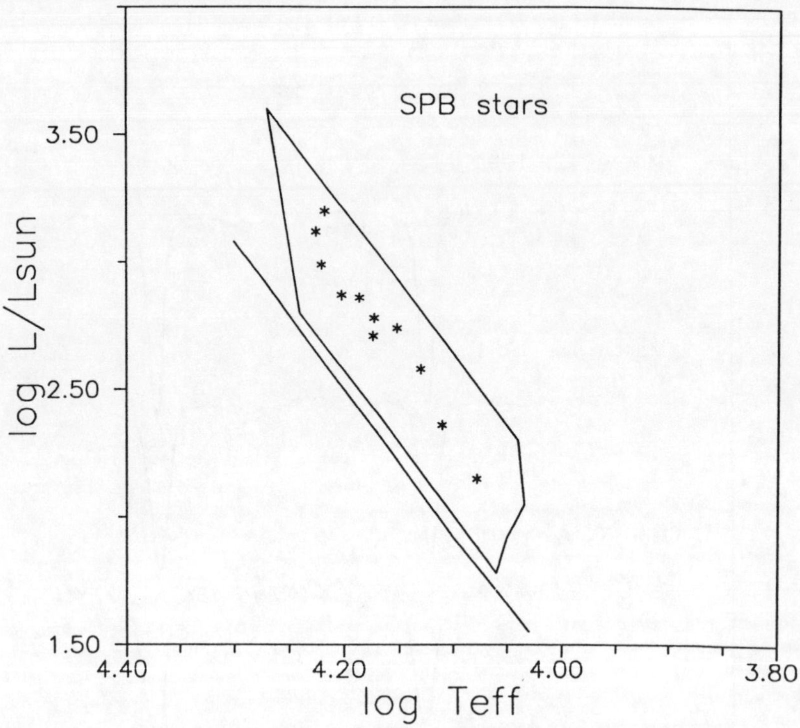

Fig. 5. The position of the SPB stars from Table 1 in the HR diagram: the physical parameters are taken from North and Paltani (1994). The box defines the theoretical instability strip as computed by Dziembowski et al. (1994).

With the advent of the new opacities, it now appears that all over the HR diagram stellar pulsation can essentially be attributed to a single (opacity) mechanism. In the coolest variables, the opacity in the hydrogen-ionisation zone drives the oscillations, the pulsations of stars in the Cepheid strip are driven in the second helium-ionisation zone, and the metal opacity drives the oscillations in the hottest pulsators. In this context, it is natural to label the zone in the HR diagram defined by the β Cephei stars and the SPBs as the β Cephei instability strip (Dziembowski 1994).

The reason that the g-mode pulsators (SPBs) are cooler than the p-mode pulsators (β Cephei stars) is related to the fact that g-modes propagate deeper into the stellar interior than do p-modes: in cooler stars the envelope layers at the temperature where the opacity maximum occurs, are situated at larger depth. It would be interesting to check whether g-mode pulsators also exist near the red edge of the Cepheid instability strip. One such star may be HD 96008 (Lampens 1987, Matthews 1990).

6. Prospects for Asteroseismology

Theory predicts that many modes are excited in SPB stars. Moreover, these modes penetrate deep into the interior of these stars. The SPB stars are, therefore, promising targets for asteroseismology. The potential may be as large as it is for the pulsating white dwarfs, which also oscillate with a large amount of g-modes of high radial order.

The information that one can expect to obtain from SPB asteroseismology is threefold (Dziembowski et al. 1993):

- In the asymptotic limit for high values of the radial wavenumber k, the period differences between modes with the same ℓ but subsequent k should be nearly constant, and this constant is essentially determined by the mass and the effective temperature. One may thus hope to obtain a precise estimate for the stellar mass from pulsation theory.
- The period differences discussed in the previous item are not exactly constant, but show an oscillatory behavior with increasing k, the reason being that the eigenfunctions have to be adjusted so that nodes occur at the composition discontinuities inside the star. The study of this behaviour should then contain information about the μ-gradient zone inside the star (see also Bertomieu & Provost 1988).
- The rotational splitting of modes depends on the rotation frequency of the layers where the modes attain large amplitudes; comparison of the splitting pattern for different modes may then test the rotation law within the star.

In fact, the asteroseismologic potential of SPB stars is very similar to that of the hot pre-white dwarf pulsators such as PG1159, for which the WET network provided such impressive results (Winget et al. 1991). It would certainly be important if the same kind of information that can now be obtained for pulsating degenerate stars were also to become available for massive main-sequence stars, eventually answering long-standing questions on the internal structure of these stars, such as those about the *mu*-gradient zone, the extent of convective overshooting, internal mixing due to differential rotation, ...

The observational task is huge, however. Indeed, the time-scale of the variations is two orders of magnitudes longer than in the white dwarfs, so that not two weeks, but up to ten years of dedicated observations may be necessary before all important frequencies are detected! On the other hand, many bright programme stars can be found, so that small telescopes are suitable for the photometric observations and spectroscopic follow-up is possible.

Interesting targets for such a programme would be o Velorum and HR 3467, the two brightest members of the galactic cluster IC 2391. Besides being bright, these stars offer the advantage of being nearly circumpolar at the major southern-hemisphere sites. Their occurrence in a cluster, with a distance modulus that is precisely known, is of course an additional bonus. Also, it may be that the only realistic way of carrying out this kind of programme is by means of CCD photometry, in which case cluster members are ideal targets.

Acknowledgements

This project could be realised thanks to the generous awarding of telescope time at the southern Swiss Telescope by the Geneva group. My warmest thanks go to Professor Golay and Professor Rufener, and also to the team that took over the Geneva Photometry. Financial support from the Belgian Fund for Collective Basic Research (grants 2.0119.83, 2.0098.87 and 2.0060.91) is gratefully acknowledged.

References

Balona, L.A., 1992, *Monthly Notices of the RAS*, 256, 425

Bertomieu, D., Provost, J., 1988, in *Advances in Helio- en Asteroseismoloy*, Proceedings of IAU Symposium 123, Reidel, Dordrecht, 121

Burki, G., 1983, *Astronomy and Astrophysics*, 121, 211

Buta, R.J. , Smith, M.A., 1979, *Astrophysical Journal*, 232, 213

Chapellier, E., 1990, Thèse de Doctorat, Université de Nice

Cox, A.N., Morgan, S.M., Rogers, F.J., Iglesias, C.A., 1992, *Astrophysical Journal*, 393, 272

Cramer, N., Maeder, A., 1979, *Astronomy and Astrophysics*, 78, 305

Dziembowski, W.A., 1994, Review presented at the IAU Symposium 162, *Pulsation, Rotation and Mass Loss in Early-Type Stars*, Juan-les-Pins, 5-8 October 1993

Dziembowski, W.A., Moskalik, P., Pamyatnykh, A.A., 1993, *Monthly Notices of the RAS*, 265, 588

Dziembowski, W.A., Pamyatnykh, A.A., 1993, *Monthly Notices of the RAS*, 262, 204

Gautschy, A., Saio, H., 1993, *Monthly Notices of the RAS*, 262, 213

Grenon, M., 1993, *Astronomical Society of the Pacific Conf. Ser.*, 40, 693

Heynderickx, D., 1991, PhD Thesis KU Leuven

Lampens, P., 1987, *Astronomy and Astrophysics*, 172, 173

Manfroid, J., Renson, P., 1989, *Astronomy and Astrophysics*, 223, 187

Matthews, J.M., 1990, *Astronomy and Astrophysics*, 229, 452

Moskalik, P., Dziemowski, W.A., 1992, *Astronomy and Astrophysics*, 256, L5

North, P., Paltani, S., 1994, *Astronomy and Astrophysics*, , in press

Pigulski, A., Boratyn, D.A., 1992, *Astronomy and Astrophysics*, 253, 178

Rogers, F.J., Iglesias, C.A., 1992, *Astrophysical Journal, Supplement Series*, 79, 507

Seaton, M.J., 1993, in *New Perspectives on Stellar Pulsation and Pulsating Variable Stars*, IAU Colloquium 139, edrs. K.M. Nemec and J.M. Matthews, Cambridge University Press, p. 231

Sterken C., 1993, in *New Perspectives on Stellar Pulsation and Pulsating Variable Stars*, IAU Colloquium 139, edrs. K.M. Nemec and J.M. Matthews, Cambridge University Press, p. 171

Sterken, C., Jerzykiewicz, M., 1988, *Monthly Notices of the RAS*, 235, 565

Sterken, C., Jerzykiewicz, M., 1993, *Space Science Reviews*, 62, 95

van Genderen, A., et al., 1992, *Astronomy and Astrophysics*, 259, 574

Waelkens, C., 1991, *Astronomy and Astrophysics*, 245, 453

Waelkens, C., Van den Abeele, K., Van Winckel, H., 1991, *Astronomy and Astrophysics*, 251, 69

Waelkens, C., Rufener, F., 1983, *Astronomy and Astrophysics*, 119, 279

Waelkens, C., Rufener, F., 1985, *Astronomy and Astrophysics*, 152, 6

Watson, R.D., 1987, Proc. Astron. Soc. Aust. 7, 38

Winget, D.E., et al., 1991, *Astrophysical Journal*, 378, 326

DISCUSSION

Kurtz: Perhaps you shouldn't take too much comfort in the beliefs of theoreticians. A few years ago at a meeting in Japan Wojtek Dziembowski was sitting next to me when a controversial observational result was presented. His eyes glowed with new-forming theories to explain these observations, and he leant over to me and asked "Don, what do you think of this result?" I said "I don't believe it". To which he said "Oh! Well, if you don't believe it, then neither do I". Theoreticians have a tendency to believe everything observers claim.

Waelkens: In this particular case, I am really struck by the extent the theoretically determined instability box neatly matches the observed one.

Le Contel: Spectrographic observations could be obtained through MUSICOS (MUlti-SIte COntinuous Spectroscopy) campaigns ($R \sim 30000$, high S/N ratios, identical spectrograph at several sites, ...)

Waelkens: I doubt somewhat whether it will ever be possible to get sufficient spectroscopic coverage: larger telescopes are not available on such a long time scale. Surely, some spectroscopic observations, in addition to photometry, would be useful, for instance in order to detect higher-l oscillations.

Le Contel: The p-mode zone and the g-mode zone in Dziembowski's HR diagram seem to overlap. And indeed I observed two stars which confirm this, viz.

1. ι Her has not only a long period, but also a short stable cycle ($\pm 0\overset{d}{.}12$) which is probably a p-mode
2. ζ Cas, which is a B2IV star and, so, should be a classical β Cep star, presents no variation in photometry (at the millimag level) nor in radial velocity (± 0.5 km s^{-1}) on a short time scale. The only variable parameter is FWHM with at least two periods, $0\overset{d}{.}27$ and a period over 1 day, which are probably g-modes.

Waelkens: So far instability strips have only been computed for $l = 0, 1$ and 2; the picture may be slightly different for higher l-values. Also, stability analyses are performed on equilibrium models. Maybe a p-mode, which would not normally be excited in a star with given parameters, can be excited if the star is already pulsating with a g-mode.

Sterken: Your proposed project would be very interesting for studying supergiants and LBVs as well. At the Val Morin meeting I presented observational evidence and suggested that the β Cep instability strip should be extended towards higher luminosities (Sterken 1989, Photometry of Luminous Blue Variables in Quiescent state, IAU Coll. 113, Eds. K. Davidson, A.F.J. Moffat, H.J.G.L.M. Lamers. Kluwer Ac. Publ., 59–66). From what I understood from the first papers of Dziembowski and co-workers, it seems that a wider β Cephei region is indeed possible, but in the picture you showed the region towards the more luminous stars is not closed. How far does it extend?

Waelkens: Calculations by the Warsaw group show that the instability strip widens toward higher luminosities, eventually merging with the Cepheid instability strip. In fact, it has already been suggested several years ago by Lamers and Fitzpatrick that LBVs may owe their instability to metal-line opacities, in the sense that the Eddington luminosity is exceeded in some layers. Clearly, with opacities three times higher this kind of scenario is attractive.

Kroll: What is the photometric accuarcy aimed at in your project?

Waelkens: The Geneva data have a typical scatter of some $0^{\mathrm{m}}.004$. Amplitudes much smaller than that could be detectable, however, if the observations are sufficiently numerous. The critical issue is long-term stability of the photometric equipment.

Kroll: What would be the observational schedule: to observe one star the whole night, or one observation per night?

Waelkens: Five points per night, spread over some 6 hours, are enough, given the length of the periods, and if a high precision is guaranteed. Ideally, of course, one would like to have more data points, but it is not likely that such an observational effort can be continued for years. This kind of programme has most chances to be carried out if it is combined with others.

DISCUSSION No 4: PULSATING STARS

Duerbeck: Can you comment on the Maia (27 Tau) variables with respect to the SPBs?

Waelkens: Maia, if its variability is real, may in fact be a SPB. It falls inside the instability strip. Another such star is 53 Per, which was first recognized as a line-profile variable. I am afraid that too many confusing nomenclatures exist for variable B stars, but I am confident that the recent theoretical breakthrough will help us to clarify the picture.

Kurtz: If I remember correctly, Myron Smith and Ron Buta found photometric variability on a time scale of 1-2 days in 53 Per in the late 1970s. I think that they should be recognized for early work on these kinds of stars.

Waelkens: Multiperiodic photometric variations indeed were found in 53 Per first and, most probably, it is also a SPB. It may then look fair to call these stars "53 Per stars". But, in the literature, the label 53 Per star is used for a broader class of line-profile variables, many of which may be unrelated to the SPBs.

Breger: (to Waelkens) Several small variability surveys have examined the cooler stars between the β Cep and δ Sct stars. Little convincing variability among the "mythical Maia variables" was detected. Among the B4-A0 stars you have both g-mode and p-mode pulsators (ET And). A division by type of pulsator appears more relevant to me than the discussions of whether one should call the stars "slowly-variable B stars", "53 Per variables" or "Maia variables". All three names can be justified, but would not a name with mode description be better?

Waelkens: Slowly pulsating B stars means exactly that: B stars pulsating in g-modes.

Hall: In my recent work (Hall 1991, IAU Coll. 130, 353) which defined the limits which guarantee that a star will be "heavily spotted", I discovered two stars which were exceptions. One was the F0V star in the detached eclipsing binary VZ CVn (Popper 1988, *Astronomical Journal*, 95, 190) and the other was AI Hya dealt with in that same paper. Both are variable but yet not understood by any known mechanism: solar-type spots, Ap-type spots, δ Scuti pulsation, or something else. At this meeting Ennio Poretti told me he has discovered two other variables which also are difficult to understand in this same way. And Kevin Krisciunaz, in a recent

C. Sterken and M. de Groot (eds.), The Impact of Long-Term Monitoring on Variable Star Research, 267–268.

paper in *Monthly Notices of the RAS*, has drawn attention to 9 Aurigae, as a fifth example. We might be witnessing the discovery of a totally new class of variable, which would be exciting. Or maybe not. Anyway, I just wanted to share this item of interest with the participants of this Workshop, and perhaps to involve them in working towards solving the mystery.

Poretti: In recent years we discovered an increasing number of F0-F2 variable stars (for a reference list see Mantegazza & Poretti (1991, *Information Bulletin on Variable Stars*, 3690). The photometric behaviour of the best observed objects is difficult to interpret: light curves are changing and there is no clear evidence for periodicities. We are investigating possible explanations: *g*-mode excitation (pulsating variables) and inhomogeneous surfaces (rotational variables).

Kurtz: Wait a moment, spotted F0V stars are Ap stars—we talked about them earlier today.

Dravins – would Hall be right?

INSTRUMENTAL EFFECTS IN STELLAR SPECTROSCOPY

Dainis DRAVINS
Lund Observatory
Box 43
S-22100 Lund
Sweden

ABSTRACT. Although in some aspects spectroscopy may be simpler than photometry (e.g. no need for "photometric" observing conditions), in other respects it is more difficult, because a complex spectrum is easily degraded by imperfections in instrumentation and/or data-reduction procedures. An understanding of instrumental effects is required to fully exploit archival spectra from different instruments at various epochs, or for using recorded spectra to compute accurate values of parameters such as spectral-line intensities, equivalent widths, or wavelength positions. In the literature, the term "signal-to-noise" is too often used to denote merely the *precision*, i.e. the reproducibility of a (possibly erroneous) observation, rather than the *accuracy*, i.e. the proximity to the true spectrum. In this review, such differences (and ensuing pitfalls) due to common spectrometer properties are exemplified, effects are described for electronic and photographic detectors, and issues discussed for the optimum use of archival data.

Keywords: stars – spectroscopy – instruments – data reduction – data archives – measuring accu- racy – signal-to-noise – detectors

1. Low-Noise, High-Resolution, High-Precision, or High-Accuracy?

During recent times, low-noise detectors have become generally available, and the literature now abounds with claims of astonishingly high signal-to-noise ratios in stellar spectroscopy, perhaps conveying a belief that one would now have reached a "noise-less" state. In many cases, however, these claims are misleading. Often, the term "signal-to-noise" is not well defined, or it is used to merely denote the *precision*, i.e. the reproducibility of a (probably erroneous) observation, rather than the *accuracy*, i.e. the proximity to the true stellar spectrum. Irrespective of the detector system used, it actually is exceedingly difficult to reach *true* signal-to-noise ratios (in the accuracy sense) in excess of about 100, if using any ordinary stellar spectrometer. The limits are set both by optical stray light, the finite spectral resolution, and the finite sampling steps of the detector. Stray light lowers the spectral continuum and fills in absorption lines, thereby causing systematic errors in spectral-line widths, depths, and shapes. Since the focal surfaces in most instruments are curved, measurements with plane multi-element detectors further cause the focus and optical aberrations to vary across the spectrum.

C. Sterken and M. de Groot (eds.), The Impact of Long-Term Monitoring on Variable Star Research, 269–290.

In this paper, a number of examples of such instrumental effects will be shown, however, without the ambition of giving a complete review. For more detailed discussions, together with numerical simulations of spectral-line degradation in different spectrometers, see e.g. Griffin (1969a), Dravins (1978a), or Andersen and Dravins (1982). A discussion of the trade-offs between resolution and noise is in Gray (1988).

2. Early High-Fidelity Spectra: The Utrecht Solar Atlas (1940)

As a point of departure for our discussion, we take the classical *Utrecht Solar Atlas* (Minnaert et al. 1940), which was recorded photographically at the Mt.Wilson Observatory. This was the first high-resolution atlas of the entire visual spectrum of a star, and remained a standard reference for several decades. For example, solar equivalent widths, as derived from this atlas, have been extensively used (not least for stellar comparisons) up to the present time, as listed in the revised Rowland wavelength table by Moore et al. (1966). The Utrecht Atlas is particularly illustrative since it is still better than almost all non-solar spectra up to this day while, at the same time, there exist modern and substantially better solar atlases which permit an illustration of systematic errors of types that are likely to exist in practically all stellar spectroscopy.

The Utrecht Atlas gives the spectrum of the *solar disk centre*, which (due to solar rotation, and line-profile changes across the solar disk) is slightly different from the spectrum of *integrated sunlight* (i.e. the solar irradiance, or flux). We start by comparing the Utrecht Atlas with a more recent disk-centre atlas, namely the one from Jungfraujoch which was completed in the late 1980s (Delbouille et al. 1973). Although there could exist some very slight physical changes in the solar spectrum between these respective epochs of observation (due to e.g. activity-cycle influences), such changes in the solar [photospheric] spectrum should be utterly negligible, compared to other effects we will be discussing. Thus, the differences between the Utrecht and Jungfraujoch Atlases reflect a few decades of progress in spectroscopy, and not any long-term changes on the Sun!

3. Different Recordings of the Same Spectrum

Fig. 1 shows the region around the familiar sodium D-lines at $\lambda\,590$ nm,as recorded in these two atlases. These lines are quite broad, and they are spectrally fully resolved by one or two orders of magnitude. Although superficially similar, there are some conspicuous differences between the atlases: (a) Different "continuum" placement in between the D-lines (probably mainly caused by different human judgment, and spectral segments of different lengths used for normalization); (b) Quite different strengths of many weak and narrow lines. These are telluric absorptions, mainly from H_2O vapor in the atmosphere, which are very much weaker (but not absent) in the Jungfraujoch Atlas, recorded at a cool and dry high-altitude site, compared with the medium-altitude site used for recording the Utrecht Atlas. Further: (c) The residual intensities in the line-bottoms of the Na I D_1 and D_2 lines are substantially different—the Jungfraujoch Atlas presents them at around 4% residual

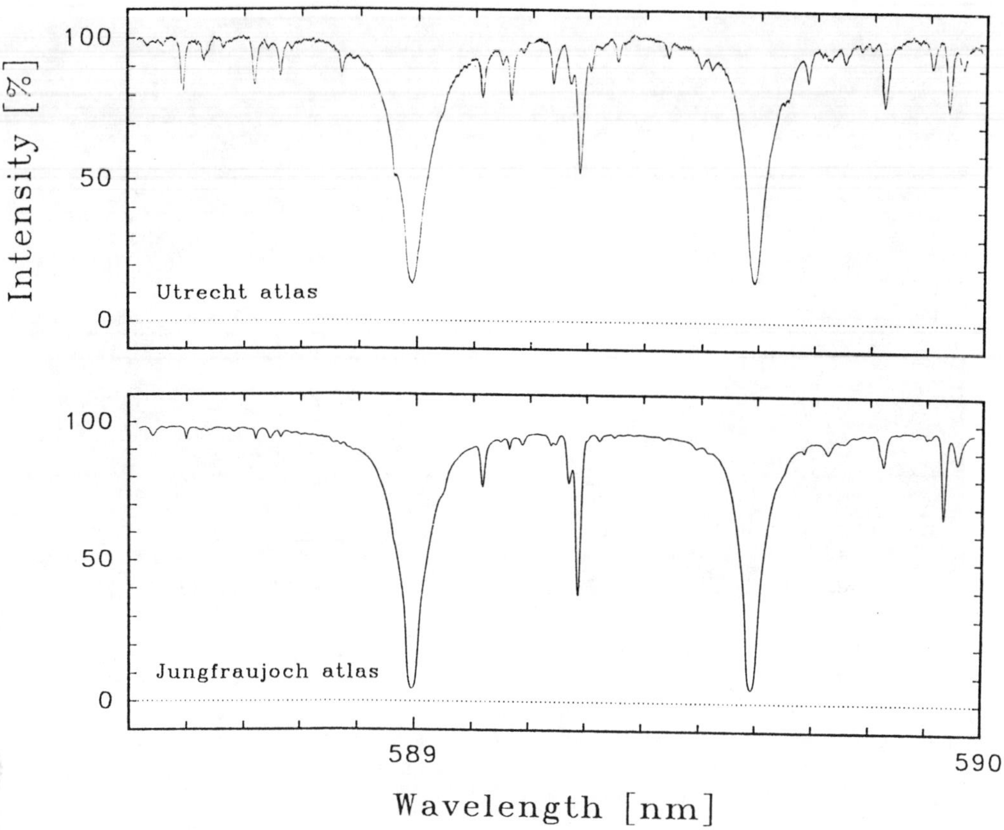

Fig. 1. Top: A region around the Na I D lines from the Utrecht Atlas of the solar disk-centre spectrum (Minnaert et al. 1940), and (bottom) in the more recent Jungfraujoch Atlas (Delbouille et al. 1973). Notable differences are seen in: (a) continuum placements, (b) strengths of the narrow telluric lines, and (c) intensities in the bottoms of the Na I D_1 and D_2 lines. The quality of the Jungfraujoch Atlas is not yet attainable for other stars, while that of the Utrecht Atlas is still better than practically all stellar data.

intensity, while the Utrecht one shows $\simeq$ 12%, after the estimated straylight level in the spectrometer has been subtracted. What causes this difference of a factor $\simeq$ 3 in measured spectral intensity?

4. Stray Light and Spectrometer Instrumental Profiles

The normally dominant source of stray light in ordinary single-pass spectrographs is the ruled diffraction grating, which causes light scattering because of its surface micro-roughness, and also exhibits some amount of grating ghosts, satellite, and grass lines. The origin of the micro-roughness can be traced back to the ruling process, where the ruling diamond is moved under mechanical pressure across a blank, thereby tearing loose surface material from the grating-to-be. Ghost lines are

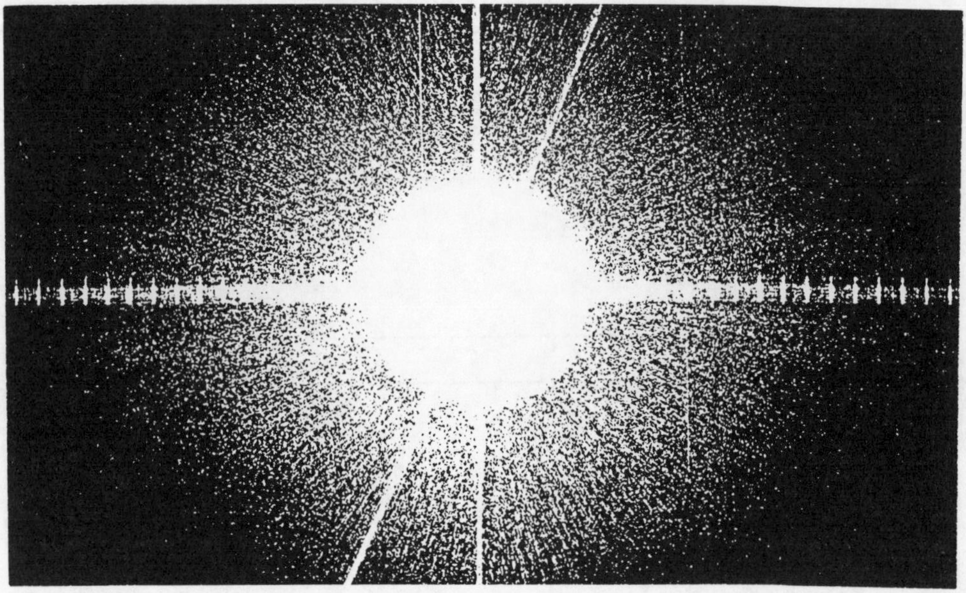

Fig. 2. Photograph of a "spectrum" of a monochromatic laser light source, recorded with a ruled diffraction grating in a coudé spectrometer. In an ideal instrument, this monochromatic light would have been focussed into a narrow line at the center of the image. One sees symmetric stray light caused by scattering on the optical surfaces and inside the detector, together with a series of grating ghosts and diffuse stray light, scattered by the grating in the direction of dispersion.

caused by periodic errors in the intergroove spacing, while random errors increase the diffuse stray light. For absorption-line spectroscopy, the major problem is diffuse stray light and not grating ghosts, even for older gratings not ruled on interferometrically-controlled ruling engines. To see such straylight patterns in the focal plane, one can illuminate the spectrometer with monochromatic light and record its "spectrum" (Fig. 2). It is, however, *not* possible to estimate correctly the level of straylight by measuring the flux on either side of a stellar exposure. Not only the grating ghosts, but also a great part of the continuous straylight is diffracted along the direction of dispersion (due to micro-roughness on the diffraction-grating grooves), and none of this will appear outside the stellar spectrogramme. To see at which optical elements each part of the stray light originates, one can image the inside of a spectrometer using the light falling at each position in the focal plane. E.g., in the light of the grating ghosts, the grating is by far the brightest source. For further discussions, see Gray (1992) and references therein.

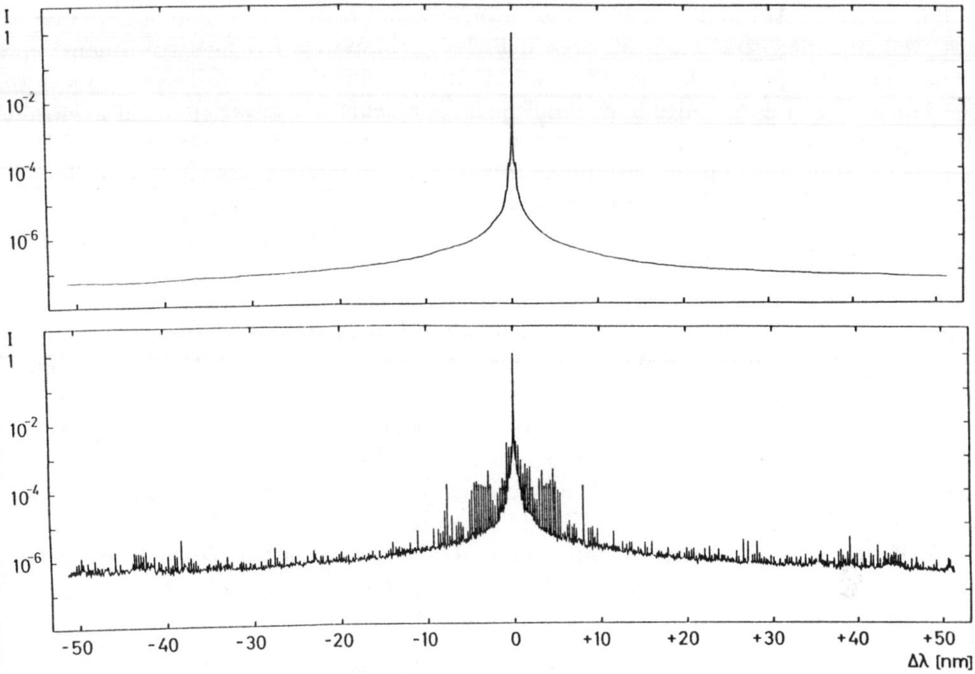

Fig. 3. Instrumental profiles for the [old] coudé spectrograph of the ESO 1.52-m spectrographic telescope, for a holographic diffraction grating (top) and a ruled one (bottom). Intensity I is plotted as function of $\Delta\lambda$, the wavelength distance from the central peak. For the ruled grating, a large number of grating ghosts and grass lines are seen. The lower stray light from the holographic grating reveals secondary maxima at ± 0.3 nm, which arise from halation rings in the photographic plate used as detector, and are not intrinsic to the spectrograph (Dravins 1978a).

Holographic diffraction gratings are made by exposing equidistant and parallel interference fringes from mode-stabilized laser light on a blank coated with photoresist. After chemical treatment, grating grooves are produced. Since the grating-surface microstructure is shaped by chemical reactions on a molecule-to-molecule level rather than by a diamond tearing loose material, there is much less micro-roughness. Since the laser wavelength is well defined, there are neither grating ghosts, satellites, nor grass lines (Fig. 3). However, the grooves of a holographic grating will retain any optical aberrations inherent in the imaging system used for the illumination with laser light. To a first approximation, the resulting grating-groove profiles become sinusoidal, and such a grating will direct light in first diffraction order only.

Thanks to the absence of grating ghosts, and low levels of scattered light, holographic diffraction gratings are being used for, e.g., emission-line spectroscopy in the laboratory. That they have not become equally widespread in astronomy, can be attributed to their somewhat lower reflective efficiencies: although more elaborate fabrication techniques do permit non-sinusoidal ("blazed") groove profiles, they do not quite reach the light efficiency of echelles, and their linear output format is not well matched to solid-state area detectors, compared to the two-dimensional cross-dispersed spectral format of an echelle spectrometer. The price to be paid for the fainter limiting magnitude of the echelle, is a generally lower spectral accuracy due to the more complex optical system. However, it should be acknowledged that modern gratings have been very much improved, and in some cases they no longer constitute the limiting factor for spectral fidelity. Rather, issues such as light leaks through order-separating filters or diffuse stray light from dusty optical surfaces could be more important.

The amount of stray light outside ±0.03 nm (= 0.3Å; outside the widths of typical spectral lines in cool stars) was introduced by Griffin as a criterion for spectral quality. For the old ESO coudé (Fig. 3), the value was 14.2% for the ruled grating (of which 2.1% in grating ghosts) at a reciprocal dispersion of 0.32 nm mm^{-1}(3.2Åmm^{-1}), and 2.7% for the holographic grating at 0.25 nm mm^{-1} (Dravins 1978a). In higher-dispersion instruments, the Mt.Wilson 100-inch telescope coudé spectrograph (as used for e.g. the Arcturus and Procyon spectral atlases), had 5.3% of which 0.5% in ghosts (Griffin 1969a), and for the Utrecht Solar Atlas it was 7.9% of which 3% in ghosts (Griffin 1969b). Only double-pass or Fourier-transform spectrometers (such as used for modern solar-spectrum atlases) have been shown to reach values less than one percent.

5. Instrumental Effects on Line Profiles and Equivalent Widths

To study instrumental effects in measured spectra, observations may be simulated by convoluting idealized or real astronomical spectra with different instrumental profiles. Fig. 4 shows an example.

Suppose, that one uses observed spectra for the purpose of determining equivalent widths (the total absorption produced by a line). Not seldom, its value is alleged to be independent of instrumentation in the erroneous belief that a convolution with an instrumental profile would not change the total measured energy in a spectral line. Now, consider the spectrum in Fig. 4. The stray light in the instrumental profile has redistributed light between the spectral continuum and the absorption lines, somewhat lowering the former and filling in the latter. Systematically erroneous equivalent widths will be obtained due to two causes. First, part of the absorption of each spectral line has been displaced so far away that it can no longer be identified as being part of that line. For real spectra with noise present and other lines blending the line of interest, an optimistic estimate is that absorption within ± 0.03 nm from the line center will be attributed to that particular line, but what has been displaced beyond this limit will be lost for measurement. Second, measurements are made relative to the spectral continuum, which is drawn through the highest-intensity points, which are then assigned the value "100%". However, the highest points midway between successive spectral lines may reach only 97%, say, of the original value, and in the absence of detailed instrumental-profile studies, one will place the continuum at that level, losing also that part of the absorption that has remained within ± 0.03 nm, but is between the apparent continuum and the true (but unknown) 100% level. A slight compensatory effect enters because

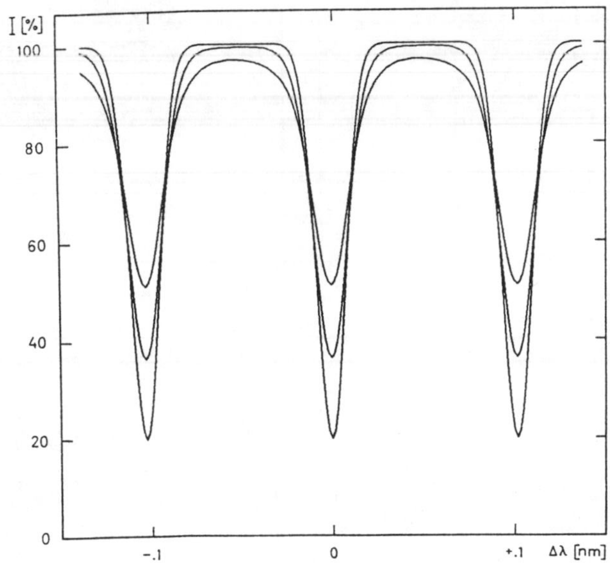

Fig. 4. Instrumental effects on an idealized spectrum with one narrow Gaussian absorption line every 0.1024 nm. Curves with successively smaller amplitudes are: (a) original spectrum, (b) convolution of (a) with the holographic instrumental profile of Fig. 3, and (c) convolution with its ruled-grating profile. The spectral resolutions of the holographic and ruled-grating setups are identical (full width at half-maximum = 5 pm = 50 mÅ), and the differences between curves (b) and (c) arise from instrumental-profile wings only (Dravins 1978a).

the apparent amount of line absorption is now measured in the smaller units of a lowered continuum.

For the two cases shown in Fig. 4, such equivalent widths will be underestimated by 4.2% (holographic) and 20.0% (ruled grating) from their true values. The corresponding value for the Utrecht Solar Atlas is 11.0% (Griffin 1969b).

From time to time, such systematic, instrumentally-induced, differences in equivalent widths or other line parameters are brought to our attention, most often when the spectrum of a previously-studied object becomes observable with superior quality. For example, in the new ESO coudé echelle spectrometer it has been found that *"there are substantial differences between the equivalent widths measured with the short camera plus CCD and those measured with the long camera plus Reticon. ... These differences appear to be wavelength dependent being larger in the blue and can be as large as 20–30% around 3880 Å."* (Crane 1988). From our discussion, we see that this should not really be surprising: e.g., for cool stars with rich absorption-line spectra, effects of such a magnitude could normally be expected in any ordinary high-resolution instrument.

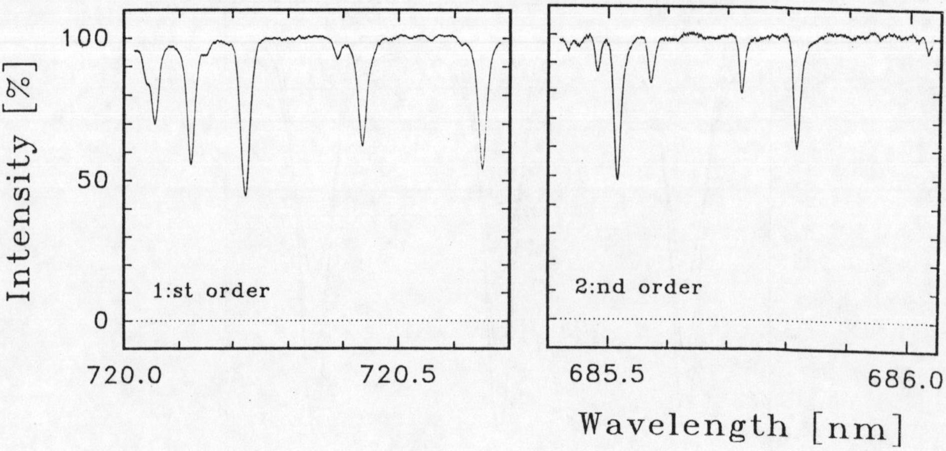

Fig. 5. Instrumentally-induced line asymmetries in the Utrecht Solar Atlas (Minnaert et al 1940). While absorption lines around λ 720 nm are symmetric, asymmetries are seen around λ 685 nm, in the sense of a depressed longward wing (as if a dense solar wind were accreting onto the Sun!). This effect can be traced back to a change of diffraction orders in the spectrograph: the symmetric lines were recorded in first diffraction order, while the asymmetries appeared in the second order.

6. Instrumentally-Induced Line Asymmetries

So far, the discussion has been about instrumental signatures that are *symmetric* about the central wavelength. Obviously, if the spectrometer's instrumental profile is *asymmetric*, it will transform intrinsically symmetric lines into asymmetric ones, correspondingly affecting e.g. radial-velocity measurements. Many circumstances may contribute to causing asymmetric instrumental profiles, e.g. optical off-axis aberrations in the focused spectral image, or out-of-focus aberrations caused by a geometrically-plane solid-state detector intersecting the curved focal surface in a spectrometer. One example of effects from changes in the asymmetry of the instrumental profile between different spectral portions in the same spectrometer is given in Fig. 5.

7. Diffraction Gratings: Efficiency Changes with Wavelength

7.1. EFFICIENCY CHANGES ACROSS SPECTRAL ORDERS

In order to obtain high angular dispersion, diffraction gratings are often used in high diffraction orders m. Various wavelengths $\lambda_1, \lambda_2, \lambda_3$ etc. will all be diffracted in the same direction if the products $m_1\lambda_1, m_2\lambda_2, m_3\lambda_3$ are equal. Ordinary ruled diffraction gratings may have m around 5–10, while *echelle* gratings are normally used in very high diffraction orders ($m \simeq 30$ or more), and then a large number of different wavelengths from successive overlapping orders result. To separate or isolate the wavelengths of interest, an optical device is required, such as a narrow-band filter, or a "cross-dispersing" grating or prism.

Besides the added stray light originating in the extra optical elements, this implies a complication for accurate spectrophotometry, since the optical transmission efficiency of the spectrometer is now a rapidly-varying function of wavelength. The continuum distribution of the measured spectrum will be the true value convolved with this efficiency ("blaze") function. To recover the true energy distribution, the measured spectrum must be divided by this function, which thus must be known to sufficient accuracy. Although this function can, in principle, be obtained by measuring a white-light source, in practice the attainable accuracy is limited by the difficulty to (a) find a true white-light source of the same polarization as the telescopic image of the star, and (b) illuminate the spectrometer under exactly the same conditions as those for stellar measurements. This correction becomes more problematic at positions further from the central wavelength. Suppose, that one measures the efficiency curve from 100% at the centre of an order to perhaps 10% at the edge. An error of 1%, say, in the determination of the 25% efficiency level, will transform to a 4% error in the "rectified" continuum level between the centre of the order and this 25% position. Errors of such a magnitude actually appear quite common. In the near ultraviolet, the *International Ultraviolet Explorer* satellite observes in diffraction orders $m \simeq 100$, and the numerous efforts made during the *IUE* lifetime to improve the echelle "ripple" calibration testify to the practical difficulties of doing this to the desired accuracy.

If the stellar spectrum itself has a clean continuum, that may be used as a calibration. However, if no obvious continuum exists, the problems become more serious, as in the case of broad absorption lines (the hydrogen Balmer lines, Ca II H & K, and similar). One then has to determine which part of the lowered flux away from the centre of a diffraction order is due to lesser flux in the star, and which is due to lower efficiency in the instrument. There are cases where the intrinsic width of an absorption line is comparable to the entire wavelength width of one echelle order (e.g. the Mg II h & k lines in cool stars around λ 280 nm, as observed in echelle orders m = 82 and 83 with *IUE*). Although one attempts to separate the "ripple" in the echelle-blaze efficiency from the analogous "ripple" in the stellar spectrum by observing suitable stars with hopefully featureless spectra, it seems unlikely that any such echelle spectra would generally produce line profiles that are more accurate than some percent at best.

7.2. EXAMPLES OF UNCERTAINTIES IN STELLAR CONTINUUM LEVELS

It is symptomatic that *addenda* or *errata* have been issued to even the highest-quality spectral atlases, after systematic errors in the continuum levels have been realized in the vicinity of broad absorption lines with extended wings, where the true continuum level is never reached.

In Fig. 6, we show an example of such a *corrigendum*, issued for the Jungfraujoch Atlas. Replacement sheets were published some years after first issue, covering spectral portions around the strongest absorption lines. A better positioning of the local continuum around Hα and Hβ had then been achieved by obtaining compressed tracings at high resolution but low dispersion. This example should absolutely not be seen as any criticism of the authors of this particular atlas—their work represented the state of the art at the time—rather, this should be seen as an illustration of the general difficulties inherent in the fitting of continua of spectra recorded with any grating spectrometer used in higher diffraction orders (e.g. here $m = 12$ around Hβ).

7.3. LIGHT LEAKS THROUGH ORDER-SEPARATING FILTERS

The challenge of high-order echelle spectroscopy can also be considered from the need to use a cross disperser and/or filter(s) in order to separate the typically thirty or more overlapping orders. Suppose that one is to measure the flux in a deep absorption line (5% of the continuum, say) in a spectral region where the flux is lower than the average. The total illumination from the [mostly near-continuum flux] in the $\simeq 30$ overlapping orders may well be more than 1000 times greater than the signal to be measured. To measure this spectral line flux to within 1% of its value, would then demand a cross-disperser blocking efficiency of 10^{-5} against stray and parasitic light. Nothing even approaching such values appears ever to have been demonstrated in light-efficient instruments.

The problems are analogous for *Fourier-transform* spectrometers used in the visible, where the measuring precision is limited by photon noise. A narrow-bandpass filter must be used to constrain the spectral interval falling onto the detector at any one time (to limit the photon noise), and the correction for the wavelength transmission profile of such a filter is analogous to the correction for echelle "ripple". (And an *FTS* has an additional complication in common with all scanning instruments: the need to monitor and compensate brightness changes of the object.)

7.4. GRATING ANOMALIES

In contrast to the reflectance of light from aluminized mirrors, the reflective efficiency of diffraction gratings may show sudden variations with wavelength that are different for light of different linear polarizations. As a result, measurements made with grating spectrometers of the energy distribution in stellar spectra may show fortuitous shallow spectral features, or even sharp discontinuities, making accurate energy-distribution measurements quite difficult, especially for partially-polarized starlight.

The presence of such diffraction-efficiency *anomalies* ("Woods" anomalies, and others) does not result from any deficiency in the grating manufacture, but rather is a fundamental property of the interaction between light and matter. The mechanism is the coupling between the electromagnetic field of the incident light wave, with guided plasma waves in the electrically conducting metal overcoating of the grating. Such anomalies can be minimized by careful choice of various grating parameters, and the optical layout of the spectrometer. For a further discussion, see e.g. Dravins (1978b) and Gray (1992).

7.5. HOW CAN THESE PROBLEMS BE AVOIDED?

These comments should not necessarily be read as criticisms against the current

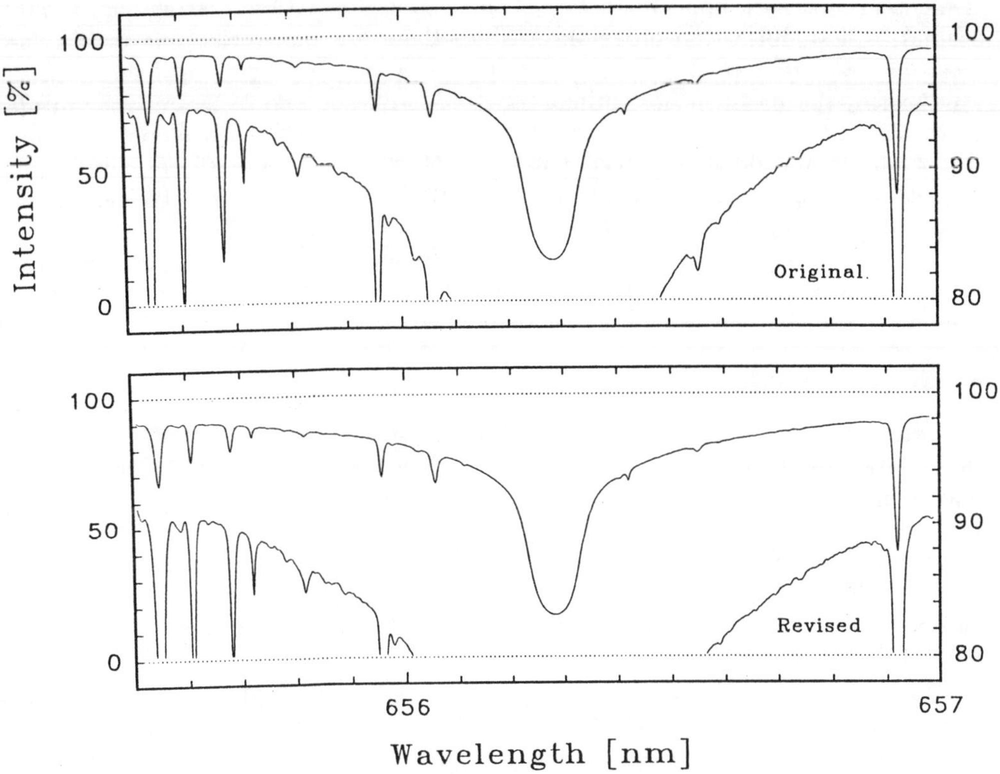

Fig. 6. A spectral region around Hα in two versions (original and revised) in the Jungfraujoch Solar Atlas (Delbouille et al.: original 1973; revision 1979). The differences arose because of the difficulty to calibrate the continuum level in a grating spectrometer used in a high diffraction order ($m \simeq 10$). The widths of broad absorption lines are comparable to the intervals over which instrumental efficiency changes, and then it is no longer possible to use simple continuum-fitting techniques. Errors of comparable magnitude are likely to be present in most stellar spectra, and data from higher-order echelle spectrometers could be considerably worse. The intensity scale at left is for the full-range plots, while that at right is for the expanded lower curves.

trend of equipping telescopes with echelle spectrometers. However, it should be understood that the reasons for doing so are to (a) utilize their high blaze efficiency in order to reach fainter objects; (b) utilize their high angular dispersion to reach high spectral resolution with moderate-size instruments; and (c) obtain the output spectra in a 2-dimensional format that is well adapted to current solid-state array detectors.

However, if the aim would instead be to obtain *accurate* stellar line profiles (to use for hydrodynamic modeling of stellar atmospheres, accurate synthetic photometry, etc.), a different approach should be taken. The best spectrally- dispersive element in this case appears to be a holographic diffraction grating, used (and usable) in first diffraction order only. Since there are no overlapping orders, no cross-disperser is needed. Since the grating efficiency changes only slowly with wavelength, the problem of continuum fitting is one or two orders of magnitude easier than in echelle spectrometers; thanks to the microscopically smooth grating surface, its stray light is no higher than that from any other smooth optical component; there are in principle no grating ghosts at all, correspondingly decreasing the straylight contamination in the spectrum. However, the light efficiency will probably be somewhat lower, and the image of the focussed spectrum will be stretched out in a long strip and, using ordinary-size optical elements, probably only a modest fraction of the spectrum could be recorded at any one time.

8. Effects of Limited Spectral Resolution

The discussion up to now has concerned effects which are largely independent of the spectral resolving power, $\lambda/\Delta\lambda$. This is often taken as the full width at half-maximum of the instrumental profile. It is a measure of how closely adjacent spectral features can be separated, but it need not be directly connected with the spectrophotometric accuracy. For example, the problems in continuum normalization (and consequent errors in, e.g., equivalent widths), exemplified in Fig. 6, arose at a resolution $\lambda/\Delta\lambda$ approaching 10^6, much higher than that encountered in any normal stellar spectroscopy.

However, in most stellar applications, the spectral resolution achievable is much more limited, and actually determines the appearance of the measured spectrum. This has effects both for the spectral-line profiles as such (e.g. equivalent widths), and for the quantities deduced from them (such as radial velocities).

8.1. EFFECTS ON LINE PROFILES

As an example of the effects of observing a normal star under more ordinary resolutions, simulated observations of the Sun are shown in Fig. 7. The instrumental profiles are those measured several years ago for cameras of different focal lengths in a coudé spectrograph at Observatoire de Haute-Provence (Andersen & Dravins 1982), but are probably representative for many other spectrometers of moderate resolution.

Fig. 7 shows that, even using a higher resolution of $\lambda/\Delta\lambda \simeq 23,000$, the degradation of the true spectrum is quite severe, and for a lower resolution it is quite profound. At such resolutions, no lines (at least in cooler stars) remain unblended, and quantities like "equivalent widths" and "linewidths" bear little relation to those in the true spectrum.

Since the measured features (e.g. bottom panel in Fig. 7) easily give the impression of a resolved spectrum, it is a common pitfall to believe that these bumps are the actual spectral lines, rather than merely those low spatial-frequency components of the spectrum that have survived the spectrograph. Obviously, great caution must be exercised in deducing astrophysical information from such resolution-degraded spectra.

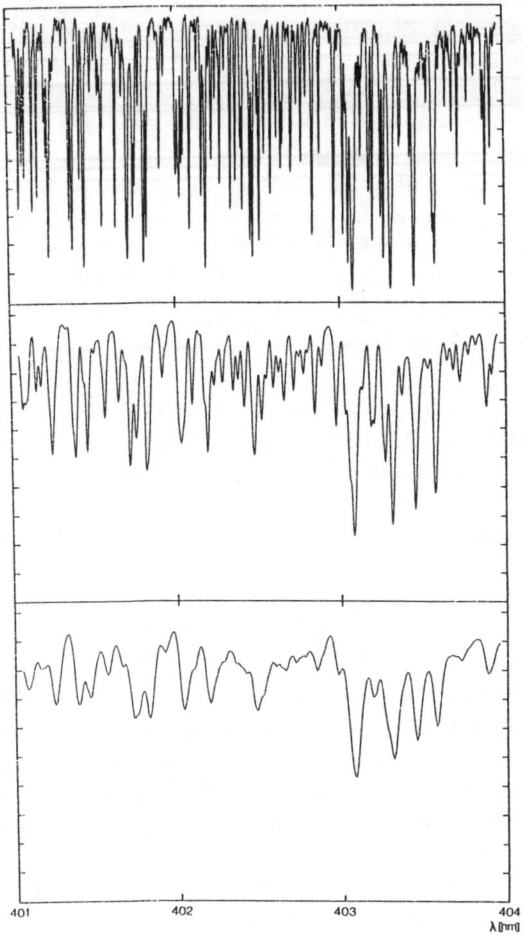

Fig. 7. Numerically-simulated observations of a normal star (Sun, G2 V) under different spectral resolutions. Top: "fully" resolved spectrum ($\lambda/\Delta\lambda \simeq 300,000$). Center: convolution with the instrumental profile for a resolution $\lambda/\Delta\lambda \simeq 23,000$. Bottom: same for $\lambda/\Delta\lambda \simeq 8,000$ (Andersen & Dravins 1982).

8.2. EFFECTS ON APPARENT RADIAL VELOCITIES

Perhaps less obvious are the effects of finite resolution on secondary parameters deduced from the observed line profiles. For example, changing the spectral resolution may change the *apparent radial velocity* of a star, or mimic a velocity dispersion in a star cluster, with an amplitude that depends on the resolution. This holds even for completely unblended lines, and using a spectrometer with a perfectly symmetric instrumental profile.

Such effects arise because stellar spectral lines are intrinsically *asymmetric* to some degree: the instrumental convolution of an asymmetric stellar line with even a perfectly symmetric instrumental profile, results in a line of different asymmetry. Intrinsic line asymmetries may arise from velocity-brightness correlations in convective motions (granulation), expanding stellar winds, oscillations, or other causes. The magnitude and sign of such asymmetries changes between different stars, and between lines of different strengths in the same star. Various lines will be differently affected, possibly mimicking some astrophysically-expected behavior. In the solar spectrum, such asymmetries are quite small but, nevertheless, instrumentally-induced wavelength shifts reach $\simeq 100$ m s^{-1} already at high resolutions $\lambda/\Delta\lambda$ around 100,000 (Bray & Loughhead 1978; Dravins 1987). In many other stars, the effects must be at least one order of magnitude greater.

9. Photographic Spectroscopy

If one is to search for stellar spectral variability over several decades, one will likely have to use archival material in the form of photographic spectrogrammes. The analysis of such recordings carries additional complications due to the nature of the photographic process. First, one must calibrate the non-linear response (photographic density versus exposed intensity), and one may need to estimate the instrumental response function for the whole system of spectrograph + photographic emulsion + microphotometer.

Non-linear effects and non-uniformities in photographic processing cause problems for the photometric calibration. For the conversion from photographic density to intensity, characteristic curves are obtained from suitable calibration exposures. However, even if these are obtained under apparently the same conditions (same wavelength, same exposure duration, same latent image decay time, etc.) as the stellar exposures, the calculated intensities will not, in general, be the true ones due to *adjacency effects* and other chemical influences in photographic processing. These result in the same exposed intensity giving different densities, depending on the spatial scale of the image and the exposure of adjacent emulsion areas. Such effects may cause the measured equivalent widths to depend on, e.g., the photographic density.

The number of developed photographic grains in a particular area of the emulsion depends not only on the exposure of that area, but also on the number of developable grains in adjacent areas. Fresh developer diffuses from areas with few developable grains to areas with many, making the latter more effectively developed than the former. How severe this effect will be depends on several parameters, and it is normally counteracted through mechanical agitation during development. The resulting *chemical modulation transfer* normally has the character of enhancing the contrast in sharp spectral lines ("edge effect"), thus increasing the measured equivalent widths of absorption lines. This effect, thus, may partially compensate the spectral degradation that occurs through stray light, and leads to the (somewhat paradoxical) situation, that the change from photographic to electronic recording in the same spectrograph probably *decreases* the *accuracy* of spectral measurement (although the *precision* improves): in the absence of photographic adjacency effects, the detector no longer "compensates" for the optical degradation in the spectrometer. For more extensive discussions of photographic effects in stellar spectrophotometry, see Lind & Dravins (1980) and references therein.

10. Solar Spectrum as Calibration

For any long-term monitoring programme, a calibration source is required that stays at least approximately constant for the duration of the programme. Although various calibration lamps can be thought of, none appears better than *integrated sunlight*. It is the most stable source that is, and has been, generally available, and it should remain essentially constant for decades or centuries. High-fidelity recordings are available in published atlases (in both graphical and digital forms; Beckers et al. 1976; Kurucz et al. 1984), and thus are readily available for comparison (but do *not* use atlases of the *disk-centre* spectrum, which can be different on a level of several percent from integrated sunlight). Further, the wavelength structure of the solar spectrum (a mixture of absorption lines of varying strengths) much better approximates that of other stars, than any laboratory emission-line source. Therefore, effects from optical straylight in the spectrometer, or detector peculiarities can be determined in a more realistic manner.

To obtain truly-integrated sunlight is non-trivial but, usually, a good approximation can be obtained in the form of *daylight* or *moonlight*. These are not exactly identical to integrated sunlight: the light of the sky has a component that has undergone incoherent, wavelength-redistributing scattering against air molecules; some light may have undergone incoherent scattering from high-albedo ground portions (e.g. crystalline snow) and subsequent rescattering by molecules or aerosols, and there is a superposed spectrum of dayglow emission. Under ordinary meteorological conditions, however, such effects in daylight normally amount to a filling-in of absorption lines of no more than 1–2% of the continuum level, an effect much smaller than typical instrumental degradations. Moonlight is physically more similar to integrated sunlight. Such spectra can be readily obtained, and already exist in practically all spectral archives. A comparison of such data with solar-spectrum atlases may then enhance considerably the understanding of the data quality. For a more detailed discussion of the differences between integrated sunlight, skylight, and moonlight, see Lind & Dravins (1980) and Kattawar et al. (1981).

11. Pitfalls in Archival Research

Efforts are underway at various ground-based and space observatories to produce catalogues and archives of observations, in order to make the data available for retrieval and analysis also by scientists who never took part in the original observations.

Obviously, this can represent a significant enhancement of the scientific return from an observatory. At the same time, however, it introduces new types of problems if the data are incompletely or incorrectly documented, or contain instrumental signatures that users, who are not familiar with the precise way the data were recorded, may find awkward to identify. The currently largest archive for stellar spectroscopy is the data base of $\simeq$ 100,000 spectra recorded by the *International Ultraviolet Explorer* since its launch in 1978. A score of pitfalls that users may encounter were collected in a series of short papers published under the title *Science Fiction with IUE* (Imhoff & Grady 1985; Grady & Imhoff 1985; 1986), from which we select the following sample of "discoveries":

(a) *Strong V II emission in the spectrum of a T Tauri star*: Strong diffuse emission was discovered in the spectrum of a star having strong Mg II emission. The explanation, however, was an oblique-incidence cosmic-ray hit on the detector, intersecting the exposed spectrum. Long exposures are commonly contaminated by sharp "emission features" due to cosmic-ray hits and radioactive decays in the camera phosphor. These are normally identified by the software which looks for sharp and localized maxima. However, cosmic rays that hit obliquely to the camera faceplate produce extended "comets" of enhanced exposure levels, in this case mimicking a diffuse emission.

(b) *B-shell star found to be binary*: Archival spectra of a bright B-shell star were found to show variations in the UV continuum fluxes by a factor of 4, suggesting the star to be an eclipsing binary. The explanation, however, is that the series of archival spectra is a mixture of observations made through two different spectrometer apertures. Due to the optical layout of the spectrometer, its throughput depends upon precisely where onto the apertures the starlight enters, in some cases leading to flux differences by this factor of four. Thus, such data are not sufficient to tell whether this object is indeed a binary.

(c) *Variability in Mg II line profiles*: Different spectra of a T Tauri star showed changes in its Mg II emission lines. The emission peaks changed from rounded, flat-topped profiles to sharp emission peaks at higher flux levels. However, this "variability" could be traced back to some software routines, where the intensity calibration was truncated at some exposure level corresponding to the deepest flat-field exposure, although the detector had not reached its saturation exposure level. Since the data-reduction software for most users has the character of a "black-box", and this particular situation was not being flagged, the cause could not be identified by the data user.

(d) *A transient stellar wind detected in Cr III*: In a monitoring programme of the star 32 Cyg, a P Cygni profile in one Cr III line appeared. Since it was absent both one month earlier and later, this suggested a mass-ejection episode. The explanation, however, could be traced to a data-telemetry dropout. The data are transmitted for consecutive pixels in successive horizontal rows: a dropout impairs the data in some horizontal row. The exposed echelle spectrum, however, has its spectral orders at an angle with respect to the data-scan direction. The spectrum extraction involves subtracting a background level along an exposed echelle order. Because of the angle between the echelle orders, and the data-scan direction, a disturbance intersects the background and the stellar spectrum at somewhat displaced wavelength positions, and the subtraction of one from the other leads to a P Cygni type profile with an apparent absorption on one side, and an emission on the other.

What is common for these four examples (and others), is that these instrumental signatures could not be identified from data quality flags or other information in the archive for *reduced data*. However, all of them could be traced by having an experienced observer examine the raw data, in particular images of the spectrogrammes. There is a conclusion to be drawn for future data archiving: the need to keep also the *raw data* available for future inspection (besides data in reduced formats). It may be next to hopeless to determine the reality of an unusual feature in a spectrum, unless one can examine whether there was a mechanical scratch in the original photographic emulsion, or the CCD had a cosmic-ray hit. The effort likely to be involved at major observatories in storing, e.g., all the raw CCD images containing exposed spectra (together with all their calibration exposures), and keeping them available for a long time in the future, may appear forbidding: however the alternative—archives containing reduced spectra only—will likely produce many fortuitous "discoveries" stemming from instrumental or observational coincidences.

Generally, "instrumental" effects in archival research are increasingly likely to originate from complex software packages (which almost certainly must contain errors), which are most often not completely documented, and are commonly being used as "black boxes" by persons with limited previous experience. One example in the field of radial-velocity software is given by Fairall (1992).

12. Long-Term Spectral Monitoring: Stellar-Activity Cycles

One field which calls for long-term spectroscopic observations concerns the monitoring of stellar-activity cycles. Such studies began in the early 1960s by observations of Ca II K chromospheric fluxes in cool main-sequence stars. A large number of stars have now been monitored for many years, and a fair number of cycles have been identified, the shortest of less than 3 years (Baliunas & Vaughan 1985).

Besides Ca II H & K, the *h* & *k* lines of the Mg II resonance doublet near λ 280 nm are sensitive indicators for chromospheric activity. In cool stars, the lower flux in the photospheric continuum in the ultraviolet increases the visibility of the Mg II emission, and permits studies also of very low activity levels, e.g. those in old stars. To study the secular evolution of solar-type chromospheres, a comparison of the present *Sun* (G2 V) was made with the very old ($\simeq$ 9.5 Gy) star β Hydri (G2 IV). Its Mg II *h* & *k* variability was monitored with the *IUE* satellite for about 12 years, recording some 100 high-resolution line profiles.

At first sight it could appear that such a space programme would be ideal for long-term monitoring: exactly the same instrument is used continuously (and there is *no* human intervention in the spectrometer!). However, when examining the data series, significant "variability" was found, largely originating from innumerable changes in the *IUE* data-reduction software. One year, the photometric calibrations were improved, another year the spectrum -extraction algorithms, a third year the wavelength calibration, a fourth year the method of smoothing the background level, etc.

Although each had been introduced with the purpose to improve the data, these perennial software changes precluded any meaningful identification of astrophysical changes in the feeble chromospheric emission of β Hyi. The chromospheric Mg II changes were small, and indeed comparable in magnitude to the inaccuracies of the standard *IUE* data-reduction software. To enable a study of real stellar variability then required a reexamination of the entire processing of the *IUE* spectral images, and the subsequent extraction of the spectra, starting with the "raw" data.

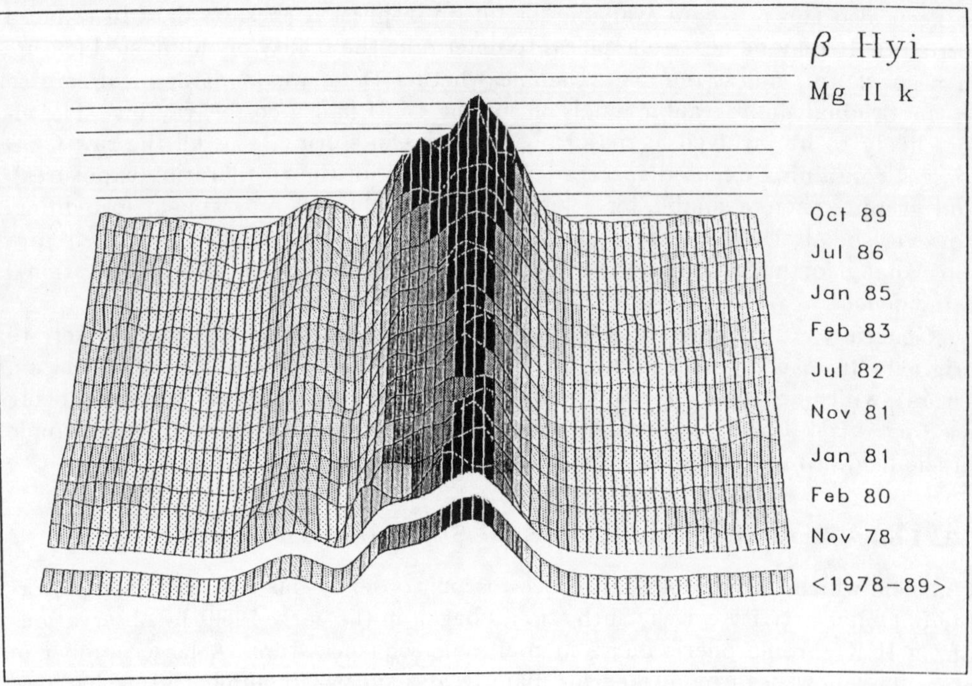

Fig. 8. Long-term variability of Mg II *k* chromospheric emission in the old star β Hyi (G2 IV). Data from 17 epochs build up a surface of intensity vs. wavelength. This plot illustrates the slight decrease of activity for a few years after 1978, followed by the build-up to maximum in 1986. The data strip in front is the average profile. The variability is consistent with an activity cycle somewhat longer ($\simeq$ 15–18 years) than the solar one. Assuming the Ca II emission to follow that of Mg II, this represents the lowest-amplitude stellar cycle so far detected (Dravins et al. 1993).

After this had been done, the signal-to-noise ratios could be improved to reveal the astrophysical variability (Fig. 8). As a by-product of this particular project, a number of *IUE* data-processing deficiencies could be identified, and the reduction methods were subsequently adopted in the new software currently being used by NASA and ESA for the reprocessing of the entire *IUE* data archive (Dravins et al. 1993).

13. Conclusions: Cause for Optimism

The reader should not have been discouraged by the above display of instrumental imperfections. A common property of them all, is that they are systematic in nature, and do not have the character of [random] noise. Further, virtually all such types of instrumental effects seem understandable, at least in principle (in contrast to our often much deeper ignorance of astrophysical systems). Thus, their magnitudes can be determined, and their effects corrected for. The conclusion to be drawn is that it is indeed profitable to examine carefully instrumental effects, and a subsequent correction for these may then yield spectra of substantially higher astrophysical credibility.

Acknowledgements

This work is supported by the Swedish Natural Science Research Council and the Swedish National Space Board.

References

Andersen, J., Dravins, D.: 1982, *Publications of the ASP*, 94, 390
Baliunas, S.L., Vaughan, A.H.: 1985, *Annual Review of Astronomy and Astrophysics*, 23, 379
Beckers, J.M., Bridges, C.A., Gilliam, L.B.: 1976, *A High Resolution Spectral Atlas of the Solar Irradiance From 380 to 700 Nanometers*, Sacramento Peak Observatory
Bray, R.J., Loughhead, R.E.: 1978, *Astrophysical Journal*, 224, 276
Crane, P.: 1988, *The Messenger*, 52, 52
Delbouille, L., Roland, G., Neven, L.: 1973, *Photometric Atlas of the Solar Spectrum From λ 3000 to λ 10000*, Institut d'Astrophysique de l'Université de Liège ["Jungfraujoch Atlas"]
Dravins, D.: 1978a, *Applied Optics* 17, 404
Dravins, D.: 1978b, in *High Resolution Spectrometry*, ed. M. Hack, Trieste, 221
Dravins, D.: 1987, *Astronomy and Astrophysics*, 172, 200
Dravins, D., Linde, P., Fredga, K., Gahm, G.F.: 1993, *Astrophysical Journal*, 403, 396
Fairall, A.P.: 1992, *The Observatory*, 112, 286
Grady, C.A., Imhoff, C.L.: 1985, *IUE NASA Newsletter* 28, 140
Grady, C.A., Imhoff, C.L.: 1986, *IUE NASA Newsletter*, 29, 46
Gray, D.F.: 1988, in *The Impact of Very High S/N Spectroscopy on Stellar Physics*, eds. G.Cayrel de Strobel, M.Spite, IAU Symp. 132, p. 185, Reidel, Dordrecht
Gray, D.F.: 1992, *The Observation and Analysis of Stellar Photospheres*, 2:nd ed., Chapter 3, Cambridge University Press
Griffin, R.F.: 1969a, *Monthly Notices of the RAS*, 143, 319
Griffin, R.F.: 1969b, *Monthly Notices of the RAS*, 143, 361
Imhoff, C.L., Grady, C.A.: 1985, *IUE NASA Newsletter*, 26, 66
Kattawar, G. W., Young, A. T., Humphreys, T. J.: 1981, *Astrophysical Journal*, 243, 1049
Kurucz, R.L., Furenlid, I, Brault, J., Testerman, L.: 1984, *Solar Flux Atlas From 296 to 1300 nm*, National Solar Observatory
Lind, J., Dravins, D.: 1980, *Astronomy and Astrophysics*, 90, 151
Minnaert, M., Mulders, G.F.W., Houtgast, J.: 1940, *Photometric Atlas of the Solar Spectrum*, Sterrewacht Sonnenborgh Utrecht, Amsterdam ["Utrecht Atlas"]
Moore, C.E., Minnaert, M.G.J., Houtgast, J.: 1966, *The Solar Spectrum 2935 to 8770 Å*, Nat. Bur. of Standards Monograph 61, Washington

DISCUSSION

Young: The $1/r^2$ wings due to surface scattering are as important in telescopes as in spectrographs; a considerable percentage of the light in a star image is in these wings. Telluric lines are unresolved at $R = 500\,000$ because the resolution required is the ratio of the speed of light to the speed of sound, which is 10^6.

Although you discounted the differences between the light of the Sun, sky, and clouds, nonconservative molecular scattering (due to translational and rotational motion of air molecules) can produce variations of tens of per cent at high resolution, depending on polarisation state and viewing geometry, resulting in large line-filling in by moonlight. So, old plates are not necessarily reliable comparison spectra.

Dravins: When observed under spectral resolutions that are commonly used in stellar spectroscopy, the differences between integrated sunlight, skylight and moonlight, are very small compared to other likely sources of error (especially if one is using inadequately calibrated spectra from old archives). The physical differences between sunlight and skylight can be seen under very high resolutions as a slight filling-in of line-bottoms in the latter (due to e.g. inelastic Raman scattering of sunlight against air molecules). For a further discussion of these differences, and the use of skylight as stellar comparison, see Lind & Dravins (1980, *Astronomy and Astrophysics*, 90, 151).

Harmanec: You recommended to correct the instrumental effects of the spectrograph using the solar spectrum. At the same time you showed us that even the best solar-spectrum atlases differ from each other. Which one, then, should we use?

Dravins: One can safely use any of the recent atlases of the solar irradiance (i.e. of the integrated flux, and **not** the solar disk-centre spectrum). There are two atlases, each of which is fully adequate for stellar comparisons:

1. Beckers, J.M., Bridges, C.A. and Gilliam, L.B. 1976 *A high resolution spectral atlas of solar irradiance from 380 to 700 nm*, Sacramento Peak Observatory.
2. Kurucz, N.L., Furenlid, I., Brault, J. and Testerman, L. 1984, *Solar Flux Atlas from 296 to 1300 nm*, National Solar Observatory, Sunspot, N.M.

Harmanec: If instrumental effects affect the radial velocity for something like a few tenths of km s^{-1} for cool stars, how is it possible that, for instance, our Canadian colleagues (Walker, Young, Campbell,...) are able to achieve an accuracy of about 15 m s^{-1} in their search for planetary systems?

Dravins: Although it is not directly possible to obtain the true velocity of the stellar centre of mass with an accuracy better than perhaps 0.5 km s^{-1}, measurements of *relative* displacements of spectra are possible on the level of some m s^{-1}. In the case of the Canadian group, the shift of the stellar spectra is measured relative to the sharp lines of a laboratory source.

Sterken: The various instrumental effects you have shown clearly illustrate that in spectroscopy—as in photometry, see Young 1984 (in *Proceedings Workshop Improvements to Photometry*, NASA Conf. Publ. 2350, 8)—the accuracy routinely achieved today is of the same level as the expert work which was done decades ago. This phenomenon has a very dangerous potential, since nowadays the shift to use mainly large telescopes causes the importance and popularity of an auxiliary instrument to be measured by its sensitivity, rather than by its instrumental fidelity.

Dravins: Yes, spectrometers using echelle gratings in very high diffraction orders are being chosen because they offer a high angular dispersion (which matches the stellar image sizes in large telescopes) and because their output format matches two-dimensional electronic detectors. However, from a spectral fidelity point of view, they are a poor choice and are unlikely to permit any line-profile measurements at levels better than ~1%. The best spectral fidelity would result from holographic gratings used in first diffraction order.

Dravins, Griffin

de Groot and Wolf – talking LBVs?

INVESTIGATION OF LUMINOUS BLUE VARIABLES OF THE MAGELLANIC CLOUDS DURING THE PAST DECADE WITH LTPV, CASPEC AND IUE

B. WOLF
Landessternwarte
Heidelberg-Königstuhl
Königstuhl 12
69117 Heidelberg
Federal Republic of Germany

ABSTRACT. The Luminous Blue Variables (LBVs) R40, S Dor, R71, and R127 of the Magellanic Clouds have been photometrically and spectroscopically monitored by LTPV, and CASPEC and IUE, respectively, during the past decade. From these coordinated observations considerable insights in the physical nature of LBVs have been gained. A few highlights will be discussed in this review. Due to the good time resolution of the LTPV observations, microvariations of the order of $0\overset{m}{.}1$ to $0\overset{m}{.}2$ in time scales of months have been found to be a general feature of LBVs. Particularly fascinating envelope-contraction episodes of S Dor and R127 have been caught with LTPV, CASPEC and IUE. R40, the first LBV of the SMC, has been discovered and thoroughly investigated. Since the metallicity of the SMC is widely believed to be low, R40 is particularly important in connection with late phases of the evolution of massive stars in a low-metallicity environment.

Keywords: Luminous Blue Variables – Magellanic Clouds

1. Introduction

Almost exactly ten years ago the issue of Sep. 1983 of "The Messenger" contained three articles closely related to my talk. D'Odorico et al. (1983) presented a description of the ESO echelle spectrograph CASPEC for the Cassegrain focus of the 3.6 m telescope (i.e. one of the first echelle spectrographs equipped with a CCD) and showed first results. Sterken (1983) reported about LTPV which he had initiated one year before (see also Sterken, 1994). Wolf and Stahl (1983) contributed the story of the discovery of an S Dor-type outburst of the Of star R127 of the LMC where the combined potential of LTPV and high-resolution IUE observations for the investigation of LBVs was demonstrated. Thus, that issue of "The Messenger" virtually marks the beginning of a decade of successful investigations of LBVs of the MCs with new methods and new instrumentation. A few highlights obtained during this period with LTPV, CASPEC and IUE are presented here.

C. Sterken and M. de Groot (eds.), The Impact of Long-Term Monitoring on Variable Star Research, 291–302.

2. Main Characteristics of LBVs

The main properties of LBVs have been described in various reviews (cf. e.g. Hillier 1991, Stahl 1993, Wolf 1992). The LBVs of the Magellanic Clouds are particularly useful since their distances are reliably known and, thence, their location in the HRD. Apart from supernovae during outburst, LBVs are the most luminous stars of the Universe with visual magnitudes of typically -9 to -11. At quiescence, the LBVs define an inclined instability strip (spectral type OB) in the sense that the hottest members are the most luminous ones. At maximum the spectral type typically is middle A. The photometric variations between minimum and maximum are typically of the order of 1 to 2.5 magnitudes in the visual range and occur under the condition of practically constant bolometric magnitude. The time scales of these major outbursts are years, decades or longer. In addition, photometric microvariations of about $0\overset{m}{.}1$ to $0\overset{m}{.}2$ on time scales of months are observed (see below). The mass-loss rates are high (typically 10^{-5} to 10^{-4} $M_\odot$ yr^{-1}) and the expansion velocities are low (typically 100 to 150 km s^{-1}) during outbursts so that the wind densities are very high. LBVs represent ideal laboratories of unique wind characteristics, i.e. slow, dense and cool winds around hot stars.

3. The case of R71

Mainly for historical reasons, I want to report briefly some results obtained for R71 of the LMC. R71 was the first LBV observed during both maximum (B4Iae) and minimum (middle A) by high-dispersion spectroscopy (Wolf et al. 1981). It was also the first LBV of the LMC observed in the long-wavelength range of IUE in the high-resolution echelle mode. The light curve shown in Fig.1 demonstrates immediately the advantage of continuous monitoring. Whereas the snapshot observations obtained for this star before 1975 are sufficient to describe the large amplitude variations of LBVs, the microvariations of R71 of the order of $0\overset{m}{.}1$ to $0\overset{m}{.}2$ on time scales of months have only been detected through continuous monitoring by van Genderen (1978) and by the LTPV group. These campaigns have shown for the first time that, apart from the large outbursts, pulsation-like motions in the expanding envelopes of LBVs do occur.

The high-resolution observations with IUE of 1981 (Wolf et al. 1981) have shown the wind of R71 to be cool (below 10 000K) and slow (v_∞=130 km s^{-1}). An early CASPEC observation of R71 in the red spectral range belongs to the first high-resolution (about 20,000) spectrogrammes obtained of any hot LMC star in this spectral range with a high S/N ratio. Very narrow (40 km s^{-1}) nebular lines of [N II] 6548, 6583 Åwere detected (Stahl and Wolf, 1986). These observations have proved IUE and CASPEC to be very efficient for investigating the physical conditions in the envelopes of LBVs. Subsequently, R71 has even been identified as an IRAS point source evidencing the existence of a very loosely-bound envelope with a temperature of only 140 K (Wolf and Zickgraf, 1986).

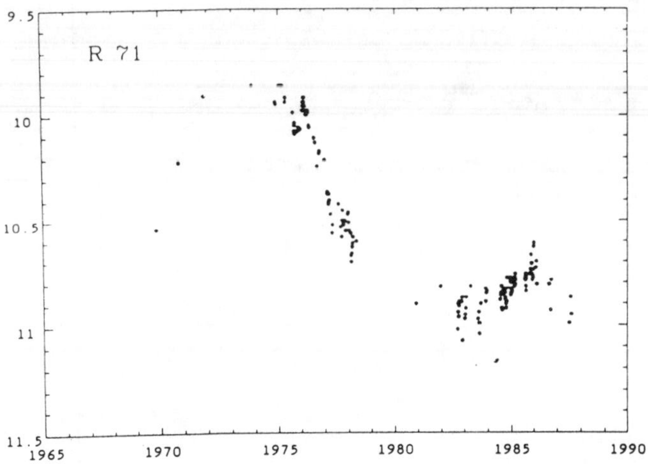

Fig. 1. Light curve of the LBV R71 in the LMC.The observations prior to 1978 were taken from Van Genderen (1978) and the sources quoted by him. Observations after 1982 are from the LTPV group, initiated by Sterken. Arrows denote epochs at which high-resolution spectrogrammes were taken.

4. Envelope-Contraction Episodes of S Dor and R127 of the LMC

S Dor has been photometrically monitored since the beginning of the LTPV campaign. Likewise, it has been monitored with CASPEC and IUE during the same period with a frequency of typically one spectrum per year. Leitherer et al. (1985) have carried out a thorough discussion of the coordinated LTPV, CASPEC and IUE observations obtained before 1985. The light curve from the LTPV observations after 1985 is shown in Fig. 2: S Dor has become brighter than magnitude eight in 1988, which represents a historical maximum. It has been declining since January 1989. Fortunately, this interesting phase is well covered by coordinated CASPEC observations, indicated by arrows in Fig. 2. The spectra shown in Fig. 3 provide, for the first time, the kind of information, necessary for a better understanding of the physical processes correlated with outburst maxima of LBVs. As shown by the sequence, S Dor exhibited extraordinarily strong P Cygni-type profiles in 1987 indicating a particularly high mass-loss rate during this maximum phase. The CASPEC spectrum taken in January 1989, shortly after the maximum, exhibits tiny inverse P Cygni-type absorption features which became progressively stronger and were present for about two years. In December 1991, when S Dor had already become fainter by $0\overset{m}{.}7$, the inverse P Cygni-type absorption components had practically disappeared again. Interestingly, the spectral appearance of S Dor in the satellite-UV range has not changed strongly during the past decade

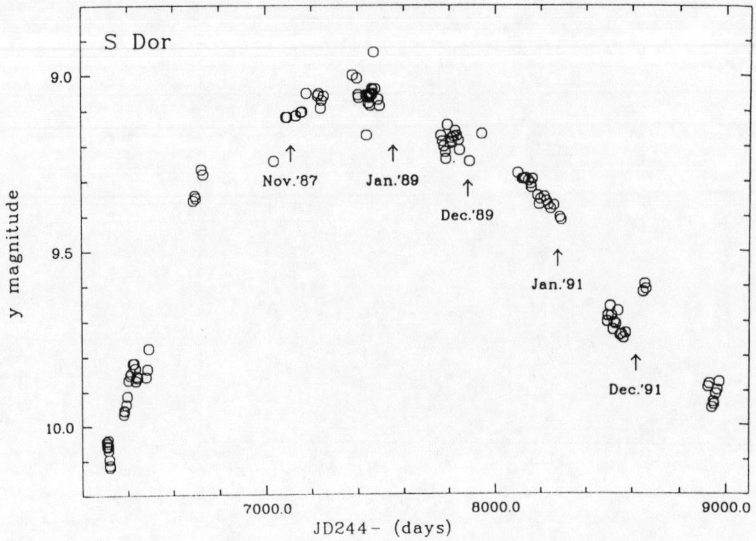

Fig. 2. Light curve as observed by the LTPV group during the the historical maximum of S Dor of the LMC. Coordinated CASPEC observations are indicated by arrows.

(cf. also Wolf, 1990). A plausible interpretation of these observations is provided by the simple model sketched in Fig. 4. During the ascension to the historical maximum S Dor has pushed so much matter into its envelope that it could no longer drive it to infinity. Part of the matter of the inner part of the envelope moved back as evidenced by the inverse P Cygni absorption profiles. The receding envelope is extended enough to account for the long duration (about two years) of the contraction episode. The strong UV absorption lines observed by IUE are supposed to be formed mainly at very large distances from the stellar core so that gravitational forces can no longer decelerate this matter. This would explain why no major changes in the satellite-UV spectrum have been observed so far. During maximum the massive dense envelope cools very efficiently so that hydrogen recombination occurs, i.e. the opacity goes down, and the pressure exerted to the envelope diminishes, allowing the inner part of the envelope even to recede. This interpretation is generally in accordance with a previous theoretical interpretation of the so-called wind limit of LBVs of Davidson (1987).

Interestingly, a very similar envelope-contraction episode was observed shortly after in the case of R127 which reached its historical maximum in 1989 as evidenced by the LTPV light curve shown in Fig. 5. The coordinated CASPEC observations (see Fig. 6) with the appearance of inverse P Cygni-type profiles in January 1991 clearly show that very similar physical envelope processes as found for S Dor have occurred.

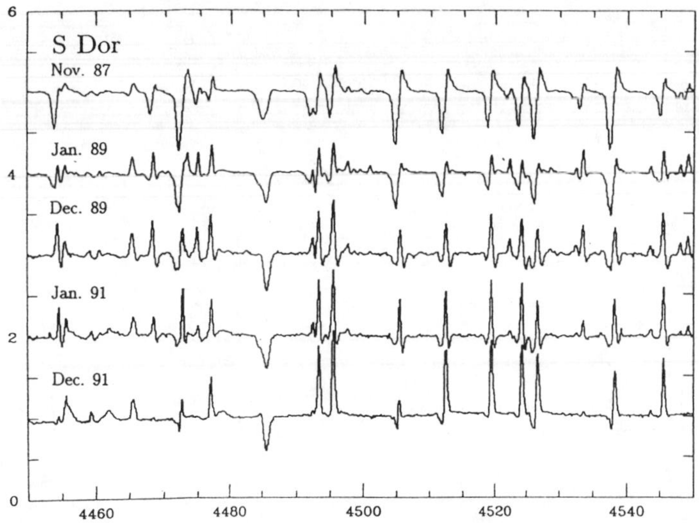

Fig. 3. Section of CASPEC spectra of S Dor taken around the historical maximum. Whereas strong P Cygni-type profiles of singly ionized metallic lines were observed in Nov.1987, intriguing inverse P Cygni profiles were observed in Dec.1989 for the first time. Obviously an envelope contraction phase lasting for about two years was observed (see text)

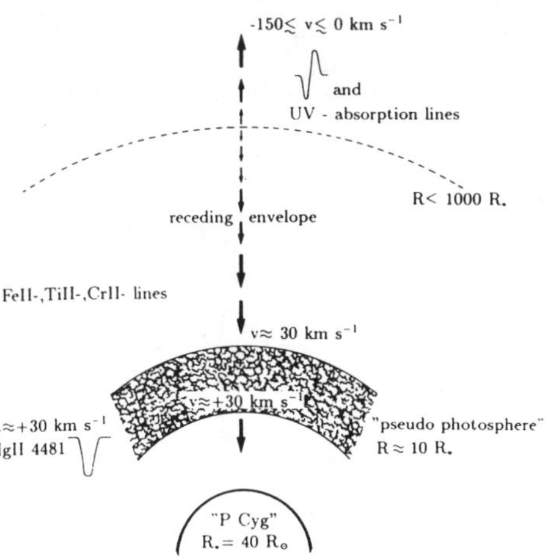

Fig. 4. Model of the outer configuration during the envelope contraction episode around the historical maximum of S Dor. In this simple model the UV absorption lines are formed at great distance from the stellar core where the matter can not be decelerated anymore, whereas the matter of the inner part of the envelope is sufficiently decelerated to recede (for more details see text)

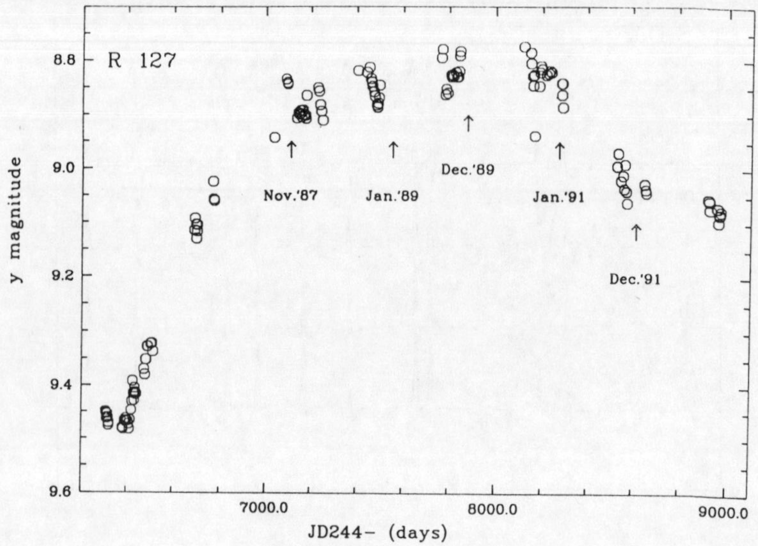

Fig. 5. light curve observed by LTPV during the historical maximum of R127 of the LMC. Coordinated CASPEC observations are indicated by arrows

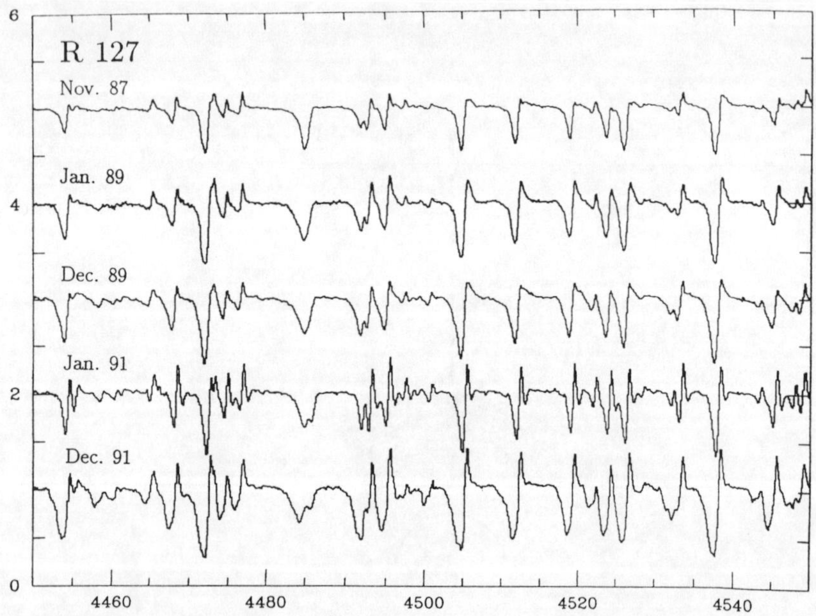

Fig. 6. CASPEC spectra of R127 of the LMC taken at different epochs. Like for S Dor inverse P Cygni-type absorption components have been observed around the historical maximum. Note the dramatic profile variations of the pseudo-photospheric line MgII 4481 with occasional line splitting

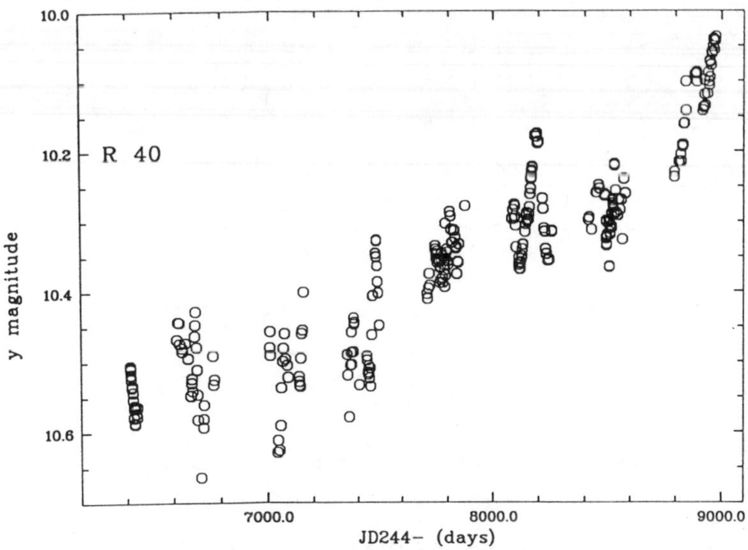

Fig. 7. Light curve of R40 in the SMC. R40 has been identified as an LBV on the basis of LTPV data. Superimposed on the increasing brightness by $0\overset{m}{.}5$ over the past seven years are microvariations of about $0\overset{m}{.}2$ on time scales of moths.

LBVs are well known to show typical equivalent spectral types of middle A during maximum (cf. Wolf, 1989). Due to the coordinated photometric and spectroscopic observations of S Dor and R127 of the LMC with LTPV and CASPEC, we now have some idea under which conditions this limiting spectral type of the opaque envelope is formed.

5. R40: the First LBV in the SMC

R40 in the SMC has been discovered on the basis of LTPV obsevations. About 40 observers have contributed to the light curve shown in Fig. 7. A detailed discussion of R40's properties is given in a forthcoming paper by Szeifert et al. (1993). R40 had been classified as B8Ie by Feast et al. (1960). The spectrum taken in December 1991 exhibits the characteristics of an almost normal A0Ia-O supergiant if compared with R45, the brightest star of the SMC (Fig. 8). R40 is still getting brighter and is now the brightest star of the SMC in the visual range. In the spectrum secured in January 1993 the strengthening of the Ti II lines is particularly obvious; the spectral appearance is now pretty much the same as that of the A3Ia-O hypergiant R76 of the LMC (see Fig. 8). Thus, this spectral development clearly shows the well-known property of LBVs that they become cooler when visually brighter. Further evidence that this occurs under the condition of constant bolometric luminosity is provided by the continuum observations (IUE low dispersion and LTPV). Note that the IUE flux of the outburst spectrum is considerably below the pre-outburst spectrum (cf. Fig. 9).

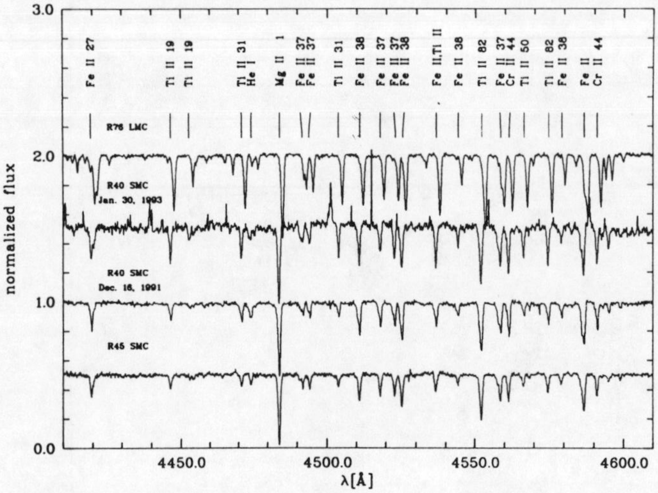

Fig. 8. CASPEC spectrogrammes of R40 for two epochs along with spectrogrammes of the hypergiants R45 (A0Ia-O) and R76 (A3Ia-O) in the SMC and LMC, respectively. The hypergiant spectra fit the spectral type of R40 for the two epochs, i.e. the spectral type gets later when the star becomes visually brighter.

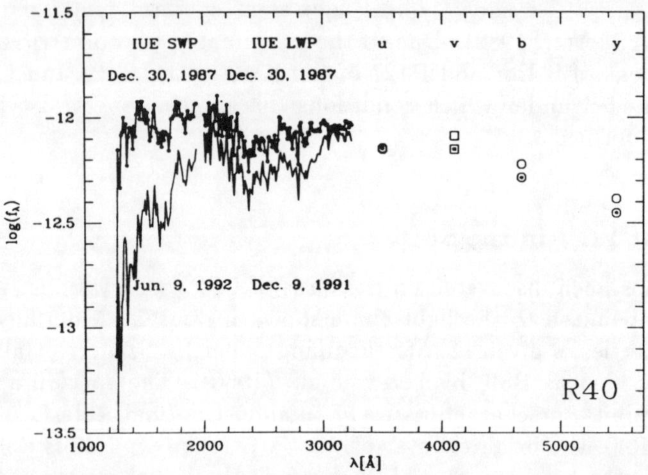

Fig. 9. Spectral-energy distribution of R40 in December 1987 and December 1991 (LWP) and June 1992 (SWP), respectively, as observed with IUE along with the Strömgren photometry of the LTPV. While R40 has become brighter in the visual the UV has decreased so that the luminosity remained approximately constant.

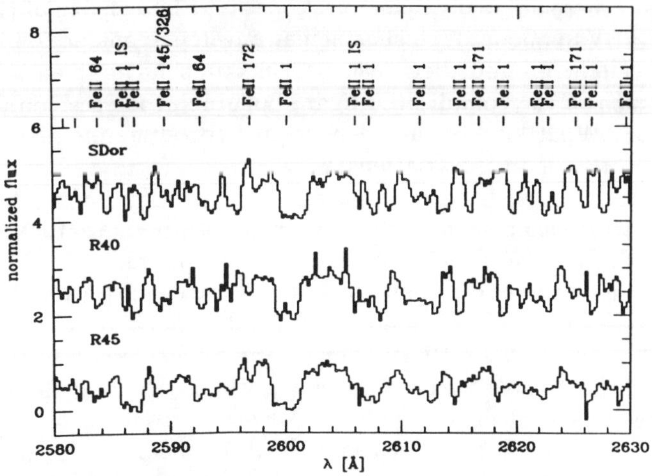

Fig. 10. Section of the high-resolution IUE spectrum of R40 along with a spectrum of S Dor and of the A0Ia-O star R45 in the SMC. Particularly remarkable is the close agreement between R40 and S Dor, bearing in mind the striking differences in the optical range.

Since it is now widely accepted that the metallicity of the SMC is down by about 0.6 dex or more compared with the Galaxy (cf. e.g. Spite & Spite 1993, Jüttner 1993), R40 is particularly important for the investigation of the dependence of the LBV phenomenon on metallicity. Obviously, low metallicity does not preclude LBV outbursts. This is interesting, since LBV outbursts have been suggested to be due to a mass-loss instability effect (Appenzeller 1986, Lamers 1986) and mass loss of early-type stars is supposed to be affected by their metallicity (Kudritzki et al. 1989).

Interestingly, the high-resolution spectrum of R40 is very similar to that of S Dor (Fig. 10). This agreement is very surprising bearing in mind the striking differences in the optical spectrum. Obviously, the strong UV lines are saturated in both stars so that the metallicity difference is of minor importance here.

6. Conclusions and Future Prospects

The triad LTPV, CASPEC and IUE has contributed considerably to a better understanding of the nature of the LBVs. A number of properties given in the introduction have been recognized through it, and new cases of LBVs have been discovered (R40, R110, R127 and HD160529). We have ideas about how to arrange LBVs in evolutionary sequences. A dichotomy is indicated in the sence that the most luminous members of the class ($M_{bol} \leq -9.5$) remain in the hot part of the HRD during their evolution along the sequence: O–Of–LBV–WR. The less-luminous members are supposedly in a post–RSG evolutionary stage.

On the other hand, there are many open questions. Our knowledge of the instability mechanism of LBVs is poor. The spectacular envelope-contraction episodes of S Dor and R127 caught around their historical maxima have given a particularly striking impression of the complicated hydrodynamic processes connected to the LBV phenomenon. What kind of time-dependent hydrodynamic processes are connected to the photometric microvariations?

Certainly, spectroscopic snap-shot observations can not provide satisfactory answers to questions of this nature. Continuous spectroscopic monitoring during time spans of several months with good resolution, both in wavelength (about 20 000) and time (typically one spectrum per night), and good S/N ratio (100 or better) is a prerequisite. What we need is a LTSV (Long-Term Spectroscopy of Variables) Group in addition to the LTPV. Such campaigns no longer belong to the realm of dreams. The development of fiber-linked echelle spectrographs as described by Mandel (this volume) which can be attached to small or medium-size telescopes, makes it possible. A campaign of four-months duration with Mandel's spectrograph attached to the 50-cm ESO telescope has been carried out already and has been described in "The Messenger" (Wolf et al. 1993). Results obtained for the galactic LBV AG Car which give an excellent impression of what one can obtain from such campaigns are presented by Stahl and Kaufer (this volume).

References

Appenzeller, I., 1986, IAU Symp. 116, 139

Davidson, K., 1987, *Astrophysical Journal*, 317, 760

D'Odorico, S., Enard, D., Lizon, J.L., Ljung, B., Nees, W., Ponz, D., Raffi, G., Tanne, J.F., 1983, *The Messenger*, 33, 2

Feast, M.W., Thackeray, A.D., Wesselink, A.J., 1960, *Monthly Notices of the RAS*, 121, 344

Hillier, D.J., 1992, in *The Atmospheres of Early Type Stars*, 105, Eds. Heber and Jeffery, Lecture Notes in Physics 401, Springer Verlag.

Jüttner, A., 1993, in *New Aspects of Magellanic Cloud Research*, 301, Eds. Baschek, Klare, Leqeux, Lecture Notes in Physics 416, Springer Verlag.

Kudritzki, R.P., Pauldrach, A., Puls, J., Abbott, D.C., 1989, *Astronomy and Astrophysics*, 219, 205

Lamers, H.J.G.L.M., 1986, IAU Symp.116, 157

Leitherer, C., Appenzeller, I., Klare, G., Lamers, H.etc., Stahl, O., Waters, L.B.F.M., Wolf, B., 1985, *Astronomy and Astrophysics*, 153, 168

Mandel, H., 1993, this volume

Spite, F., Spite, M., 1993, in *New Aspects of Magellanic Cloud Research*, 313, Eds. Baschek, Klare, Leqeux, Lecture Notes in Physics 416, Springer Verlag.

Stahl, O., 1993, in *New Aspects of Magellanic Cloud Research*, 263, Eds. Baschek, Klare, Leqeux, Lecture Notes in Physics 416, Springer Verlag.

Stahl, O., Kaufer, A., 1993, this conference

Sterken, C., 1983, *The Messenger*, 33, 10

Sterken, C., 1994, this conference

Szeifert, Th., Stahl, O., Wolf, B., Zickgraf, F.-J., Bouchet, P., Klare, G., 1993, *Astronomy and Astrophysics*, (in press)

van Genderen, A.M., 1978, *Astronomy and Astrophysics, Supplement Series*, 38, 381

Wolf, B., 1989, in *Physics of Luminous Blue Variables* Eds. Davidson, Moffat, Lamers, 91, Kluwer Academic Publishers, Dordrecht.

Wolf, B., 1990, ESA Sp 310, 33
Wolf, B., 1992, in *Nonisotropic and Variable Outflows from Stars* Eds. Drissen, Leithere, Nota, A.S.P. Conf.Ser. Vol.22, 327
Wolf, B., Appenzeller, I., Stahl, O., 1981, *Astronomy and Astrophysics*, 103, 94
Wolf, B., Stahl, O., 1983, *The Messenger*, 33, 11
Wolf, B., Zickgraf, F.-J., 1986, *Astronomy and Astrophysics*, 164, 435
Wolf, B., Mandel, H., Stahl, O., Kaufer, A., Szeifert, Th., Gäng, Th., Gummersbach, C.A., Kovacs, J., 1993, *The Messenger*, (in press)

DISCUSSION

Moffat: Using an echelle spectrograph at a 50 cm telescope will limit you to bright stars (you say you can get $S/N \sim 100$ per resolution element at $R \sim 20000$ in about 2 hours (for $V \sim 7$). Most LBVs are fainter than this!

Wolf: That is right. But you may perhaps know that ESO plans to build FEROS, a fiber-linked extended-range spectrograph, supposed to be permanently attached to the 1.5 m telescope. The limiting magnitude is supposed to be 15th magnitude, so that the LBVs of the Magellanic Clouds can easily be monitored with this spectrograph.

Le Contel explaining architecture to Walraven

Olsen explaining levitation — sceptic Hensberge

FIBER-LINKED ECHELLE SPECTROGRAPHS: THE IDEAL TOOL FOR HIGH-RESOLUTION SPECTROSCOPIC LONG-TERM MONITORING

H. MANDEL
Landessternwarte
Heidelberg-Königstuhl
Königstuhl 12' 69117 Heidelberg
Federal Republic of Germany

ABSTRACT. The concept to build a compact fiber-linked, portable, high-resolution echelle spectrograph was conceived in 1984 and looked very promising to fill a significant instrumentation gap of nearly all small or modest-aperture telescopes. The spectrograph, named FLASH (fiber-linked astronomical spectrograph of Heidelberg), has been designed and constructed at the Landessternwarte Heidelberg and has now successfully been used for spectroscopic monitoring campaigns at different sites and telescopes. A short characterization of the instrument together with some aspects of long-term spectroscopic observations and data-handling procedures are presented here.

Keywords: echelle-spectrographs – optical fibers – high-dispersion spectroscopy – observing techniques – administration – data handling – publication

1. Introduction

More than 70 optical telescopes with apertures between 0.5 and 1.5 meters are located somewhere in Europe, three times more worldwide. Most of them are located near universities or research institutes, near big cities with bright skies and polluted air. Compared with southern European sites with at least 200 usable nights per year or top sites like northern Chile with more than 300 nights, some 100 nights for central Europe are looking rather scarce. On the other hand, during the last decades especially, the dramatic advances in detector technology provided a much better performance of even a 50-cm telescope compared with the 5-m Mt. Palomar telescope in the 1950s.

We now have the grotesque situation that there are plenty, in principle very efficient, telescopes without any significant auxiliary instrumentation. For some of them there may be a problem with the weight of modern back-end instruments, but the conditions have changed dramatically with the rise of high-quality optical fibers. Therefore, portable fiber-linked instruments connected to small telescopes with low pressure factors for observing time are an ideal tool to study bright stars during long-term programmes, even in modest climatic conditions. With a homogenous set of data they can produce better scientific results than several too short stays

C. Sterken and M. de Groot (eds.), The Impact of Long-Term Monitoring on Variable Star Research, 303–310.

with larger instruments at better sites. Spectroscopic monitoring campaigns with FLASH are an example for those "lean observations". Don't wait to get long blocks of observing time at an overcrowded guest observatory. Build your own instrument, look around for suitable naked telescopes and get the data you'll never obtain during snapshots at well-equipped observatories.

2. Instrumentation

Over at least the last decade or so it has become quite obvious that high-dispersion spectroscopy is no longer a domain of big coudé spectrographs attached to large telescopes, but can be done with echelle spectrographs at smaller instruments also. Their well-known sensitivity to bending effects, avoidable by mechanical stiffness at the expense of great weight, is no longer a problem when using single fused-silica fibers and mechanical separation of the spectrograph from the telescope (Mandel 1988). While eliminating telescope-related flexure, a light-weight instrument can be housed in a fixed place in a temperature and humidity-controlled room. The light-scrambling property of the fiber results in a homogenous illumination of the echelle grating independent of guiding errors and seeing variations which simplifies also the merging of the reduced different echelle orders. During observations there are two connections to the outside world only: one incoming fiber and one outgoing coaxial cable to the CCD-control computer.

Our equipment, containing the spectrograph, a TV-guiding system, mechanical interfaces to the telescope, a computer-system with a magnetic tape, monitors and power supplies, weighs about 600 kg and fits into eight medium-size containers. Two well-trained persons are able to install and adjust the system in less than half a day.

With an efficient detector, stars down to a magnitude of 7 are within reach of even a 50-cm telescope (Wolf *et al.* 1993). Using an EEV CCD with 1252 × 770 pixel of 22 μ we get a spectral range of 2700 Å in one exposure. As standard setup we selected the wavelength range from 4050 Å to 6780 Å. This setting allows us to observe 57 echelle orders simultaneously, with a generous overlap between the orders. With a 100 μ fiber, which corresponds to a few arcseconds at smaller telescopes, the spectral resolution is about 20,000. During observation, up to 100 CCD frames (including Th-Ar and flatfield spectra) can be stored on the hard disk of the CCD-control computer. The frames are transferred to magnetic tapes each morning and copied to DAT-tapes later for sending them home periodically. It should be noted that the efficient handling of the huge amount of spectroscopic data was only possible thanks to an already existing far-reaching automatisation of the reduction procedure which is a modified version of the ESO-MIDAS echelle-reduction package running at the UNIX workstations of our institute (Stahl *et al.* 1993).

The major components of the TV-guiding system and the spectrograph are shown in Fig. 1 and Fig. 2. Inside the guiding-system the infalling beam from the telescope is reflected by several mirrors and focussed on the photo-cathode of a TV-camera. A motor-driven mirror and prism allow to select between starlight or light from the comparison lamps. All light is fed into the same fiber. Therefore,

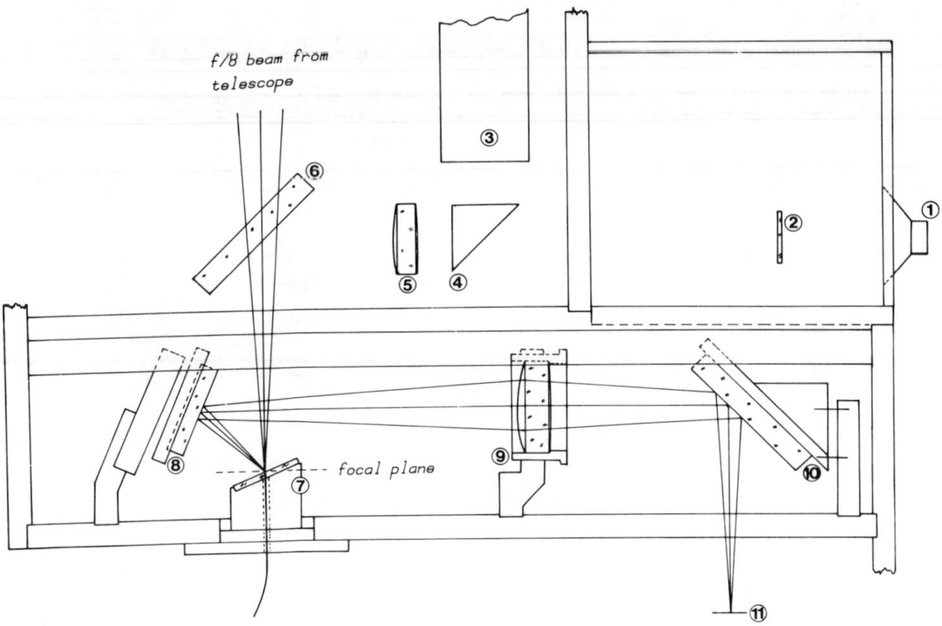

Fig. 1. Optical layout of the TV-guiding system: (1) flat-field lamp; (2) diffuser; (3) Th-Ar spectral lamp; (4) motor-driven prism for selection of one lamp; (5) lens f/2.5, f=80 mm; (6) motor-driven mirror; (7) fiber coupler; (8) plane mirror; (9) lens f/2.25, f=90 mm; (10) plane mirror; (11) focal plane of TV camera.

the whole system works as differential as possible and the precision of the data, especially for radial-velocity measurements ($\approx$ 1 km/s), is much higher compared than for a conventional slit spectrograph. The fiber coupler is a mirror with a pinhole of 120 μ. The glued-in fused-silica fiber, which represents the slit of the spectrograph, has an outer diameter of 110 μ and a core diameter of 100 μ. Because of degradation effects and signal attenuation, each individual fiber must be tested carefully. In practice, a well-prepared set of fibers with different core diameters allows a fast exchange and reaction to all unforeseen situations.

3. Long-Term Monitoring Campaigns with FLASH

During the last years FLASH has been used for different observations and monitoring campaigns at the following sites: 7 $\times$ at Heidelberg (0.7 m, 0.72 m and 0.75 m), 5 $\times$ at Calar Alto (1.23 m and 2.2 m), 3 $\times$ at ESO (0.5 m and 2.2 m) and one 5-month stay at Tautenburg (2.0 m) with two-week observing blocks in exchange with the Tautenburg Schmidt camera.

There have been up to eight team members in our "Wolf pack" simultaneously who shared the administration, transportation, installation, observation and technical service. Except for the ESO campaigns, transportation and custom regulations

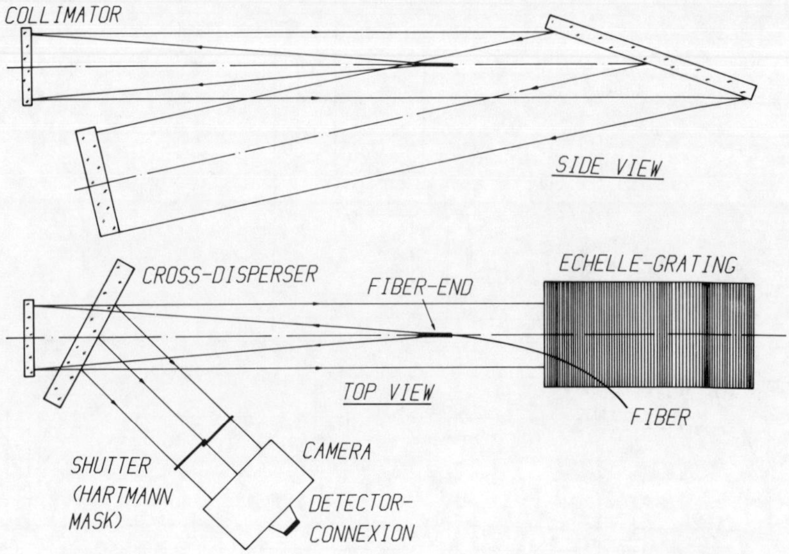

Fig. 2. Optical layout of the echelle spectrograph: collimator: f/5, f=400 mm or f/4.5 f=360 mm; echelle grating: 110 × 220 mm, 31.6 grooves/mm, blaze: 65°; cross disperser: 165 × 135 mm, 300 grooves/mm, blaze wavelength = 5000 Å; camera: f/2.8, f=300 mm.

have been organized by ourselves. Normally, FLASH is permanently ready for operation at one of the telescopes in Heidelberg. With our small transporter, a well-prepared and tested instrument can be transported and installed at a new site in a very short time (typically 2-3 days). This method, which reduces also the risk of damage, is much more efficient than using a national or commercial transportation company. In this way the loss of observing time during campaigns at different telescopes can be minimized.

Our group consists of three master scientists and a changing number of students and guests. At the beginning of each campaign the equipment is installed and adjusted by an experienced astronomer, assisted by the first observer. However, each regular team member is at least roughly familar with the instrument and data handling in such a way that autonomous observations can start without any extensive introduction. Typical observation blocks for one single observer are 3-4 weeks. Our latest ESO run lasted 4 months. According to the personal experience, the overlap time between two successive observes can be adapted. Our campaigns at Calar Alto and ESO have shown that even guests who never have seen the instrument before, are able to work completely independent after two or three nights of joint observations. Because of the overlap of up to one week, a permanent e-mail contact between the observers and our institute and a fast standard reduction of the incoming data at Heidelberg, no significant bias between different observers has been recognized.

Up to now we collected several thousand spectra. Some data have been used directly to drive the work of the participating students. An equal amount of data flows into long-term monitoring key projects. Therefore, the motivation of all participants is high. All team members of the campaign are present as co-authors of basic papers (Stahl et al. 1991).

Summarizing, one can say that the infrastructure of our home institute played an important role during building, testing and introducing the instrument. The prime costs of some 200,000 DM are of course not negligible. But now FLASH has grown-up to a flexible expedition spectrograph which produces valuable scientific results at relatively low costs. Only a few dozen stars have been observed so far. Therefore, the overwhelming majority of unobserved objects opens a wide field for further programmes.

4. Results

As an example, only a few highlights from our P Cygni campaign will be presented. For further results, please, see the contribution of O. Stahl in these procedings.

An unexpected result has been the detection of forbidden [FeII] emission lines in the spectrum of P Cygni. At standard set-up we observed this star in the wavelength range from 4050 Å to 6780 Å with a 200 μ fiber, corresponding to a spectral resolution of 12,000. Although we used only a 75-cm telescope, the simultaneous observation of more than 2700 Å allowed us to detect many flat-topped lines, never seen before, which are formed in a region with a constant expansion velocity of 230 km s^{-1} (Stahl *et al.* 1991). To create a spectral atlas of P Cygni we averaged 119 spectra and obtained a S/N-ratio of $\approx$ 2000 at 5400 Å. By turning the cross-disperser grating we obtained a red to near-IR spectrum from 7 short-exposure spectra, taken at the Calar Alto 2.2-m telescope. Its S/N ratio at 8000 Å is 350. The complete atlas covers the wavelength range from 4050 Å - 9050 Å (Stahl *et al.* 1993). Only the stronger lines have been identified so far. From Fig. 3 it can be seen that much more work is necessary to investigate the numerous weak lines.

Another example is shown in Fig. 4. To study the time and depth-dependent atmospheric-velocity field of P Cygni, 38 spectra covering 2700 Å have been obtained during one campaign of 127 days. The presented plot of the HeI 6678 line shows a P Cygni-type profile with strong intensity and velocity variations on time-scales of weeks. Such spectra—highly resolved in wavelength and time—are an important tool for studying the depth-dependent velocity field and allow identification of the relevant hydrodynamic wind-driving mechanisms.

5. Plans for the Future

The S/N-ratio of our spectra depends on the wavelength and is lowest in the blue spectral range. This unavoidable effect is present due to the use of a commercial camera objective and depends only slightly on the detector efficiency and the fiber. Below 4000 Å the overall efficiency drops rapidly. 10 years ago this version was the cheapest solution and nobody thought about near-UV observations. To use the wavelength region below this limit, the camera must be replaced by one with special UV optics. By using a dichroic filter, FLASH will be extended into a two-

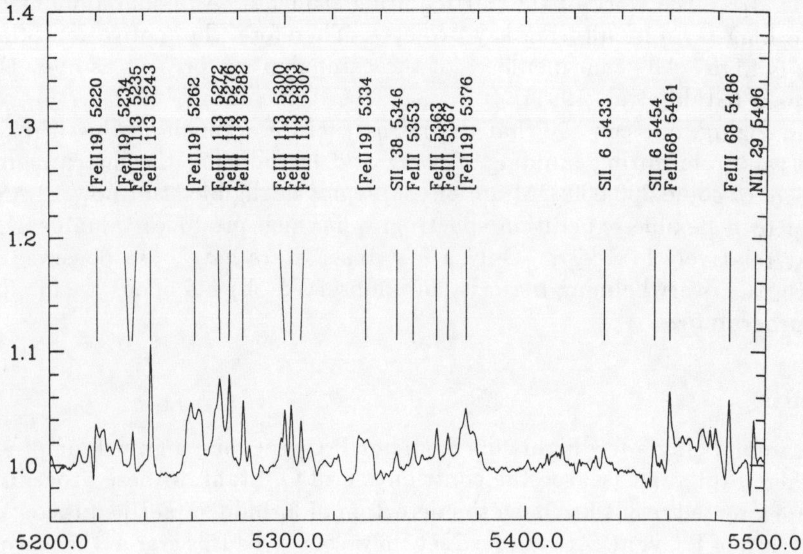

Fig. 3. A spectral range of the P Cygni atlas where only relatively weak lines are observed. Note the many faint lines which are present.

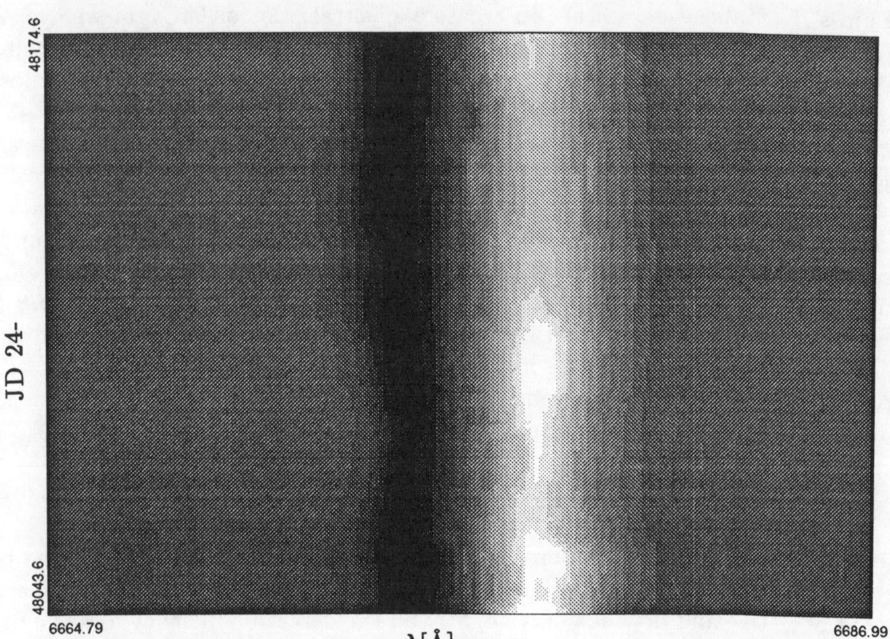

Fig. 4. Time series of the P Cygni HeI 6678 spectral line over 127 days; velocity range $-550, +350$ km s^{-1}.

channel spectrometer in an on-going programme. This rebuilding requires a second CCD-camera also, but then our spectrograph will be able to cover the wavelength range from ≈ 3500 Å to ≈ 8500 Å simultaneously.

6. Dreams

At this point one can ask for ideal input data to model stellar atmospheres. Spectroscopic observations are certainly an important tool, but photometric data are not less important. For that reason a merging of both observation techniques will be a perfect match to get the most homogenous set of data. I do not speak about observations at different telescopes, sites or time! Normally, there is only a small chance to obtain two telescopes at the same observatory. Each of us knows the struggle to get observing time, especially for long-term programmes. I want to discuss the combination of two instruments on one single telescope mounting. By adding a small light-weight piggy-back telescope with its own CCD-camera and filter wheel, it will be possible to obtain spectroscopic and photometric data simultaneously.

The design of and prospects for such an instrument must be considered carefully. The major problems, without considering the financial aspects, are the instrument's weight and its adaption to a standard CCD-pixel format. In principle, small light-weight and relatively cheap amateur instruments like flat-field or Schmidt-Cassegrain telescopes can be used. But most of them have the disadvantage of too short focal lengths. With such telescopes one arcsecond corresponds to a few micrometers in the focal plane. Therefore, only CCD chips with very small pixels can be accepted. Such chips are on the market, but they are relatively small and not suitable for CCD photometry which requires a large field of view. I see two solutions: using a standard CCD-camera with a large chip and pixels of 15-25 μ, one can either build a light-weight piggy-back telescope with a focal length of more than 4 m, or one can use a medium-size amateur telescope in combination with a special Barlow lens which doubles its focal length. Together with the CCD and computer system, filter-wheel, monitors and DAT-tapes some 200,000 DM will be needed. A huge amount of money – but in principle it is possible to obtain complementary photometric data during on-going spectroscopic exposures. The future of variable-star research looks quite promising.

Acknowledgements

This work was supported by the Deutsche Forschungsgemeinschaft (Wo 269, 9-1).

References

Mandel, H., 1988, *IAU Symp.* 132, 9

Stahl, O., Mandel, H., Szeifert, Th., Wolf, B. and Zhao, F., 1991, *Astronomy and Astrophysics*, 244, 467

Stahl, O., Mandel, H., Wolf, B., Gäng, Th., Kaufer, A., Kneer, R., Szeifert, Th. and Zhao, F., 1993, *Astronomy and Astrophysics, Supplement Series*, 99, 167

Wolf, B., Mandel, H., Stahl, O., Kaufer, A. Szeifert, Th., Gäng, Th., Gummersbach, C.A. and Kovacs, J., 1993, *The Messenger*, 74, in press

DISCUSSION

Le Contel: What kinds of tests did you perform with respect to mechanical stability and termal stability in the laboratory and at the telescope itself?

Mandel: First of all: the TV-guiding system and the spectrograph are physically separated and, therefore, the echelle spectrograph shows no bending effects due to telescope-related flexure. When the instrument was housed in a humidity- and temperature-controlled environment, no drifts were found. While using the spectrograph in the dome we have seen systematic drifts which can be verified with water-vapour lines. But when we took the instrument downstairs to the basement of the telescope (which is highly temperature-stabilised) no effects were seen.

Moffat: In your future dreams, please consider the possibility of including a polarisation (linear and circular simultaneously) module to allow one to carry out spectropolarimetry! This could be very important to improve our understanding especially of LBVs, which often reveal disks and other asymmetries for which spectropolarimetry is an excellent discriminator.

Mandel: It is an interesting alternative. I will think about this suggestion. But, of course, instrumentation in any case is a question of money and we have to find economically feasible solutions.

Young: There are difficulties in doing spectroscopy and photometry at the same telescope. The auxiliary measurements (sky, comparison stars, extinction stars) needed to do photometry take telescope time away from the spectroscopy.

Mandel: In principle, I agree, but the technical problems haven't been discussed to an end. There are different solutions. You can take a small, let's say 30-cm amateur telescope, as a piggy-back instrument to a larger telescope and do some CCD photometry simultaneously. Normally, you will have problems to find an adequate CCD-pixel format. Therefore, you have to increase the focal length. Another possibility is to add the photometer to the TV-guiding system. In this case you can use the primary telescope only and, indeed, in this way the photometry will take some time from the observing programme. Maybe we have to find a compromise.

SPECTROSCOPIC MONITORING OF OBA STARS

O. STAHL, and A. KAUFER
Landessternwarte
Heidelberg-Königstuhl
Königstuhl 12
69117 Heidelberg
Federal Republic of Germany

ABSTRACT. We have been using the fiber-linked echelle spectrograph described by Mandel (these proceedings) at various telescopes, both in the conditions of central Europe and in the excellent conditions of La Silla. Our main objects of study were early-type stars, mainly supergiants. Our studies show that the variability of early-type stars displays a large variety of phenomena. The variations are generally irregular. There are exceptions, however, as in the surprising case of θ^1Ori C. Most stars show variations on several widely-different time-scales, which makes their study particularly difficult. Spherical symmetry may be violated in many cases, as suggested, e.g., by variable blue and red-shifted emission components.

Keywords: spectroscopic monitoring – echelle spectrographs

1. Introduction

Time scales of the order of weeks or months are difficult to study photometrically and even more difficult spectroscopically. For spectroscopy the main reason is that spectrographs, particularly high-resolution spectrographs, are available mainly at large telescopes. Unfortunately, a typical observing run at a large telescope is only a few nights long. That means that only variations on short (or very long) time scales can be studied easily. The time scales from weeks to months are, therefore, almost completely unexplored for stars of any kind. However, small telescopes of the 50-cm class, equipped with modern equipment, are sufficiently large for spectroscopy of bright stars, even at moderately-high resolution. We realized that this is an excellent opportunity for studies with small telescopes and a fiber-linked echelle spectrograph. Our main interest is in early-type supergiants, where high spectral resolution is needed to resolve the lines. We, therefore, used the echelle spectrograph described by Mandel (these proceedings).

The variability in light and spectrum of some of the most luminous early-type supergiants has been known for a very long time (e.g. η Car, P Cygni). The range of variability is large: Normal OB supergiants show small stellar-wind variations only, which are mainly observed in the strong resonace lines in the UV. Some Luminous Blue Variables (LBVs), on the other hand, exhibit huge variations (e.g. the

C. Sterken and M. de Groot (eds.), The Impact of Long-Term Monitoring on Variable Star Research, 311–328.

η Car eruption) of several magnitudes in V and also show spectacular spectroscopic changes. The range of time scales of variability is from hours to days for normal O stars and up to decades for the large variations of LBVs.

There is no generally accepted theory for all these variations and it is not clear whether these variations are physically related. The stellar-wind variations may be triggered by variations of the star (pulsation?) or be due to intrinsic instability of the stellar wind. It is, therefore, important to study many lines simultaneously, which makes echelle spectrographs attractive.

2. Observations

We have monitored P Cygni (and a few other target stars) with our instrument from 1990 to 1992, using mainly the 70-cm telescopes of the Landessternwarte Heidelberg. Some of the spectra have been obtained with the 2-m telescope of the Landessternwarte Tautenburg and the 2.2-m telescope at Calar Alto, Spain, during runs for other projects. Exposure times for this star were typically 30 minutes to one hour at a 70-cm telescope. An average spectrum of the star has been published by Stahl et al. (1993a). It turned out that the climatic conditions of central Europe are sufficiently good to cover the variability on time scales of weeks to months, although some gaps are unavoidable (see below).

More recently, we had two long runs at La Silla. The obvious advantage of La Silla is that the stable weather provides better sampling for variability on shorter time scales, and simultaneous photometry is possible in principle. In addition, there are many interesting objects in the southern sky.

Among others, we have observed the following objects: HD 160529 (a recently-detected galactic LBV), HD 152236 (ζ^1 Sco, a very luminous B1Ia supergiant), the well-studied galactic LBV AG Car, the famous peculiar star η Car, θ^1Ori C (a young main-sequence O star), β Ori (B8Ia), the brightest OB supergiant in the sky, and HD 92207 (A2Ia), an extreme A-type supergiant.

In our two runs at La Silla, we have used the ESO 50-cm telescope on 60 nights from June to August 1992 (287 object spectra) and on 120 nights from February to June 1993 (1001 object spectra).

The reduction of the spectra has been automated as far as possible and, so, all spectra have been reduced within a few weeks. Only a few objects have been analyzed in some detail so far.

3. P Cygni

P Cygni is, with a visual magnitude of $V = 4.8$, the brightest LBV in the sky. It had a large outburst in 1600, large variations (3 mag) for a period of about 200 years, and has been relatively stable for the last 200 years. It exhibits a peculiar spectrum with almost all lines showing P Cygni profiles (undisplaced emission and blue-shifted absorption). P Cygni is photometrically slightly variable ($\approx 0^{m}.2$). The history and an extensive discussion of the older literature have been presented by de Groot (1969).

Already in the older literature, a splitting of the spectral lines of P Cygni has been reported by several authors (see, e.g., the discussion by de Groot, 1969). Lamers et al. (1985) have found variable absorption components in ultraviolet lines of FeII and FeIII as well. The splitting of absorption components of optical lines has been studied extensively by several authors at a later date; e.g. Markova (1986), Markova (1988). Sometimes, even periodic variations of the radial velocity of certain lines or periodically-appearing components have been claimed. A re-analysis of all published data by van Gent (1986) did not recover any of the claimed periods.

The study of the moving sub-components of some lines and the clarification of the possible periodicities was our main aim in starting our monitoring project of P Cygni.

We found from our observations variations in emission-line strength of ($\approx$ 30%) and in the expansion velocity measured from the lines of ($\approx$ 30 km s^{-1}). However, *no* line-splitting has been observed by us at any time. The observed variations of e.g. the He I λ6678 line do not give the impression that the variable radial velocities, which we observe, are due to moving components. Rather, the impression is, that the whole absorption component is shifting towards shorter wavelenghts. This is in contrast to results published in other papers, where line-splitting has been reported to be very common; e.g. Markova (1988). It should be mentioned that the spectral resolution of our data ($\lambda/\Delta\lambda \approx 12,000$ for most spectrogrammes) is somewhat lower than the resolution of the coudé spectra of Markova (1988), but this is at least partly compensated by the much higher S/N ratio of our spectra. There is no doubt that at some phases line-splitting is present in P Cygni, at least in the higher Balmer lines; e.g. Adams and Merrill (1957). Our result indicates, however, that the reported line-splitting is not as ubiquitous as sometimes reported in the literature. Some of the reported line splittings appear to be over-interpretations of noisy photographic data.

We searched for correlations of the variations which we found, with photometric observations kindly provided by Mart de Groot, but could not detect any clear correlation. This is partly due to the absence of prominent features in the light curve during our spectroscopic campaign.

The typical time scale of the variability is from weeks to months, which is consistent with the dynamical time scale of the envelope. The variations in different lines are correlated (see Figs. 1 – 4) and can be used to constrain models of the variations. There is also a long-term component of the variations, which is especially clear from the observations of Hα. Variations with a much shorter time scale (less than a few days), cannot be studied very well with our data, but appear unlikely because of the smooth variations observed on a time scale of weeks.

A real understanding of the spectroscopic variations requires detailed modelling, which has not yet been done. Qualitatively, some relations between the variations of the He I lines (supposed to be formed deep in the envelope) and the lower Balmer lines (supposed to be formed further out in the wind) can be understood in terms of density enhancements moving through the envelope.

4. The peculiar O star θ^1Ori C

θ^1Ori C is the brightest and hottest star in the Orion trapezium and the main source of ionization of the Orion nebula. It is an extremely young O7 main-sequence star.

A few peculiarities of the star have been known for a long time. Conti (1972) found that He II λ4686 is highly variable on a time scale of days, sometimes showing an *inverse* P Cygni profile. One the other hand, he found the radial velocity to be constant within the errors of measurement ($\approx$ 10 km s^{-1}). He suggested infall of circumstellar matter as a possible cause for the variations. Walborn (1981) found the ratio He II λ4541/ He I λ4471 to vary on a time scale of days, suggesting variations of the effective temperature of the order of 10,000 K! Surprisingly, van Genderen et al. (1989) found *no* significant photometric variations. In spite of these clear indications of peculiar behaviour, no long-term spectroscopic monitoring has been reported so far.

We obtained 16 spectrogrammes in 1992 and 78 in 1993 with our spectrograph and the ESO 50-cm telescope. The data around Hα obtained in 1993 are shown in Fig. 5.

From this figure, a periodic modulation is obvious. A more detailed analysis gives a period of $15\overset{d}{.}43 \pm 0\overset{d}{.}03$. In Fig. 6, we show the phase diagram of the equivalent width of Hα, constructed with a period of $15\overset{d}{.}43$. The phase diagram shows two peaks, separated by about 0.5 in phase. Knowing the period, phase-averaged spectra can be obtained. In Fig. 7 we show phase-averaged spectra of the He II λ4686 line. It clearly shows the same qualitative behaviour as Hα. This is true for other strong lines, e.g. Hβ, Hγ and He I λ5876. The radial velocity of the C IV λ5801 line is shown in Fig. 8 (most other lines are disturbed by emission from the Orion nebula). The radial velocity is variable, but not in phase with the emission-line variations. The ratio He II λ4541/He I λ4471 is about constant.

From the width of the lines we estimate a small value of $v \sin i$ ($<$ 50 km s^{-1}), which is consistent with $15\overset{d}{.}43$ as the rotation period of the star.

After the publication of our detected period, Walborn (1993) found that archival IUE spectra obtained ten years ago show a modulation of the C IV stellar-wind line with the same period. Unfortunately, the time gap between the two data sets is too long to allow an unambiguos phasing of both data sets. The finding shows, however, that the period has been stable for at least ten years.

These results suggest the following interpretation: Since the period is too long for pulsation and the radial-velocity variations are not in phase, the periodic variations are probably not due to binary motion. A rotational origin seems more likely. Then, an oblique magnetic-rotator model may be applicable for this star (Stahl 1993b). Such a model could also account for the two peaks in the phase diagram.

5. The galactic LBV AG Car

AG Car is the brightest LBV in the southern sky. In contrast to P Cygni, it is relatively active, with variations of about two magnitudes in the visual range (roughly between 6 and 8 mag) on a time scale of years. The latest maximum of AG Car has been around 1980. After a minimum phase, which lasted from

1985 to 1990, the star is now brightening again. In minimum phase, the star has the spectrum of an extreme Of or an Ofpe/WN9 star, in maximum the spectrum resembles an early A-type star with a strong stellar wind—a typical behaviour of active LBVs. We were lucky to catch AG Car during an active phase of increasing brightness ($V \approx 6.7$).

We observed several sub-components in the Hα line. This is shown in Fig. 11, where the slow disappearance of one component can be seen. This component accelerated from about -120 to about -136 km s^{-1} during a time-interval of about 300 days. The equivalent width of this component decreased strongly during this time. This behaviour is consistent with a discrete expanding shell, probably ejected during the beginning of the present outburst. The high-velocity component at about -270 km s^{-1} did not show any significant variations in radial velocity. Its expansion velocity suggests that it is a remnant of the stellar wind before outburst.

AG Car is surrounded by a well-known ring nebula, which shows peculiar bipolar sub-structure (see e.g. Nota et al. (1989). The bipolar structure suggests an axisymmetric wind from the star, not a spherically symmetric one. Further support for a non-spherical wind from LBVs comes from the recent detection of intrinsic continuum polarization (1 – 1.5%) in the LBV R 127 of the LMC (Schulte-Ladbeck 1993), which is surrounded by a nebula similar to AG Car. In AG Car we observed for the first time Be-like profiles (i.e. unshifted absorption, flanked by a red *and* a blue-shifted emission component) in several lines, e.g. the He I λ6678 line (see Fig. 12), at least on some observing dates. This kind of profile is usually interpreted as evidence for a ring or disk around the star. The observation of such a profile strengthens the suggestion for non-spherical winds in LBVs.

AG Car shows strong variability in all spectral lines. This is shown for the He I λ6678 line in Fig. 12. Both the central absorption and the blue-shifted wind absorption vary. All kinds of profiles are observed, double-peaked emission with central absorption, classical P Cygni and also inverse P Cygni profiles.

The Si II λ6347 line (Fig. 12) shows a different behaviour with a central emission and a blue-shifted absorption, while sometimes red-shifted absorption appears as well.

AG Car shows an extremely complex spectroscopic behaviour—the profiles change *qualitatively* within a few days. In addition, different lines, even from the same ion, can show completely different behaviour (e.g. one He I line shows a P Cygni profile, an other one inverse P Cygni).

6. Normal B and A supergiants

On La Silla we also observed several normal B and A supergiants of high luminosity, among them β Ori and HD 92207.

β Ori is the brightest OB star in the sky and a well-known spectroscopic and photometric variable. Strong variations of Hα and in radial velocity have been known for a long time. Hα, with both P Cygni and inverse P Cygni profiles, has been observed by Sanford (1947), but no long-term monitoring has been available so far. Photometrically, the star shows variations with an amplitude of about $0^{m}_{\cdot}15$ (Guinan et al. 1985).

The Hα and Hβ observations are shown in grey-scale representation in Fig. 13. Our measurements of β Ori also show strong radial-velocity variations on a time scale of days (see Fig. 14). These variations are probably due to pulsations, or at least some pulsation-like motions, in the atmosphere. The amplitude is similar in all lines, but the absolute value of the radial velocity differs from line to line. This can probably be explained by stratification in an expanding atmosphere. Whether these variations are correlated with the emission-line activity is not yet clear. The emission-line profiles are very much "Be-like", suggesting deviations from spherical symmetry.

HD 92207 is one of the most extreme A-type supergiants known. The variability of this star has been known for a long time, but no detailed study of its variations has been published. Sterken (1977) has monitored the star photometrically and found a semi-period of 22 days and an amplitude of $0^{m}.035$.

The radial-velocity variations of this star are similar to the variations of β Ori, but the time scale is longer. The Hα variations (Fig. 15) look different from the variations observed for β Ori—note particularly the appearance of enhanced absorption around JD 2,449,090 and the preceding sudden shift in the radial velocity of the Hβ absorption (Fig. 16). The radial velocity of Hα also has a long-term trend, demonstrating that three months of monitoring are not enough!

7. Conclusion

Our monitoring of luminous OBA stars has shown that small telescopes can be useful for high-dispersion spectroscopy. If you one concentrates on a few objects and look closely for a long time, small telescopes can produce results which can not be obtained with a large telescope. Even for bright stars, which are often considered to be well studied and not worthwhile to observe anymore, you find unexpected surprises when you look closely. In many cases, however, you shall not find what you expect or what you are looking for, but something completely different.

Acknowledgements

This work has been supported by the Deutsche Forschungsgemeinschaft (Wo 269, 9-1). We thank the European Southern Observatory for the generous allocation of observing time.

References

Adams, W.S., Merrill, P.W., et al., 1957, *Astrophysical Journal*, 125, 102

Conti, P., 1972 *Astrophysical Journal*, 174, L79

de Groot, M., 1969, *Bulletin of the Astronomical Institutes of the Netherlands*, 20, 225

Guinan, E.F., McCook, G.P., Harris, W.T., Speranzini, D., Wacker, S.W., 1985, *Information Bulletin on Variable Stars*, 2762

Lamers, H.J.G.L.M., Korevaar, P., Cassatella, A., 1985, *Astronomy and Astrophysics*, 149, 29

Markova, N., 1986, *Astronomy and Astrophysics*, 162, L3

Markova, N., Kolka, I., 1988, *Astrophysics and Space Science*, 141, 45

Nota, A., Paresce, F., 1989, IAU Coll. 113, Kluwer, eds. Davidson, Moffat and Lamers, 159

Sanford, R.F., 1947, Contrib. Mount Wilson, 729

Schulte-Ladbeck, R.E., Leitherer, C., Claytron, G.C., Robert, C., Meade, M.R., Drissen, L., Nota, A., Schmutz, W., 1993, *Astrophysical Journal*, in press
Stahl, O., Mandel, H., Wolf, B., Gäng, Th., Kaufer, A., Kneer, R., Szeifert, Th., Zhao, F., 1993a, *Astronomy and Astrophysics, Supplement Series*, 99, 167
Stahl, O., Wolf, B., Gäng, Th., C.A. Gummersbach, Kaufer, A., J. Kovacs, H. Mandel, Th. Szeifert: 1993b, *Astronomy and Astrophysics*, 274, L29
Sterken, C., 1977, *Astronomy and Astrophysics*, 57, 361
van Genderen, A.M., Bovenschen, H., Engelsman, E.C., Goudfrooy, P., van Haarlem, M.P., Hartmann, D., Latour, H.J., Ng, Y.K., Prein, J.J., van Roermund, F.H.P.M., Röttgering, H.J.A., Steeman, F.W.M., Tijdhof, W, 1989, *Astronomy and Astrophysics, Supplement Series*, 79, 263
van Gent, R.H., Lamers, H.J.G.L.M., 1986, *Astronomy and Astrophysics*, 158, 335
Walborn, N.R., 1981, *Astrophysical Journal*, 243, L37
Walborn, N.R., 1993, *Astrophysical Journal*, in press

DISCUSSION

Delgado: Would you comment on the observed variations in η Car and, in particular, on the correlation, if any, between the spectral and light variations?

Stahl: We have a few spectra of η Car that were obtained last year which show clear variations. Our long series of spectra obtained this year does, at least at first glance, show hardly any variations. Apparently, η Car was also photometrically rather quiet during this period.

de Groot: Your graphs of emission-line strength and absorption radial-velocity in P Cygni shows larger velocity shifts when the emission is stronger. Couldn't this be due to the emission-line effect?

Stahl: An emission-line effect appears very unlikely to me, since the absorption-line variations are strongest at the blue edge and the emission-line variations appear too small to account for this.

Harmanec: I think it could be due to the emission-line effect because there is a very steep flank to the emission line and small intensity variations can produce relatively large changes in absorption radial velocity.

Moffat: The variability in the blue absorption edges of the lines you showed (e.g. He I 6678 in P Cygni) are often seen in such lines of hot stars, especially in the UV where resonance lines are most prominent. These are related to "DACs" (discrete absorption components) which are wind perturbations spiralling out in the wind at sub-wind material speeds. They may be some kind of density wave triggered at the surface of the underlying rotating star.

Stahl: I agree that the variations we see may be related to the DACs seen in UV resonance lines.

Kurtz: Are there any polarimetric, magnetic or photometric observations of θ^1 Ori C?

Stahl: There are some polarimetric and photometric data available, but there is no evidence for variability; but nobody seems to have looked carefully so far.

Wolf: It might be interesting to note that, having the precise period of the spectroscopic variations of θ^1 Ori C, Walborn and Bohlin were able to find the same periodic variation for the CIV line in He IUE spectra. They used archive data of spectra taken about ten years ago. So, the period must be very stable.

Harmanec: Regarding the lack of an apparent correlation between various photometric and spectral characteristics of P Cyg, I would like to remind you that I suggested a few years ago at the ESO Workshop on *Rapid Variability of OB-Stars* (1991, Ed. D. Baade, ESO Conf. Workshop Proc. No. 36, p. 265) that this could possibly be understood in terms of *local* outflow of material from the stellar equator. In that case, the material would spiral out when expanding away from the star and different material will be projected against the stellar disk at different times. Therefore, for instance, high-ionization and low-ionization lines, formed at different parts in the outflowing stream, will not show the same line variation.

Stahl: The possibility you suggest can probably account for the effect we observe. It will be difficult to prove, however, without a detailed physical model.

Moffat: Many of the problems of variable line profiles in hot stars, as discussed here, were the main topic of interest at a recent workshop on "Instability and Variability in Hot-Star Winds" held at Isle-aux-Coudres, Quebec, in late August 1993. This will be published in 1994 by Kluwer Academic Publishers (Astrophysics and Space Sciences Series).

Coffee time for Breger, Cuypers, Young, Moffat
Burki goes for apéritif

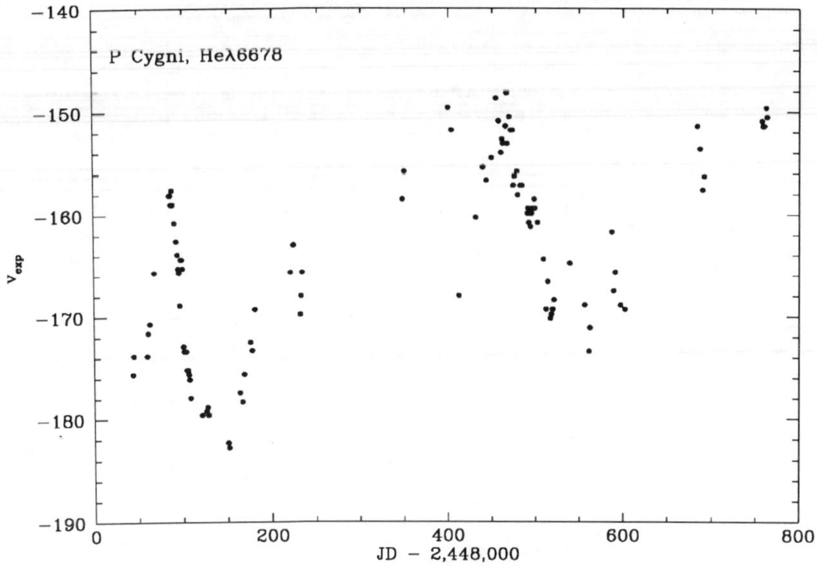

Fig. 1. The expansion velocity as measured from the blue-shifted absorption component of the HeIλ6678 line of P Cyg during our monitoring campaign.

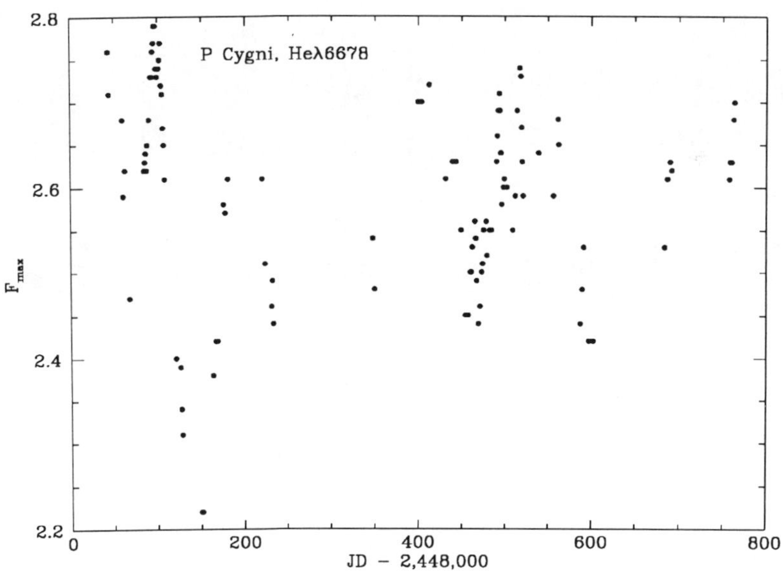

Fig. 2. The peak intensity of the HeIλ6678 line of P Cyg during our monitoring campaign.

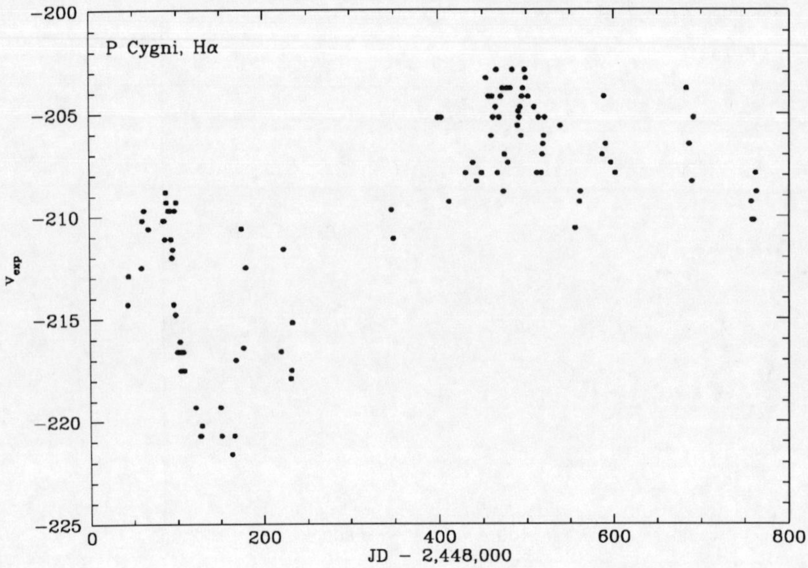

Fig. 3. The expansion velocity as measured from the blue-shifted absorption component of the Hα line of P Cyg during our monitoring campaign.

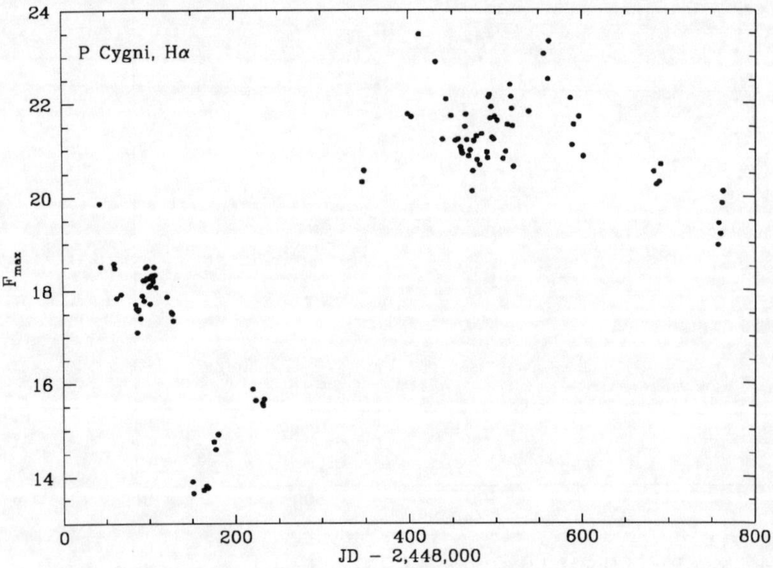

Fig. 4. The peak intensity of the Hα line of P Cyg as observed during our monitoring campaign.

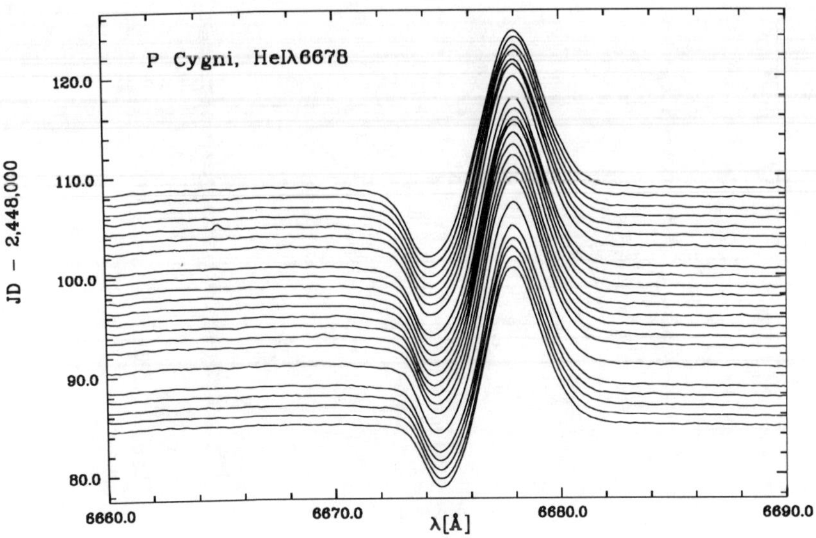

Fig. 5. A series of HeIλ6678 spectra obtained during the summer of 1990. Note the smoothly varying blue edge of the absorption component of this line.

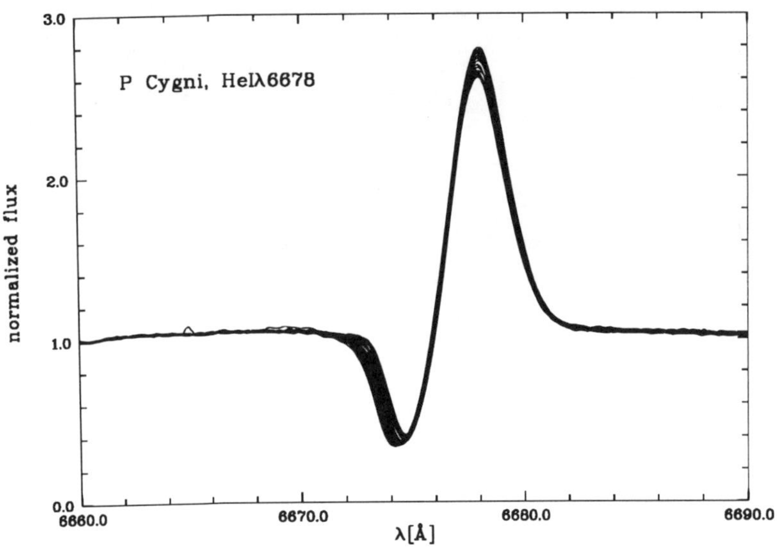

Fig. 6. Same as Fig. 5, but without offsets between the individual spectra. In this plot, it can be clearly seen, where in the line profile the variations are located.

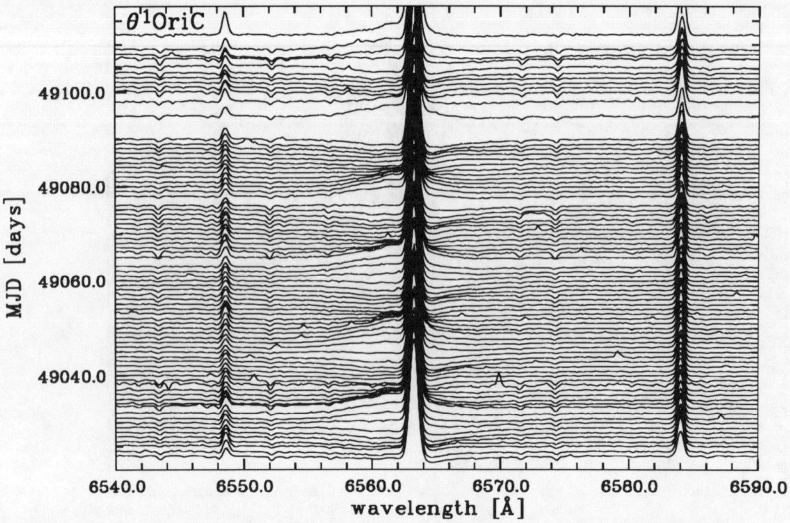

Fig. 7. Spectral variations of the Hα line of θ^1 Ori C as observed in 1993. The ordinate (at continuum level of the spectra) gives directly the modified Julian date (JD$-$2,400,000.5 days). Note the periodically appearing bump at the blue side of the nebular emission line and the fainter red bump at the times in between. The nebular [NII]-lines at 6548 and 6583 Å are also clearly seen. Other sharp absorption and emission features are terrestric water vapor absorptions and cosmic ray hits, respectively.

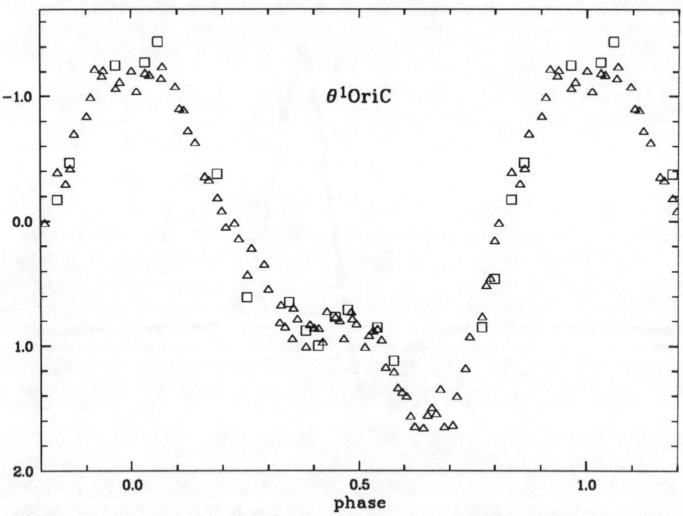

Fig. 8. Equivalent width of the Hα line of θ^1 Ori C versus phase (for an adopted period of 15.43 days and a zero point of MJD = 48832.5). Phase zero corresponds to maximum strength of the emission. The data from 1992 are plotted as square symbols, the data from 1993 as triangles. Note the good agreement between the two observing runs. The total time span is about 300 days or more than 15 cycles. This long time span results in a small error of about 0.03 days for the period.

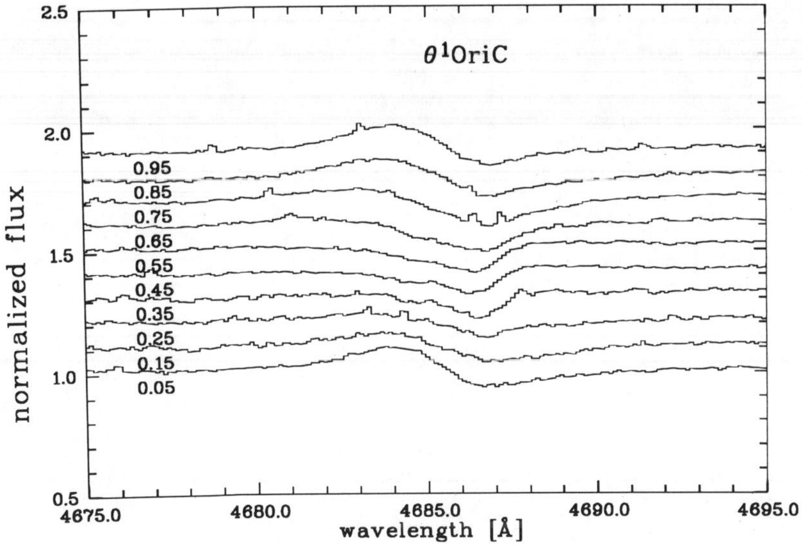

Fig. 9. Phase averaged spectra of θ^1 Ori C around HeIIλ4686 obtained by averaging all frames within phase bins of 0.1. From this plot it can be seen that this line behaves qualitatively very similar to Hα. The variations in both lines are in phase.

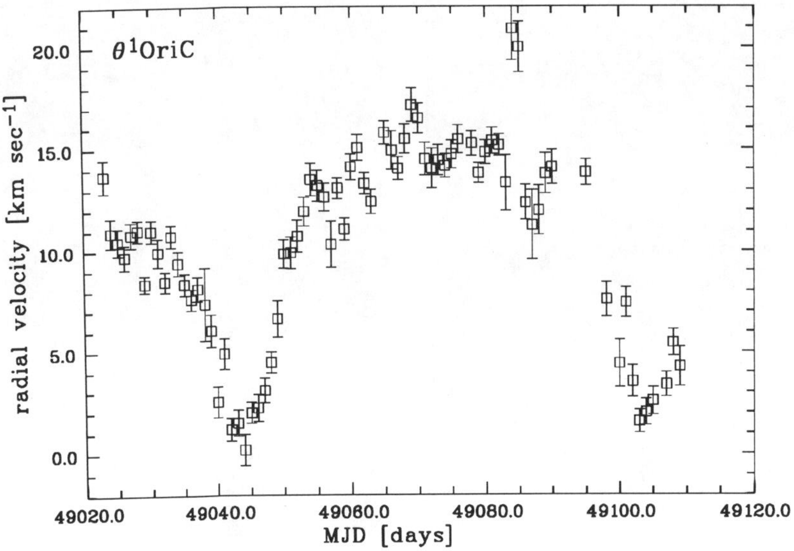

Fig. 10. Radial velocity of the CIVλ5801 line of θ^1 Ori C versus time. Note the absence of the 15.4 day period. The wavelength of the line has been determined by fitting a gaussian profile. The error bars have been determined from this fit and are the interior errors. Systematic errors have been checked with the nebular lines and are smaller than 1 km sec^{-1}.

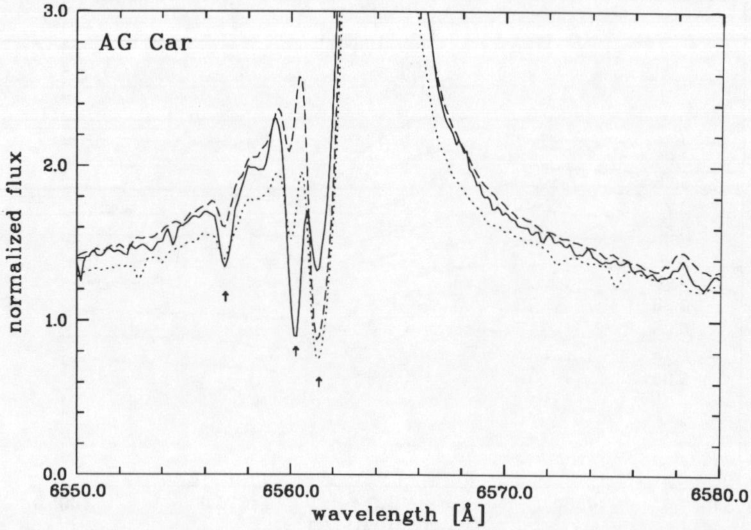

Fig. 11. Variations of the Hα profile of AG Car in 1992 and 1993. The three profiles shown have been obtained at JD 2,448,822 (full line), JD 2,449,023 (dotted) and JD 2,449,139 (dashed). The three discrete absorption components at -60, -125 and -275 km sec^{-1} are marked with arrows. Note the fading of the intermediate component during the time of almost one year.

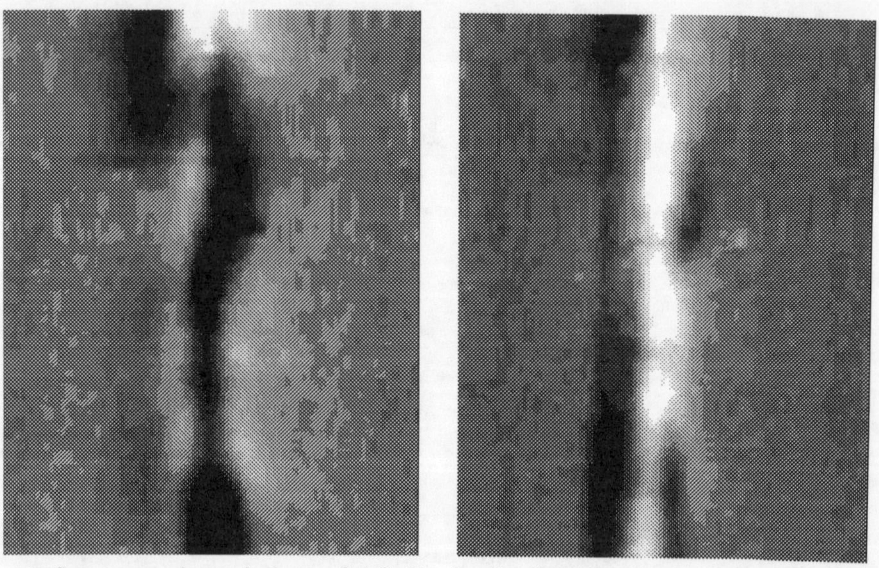

Fig. 12. Spectroscopic variations of AG Car in the lines HeIλ6678 (left) and SiIIλ6347 (right), as observed with our echelle spectrograph attached to the ESO 50cm telescope. In this gray-scale representation of all spectra obtained in 1993 we plot the spectra with a velocity range of ± 300 km sec^{-1} versus Julian date (from JD 2,449,023 (bottom) to JD 2,449,139 (top)). Missing data are interpolated. Note the dramatic variations within a timescale of days and the double emission with a central absorption observed around JD 2,449,070 in the HeI line.

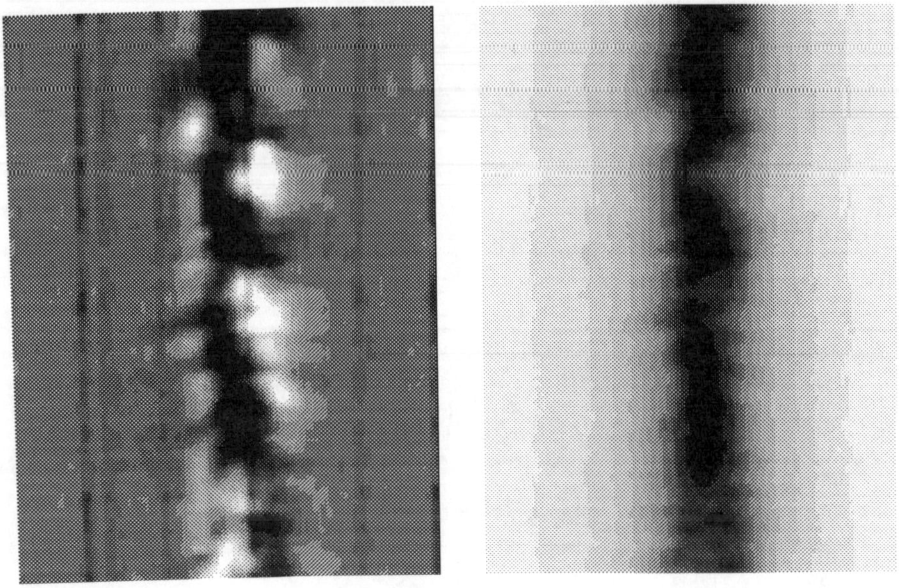

Fig. 13. Grey-scale representation of the spectrogrammes of β Ori obtained in 1993 around Hα (left) and Hβ (right). Velocity range: ± 410 km s^{-1}, time-span: JD 2,449,023 to JD 2,449,127. Obviously, the Hα profile is strongly variable within a few days. The behaviour of this line is very similar to the variations of Be stars.

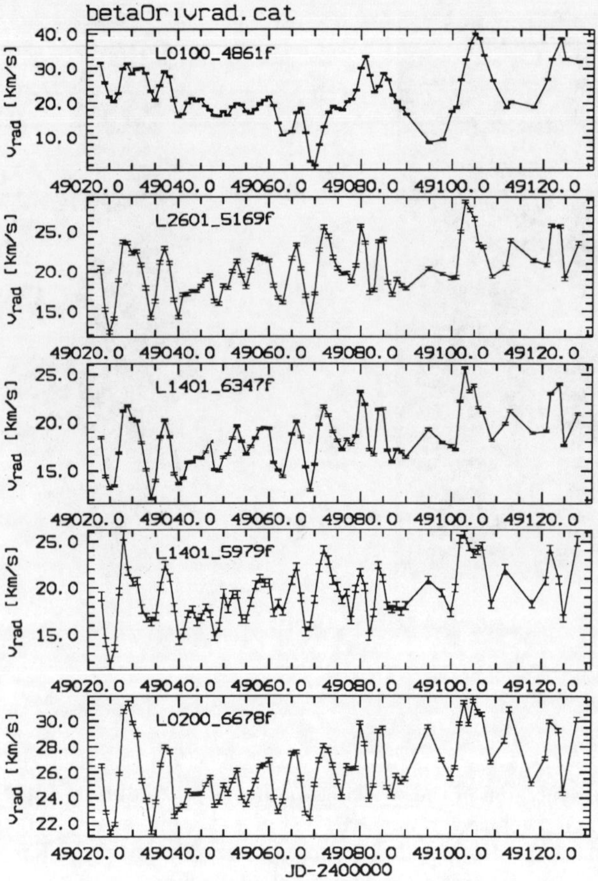

Fig. 14. Radial-velocity variations of β Ori as observed from the lines Hβ, Fe II $\lambda5169$, Si II $\lambda6347$, Si II $\lambda5979$ and He I $\lambda6678$. The lines are sorted by their depth of formation from the top to the bottom.

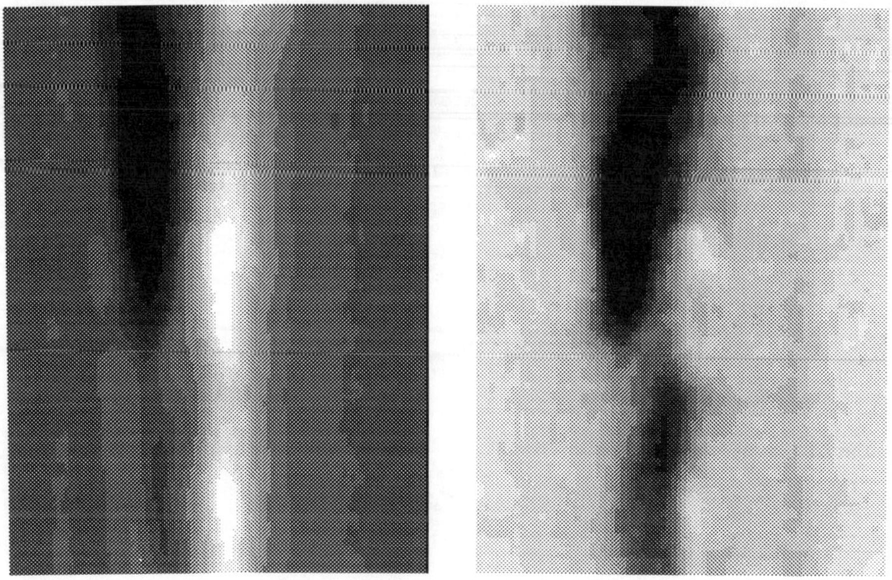

Fig. 15. Grey-scale representation of the spectrogrammes of HD 92207 obtained in 1993 around Hα (left) and Hβ (right). Velocity range: ± 300 km s^{-1}, time-span: JD 2,449,023 to JD 2,449,137. Obviously, the Hα profile is strongly variable. The time scale is much longer than for β Ori. Note the sudden shift in Hβ.

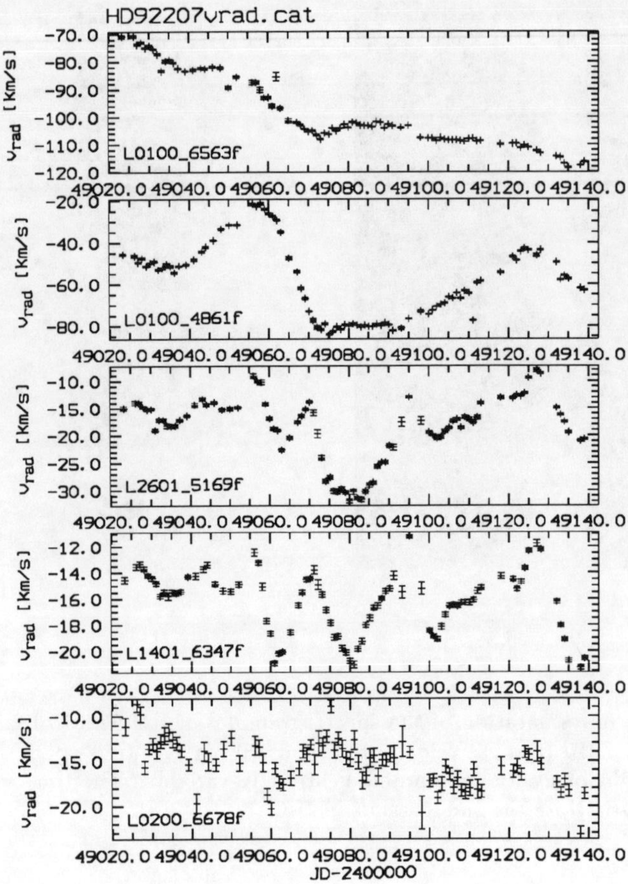

Fig. 16. Radial-velocity variations of HD 92207 as observed from the lines Hα (blue-shifted absorption component), Hβ, Fe II λ5169, Si II λ6347 and He I λ6678.

MULTICHANNEL MULTICOLOUR PHOTOMETRY: EXPERIENCES FROM A LONG-TERM MONITORING CAMPAIGN

K.H.MANTEL, and H.BARWIG
Universitaets-Sternwarte Muenchen
Scheinerstrasse 1
81679 Muenchen
Federal Republic of Germany

ABSTRACT. We present the design of a multichannel, multicolour photometer developed for the study of rapidly variable objects like Cataclysmic Variables, Low Mass X-Ray Binaries, Burst Sources or Flarestars and discuss experiences from a long-term observing campaign. It is demonstrated that high-speed multicolour photometry with an accuracy on the millimagnitude scale can be achieved. We present the design of a new four-channel spectrophotometer which allows to treat correctly higher-order extinction effects and which guarantees precise transformations to other standard photometric systems.

Keywords: multichannel photometer – high-speed photometry – MCCP – MEKASPEK

1. Introduction

The study of variable stars requires spectral- and time-resolved photometry. In order to gain insights into the underlying physical mechanisms responsible for the variability, information about the spectral behaviour on time scales typical for the systems being studied is needed. Depending on the nature of the variability (e.g. orbital motion, dwarf-nova outbursts, quasi-periodic variations, period changes, amplification events, flares, bursts, etc.), the typical time scales of light-variation phenomena range from years or days to minutes and seconds or even to milliseconds. This implies that the equipment used must be stable and supply the observer with high-accuracy multicolour data not only on short time scales, but also in the long run between different observing campaigns (which in some cases make use of different telescopes, e.g. for objects with periods in the range of multiples of a day). Therefore, high accuracy of the single photometric measurement is needed as urgently as is a stable and calibratable photometric system throughout the whole process of observations, data reduction and data processing. This includes not only control of all relevant instrumental parameters like quantum efficiency, dead time, etc., but also careful monitoring of the sky background and the varying atmospheric transparency as well as optimized reduction procedures.

During the last years extended efforts, including instrumental development and

C. Sterken and M. de Groot (eds.), The Impact of Long-Term Monitoring on Variable Star Research, 329–344.

implementation of specially adapted reduction procedures, have been made by our working group at the Universitaets-Sternwarte Muenchen with the aim to install photometric equipment which allows precise high-speed multicolour photometry of rapidly variable objects. As a result, two multichannel photometric instruments have been developed, together with comprehensive reduction software. A brief description of the Multichannel Multicolour Photometer (MCCP) can be found in Section 2 together with experiences from long-term observing campaigns. It is demonstrated that millimagnitude accuracy can be achieved with a multichannel fiberoptic photometer. In Section 3 the design of the new Multichannel Spectrophotometer (MEKASPEK) is described. A 33-channel instrumental-colour system is presented, which allows to treat higher-order colour effects correctly. Results are summarized in Section 4.

2. The Multichannel-Multicolour Photometer (MCCP)

This instrument was designed for high-speed *UBVRI*-photometry of rapidly variable objects. A block diagram of the instrument is presented in Fig. 1; details can be found in Barwig et al. 1987. In the following a brief description of the instrument will be given. The main features can be characterized as follows:

- Simultaneous measurements of three different light sources (the object, a nearby comparison star and the sky background)
- Simultaneous *UBVRI* measurements of each light source covering the whole optical wavelength range with high efficiency
- High time-resolution up to 20 milliseconds

2.1. INSTRUMENTAL DESIGN

Three optical fibers which can be positioned in the focal plane of the telescope within an area of 10x10cm transfer the signals of the three input channels to the entrance slit of three identical prism spectrographs. The fibers are fixed by an acryl plate with holes drilled at the positions of the selected objects. The size of the measuring diaphragms can be selected between 0.6 and 1 mm. The resulting spectra cover the region 3400-9000 Åand are imaged onto an array of mono fibers. These arrays select 5 broad passbands and are optimized to match the Kron-Cousins *UBVRI* system as closely as possible. 15 Peltier-cooled photomultipliers are used to record simultaneously the signals of the 5 *UBVRI* passbands of the three input channels. The multiplier signals pass the preamplifier and discriminator unit and are recorded and stored on disk by a microcomputer system. A CCD-camera in combination with an image intensifier can be positioned in the focal plane of the telescope for autoguiding.

2.2. MEASURING METHOD AND STANDARD REDUCTION PROCEDURE

During each integration the individual channels (1=object, 2=comparison star, 3=sky background) provide a combination of the following quantities for each of the 5 *UBVRI* passbands:

$$\text{Channel 1 (object)}: TC(1) = O(1) + S(1) + D(1)$$

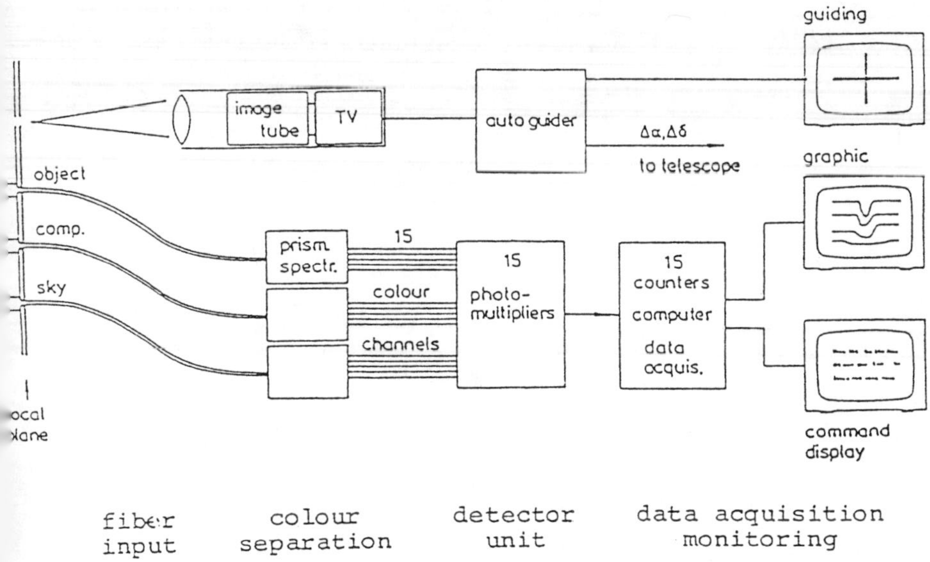

Fig. 1. Block diagram of the Multicolour-Multichannel-Photometer (MCCP).

Channel 2 (comparison star): $TC(2) = C(2) + S(2) + D(2)$
Channel 3 (sky background) : $TC(3) = S(3) + D(3)$

where $TC(i)$ is the signal measured in channel i, $O(i)$, $C(i)$, $S(i)$ and $D(i)$ are the intrinsic signals of object, nearby comparison star, sky background and dark current in channel i respectively (i=1,2,3). To a first approximation (so called *standard reduction*, see Barwig et al. 1987) changes in atmospheric transparency and sky-background contribution are the same for object and comparison star. Therefore, the ratio $O(1)/C(1)$ gives the object signal relative to the nearby comparison star with extinction effects and contributions from the sky background removed. The simultaneous measuring method allows the determination of $S(1)$ and $C(1)$ from the signals of the comparison star measured in channel 2 and the sky background measured in channel 3.

Due to different quantum-efficiency of the different input channels additional calibration measurements are required. This can either be done with special calibration lamps or with the nearby comparison star and/or an additional standard star under photometric conditions. The measurements with calibration lamps have the advantage of not consuming valuable observing time and may be performed even during day time.

From these measurements calibration coefficients are determined which allow the transformation of the measured quantities from one channel to the other. These calibration coefficients also reflect the stability of the instrument and instrumental changes (transmission, efficiency). Figure 2 shows the calibration coefficient for the transformation of the comparison signal from channel 2 to channel 1 in the $UBVRI$

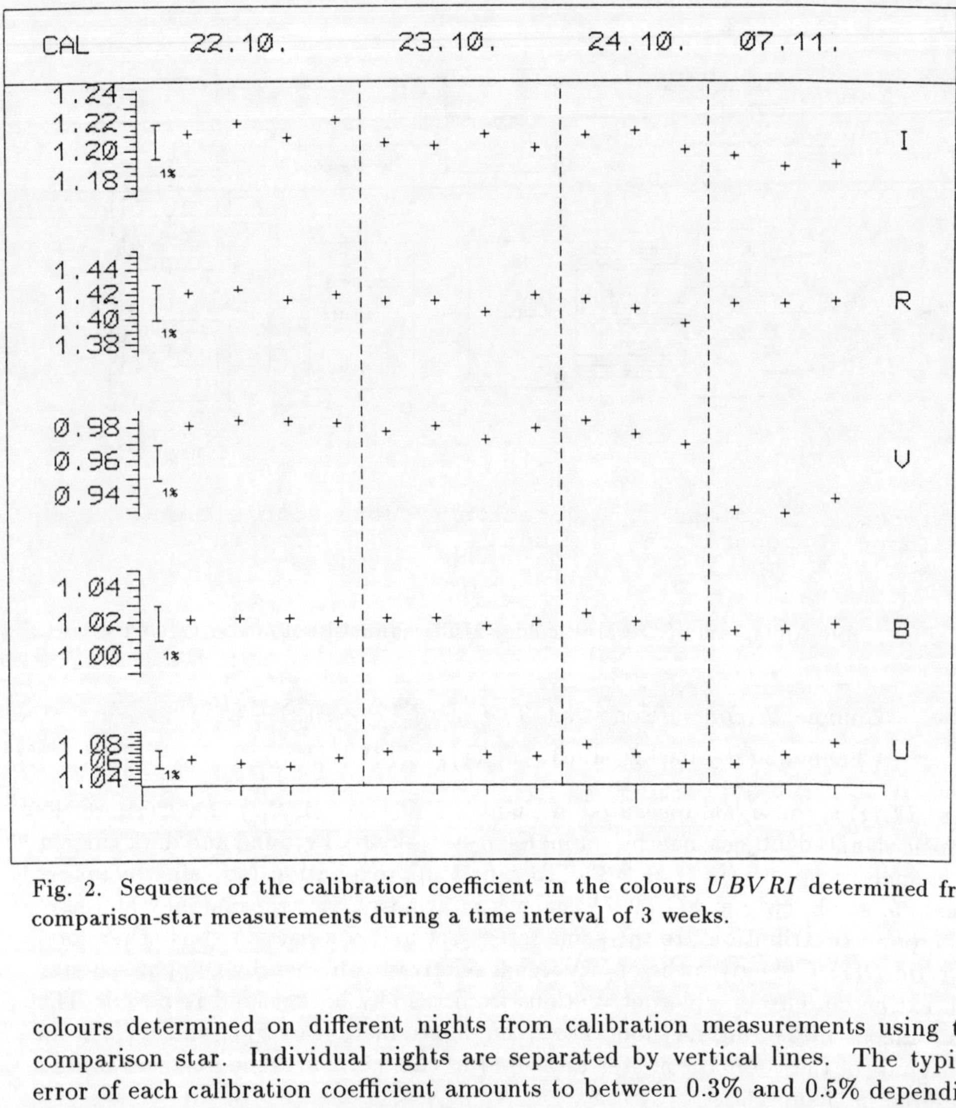

Fig. 2. Sequence of the calibration coefficient in the colours *UBVRI* determined from comparison-star measurements during a time interval of 3 weeks.

colours determined on different nights from calibration measurements using the comparison star. Individual nights are separated by vertical lines. The typical error of each calibration coefficient amounts to between 0.3% and 0.5% depending on colour and is marked as a vertical error bar. These errors are probably due to improper centring of the comparison star in the entrance diaphragm. Except in the V passband, the individual variations are smaller than the typical errors. Therefore, mean values of the calibration coefficients can be used for the reduction to improve the accuracy. In this case the error of the coefficients can be reduced to 0.05% in B and less than 0.2% in the other colours. Between night 2 and 3 a power-failure caused a change in the efficiency of the multiplier used for the V passband in channel 2 which resulted in a change of the transformation of about 4.2%. The data clearly demonstrate that the calibration of a multichannel photometer with an accuracy of better than 0.2% is feasible. The standard reduction is a powerful tool to reduce atmospheric extinction effects: transparency variations up to 80% can be corrected (for an example see Section 3, Fig. 6). Details of the procedures as well as an extensive discussion of thc standard reduction, the assumptions involved, and crrors depending on brightness of object, comparison star and sky background can be found in Barwig et al. 1987.

2.3. ITERATIVE REDUCTION PROCEDURE

Besides applying the standard reduction, data can also be reduced in a more conservative way: sky-background and dark-current contribution are subtracted from the object and the comparison channel. From the comparison star data measured during photometric conditions, extinction coefficients and extra-atmospheric count-rates for the comparison star are derived. The extinction coefficients are used to reduce all object and comparison-star data to zero airmass. Residual variations in the comparison-star data are used to correct transparency variations in the object measurements. The transformation to the *UBVRI* system is determined from photometric standard-star measurements. This reduction procedure requires photometric conditions; otherwise, too small extinction coefficients and extra-atmospheric count-rates are determined. In order to solve this problem the so-called *iterative reduction procedure* was developed, which allows to determine extinction coefficients and extra-atmospheric count-rates with high accuracy, even in nights with varying transparency (Mantel 1993).

The interactive reduction procedure starts with a first guess for extinction coefficients and extra-atmospheric count-rates (determined in the classical way from all measurements, including those made under non-photometric conditions). For these values, a theoretical ideal data set can be computed. Differences between the theoretical and the measured data set are caused by atmospheric transparency variations, which prevent a correct determination of the extinction coefficents and the extra-atmospheric count-rates. For each datapoint the difference between the ideal and the measured data set is used to assign a weight to each measured data point depending on whether it lies above or below the theoretical value. In the next iteration step, new values for the extinction coefficients and the extra-atmospheric count-rates are derived from the weighted measured data set. In this way measurements obtained under photometric conditions are extracted from the data and, even in nights with only relatively few measurements under photometric conditions, reliable extinction coefficients and extra-atmospheric count-rates can be determined. Figure 3 shows a set of measured data of the comparison star corrected for sky background and dark current, obtained in a non-photometric night, together with the classical Bouguer-line determined from data obtained during photometric conditions, as well as the ratio between ideal and measured data.

Table 1 gives the results of extra-atmospheric magnitudes of the comparison star determined by using the iterative reduction procedure during five non-photometric nights.

As a result, even in *non-photometric* nights, extra-atmospheric count-rates of the comparison star and extinction coefficients can be determined with an accuracy between 0.5% and 2% *without* any additional standard-star measurements. The lower accuracy in the *R* and *I* band is caused by the diminished quantum efficiency

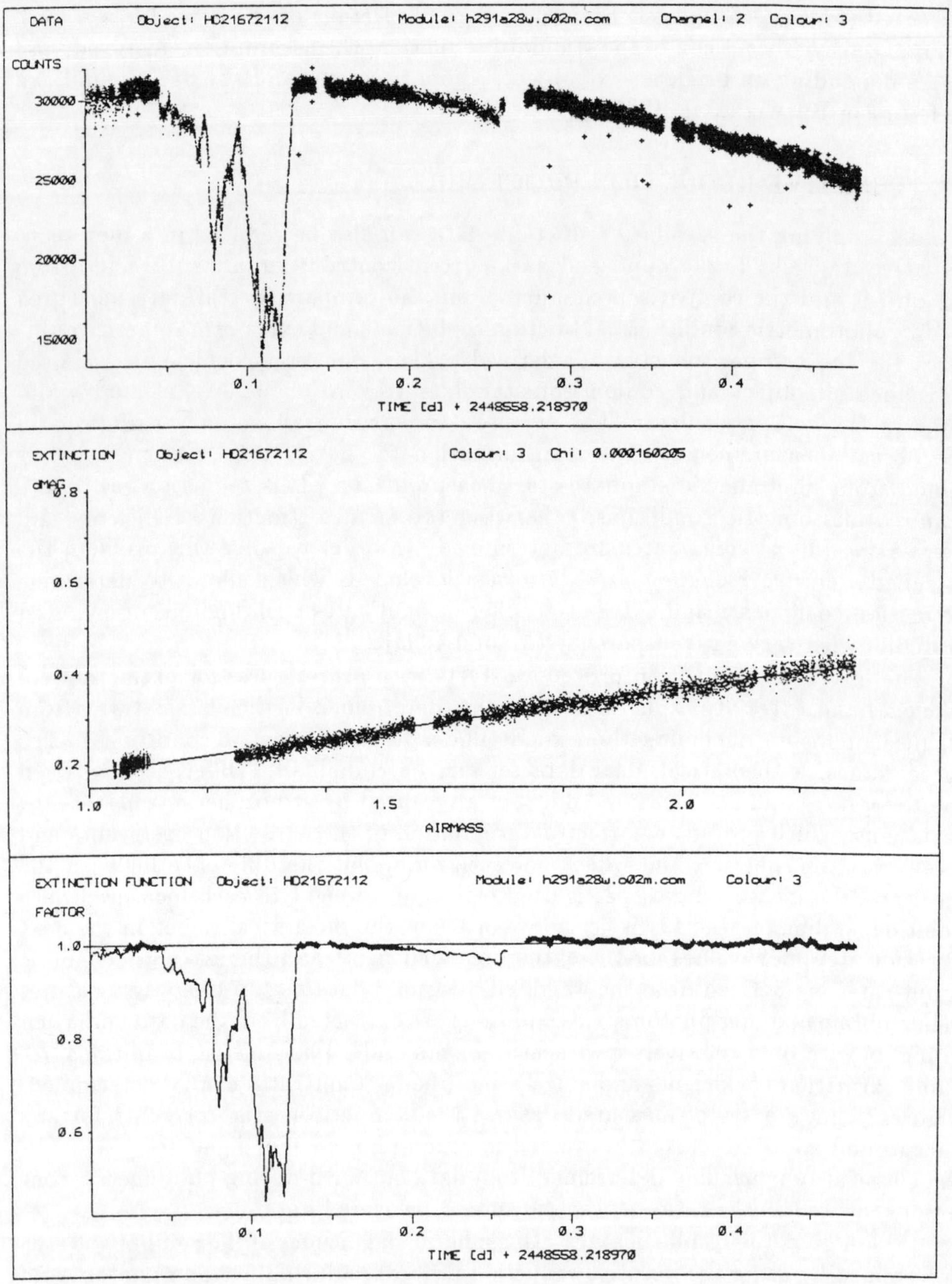

Fig. 3. Measurement set of the comparison star (sky background and dark current subtracted) during a non-photometric night (above); data obtained during photometric conditions are marked by bold points. The centre panel shows the Bouger-line fit for the data under photometric conditions. Residuals between ideal and measured data are shown below.

TABLE I

Extra-atmosperic magnitudes of the comparison star in different nights.

Night/Colour	*U*	*B*	*V*	*R*	*I*
27.10.90	9.2734	7.9307	8.6209	9.1926	10.5371
28.10.90	9.2850	7.9298	8.6188	9.2067	10.5647
29.10.90	9.2808	7.9343	8.6280	9.1940	10.5408
30.10.90	9.2605	7.9302	8.5931	9.1512	10.5432
31.10.90	9.2881	7.9418	8.6409	9.2046	10.5920
mean	9.2776	7.9316	8.6226	9.1978	10.5475
error	±0.0110	±0.0050	±0.0175	±0.0225	±0.0230

of the photomultipliers in these wavelength regions.

2.4. ACCURACY IN LONG-TERM MONITORING

In order to demonstrate the accuracy obtainable with a multichannel multicolour instrument, the final reduced light curve from a long-term observing programme is shown. The aim of this programme was to determine the physical parameters of the hot binary system DN Cas. Due to the $2\overset{d}{.}3$ orbital period of DN Cas, data of different nights had to be combined in order to cover the most important parts of the light curve. Over 3 years 14 nights of data were obtained with the MCCP at the 0.8-m telescope of the Wendelstein Observatory with a time resolution of 4 s. Figure 4 shows the final reduced light curve together with a best fit derived from theoretical models. Data were binned to time intervals of 8 minutes resulting in count rates of more than 5×10^7 counts per time interval. Therefore, the Poisson statistics should not contribute to the error of individual data points. The remaining error of individual measurements is $0\overset{m}{.}001$ to $0\overset{m}{.}002$ (Fiedler & Simon 1992, Simon et al. 1993). Extensive investigations were made in order to determine the source of this remaining error. As the most probable error source the size of the input diaphragm was identified. During the observations a size of 16″was used, which, in case of slightly inaccurate centring might cause an error at the millimagnitude level.

Although the multichannel multicolour photometer proved to be a reliable and stable instrument the following problems remain:

- the size of the entrance diaphragms is limited to 1 mm in diameter
- second-order extinction effects are neglected by the reduction procedure
- fiber positioning requires the exchange of a mask and is therefore time-consuming
- transformations to the standard system involve errors of the order of 3-5%

Therefore, in 1987 the MEKASPEK project was started, which aimed to develop an instrument which overcomes these problems.

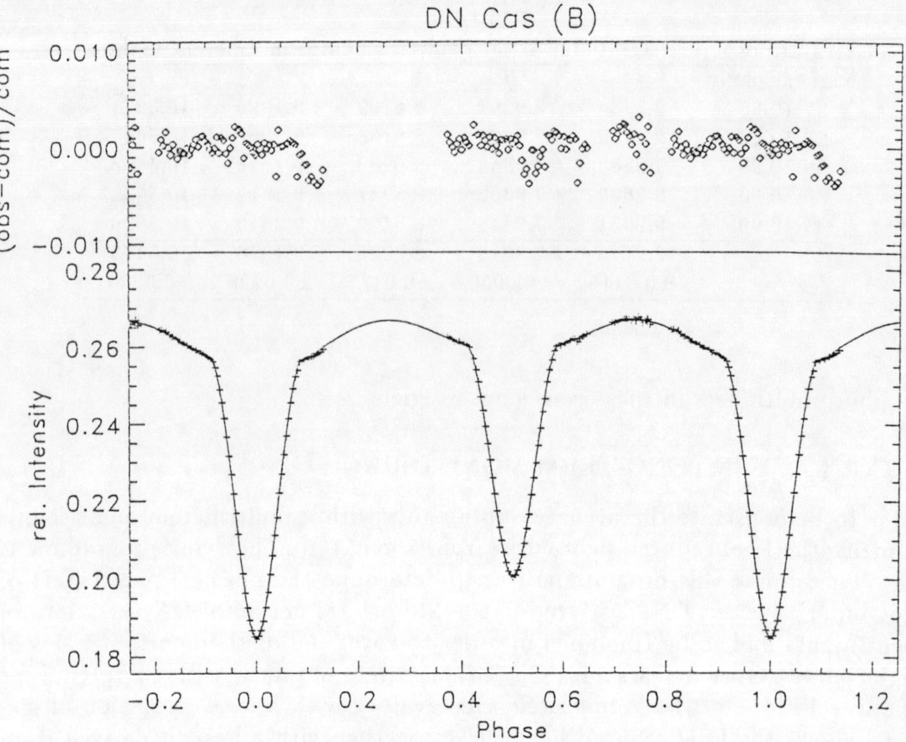

Fig. 4. Reduced light curve of DN Cas (marked by '+') together with the best fit derived from theoretical models (straight line). The error of the individual data points is $0\overset{m}{.}001$ to $0\overset{m}{.}002$.

3. The Multichannel Spectrophotometer (MEKASPEK)

3.1. INSTRUMENTAL DESIGN

The MEKASPEK instrument is a four-channel fiber-optics spectrophotometer which allows the measurement of 4 sources simultaneously with a spectral resolution of about 50 in the optical wavelength range between 3500 and 9000 Angstroem. Fig. 5 shows a block diagram of the instrument. Four fiber bundles can be positioned in the focal plane of the telescope by computer control. A Fabry lens at the entrance of the fiber bundles projects the entrance pupil of the telescope onto the circular fiber bundle with a diameter of 0.2 mm. The diameter of the entrance diaphragm can be changed between 0.6 and 2mm. The rectangular exit of the fiber bundles forms a homogenously-illuminated entrance slit of a prism spectrograph. The spectra of the four sources are imaged onto the photocathode of a two-dimensional, photon-counting detector (MEPSICRON, Firmani et al. 1983) and are recorded by a real-time multitasking VME-Bus computer system. A maximum time resolution of 20 ms can be achieved. Instrumental details can be found in Mantel et al. 1993.

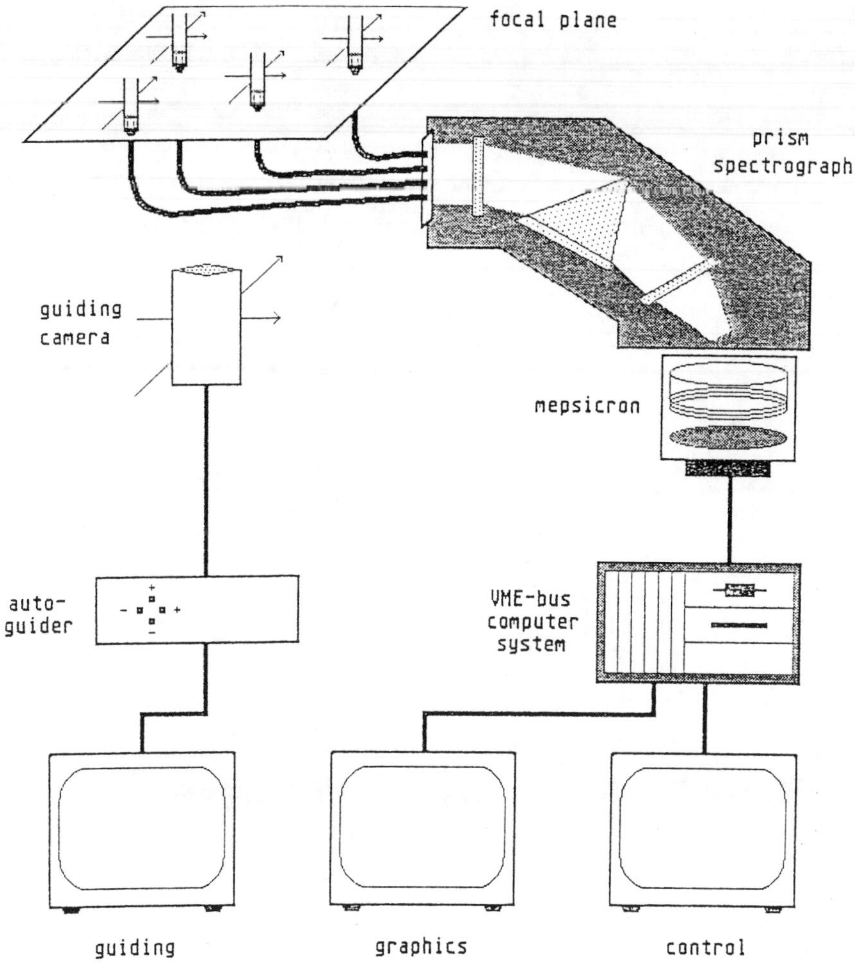

Fig. 5. Block diagram of the Multichannel Spectrophotometer (MEKASPEK).

3.2. MEASURING METHOD AND DATA REDUCTION

The measuring method is similar to that used for the MCCP apart from the fact that with MEKASPEK an additional fiber is available, which can be used to monitor a second comparison star (e.g. with a different spectral distribution) for improved description of colour-dependent extinction effects, or, in cases of dominant stray-light (Moon), can be used to monitor the sky background at a second position to compensate for spatially varying background contribution.

An example of the elimination of atmospheric-transparency variations is shown

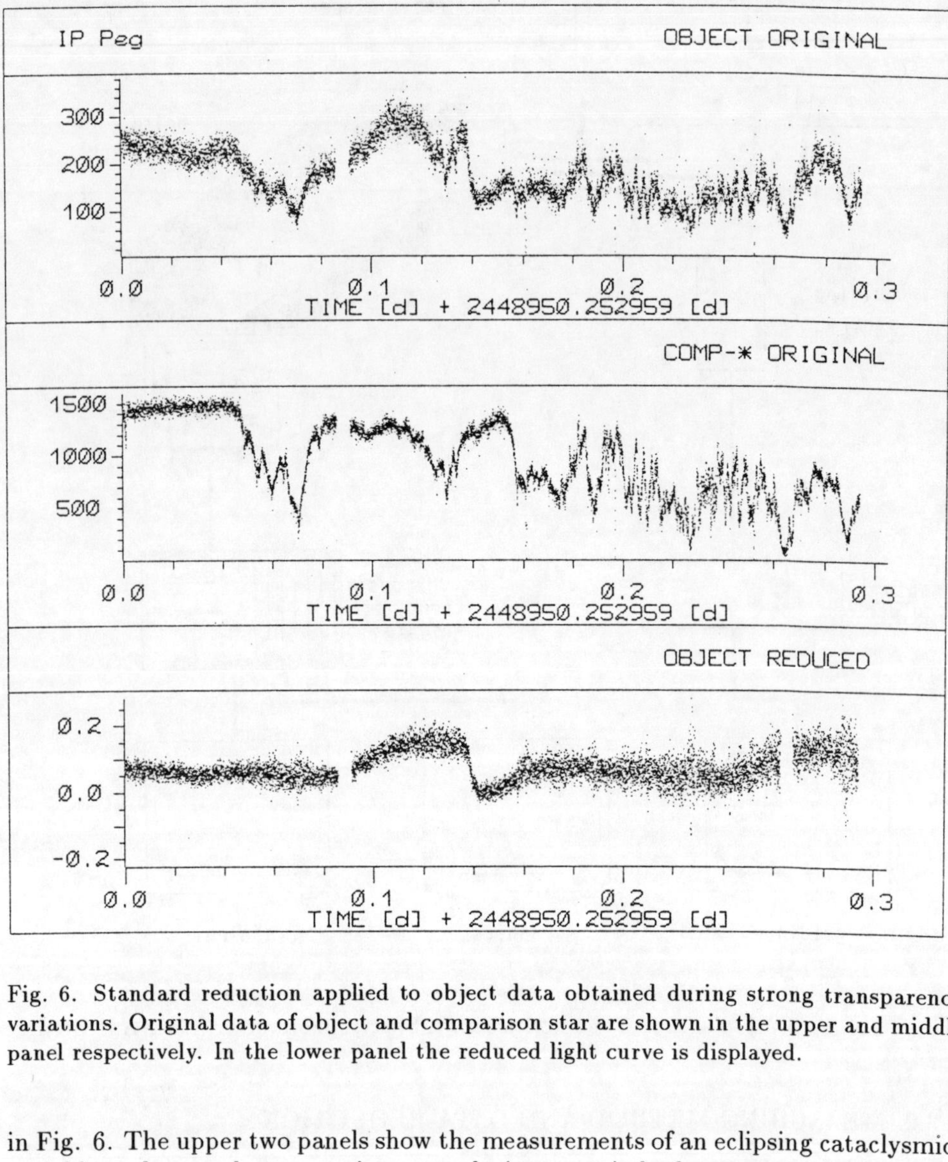

Fig. 6. Standard reduction applied to object data obtained during strong transparency variations. Original data of object and comparison star are shown in the upper and middle panel respectively. In the lower panel the reduced light curve is displayed.

in Fig. 6. The upper two panels show the measurements of an eclipsing cataclysmic-variable and a nearby comparison-star during a period when strong cirrus caused large transparency variations up to 90%. In the lower panel the reduced lightcurve (standard reduction applied) is shown. Most of the transparency variations could be reduced completely; only during times of verly low transparency the scatter in the reduced lightcurve is increased due to photon noise.

3.3. DETERMINATION OF HIGHER ORDER COLOUR EFFECTS

All photometric reduction procedures suffer from insufficient knowledge of the spectral distribution of the observed stars. This implies that second-order extinction effects can be treated only in a very crude way and limit the accuracy obtainable so far (for an excellent review of this problem see Young 1988). This problem is solved with MEKASPEK: flux-calibrated low-resolution spectra provide the necessary information on the spectral distribution of the observed objects and allow to treat higher-order extinction corrections. In general, the influence of atmospheric extinction on the measurement within a passband can be described by (cf. Mantel 1993):

$$C(t) = I(t)A(t)(1+x)F$$

where:

$C(t)$ recorded count-rate

$I(t)$ extra-atmospheric flux of object

$A(t)$ atmospheric extinction

x correction term for higher-order colour effects

F integral of filter curve of passband used

All the higher-order extinction effects are described by the x-term which includes airmass, moments of the passband used and also the first and second-order derivatives of extinction coefficients and spectral intensities. Typical values of this term depend on the combination of the spectral distribution of the object and the filter pass bands used and may range between 0.0001 and 0.1. Small values of x indicate a well-adapted description of colour effects by the photometric system used.

Due to the limited spectral information provided by photometric standard systems, so far only crude approximations could be derived for the derivatives involved in the x-term. Because of the spectral resolution, the MEKASPEK data now enable to solve this problem and allow a numerical determination of the derivatives involved.

This is accomplished by summing up data in widely-overlapping filter passbands, which fulfil the sampling theorem. Intensities of these overlapping passbands can be used to describe derivatives in a proper way. Therefore, higher-order extinction effects can be treated correctly. Additional software was developed which allows to determine the x-term numerically from the intensities of all neighbouring filter passbands by using an approximative spline function fitted to the intensities obtained in the overlapping filter passbands. Details can be found in Mantel (1993).

An example for this is shown in Table 2: the measured data were summed up into 33 widely overlapping passbands (see Fig. 7), which fulfil the sampling theorem. For each passband the x-term was computed. As can be seen from Table 2, the x-term is usually below 3%, only on the very blue and red side higher correction terms appear. The reason for this insufficient description of the derivatives is due to low efficiency of the instrument in these wavelength regions. The low values of the correction term show, that higher-order colour effects are well described by an instrumental system with widely overlapping filter-curves.

Transformations to other broad-band standard systems also benefit from the correct description of higher-order colour effects. Due to the widely overlapping

TABLE II

Correction term x computed for a 33-colour photometric system.

filter	cen.wavelength	x-term	filter	cen. wavelength	x-term
1	4150	1.0450	18	5803	0.9948
2	4161	1.0222	19	5920	0.9942
3	4191	1.0106	20	6048	0.9935
4	4256	1.0044	21	6186	0.9930
5	4346	1.0007	22	6342	0.9923
6	4454	0.9983	23	6512	0.9914
7	4575	0.9969	24	6692	0.9904
8	4717	0.9961	25	6892	0.9887
9	4848	0.9961	26	7113	0.9857
10	4983	0.9961	27	7366	0.9799
11	5138	0.9964	28	7638	0.9711
12	5283	0.9972	29	7933	0.9587
13	5352	0.9970	30	8227	0.9434
14	5427	0.9968	31	8528	0.9331
15	5508	0.9963	32	8874	0.9285
16	5598	0.9959	33	8953	0.9198
17	5696	0.9953			

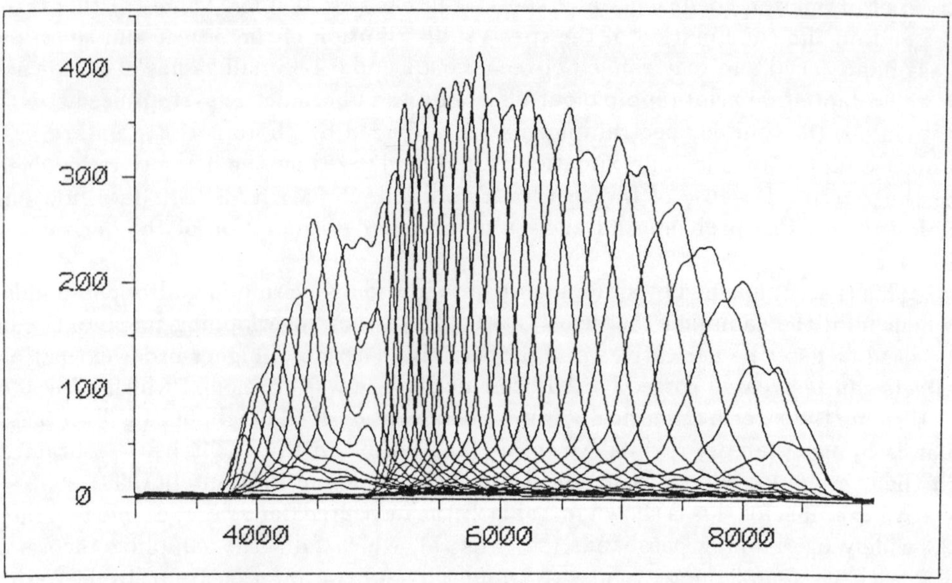

Fig. 7. Instrumental 33-colour system with widely overlapping filter-curves.

filter-curves, colour terms in transformation equations can be neglected and colour transformations can be calculated from the instrumental filter-curves. The filter curves of the 33-channel instrumental colour system have been measured using

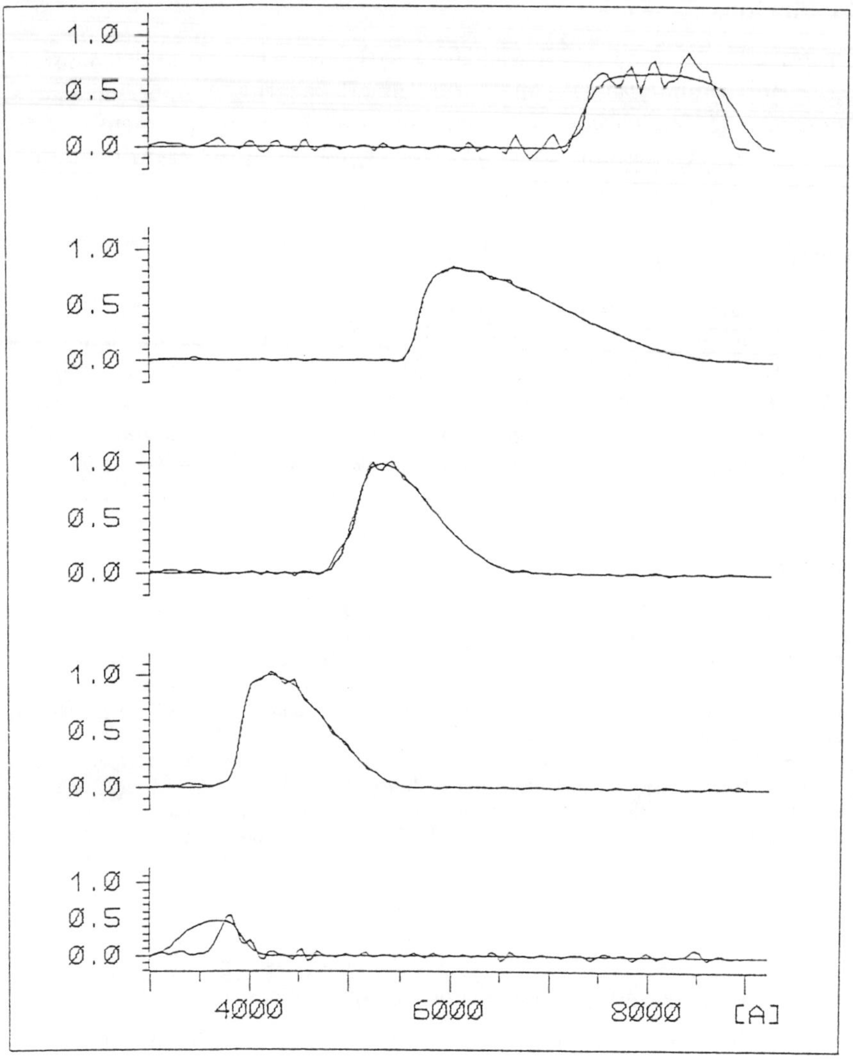

Fig. 8. Approximation of the $UBVRI$-system.

a flux-calibrated scanning monochromator (see Fig. 7). From these filter curves, transformations to the Kron-Cousins $UBVRI$, the MCCP and the Strömgren colour system have been calculated. Fig. 8 shows the filtercurves of the $UBVRI$ system together with a fit calculated from the transformation. The bad approximation in the U- and I-band is caused by the low quantum efficiency of the detector in this wavelength region. The excellent approximation in the other colours demonstrates that an instrumental colour system with widely overlapping filter-curves fitting the requirements of the sampling theorem, allows precise colour transformation to broad-band standard systems.

4. Conclusions

Millimagnitude accuracy can be achieved with multichannel multicolour photometry. Due to the simultaneous measurement of a nearby comparison star and the sky background, transparency variations up to 80% can be compensated for. The multichannel spectrophotometer allows to calculate accurate transformations to all broad-band photometric systems.

Acknowledgements

The authors wish to thank A. Fiedler for valuable discussions on the long-term stability of the MCCP and for placing his data at our disposal. They also wish to thank S. Wolf for his help in preparing the Latex source file. Observations presented in this work were performed at the 0.8-m telescope at Wendelstein Observatory, Germany and at the Centro Astronómico Hispano-Alemán at Calar Alto, Spain. The support by the Deutsche Forschungsgemeinschaft (grants $Ba867/2-1$, $Ba867/2-2$) is kindly acknowledged.

References

Barwig, H., Schoembs, R., Buckenmeyer, C., 1987, *Astronomy and Astrophysics*, 175, 327.

Mantel, K.H., 1993, Dissertation, Universitaets-Sternwarte Muenchen.

Fiedler, A., Simon, K.P., 1992, *Astron. Ges. Abstract Series*, 7, 92.

Simon, K.P., Sturm, E., Fiedler, A., 1993, *Astron. Ges. Abstract Series*, 9, 21.

Firmani, C., Gutierrez, L., Ruiz, E., Bisiacchi, G.F., Salas, L., Paresce, F., 1983, in SPIE Instrumentation in Astronomy V, 445, 192.

Mantel, K.H., Barwig, H., Kiesewetter, S., 1993, in *Stellar Photometry - Current Techniques and Future Developments*, C.J. Butler and I.Elliott (Eds.), Cambridge University Press, 172.

Young, A.,T., 1988, in *Proc. of the Second Workshop on Improvements to Photometry*, W.J. Borucki (Ed.), NASA Conference Pub. 10015, 215.

DISCUSSION

Sterken: What is the limiting magnitude on the bright side?

Mantel: The limiting magnitude on the bright side, at the 80-cm telescope, is 8th magnitude.

Sterken: In LTM, as you have seen, many programme stars are bright. Do you simply avoid monitoring bright stars, or do you have a solution?

Mantel: So far we tried two different solutions. For observations of Supernova 1987A with the MCCP we used neutral filters. The data showed large variations of the flat field accross the entrance diaphragm caused by the neutral filters. The second solution we tried was reducing the size of the light-collecting area by mounting a diaphragm on the top ring of the telescope. With this set-up we observed a $6\overset{m}{.}5$ star close to the full Moon. Due to the additional obturation at the top ring, we got additional stray light from the Moon, inducing strong variations, which, in turn, caused many problems during the reductions. So far, the experience has been very discouraging, but we hope to proceed in the future with an improved approach.

Kurtz: In response to Chris Sterken's question: I encourage you *not* to stop your telescope down to observe bright stars. The scintillation noise is greater for similar apertures, so this will degrade your observations. For brighter stars you should use narrower filters, or neutral-density filters.

Williams: What is the projected size of the fibres on the sky, and how long does it take to re-position the fibres feeding the multi-colour spectrograph if you are changing stars?

Mantel: This depends on the focal scale of the telescope and on the entrance diaphragm used. The entrance diaphragms can be changed between 0.6 and 2 mm, corresponding to 13″and 42″at the 0.8-m telescope at Wendelstein, and to 4″and 14″at the 3.6-m ESO telescope.

Young: How does the transmission of the fibres change when you move them? When you move around the sky and measure extinction, how accurately can you do photometry?

Mantel: We do not move the fibres around during measurements. Therefore, we have not tested the influence of movements of the fibres on fibre transmission. Normally, we use the instrument for photometry of rapidly varying objects like cataclysmic variables, flare stars, X-ray bursts, etc. We stay on one object during the whole night and determine the extinction from the simultaneous measurements of one (MCCP) or two (MEKASPEK) nearby comparison stars. As I have shown, precision on the millimag level is achievable even if data are obtained over a time interval of 3 years.

Le Contel: Would you please remind us of the main properties of the MEPSICRON detector?

Mantel: The MEPSICRON is a photon-counting detector. It has a sensitive area of 25 mm in diameter, corresponding to 512 x 512 pixels and is equipped with a multi-alkali photocathode with a typical quantum efficiency of 15-20% at 5000Å. Five cascaded microchannel plates behind the photocathode accelerate the electrons to a resistive anode inducing a distinct charge distribution for each registered photon. The position is determined by a position-computing system. The detector is cooled to -25°C. It has no readout noise and very low dark current (less than one count/pixel/hour) There is no overflow due to cosmic events; cosmic particles just increase the dark current by one count. The detector has a maximum time resolution of 1 ms; the maximum count-rate is limited to 1 x 10^5cts/s.

Schwarzenberg-Czerny: What effect have cosmic-ray particles on the MEPSICRON measurements?

Mantel: A cosmic-ray particle just increases the dark current by one count, because high and low-energy particles are treated by the MEPSICRON in just the same way. So, there is no influence on the measurements from cosmic-ray particles.

Young: Cosmic rays contribute negligibly because the MEPSICRON is a digital device (one count per cosmic ray event). Cosmic rays are a big noise source in CCDs because those are analog devices (hundreds of electrons produced per cosmic ray event).

Kurtz: This morning we saw the point-spread function and spikes on stellar images. It worries me that the largest aperture on your photometer is only 1 mm. I would hope our experience with the HSP on the HST would have taught us permanently *not* to build photometers with *only* small apertures.

Mantel: The maximum diameter of the entrance diaphragm at the MEKASPEK is 2 mm which, I think, is large enough in most cases. If you increase the diameter further, the probability of including a faint star increases rapidly, causing additional errors, especially if the faint star is situated at the rim of the entrance diaphragm.

Chochol: The instrument you described is of a very professional design. Did you consider offering it for commercial purposes? What would be the price for a copy of the instrument?

Mantel: I think the market for such an instrument is very small, but it is possible to build a copy, or even a few copies, of it; the know-how is available at Munich Observatory. My guess is that you would need to invest about DM 200 000, for the parts alone.

Wolf: When the cost of labour is added, the final price may be in the DM 700 000 range.

Hric: What is the minimum distance between two fiber channels?

Mantel: The minimum distance between two fiber channels is 4 mm. This corresponds to 80″at the 0.8-m Wendelstein telescope, and to 29″at the 3.6-m ESO telescope.

Vyverman: What is the time delay between the flare observations with MEKASPEK and ROSAT? Did you observe other such events? With comparable delays?

Mantel: The time delay between the optical and X-ray data of the flare is of the order of 10 s. We observed a comparable delay during a micro-flare on the flare star UV Ceti (Schmidt and Barwig 1993).

DISCUSSION No 5: SPECTROSCOPY

Le Contel: I would like to ask a rather provocative question: is an echelle spectrograph really the best solution for hot stars where you have only a few lines? We shall have an enormous amount of data to archive (comparable in quantity to Schmidt-plate data) and, maybe, for nothing.

The Audience: —

Schwarzenberg-Czerny: Long-term ($\sim$ 3 years) coverage simultaneous with ground-based photometric and spectroscopic campaigns will be provided (for bright stars) by HIPPARCOS.

Young: HIPPARCOS generally does not sample a given star uniformly in time, but in bursts of a few measurements separated by longer intervals.

Le Contel: HIPPARCOS photometry is actually very good, and we may expect some nice surprises when the data will appear. Of course, the quality of the light curves of variable stars will depend on the star. But for some cepheids I can say that very nice light curves already exist.

Chochol: I should like to ask the organisers of spectroscopic campaigns to plan always for the necessary simultaneous photometric observations of objects under study.

de Groot: The Heidelberg people have mentioned Long-Term Spectroscopy of Variables (LTSV). Can we see a way forward towards realizing this idea and make spectroscopic monitoring accessible to a wider circle of observers?

Sterken: One way forward could be duplication of the Heidelberg instrument and install it at various sites. Combined with a photometric instrument it would require only one observer to get photometry and spectroscopy simultaneously.

Wolf: A similar system exists in the MUSICOS collaboration, in which there are now three identical instruments.

C. Sterken and M. de Groot (eds.), The Impact of Long-Term Monitoring on Variable Star Research, 345–346.

Danish spoken here – Olsen and Pedersen

LONG-TERM OPTICAL MONITORING OF GAMMA-RAY BURST SOURCES

H. PEDERSEN
Copenhagen University Observatory
Øster Voldgade 3
DK 1350 Copenhagen K
Denmark

ABSTRACT. In spite of a massive experimental and theoretical effort, there is still no consensus as to the distance scale of the GRB phenomenon and, therefore, the basic physics remains unknown also. It is not clear which approach should be applied to advance our knowledge. However, multiwavelength observations seem interesting, and long-term optical monitoring may constrain some of the proposed models.

Keywords: Gamma Ray Bursts – Optical Transients

1. Introduction

More than 50 spacecraft have been used to do observations of cosmic Gamma-Ray Bursts (Castro-Tirado 1993) and the number of resulting theories exceeds 100 (Nemiroff 1993). Data from the Compton Observatory and other instruments were dealt with in detail during the recent workshop in Huntsville (Fishman et al. 1993a), but no consencus was reached regarding the distance scale; the great debate stands between "local" and "cosmological" models.

This is not the place to give a full introduction to the subject of GRB. However, being of particular relevance to the issue of long-term optical monitoring, a few points should be mentioned. Cosmic GRBs can be divided into "classical GRBs" and "soft gamma repeaters", SGRs. The celestial distribution of the classical sources is consistent with isotropy, and no association has been found with any class of persistent objects, including γ-ray objects. There is some evidence that the faintest GRBs observable with the BATSE modules on the Compton Observatory, are placed at z $\sim$ 0.8 - 1 (Fenimore et al. 1993, Norris et al. 1993), in which case host galaxies must be fainter than absolute magnitude $\sim$ -18. The classical sources show a bimodal distribution of durations, with peaks near 0.2 s and 30 s, see Fig. 1, and short and long events seem to be derived from the same source population (Mao et al. 1994). Only three members are known of the class of SGRs. It seems that they can be associated with supernova remnants in the Galaxy, or in the LMC, but details of the burst emission mechanism remain unknown.

C. Sterken and M. de Groot (eds.), The Impact of Long-Term Monitoring on Variable Star Research, 347–352.

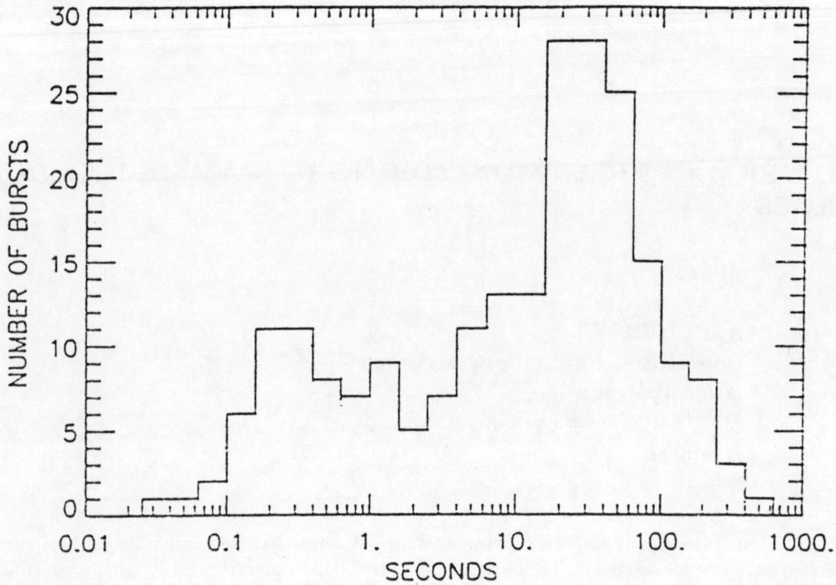

Fig. 1. The distribution of GRB durations has two peaks, near 0.2 and 30 s. The present data from BATSE (Fishman et al. 1993b) show the burst durations counted from 5% till 95% of the integrated burst fluence. Other duration definitions agree qualitatively, as do data from other spacecraft. BATSE positions, accurate to 5°- 10°, can reach ground-based observatories ~ 6 s after on-board trigger, permitting simultaneous observations while the long-lasting events are still in progress.

2. Optical Observations

In the MeV energy range, GRBs tend to dominate all other celestial sources during their brief presence. At optical wavelengths, the counterpart is not likely to be the brightest object in the error box, which in nearly all cases extends over more than one square arcminute. The number of faint stars and galaxies to be examined is so large, that spectroscopic studies remain impractical. Long-term optical studies are therefore limited to broad spectral passband observations and to the search for flaring optical counterparts of large amplitude.

3. Variability of Quiescent Optical Counterparts

That quiescent counterparts, if they exists at all, should be time-variable, and can only be surmised from analogy with other high-energy sources, such as X-ray pulsars, X-ray bursters and quasars, but that analogy is not well founded. Due to the size of most GRB error boxes, only few detailed studies have been made of their content of stars and galaxies (see Schaefer, 1992, and Harrison et al. 1994). Motch et al. (1985) specifically searched for time-variable objects, finding none, in the error box of the April 6, 1979 source, which is the smallest one known for classical GRBs.

4. Flaring Optical Counterparts

Several theories predict low-energy emission in conjunction with GRBs (e.g. London and Cominsky 1983, Woosley 1984, Melia 1988, Band and Hartmann 1992). Depending on the mechanism, the low-energy emission may be simultaneous with the γ-radiation, or it may arrive with a delay. One recent paper (Mészáros and Rees 1993) deals with a quite general fireball scenario. It concludes that the ratio of UV + optical radiation, compared to γ-radiation, will be in the range 1 to 10^{-4}, depending on model parameters. For an average γ-ray burst fluence, the implied low-energy fluence is detectable with equipment of modest size.

5. Historical Optical Transients

Based on the assumption that some GRB sources may repeat, Schaefer (1981) undertook the first investigation of wide-field exposures from the Harvard Sky Patrol. One interesting candidate event, from 1928, was found. Several similar studies have been made since then, in particular utilizing the large plate archives in Bamberg, Odessa, Ondřejov, and Sonneberg. However, none of the candidates have passed independent examination nor have been confirmed otherwise (e.g. Zytkow 1990, Greiner 1992, Greiner and Moskalenko 1993, but see also Hudec and Soldán 1993). In fact, a number of background effects can be difficult to eliminate, such as plate faults, accidental double exposures, aircraft flashes, satellite glints, and head-on meteors. The observational difficulties and the existing data have been reviewed by Hudec (1993).

The early reports on OTs prompted MIT/GSFC to construct the "Rapid Moving Telescope" (Barthelmy et al. 1992) and "Explosive Transient Camera" (Vanderspek et al. 1992, Krimm et al. 1993), now in operation at Kitt Peak, and ESO to embark on the GMS.

6. Simultaneous Detections

Simultaneous optical/γ-ray observations obviously are most suited for comparison with theory, but are technically difficult due to the unpredictable character of the events searched for. High-sensitivity, photoelectric observations can be done on the two SGR members, for which error boxes are known to better than an arcminute, SGR0526-66 (Cline et al., 1982) and SGR1806-20 (Kouveliotou et al. 1993). The first of these, the site of the famous March 5, 1979 event coincides with the direction to the young supernova remnant N49 in the LMC. It was subjected to a long observing campaign, using the ESO 50-cm telescope (Pedersen et al. 1984), but none of the three suspected optical events were noted in γ-rays. The second source, for which an association with the galactic supernova remnant G10.0-0.3 seems certain (Kulkarni et al. 1993), will probably become a target for IR and other observations during 1994, in particular if the burst activity continues.

The error boxes of several classical GRBs are being monitored by the Santa Barbara Astronomy Group (Schwartz, 1992).

Wide-angle, photographic observations are relatively easy but suffer from low time-resolution. For the first coincident exposure Grindlay et al. (1974) derived a fluence ratio $L_{opt}/L_\gamma < 10^{-2}$. Several other upper limits are known (Hudec 1993, Greiner et al. 1992a,b, Castro-Tirado 1993), primarily from fish-eye cameras used in fireball patrol networks. They are typically one to two orders of magnitude less significant, and some correspond to non-unique GRB positions.

7. Multiwavelength Observations from Space or Triggered from Space

A system is currently under development by Marshall Space Flight Center and Goddard Space Flight Center, which will allow 5°-10° accurate burst localisations to be distributed to ground-based observers only a few seconds after GRO/BATSE has triggered (Barthelmy et al. 1993). Since many GRBs last longer than 10 s, this will facilitate simultaneous observations, eventually with fast-slewing cameras.

The HETE mission, scheduled for launch in 1994/1995, includes a facility for direct transmission of data via shortwave radio and further dissemination will likely take place via Internet. HETE aims to locate High Energy Transients to arcsecond precision, through simultaneous on-spacecraft detections in γ-rays, X-rays, and UV (Ricker et al. 1992).

ESA's INTEGRAL mission, foreseen for launch in 2001, will include an X-ray detector and a wide-angle optical CCD detector, to operate in unison with the γ-ray imager. It is possible that INTEGRAL data will also be made available for immediate follow-up observations elsewhere.

8. Conclusion

More than two years into the Compton mission, it is clear that progress in the understanding of the GRB phenomenon will be slow, if it is to come from improved statistics. Multiwavelength observations, in particular optical and IR, have the potential of providing some missing links. To observers of variable-star phenomena, there can hardly be a geater challenge.

References

Band, D.L., and Hartmann, D.H., 1992, in *Gamma Ray Bursts*, AIP Conference Proc., 265, eds. W.S.Paciesas and G.J.Fishman, New York, p.342

Barthelmy, S., et al., 1992, in *Robotic Telescope of the 1990s*, ed. Alex Filippenko, *Astronomical Society of the Pacific Conf. Ser.*, 34, p.137

Barthelmy, S., et al., 1993, in *Proceedings of the 2nd Huntsville Gamma Ray Burst Workshop*, eds. G.J.Fishman et al., American Institute of Physics, in press

Castro-Tirado, A., 1993, Ph.D. Thesis

Cline, T.I., at al., 1982, *Astrophysical Journal, Letters to the Editor*, 255, L45

Fenimore, E.E., et al., 1993, *Nature*, 366, 40

Fishman, G.J., et al., 1993a, eds., Proceedings of the 2nd Huntsville Gamma Ray Burst Workshop (American Institute of Physics), in press

Fishman, G.J., et al., 1993b, in *Proceedings of the 2nd Huntsville Gamma Ray Burst Workshop*, eds. G.J.Fishman et al., American Institute of Physics, in press

Greiner, J., 1992, *Astronomy and Astrophysics*, 264, 121

Greiner, J., et al., 1992a, in *Gamma Ray Bursts*, AIP Conference Proceedings 265, Eds. W.S.Paciesas and G.J.Fishman, New York, p.327

Greiner, J., et al., 1992b, contribution to the Compton Symposium, St.Louis, Oct. 1992, preprint

Greiner, J., and Moskalenko, E.I., 1993, *Astronomy and Astrophysics*, preprint

Grindlay, J.E., Wright, E.L., and McCrosky, R.E., 1974, *Astrophysical Journal, Letters to the Editor*, 192, L113
Harrison, T.E., McNamara, B.J., and Klemola, A.R., 1994, *Astronomical Journal*, 107, 254
Hudec, R., 1993, *Astro. Lett. and Communications*, 28, 359
Hudec, R., and Soldán, J., 1993, in *Proceedings of the 2nd Huntsville Gamma Ray Burst Workshop*, eds. G.J.Fishman et al., American Institute of Physics, in press
London, R.A. and Cominsky, L.R., 1983, *Astrophysical Journal, Letters to the Editor*, 275, L59
Kouveliotou, C., et al., 1993, *Nature*, 362, 728
Kulkarni, S.R., et al., 1993, in IAU Circular, No.5879
Krimm, H., et al., 1993, in *Proceedings of the 2nd Huntsville Gamma Ray Burst Workshop*, eds. G.J.Fishman et al., American Institute of Physics, 1993, in press
Mao, S., Narayan, R., Piran, T., 1994, *Astrophysical Journal*, 420, 171
Melia, F., 1988, *Astrophysical Journal, Letters to the Editor*, 324, L21
Mészáros, P., and Rees, M., 1993, *Astrophysical Journal, Letters to the Editor*, preprint
Motch, C., et al., 1985, *Astronomy and Astrophysics*, 145, 201.
Nemiroff, R.J., 1993, *Comments on Astrophysics*, 17, No.4, in press
Norris et al., 1993, in *Proceedings of the 2nd Huntsville Gamma Ray Burst Workshop*, eds. G.J.Fishman et al., American Institute of Physics, 1993, in press
Pedersen, H. et al., 1984, *Nature*, 312, 46.
Ricker, G.R., et al., 1992, in *Proceedings of the Los Alamos Workshop on Gamma Ray Bursts, 1990*, eds. C.Ho, R.I.Epstein, E.E.Fenimore, Cambridge University Press, p.288
Schaefer, B.E., 1981, *Nature*, 294, 722
Schaefer, B.E. 1992 in *Proceedings of the Los Alamos Workshop on Gamma Ray Bursts, 1990*, eds. C.Ho, R.I.Epstein, E.E.Fenimore, Cambridge University Press, p.107
Schwartz, R., et al., 1992, in *Proceedings of the Los Alamos Workshop on Gamma Ray Bursts, 1990*, eds. C.Ho, R.I.Epstein, E.E.Fenimore, Cambridge University Press, p.155
Vanderspek, R., et al., 1992, in *Robotic Telescope of the 1990s*, ed. Alex Filippenko, *Astronomical Society of the Pacific Conf. Ser.*, 34, p.12
Woosley, S., 1984, in *High Energy Transients in Astrophysics*, AIP Conference Proc., 115, ed. S.E.Woosley, p.485
Zytkow A.N., 1990, *Astrophysical Journal*, 359, 138

DISCUSSION

Bruch: Are there any systematic differences between the characteristics of γ-ray bursts in repeating sources and in those of the general γ-ray burst background which might shed some light on different populations of objects emitting these bursts?

Pedersen: Apart from repeating, the three sources are characterized by soft spectra and by being seen in the direction of regions of high stellar density, viz. Galactic Plane (two sources) and the LMC (one source).

Waelkens: Isn't the fact that none of the 761 sources observed by BATSE are repeated an argument that none ever will repeat? And if so, why not consider random optical observations, in the hope that on some ossasion a source observed simultaneously by BATSE will be seen?

Pedersen: From consideration of the frequency of overlapping error regions, the minimum time scale between repetitions is ~ 100 years. As to the second question: yes, wide-angle, long-duration experiments are indeed of great interest. The trend is from photographic work to CCD observations.

Jorissen: Can you use the data from the MACHO project (search for brown dwarfs in the halo from microgravitational lensing of LMC stars) to look for optical transients?

Pedersen: In principle, yes, but due to the limited area covered, the chance of ever catching an event coincident with a BATSE detection is very small. On the other hand, no one can exclude the existence of optical transients which are not related to detectable γ-ray bursts.

Delgado: What does the distribution of γ-ray sources look like? Is there the same lack of concentration as in the distribution of γ-ray bursts?

Pedersen: The persistent sources include a number of AGN and quasars, conceivably of isotropic distribution, and some galactic neutron stars (high-energy pulsars). There is no indication that persistent γ-ray sources are related to the γ-ray burst phenomenon.

Dinner International – Fernandes, Duerbeck, Moffat, Kurtz, Griffin, Walraven

ARCHIVES OF VARIABLE-STAR OBSERVATIONS – HISTORY, USE AND PROSPECTS

H.W. DUERBECK
Astronomisches Institut
Westf. Wilhelms-Universität Münster
Wilhelm-Klemm-Str. 10
D-48149 Münster
Federal Republic of Germany

He [Hipparchos] left the heavens as a bequest, as it were, to all who were found capable of receiving the inheritance
C. Plinius Sec., *Natural History, Book I*

ABSTRACT. The value of old data is exemplified by astrometric observations. To make use of old published photometric data, a good bibliographic coverage is necessary; references are given. Visual (amateur) observations are useful in establishing long-term trends in luminous blue variables and novae, as well as in studies of the period behaviour of eclipsing binaries. The value of photographic sky patrols and plate libraries is illustrated by studies of dwarf novae and symbiotic stars. The technical feasibility and costs of digitizing existing plate libraries are presented.

Keywords: variable stars – bibliography of variable stars – visual observations – data archives – digitization of plate archives – sky library

1. Introduction

The use of old data has long left its mark on astronomical research. All of astronomy originates from observations on reasonably long time scales: the changing angular positions of the Sun, Moon and planets relative to the stars over the cycles of months, seasons and years are at the base of our science.

In positional astronomy, old data, collected over centuries, have always been appreciated and used—from the detection of precession by Hipparchos to the notion of stellar motions by Halley and the solar motion by William Herschel well into modern times. More recently, the analysis of solar-system phenomena, mainly ancient solar and lunar eclipses, has become a valuable tool for studying the Earth's variable rotation. While modern measurements show its erratic fluctuations to a high precision, they reveal nothing about the long-term trend: we have to rely on observations carried out over millennia to avoid the short-term jitter (see, e.g. Morrison & Stephenson 1986).

C. Sterken and M. de Groot (eds.), The Impact of Long-Term Monitoring on Variable Star Research, 353–364.

In astrophysics, technological progress has nearly obliterated from our minds the rich inheritence of old-fashioned photometric and spectroscopic observations. CCD detectors have turned astronomers' interest very rapidly away from photographic plates. The change from visual observations to photography was a much more gradual one and left room for evaluating the merits of the older technique instead of rapidly switching to the new one. The present widespread neglect of photographic and visual observations has led to the tendency of discarding not only outdated methods but old data as well. Can one think of a philologist burning all manuscripts immediately after the invention of printing? Well, this is close to what some science-policy committees are suggesting.

Variable-star researchers are more historically minded. For them, I shall review some research of the past and present activities and future possibilities, pointing out the need for detailed planning before the intended goals are reached.

2. A case study: astrometric data

Astrometrists, in need of old observations for the derivation of proper motions, started two ambitious projects almost 100 years ago. Neither of them met its goal because of the mismatch between the work to be done and the technical means available at the time, and, also, because of insufficient flexibility in adjusting to new techniques developed during the execution of the projects.

The first project is the "Geschichte des Fixsternhimmels" (GFH). When, at the end of the last century, the number of known star positions had grown beyond easy handling, a card catalogue of star positions was started, each star with available information being represented by a card to be kept up-to-date in the future. Star positions, determined between 1750 and 1900, were to be published as a General Catalogue.

The project started in 1898, the first volume was published in 1922, the last one of 48 volumes appeared in 1968. The original plan was only carried out in part. All fully-reduced star positions, which had been determined by transit observations or equivalent methods, were collected and reduced to the equinox of 1875.0. In order to avoid similar problems with the growing number of positional catalogues in the 20th century, "index catalogues" were prepared, with listings of the catalogues where star positions are found, rather than the positions themselves.

The second project which suffered from bad planning is the Astrographic Catalogue (AC), giving photographic measurements of star positions with an accuracy of about $0\farcs3$ down to mag. 13-14. The plates themselves provide a photographic record of the sky down to 15 mag. Most observations were carried out between 1891 and 1920, with the last plates taken in 1950. In spite of the fact that the time for completion was far too long, this catalogue illustrates the advantages of old catalogues when used with modern techniques.

Machine-readable positions of the Astrographic Catalogue were an important input for the new PPM Catalogue (Röser and Bastian 1993) which, in some sense, reached the goal of the Geschichte des Fixsternhimmels: the combination of several good positional catalogues, the reduction to a well-defined epoch, and the derivation of proper motions. The PPM catalogue contains 378 910 stars, thus providing a comparatively dense grid of good positions.

Combining the stellar positions derived from the AC with the positions of the "The Space Telescope Guide Star Catalogue" (GSC; Lasker et al. 1990), observed and automatically measured within a few years, is expected to yield proper motions of 4 million stars, with an accuracy of $0\farcs8$/century (Volchkov et al. 1993). Our conclusion is that the time needed to complete a project should not be longer than the time in which techniques develop which largely outdate the former efforts.

Let us now discuss three aspects pertaining to astrophysical data, essential in variable-star research: location of published data (bibliography), access to data (archives), and accessibility of unreduced data (use of plate libraries).

3. Notes on bibliography

To locate data of a star appears to be quite simple: just log in onto SIMBAD (Set of Identifications, Measurements, and Bibliography for Astronomical Data—the data base kept at the Centre de Données Astronomiques de Strasbourg = CDS), type the name of the star, and a listing of basic data from various catalogues as well as a list of the literature appears on the screen. One should know, however, that SIMBAD only scanned the literature from the 1950s onwards, and that it is detailed only since the late 1970s (e.g. inclusion of IAU Circulars), and since the 1980s for extragalactic objects.

For stars which for a long time have been known as variables one can rely on the detailed bibliographic records in "Geschichte und Literatur des Lichtwechsels der veränderlichen Sterne"(GuL). This publication covers all stars to which variable-star names were assigned up to 1958. The authors scanned the most important journals as well as observatory publications and unpublished manuscripts.

After the completion of GuL it was obvious that this slow and costly way of compiling and editing bibliographies could not be continued. Since then, a continually-updated card catalogue is kept at the Sonneberg Observatory. An effort was made to use always the best available technique for storing data and making them available. In due time the cards were made machine-readable and information became available on microfiche and magnetic tape. Now, it is in the process of becoming accessible via SIMBAD. Unlike the standard SIMBAD, however, it covers not only the well-known professional journals, but smaller ones as well, including amateur journals, which provide a rich source of variable-star observations. The intended closing of Sonneberg Observatory by the end on 1994 will put an end to these archiving efforts. Bibliographic references to GuL and the Sonneberg catalogues are given in Table I.

For eclipsing binaries another card catalogue has been kept. It is supplied as photocopied cards for single objects since a machine-readable form is not available (Wood 1993).

If the object under investigation is not (yet) a named variable or suspected variable star, one must search the pre-1950 literature oneself. Astronomy and Astrophysics Abstracts (AAA) list papers under relevant headings, and objects which are discussed in detail are indexed individually, at least in the more recent editions. Its predecessor, the Astronomischer Jahresbericht (AJB), surveyed the literature from 1899 to 1968.

An example of using data on variable stars from bibliographies is the recent study on the *photometric* evolution of P Cygni by Lamers and de Groot (1992). It is shown that this LBV has steadily become brighter by 0.15 mag/century between 1700 to 1988. The light curve is convincing, despite the fact that the authors had to struggle against the floating definitions and zeropoints of visual magnitudes in the course of centuries.

Another case of an interesting long-term trend, this time in *spectroscopic* properties, is the enigmatic binary β Lyrae. A study based on published and unpublished material was recently presented by Harmanec and Scholz (1993). Their analysis of the radial-velocity curves during the last 100 years indicates a secular change in the semi-amplitude of the spectroscopically-prominent component, K_2, compatible with conservative mass transfer in the system.

4. Archival data

4.1. SPACE DATA

Data collected by satellites are very costly, which accounts for the fact that they are generally well maintained, reprocessed and archived. Details of the status of the data archives of IRAS, IUE, EINSTEIN, EXOSAT, ROSAT are collected in the book of Albrecht and Egret (1991).

The uniformity of the data from the IUE satellite (two fixed spectrograph settings), the straightforward way of reduction, no day-to-day variations that make homogenizations of earth-bound observations so difficult, and only well-controlled long-term trends in the sensitivity of the receivers, make it both easy and effective to establish the data archive. About 55 000 low-dispersion spectra of the IUE Satellite are available for general use in the Uniform Low-Dispersion Archive (ULDA). They are wavelength- and flux-calibrated and are accessible on-line (La Dous 1993).

4.2. RADIO DATA

In mid-1992, the total number of published flux densities at various wavelengths

Table I: Bibliographical sources for variable-star research

GuL (History and bibliography of the light-changes of variable stars) G. Müller and E. Hartwig, 1st ed. (3 vols.) 1918-1922 R. Prager and H. Schneller, 2nd ed. (8 vols, incl. Harv. Ann. 111), 1934-1963
BCVS/BCSVS: W. Wenzel: Bibliographic catalogue of variable stars, CDS Bull. 20, 105 (1981) S. Rössiger: Bibliographic catalogue of variable stars (BCVS): second supplement – 1982-1990, CDS Bull. 40, 39 (1992) H. Huth, W. Wenzel: Bibliographic catalogue of suspected variable stars, CDS Bull. 35, 137 (1988) S. Rössiger, H.-J. Bräuer: Bibliographic catalogue of suspected variable stars update (BCSVS) - revised edition, CDS Bull. 42, 31 (1993)

was roughly 700 000, according to Andernach (1992). Part of the catalogues are remotely accessible.

4.3. OPTICAL DATA

Because of their sheer plenitude, optical data taken at numerous observatories are widely scattered and poorly documented. A recent effort in archiving spectroscopic (mainly coudé) photographic observations has been made by R.E.M. Griffin (1992). Unpublished photometric observations of variable stars are collected in an IAU archive (Breger 1990, Schmidt 1992). Here I shall focus on another important source of information for variable-star researchers: the amateur data.

There are several fields of ongoing amateur activities: (a) search for variable objects, mainly novae and supernovae, (b) photometric monitoring by visual estimates, photography, photoelectric photometry or by CCD observations of known or suspected variables, (c) timing of variables: minima of eclipsing binaries, maxima of pulsating variables. The value of visual data for present-day astrophysical research will be illustrated by some examples.

The first example covers the behaviour of novae long after the outburst. The question is, whether 'post-nova' magnitudes remain constant or decline on timescales of decades. The latter implies that a nova enters a stage where accretion is dramatically reduced or even ceases, which is known as hibernation. Theoretical and statistical arguments support the second scenario, but the observational evidence is largely lacking.

In this context, it is fortunate that W. Steavenson, former president of the Royal Astronomical Society, and its last president who was an amateur astronomer, estimated post-nova magnitudes from time to time over 39 years. These estimates were published in *Monthly Notices of the RAS*, making it easy to collect and analyze them. Assuming exponential decline, the fading expressed in $dm_{\rm vis}/d\log t$, should be a constant (with t measured in years). From this data set, supplemented by observers of the Variable Star Section of the RAS of New Zealand, I derived an average decline rate of nine old novae $dm_{\rm vis}/d\log t = 1.00 \pm 0.30$, while a theoretical estimate for 'radiation-induced mass transfer', for the same average time after outburst, yields $dm_{\rm vis}/d\log t = 0.70$ (Duerbeck 1992).

The previous study dealt with a homogeneous set of visual data obtained by a single observer. Large numbers of heterogeneous observations have also been obtained by many observers over long periods. They are archived in a more or less homogeneous way, sometimes also published (although not always easily accessible). When properly reduced, the data are compact, and can easily be stored and accessed by computer. The French association of variable star observers, the AFOEV, in collaboration with the Centre de Données Astronomiques (Strasbourg), has recently stored its data of variable-star observations, arranged by constellation and object, in the CDS, and the data can be copied by a file-transfer program (Schweitzer and Vialle 1993). The usefulness of these data will be illustrated in the following.

One can use the visual magnitudes of P Cyg towards the same goal as that of Lamers and de Groot (see Sect. 3). Several thousand visual estimates of P Cyg are available in the archive of the AFOEV, but their scatter is very large, up to

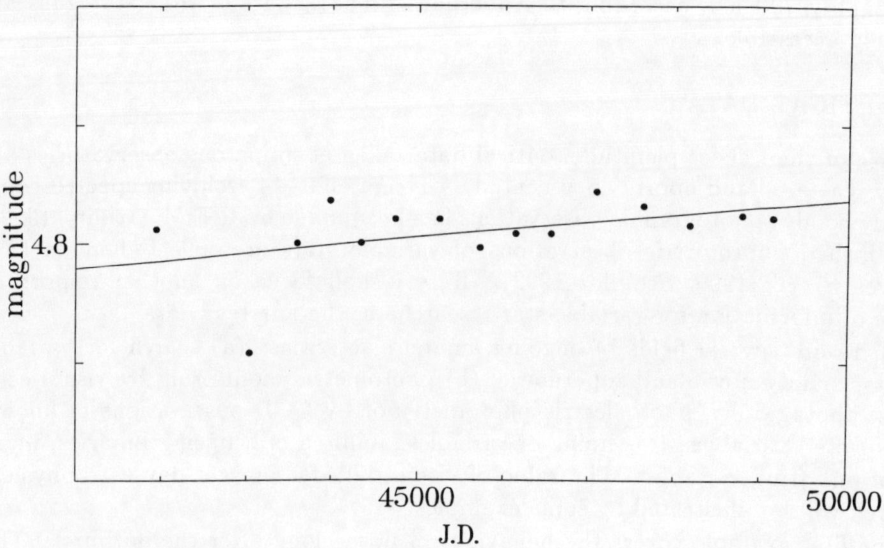

Fig. 1. Normal points of visual observations of P Cyg carried out by Vedrenne.

1 mag for a given night. In view of the good experiences with Steavenson's data, I restricted the sample to the homogeneous data set by the amateur astronomer Vedrenne. It consists of about 700 visual estimates of the years 1971 to 1993. A line is fitted to the normal points, each of which is based on 50 visual estimates (Fig. 1). The results are given in Table II.

It is clearly seen that a slope is present which amounts to a total decline of $0^{\mathrm{m}}_{.}02$ in 20 years, which shows that sufficiently many data, although given to only $0^{\mathrm{m}}_{.}1$, can still reveal this trend. It seems that the somewhat *reduced* slope found by Lamers and de Groot in the 20th-century data is also indicated in this visual data set.

The pioneering activity of the AFOEV has induced other continental amateur-astronomy groups to merge their data bases with that of the AFOEV. The data base includes at present 1 603 stars, for which almost 540 000 estimates have been made by members of the AFOEV since 1923, and 372 000 estimates by members of other continental-European and Japanese organizations (BAV and AKV, HAA, NHK, NVVW, RES-MIRA; Schweitzer 1993). The Variable Star Section of the British Astronomical Association, with about 2 million observations of more than 540 objects, also makes its archive computer-readable (McAdam 1993). The data

Table II. Brightness increase of P Cyg, magnitudes per century

mag/cty	authors / data set	time interval
-0.15 ± 0.02	Lamers & de Groot 1992	1700 – 1988
-0.11 ± 0.04	709 observations by Vedrenne	1971 – 1993
-0.12 ± 0.06	15 normal points	1971 – 1993
-0.04 ± 0.03	14 normal points (lowest removed)	1971 – 1993

Table III. Visual amateur-astronomer observations

brightness estimates				
organization	no. observations	no. objects	time interval	notes
AAVSO	7 400 000	3 600	1911 - now	comp.; list avail.
AFOEV	1 200 000	1 600	1921 - now	comp. and avail.
BAA	2 000 000	540	1900 - now	comp. in progress
BAV	600 000	1 500	1950 - now	60% comp.
VSS RASNZ	3 000 000	500	1928 - now	comp. in progress
minimum and maximum times				
organization	no. observations	no. objects	time interval	notes
BAA		200	1900 - now	comp. in progress
BAV	90 000		1950 - now	90% comp.
BBSAG	30 000		1964 - now	?

base of the American Association of Variable Star Observers with several million observations of 3000 stars is the largest collection of amateur data. Here is the status as of October 27, 1993:

"Just last week we completed the computerization of our data from 1911 to 1960, over 1.9 million observations. The AAVSO observations from 1960 to the present are already in computer-readable form. Thus, the AAVSO data archives are now completely computer-readable, and they consist of over 7.4 million observations from over 4800 observers. The data from 1960 to the present are totally accessible, and reside on 2.4 Gigabytes of hard disk in the computers at AAVSO headquaters. The data from 1911 to 1960 need to be processed and merged with the archives. We anticipate that this processing and merging will take at least two more years to complete. After that, the AAVSO archives will go back to 1911 and will be totally accessible in computer-readable form" (Mattei 1993).

An overview of the contents of archives of amateur observations and the status of their conversion into machine-readable form ("comp.") is given in Table III.

Most of these amateur organisations include groups which observe minimum times of eclipsing variables or maximum times of pulsating variables. The timings of pulsational variables have already left an imprint on our understanding of the evolution of these variables. As concerns minimum times of eclipsing binaries, we still seem to be far away from an insight into the causes of period variations. Thirty years ago, the enhanced interest of professional astronomers in period changes of eclipsing binaries induced much amateur activity in this field. Later, (a) the interest of professionals shifted to other fields, and (b) the accuracy of minimum time determinations by amateurs was found to be too low (Herczeg and Frieboes-Conde 1974).

Let me illustrate what, nevertheless, can be done with large numbers of such minimum times. The W UMa system ER Ori was noted for its irregular period since the 1950s. The (O – C) diagram (Fig. 2), based on 36 photoelectric, 423 visual, and early photographic observations is dominated by sudden period jumps. After a correction for these sudden jumps, a residual sinusoidal effect remains which can be explained by the motion of the system around the common center of gravity with a third body and a period of ≈ 60 years. The third component at an angular

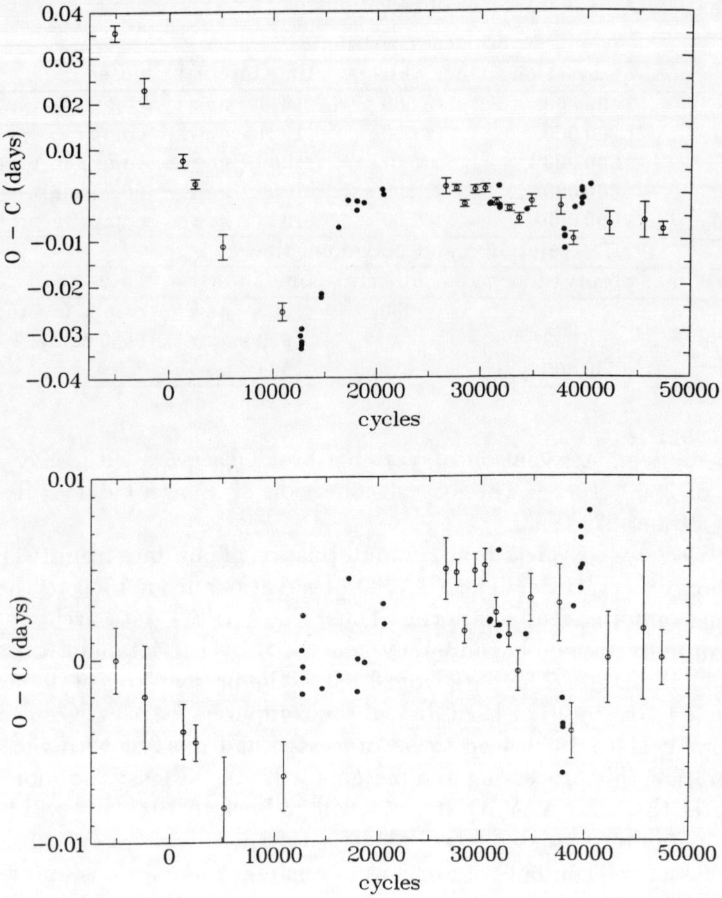

Fig. 2. The (O − C) diagram of ER Ori. Filled circles: photoelectric minimum times, open circles: normal points of visual and photographic minimum times. Upper diagram: original data; lower diagram: corrected for sudden period jumps.

separation of $0\farcs19$ was actually found by speckle studies. It appears that the orbit of the triple system is almost seen pole-on (Goecking et al. 1993).

From the scatter of the visual minimum times of ER Ori the accuracy of a single minimum time is estimated to be ±0.0055 days = 8 minutes. The photoelectric timings made by various authors yield σ's of $0\fd0016$ to $0\fd0022$, with an average of $0\fd0019 = 2\fm7$. To reach the same accuracy as a photoelectric minimum time, 8-10 visual ones have to be averaged. These numbers should be taken only as a rough guide; other binaries show eclipses of widely varying duration and depth, and different accuracies are expected. For an Algol system with deep eclipses, the scatter of visual and photoelectric minima was determined to be $0\fd0024$ and $0\fd0004$, respectively (Duerbeck 1975).

Nevertheless, it is seen that—except in the case of frequent, sudden period jumps—many data of lower accuracy can be considered a useful basis for research on period changes, and since the problem of period changes is far from being settled, a continuation of such data collection is valuable. It is obvious that no 'fine structure' can be recognized in the $O - C$ diagrams based on visual observations, and that the true nature of sudden period changes cannot be revealed by visual timings (Herczeg & Frieboes-Conde 1974).

4.4. UNREDUCED DATA

Many data exist which were collected but never reduced and published. In spite of the potential value of such 'drawer data', they are not the ones to be discussed here. This section concerns the photographic data, stored in plate archives, often taken for special purposes, evaluated and then stored away.

In the famous plate archives of Harvard and Sonneberg, with hundreds of thousands of photographic records of the sky, the plates are quite easily accessible. The photometric history of an object which catches the interest of the present-day observer can be traced back to the past up to more than 100 years. These archives are 'time machines' for inspection of the sky during different periods. The following are two examples from my experience with plate collections.

In 1929, images of a 10th-mag object were found by Schneller on three Potsdam plates. Plates of the Harvard collection also show this outburst. The object was classified as a possible nova, and received the variable-star designation HV Vir. A new brightening was observed in 1989 by amateur astronomer P. Schmeer, and intensively observed at various observatories. The star could be classified as a SU UMa dwarf nova—a type of cataclysmic variable with long outburst intervals. A subsequent search in the Sonneberg plate archive by G.A. Richter led to the discovery of three more outbursts (in 1939, 1970 and 1981). Taking into account the yearly times of invisibility, the average outburst cycle should lie between 3 and 10 years (Leibowitz et al. 1994). The star is a member of the very small class of WZ Sge-type systems.

The spectral appearance of V417 Cen, a variable previously classified as 'irregular', shows that it is a symbiotic star. The earlier classification was based on a subsample of Harvard plates. Recently B. Fuhrmann, Sonneberg, checked all available Harvard and Sonneberg plates. From the 241 brightness estimates and the knowledge of the type of system, a search in the expected time interval (100 – 1000 days) led to the discovery of the orbital period $P = 245$ days. The occurrence of a flare-like event, which stands out from the orbital variation, might have been the cause of the former classification as irregular variable (Van Winckel et al. 1994). The composite light-curve shows a large scatter of about $0\overset{m}{.}5$. Estimates of magnitudes on plates taken during the same or subsequent nights show a scatter of typically $0\overset{m}{.}1$, indicating that the overall scatter is caused by irregular variability of the system superimposed on the orbital light-curve changes.

These examples illustrate the information which can be obtained from archival photographic data. More information is given by Kroll (1994). It is certain that many more studies of this type would be carried out *if* the data were not buried in plate archives, but available in digitized form from a data bank. Steps towards establishing a wide field database are being taken (Tsvetkov 1993), so that it will be possible in the near future to find out how many plates of a given field were taken at specific epochs, and are available in plate archives all over the world. The next step could be the inspection of these plates 'on line'. Such service would be desirable, and, if offered in a convenient way, will have a major impact on variable-star research. Important aspects which decide about the realization of the project are technical feasibility and cost-efficiency considerations.

5. A catalogue of 5 billion stars?

5.1. ESTIMATED COSTS

A conservative estimate suggests that approximately \$20 million must be invested in the scanning and reduction (including internal calibration) of all survey plates available today. If service costs for maintaining the catalogue for 20 years are included, the sum should be enlarged to \$40 million. We expect that about 5 billion stellar images will be recovered from the world-wide existing approximately 2 million plates. With about 200 exposures per field (and thus for each light curve) as supported by our experience with the Harvard plate collection, 10 000 independent survey fields are available. Within these fields we expect 200 000 variables or 20 per field, if we have the same success rate as Sonneberg with 25 000 variables per 250 000 plates. Even if this number includes already known variables (1/5th of the expected total number) we may hope to find more variables than indicated by our conservative estimate. While statistical errors are the same for the experienced observer and the machine, small systematic variation over long chains might be better detectable in the impersonal machine measurements. The cost of the well-covered light curve of a variable will then be of the order of \$200. This should be compared with the cost of a galaxy spectrum taken at a large telescope, \$1000, or the cost of a single IUE spectrum at an estimated \$2000 (for the more expensive space mission prices of more than \$100 000 per observation are not unusual.)

Although these are crude estimates, it is apparent that useful information from existing survey plates is available at about 1/5th of the cost of a modern single high quality spectroscopic observation. For light curves of known short-period variables made with photoelectric devices at medium-size telescopes and light curves of long-term variables from the LTPV project at a small telescope, we pay a price for the higher quality of data. The cost per light curve is about 10 times more expensive than that from existing plate collections. We conclude that light curves obtained by modern methods from old plate collections, while not exactly a bargain, might well be worth the money.

5.2. FEASIBILITY

In view of the fact that double-slit microdensitometers are proven to be the best available machines for high-quality photometric measurements from photographic plates, we might base our estimates on the most expensive microdensitometers, the Perkin-Elmer PDS2020GM. The above cost considerations include this type of measuring device, assuming that 10 machines will be needed to construct the catalogue, including internal calibrations, within 10 years. In Section 2 this was found to be an acceptable time interval for projects of this kind. After this period one may indeed expect better techniques for sky surveys to replace the photographic surveys with small telescopes (see, e.g., Kroll & Lehmann 1993).

Another problem is that of plate transportation, because it is most efficient for the scanning and reduction processes to have the work done at one or a few locations. This requires the shipping of a substantial number of plates. In view of the fact that even large Schmidt plates are generally reduced at locations far from the telescopes, this part of the project can be well handled. Transportation costs are already included in the above estimates.

No problems arise concerning the accessibility of data once they are scanned, reduced and stored. Computer networks in progress (e.g. SuperJanet) will permit data transfer at a speed of 100 MByte/s. A potential reader of the envisioned SKY LIBRARY will have to wait only 8 seconds for the complete scan of a Schmidt plate (if s/he should prefer the raw data rather than the catalogue) or the equivalent of 400 present-day CCD frames.

At present, the best approach seems to be a pilot project to be carried out within 3 years in order to provide the basic software and to check whether the financial guesses made above and the envisioned timescales are realistic. The most severe problems, however, are not the technical ones. It will be necessary to win the astronomical community for support of this project. Let us hope that within 3 years those who do not yet see the value of our photographic heritage, will be convinced.

Acknowledgements

It is a pleasure to thank J. Mattei, AAVSO, E. Schweizer, BAFOEV, J. Hübscher, BAV, D. MacAdam, VSS/BAA, and F.M. Bateson, VSS/RASNZ, for their detailed reports on the variable-star archives of the respective organizations, for the effort that they and their colleagues put into observing, archiving and offering the results. I thank H. Andernach, C. LaDous and M. Tsvetkov for useful preprints, E. Budding for his kind help in establishing rapid communications, as well as my co-authors for the permission to quote from joint papers, and for the pleasant cooperation. W. Seitter, R.E.M. Griffin, P. Kroll and C. Waelkens gave valuable comments on earlier versions of this manuscript. Finally, I thank C. Sterken for his efforts to keep the LTPV project going, for organizing this ARW, and for his invitation to present my contribution.

References

Albrecht, M.A., Egret, D. 1991, *Databases & On-Line Data in Astronomy*, Dordrecht: Kluwer

Andernach, H. 1992, *Astronomy from large data-bases II*, ESO Proc. 43, 185

Bateson, F.M. 1993, private communication

Breger, M. 1990, *Information Bulletin on Variable Stars*, 3422

Duerbeck, H.W. 1975, *Acta Astronomica*, 25, 361

Duerbeck, H.W. 1992, *Monthly Notices of the RAS*, 258, 629

Goecking, K.-D., Duerbeck, H.W., Plewa, T., Kaluzny, J., Schertl, D., Weigelt, G., Flin, P. 1994, *Astronomy and Astrophysics*, (in press)

Griffin, R.E.M. 1992, IAU Comm. 9 WG 'Wide field imaging' Newsletter 2, p. 71
Harmanec, P., Scholz, G. 1993, *Astronomy and Astrophysics*, 279, 131
Herczeg, T., Frieboes-Conde, H. 1974, *Astronomy and Astrophysics*, 30, 259
Kroll, P. 1994, *these proceedings*
Kroll, P., Lehmann, T. 1993, IAU Comm. 9 WG 'Wide field imaging' Newsletter 3, p. 30
La Dous, C., 1993, preprint (to be published in the proceedings of the Trieste meeting on astronomical archiving)
Lamers, H.J.G.L.M., de Groot, M.J.H. 1992, *Astronomy and Astrophysics*, 257, 153
Lasker, B.M., Sturch, C.R., McLean, B.J., Russell, J., Jenkner, H., Shara, M.M., *Astronomical Journal*, 99, 2019
Leibowitz, E.M., Mendelson, H., Bruch, A., Duerbeck, H.W., Seitter, W.C., Richter, G.A. 1994, *Astrophysical Journal*, (in press)
McAdam, D. 1993, private communication
Mattei, J.A. 1993, private communication
Morrison, L.V., Stephenson, F.R. 1986, in *Earth Rotation: Solved and Unsolved Problems*, ed. A. Cazenave, Dordrecht: Reidel, p. 69
Röser, S., Bastian, U. 1993, Bull. Inform. CDS 42, 11
Schmidt, E.G. *Information Bulletin on Variable Stars*, 3733
Schweitzer, E. 1993, private communication
Schweitzer, E., Vialle, J. 1993, Bull. Inf. CDS 43, 51
Tsvetkov, M.K. 1993, IAU Comm. 9 WG 'Wide field imaging' Newsletter 3, p. 72
Van Winckel, H., Schwarz, H., Duerbeck, H.W., Fuhrmann, B. 1994, A&A (in press)
Van't Veer, F. 1991, *Astronomy and Astrophysics*, 250, 84
Volchkov, A.A., Kuzmin, A.V., Nesterov, V.V., Bastian, U., Röser, S. 1993, *Astronomische Nachrichten*, 314, 43
Wood, F.B. 1993, *Information Bulletin on Variable Stars*, 3889

Poretti explaining extinction to Kroll and Mantel

COMPUTERISED VERY–LONG–TERM MONITORING OF VARIABLES AND THE SONNEBERG PLATE ARCHIVE

P. KROLL
Sonneberg Observatory
D–96515 Sonneberg
Federal Republic of Germany

ABSTRACT. This paper uses the term "very-long-term monitoring" for monitoring longer than 100 years. There are objects with characteristics which can only be detected by observations over a very long time. Very-long-term monitoring can be done by computerised exploitation of the large plate archives of the world. As an example, current scanning activities and related problems at the Sonneberg observatory are mentioned. The digitisation of photographic plates and the future CCD-based sky patrols are not only essential for monitoring several million stars but also promise the discovery of several tens of thousands of new variables.

Keywords: Long-term monitoring – plate archives

1. Very-long-term monitoring of variables

In the very recent astronomical journals one notices a growing number of papers concerning **long-term** monitoring of variable stars, extragalactic objects such as AGNs, or quasars. However, checking through the length of time over which these observations have been achieved we find—in most cases—2 or 3, perhaps 5 years of continuing monitoring. Ten, twenty or even thirty years' monitoring is extremely rare.

In this sense we introduce the term **very-long-term monitoring** for observations lasting over a period of time of **100 years** at least. The point is that there are very specific phenomena which can only be found or studied by **very**-long-term monitoring or else remain hidden.

Let us list some objects, classes of objects and phenomena related to them:

– **Eclipsing binaries:** Some objects are known to be showing occasionally sudden or gradual changes in the length of their orbital periods, which indicates an exchange of matter between the components and/or a displacement of matter within one of the components. According to Duerbeck's paper in these Proceedings one needs very accurate photoelectric or CCD measurements to recognize subtle details in the $O - C$ diagramme. No visual nor photographic observation is able to compete. However, let us look, for example, at the $O - C$ diagramme of Algol and U Cep (Hoffmeister et al. 1990, pp. 216 and 214). Despite the scattering of data in the past—most of the

C. Sterken and M. de Groot (eds.), The Impact of Long-Term Monitoring on Variable Star Research, 365–374.

minima were detected by amateurs—we see structures of secular tendencies, sudden jumps, and periodic behaviour over decades, which can by no means be resolved by analysis of only a few years' observations. It is just these jumps and unexpected variations that are of consequence for our understanding (or explaining our not-yet-understanding of) the physics of rapid evolutionary periods of binaries.

– **Cataclysmic variables:** Every object in this class can be regarded as a unique individual. They differ from each other in cycle length, amplitude and shape of outburst light curve, flickering, spectrum, orbital period and even more. Being able to exploit a plate archive and search for outbursts in the past, yielding their frequency and amplitude, is a great advantage.
  Let us give one example: VY Aqr. This star was investigated in detail by Richter (Richter 1992). Twenty outbursts are known to have occurred between 1929 and today (Sonneberg and Harvard plates, AFOEV). Obviously, the star exhibits a growing activity. The frequency of outbursts increased during the last decades. There were only a few outbursts in the past. At the same time the amplitude was decreasing. This behaviour marvellously confirms the Kukarkin-Parenago linearity between logarithm of cycle length and amplitude of outbursts of cataclysmics. To confirm and refine such phenomena, ether one must have a large sample of objects or one needs to monitor the star over a number of decades.

– **Pulsating variables:** Like eclipsing binaries, pulsating variables are subject to period changes. This can be considered as an evolutionary process as a result of crossing the instability strip. Typically, the stay in the instability strip is about 10^6 years for δ Cephei-stars, 10^5 years for RR Lyrae-stars and 10^4 years for W Virginis-stars (Richter 1968). In this case, very-long-term monitoring will reveal much more than monitoring over just 10 years or shorter. The main task here is to separate secular from statistical effects.
  Further, there are a few Cepheids with amplitude increasing and decreasing over periods of a few years (RU Cam, Polaris). Monitoring this behaviour and searching for more objects with this peculiarity is very important. Also, the phenomenon of double-mode pulsating stars is far from being well understood. In this field observations of high accuracy as well as very-long-term monitoring is required.
  Another point of interest concerns period changes in Mira variables. There is a large variety known in the behaviour of periods (Hoffmeister et al. 1990, p. 65).

– **FG Sge:** Discovered in 1943 by C. Hoffmeister on Sonneberg plates, this star is one of the most important test cases for modern theories of stellar evolution. The star showed a secular growth in brightness over 80 years (or more) and reached maximum brightness in the *U*-band in 1962, in B in 1967 and in *V* only in 1970, indicating drastic alteration in its spectrum. This amazing object is supposed to be in a flash phase: The energy production proceeds outward in a shell around the core. The understanding of FG Sge has been rendered still more difficult by its dramatic decrease in brightness about 1 year ago.

To sum up we would like to emphasize two points:

1. There are phenomenona in variable stars which can only be detected by monitoring over several decades (50 years or longer).
2. It is not true that one year's monitoring with millimag accuracy or even better gives the same information as 100 years' rough monitoring with $0^{\rm m}.1$ accuracy. Investigation of secular effects is closely related to very-long-term monitoring.

2. How to do very-long-term monitoring ?

There are two ways of doing very-long-term monitoring: From today into the future or from today into the past. Waiting for future observations takes a very long time and we would, therefore, prefer to study recorded observations, which means to exploit plate archives.

This leads us to a question raised by H. Duerbeck (These Proceedings). Should we digitise all the plates and each plate completely or should we only study the few objects which are of interest at a given moment? One might expect that studying by eye a single object on all plates of the field covering the environment of the object is very time-consuming. Our experience at Sonneberg Observatory, however, is that an experienced observer can estimate the brightness of a star within a minute (on average) using Argelander's method. You might be tempted to mock at the primitivity of this method, but it is a real "high-speed" photometry mode compared with electronic scanning methods. For this procedure you have to set up the plates in the scanning machine, adjust the focus, find the right integration time (depending on the plate fog), start the scanning program at the computer and, finally, measure the magnitudes of object and comparison stars in the data file. It is our experience that you need at least 5 minutes per plate, which is comparatively slow.

Therefore, we suggest to scan the plates as a whole. This, of course, requires a lot of time at the beginning of a scanning project and you have much more information than you really need for studying your object. But, after months or years, another investigator will come and wish to study another object in the same field. He or she would then only have to look at the data files and need not handle the invaluable and fragile plates themselves. The advantage of such a method is obvious.

Further, one should take into consideration that a well-exposed plate contains between 50 000 and 500 000 star images. It is, therefore, advisable to apply computer programs for automatic object recognition, automatic brightness determination and, for astrometric purposes, for automatic determination of positions. This could converge into an object catalogue for every plate, containing position, brightness and other quantities for every starlike image. By collecting these object catalogues we were able to establish an object-in-time database.

To be specific, let us describe our current scanning activities at Sonneberg Observatory (for more details see Kroll 1993). It has, over almost 70 years, accumulated the world's second largest plate archive containing about a quarter of a million photographic plates. About 190 000 plates were taken with the Sky Patrol which is a system of 14 short-fucos cameras covering the whole northern sky from $-30°$

to the Pole and going down to 14th magnitude in blue, and 13th magnitude in red. About 20 000 plates are from two 40-cm-astrographs taken at selected fields along or near the Milky Way going down to 17th magnitude. Finally, there are almost 9 000 Schmidt plates of selected fields down to 18th magnitude (for details see Bräuer & Fuhrmann 1992). The log-book data of all instruments are almost completely entered into computer-readable form by B. Fuhrmann. We have also prepared a program with the help of which you are able to find out which and how many plates are available for a given celestial position.

This plate collection is an excellent stock of information to do very-long-term monitoring. In cooperation with the 'Institut für Theoretische Astrophysik' at Tübingen University we have applied a CCD line scanner for digitising plates. This scanner consists of a camera containing a CCD line with 5000 pixels movable over 7200 steps, a moving table and a backlighting unit. The scanner camera is connected with a Silicon Graphics computer. The scanning speed is about 5 cm^2 per minute with a resolution of 10 μ.

As a first step we have applied this device to the study of single objects, e.g., HZ Her, BL Lac, 3 C 273 (Maisack et al. 1993), MV Lyr, KR Aur. For the brightness determination we have developed X-window-based software. The method applied makes use of a Gaussian-shaped fit to the star image as a measure of brightness (Kroll & Neugebauer 1993). The automation of this method is currently in preparation.

Recently, we started systematic scanning of plates. For scanning a Sky Patrol plate (size 12×12 cm^2) our device needs about 40 minutes, and for an astrograph plate (size 30×30 cm^2) about 4 hours. Scaled up to the archive as a whole, digitising would take about 20 years and yield about 100 TB = 10^{14} bytes of raw data. You may consider 20 years as a very long time in our quick-changing life, but the first data are available already after a few weeks. We begin by scanning all plates of a given field, then move to the next field etc. The selection of the fields depends on the objects to be studied.

We plan to establish an object-in-time data base accessible via computer network. The advantages are:

1. Less content of data (only positions and brightness, no bit-map).
2. All of the data are immediately accessible. This requires hard-disk capabilities of several tens of gigabytes instead of terabytes for the bit-maps.
3. Recent network capabilities do not allow transferring dozens of megabytes within a convenient time, but several tens or hundreds of kilobytes can be dealt with.

Using two scanning devices would reduce the time needed to 10 years, and with four devices we would only need 5 years (for the scanning process itself). This is a reasonable time. Looking at the costs—one scanner is about US \$ 50 000—scanning big archives is almost free.

Concerning other plate archives, we encounter the problems already mentioned in Duerbeck's paper. The invaluable plates can get lost or damaged in transit. Scanning *in situ* enhances the scientific value of the "old-fashioned" plates at their location. The archiving conditions are generally rather poor. Plates are very often seen to be without scientific value for modern astronomical research. Therefore,

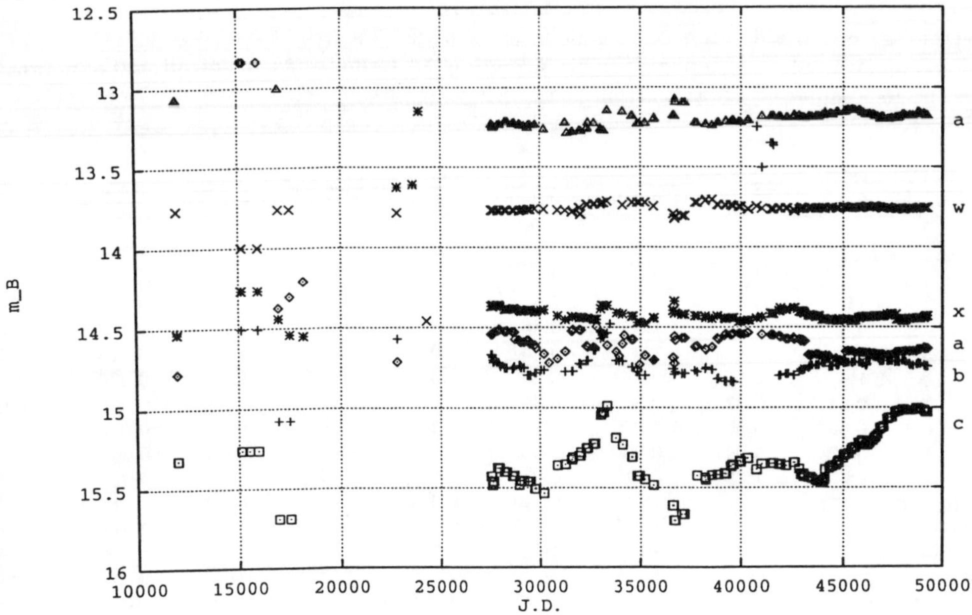

Fig. 1. Smoothed light-curves of comparison stars of HZ Her.

demonstrating the value of the plates might contribute to giving the plates a better home. The scanning could be done by a trained, small team which moves from site to site.

To illustrate what can be done with scanned data, let us consider the example of HZ Herculis. We scanned Harvard and Sonneberg plates to study the very-long-term behaviour of this prominent object.

Even more surprising for us were the light curves of the comparison stars which are commonly supposed to be "constant". They are presented in Fig. 1. Only one star (x) seems to be constant. It is strikingly evident that star c is variable. However, verifying the variability of the other stars requires a larger set of comparison stars in the brightness range of the comparison stars. This will be done in the near future.

3. Systematic search for new events and objects

This last demonstration confirmes our opinion that computer methods for large-scale digitisation and evaluation of photographic plate archives is urgently needed. This would facilitate systematic search for unknown variable objects or other events.

The existence of those many new variables is currently demonstrated by different projects of searching for gravitational microlensing effects. The MACHO project (Alcock et al. 1993) and the OGLE project (Udalski et al. 1993) study millions of stars over a few years and, already after about one year of operation, they detected, as a byproduct (!), thousands of variables.

Another indication of the number of variable stars yet to be discovered, even in

TABLE I

Percentage of variable stars known in a given maximum brightness region 'm' and of different amplitudes. (Number of stars according to formula (1) in (Kroll & Lehmann, 1993) and number of variables from the computer-readable version of GCVS (Kholopov et al. 1985)

m/mag	0.0–0.1	0.1–0.2	0.2–0.3	0.3–0.5	0.5–1.0	1.0–2.0	2.0–3.0	3.0–5.0	> 5.0
0 – 1	–	7.7	7.7	–	–	–	7.7	7.7	7.7
1 – 2	12	2.3	2.3	2.3	–	4.6	–	–	6.9
2 – 3	11	6.4	2.9	1.4	2.1	2.1	–	0.72	2.1
3 – 4	4.4	3.0	1.8	0.46	2.5	0.69	0.23	–	1.1
4 – 5	4.0	2.4	1.8	0.75	1.0	0.75	0.075	0.075	0.82
5 – 6	1.6	1.1	0.71	0.66	0.66	0.38	0.13	0.076	0.40
6 – 7	0.61	0.43	0.37	0.32	0.47	0.19	0.079	0.062	0.40
7 – 8	0.087	0.091	0.059	0.11	0.22	0.16	0.050	0.056	0.31
8 – 9	0.030	0.041	0.030	0.088	0.19	0.010	0.036	0.041	0.15
9 – 10	10^{-2}	10^{-2}	0.010	0.033	0.10	0.075	0.034	0.028	0.029
10 – 11	10^{-3}	10^{-3}	10^{-3}	0.011	0.064	0.049	0.015	0.019	10^{-2}
11 – 12	10^{-4}	10^{-4}	10^{-3}	10^{-3}	0.027	0.029	10^{-2}	0.011	10^{-3}
12 – 13	10^{-4}	10^{-4}	10^{-4}	10^{-3}	0.014	0.019	10^{-2}	10^{-2}	10^{-3}
13 – 14	–	10^{-4}	10^{-4}	10^{-3}	10^{-2}	0.016	10^{-2}	10^{-3}	10^{-3}
14 – 15	10^{-5}	10^{-5}	10^{-5}	10^{-4}	10^{-2}	10^{-2}	10^{-3}	10^{-3}	10^{-4}
15 – 16	–	10^{-5}	10^{-5}	10^{-4}	10^{-3}	10^{-3}	10^{-3}	10^{-4}	10^{-5}
16 – 17	–	10^{-6}	10^{-6}	10^{-5}	10^{-4}	10^{-4}	10^{-4}	10^{-5}	10^{-6}
17 – 18	–	10^{-6}	–	10^{-5}	10^{-4}	10^{-4}	10^{-5}	–	10^{-7}
18 – 19	–	–	–	10^{-6}	10^{-5}	10^{-6}	10^{-6}	–	–
19 – 20	–	–	–	10^{-7}	10^{-6}	10^{-7}	–	–	–

"old" plate archives, is obtained by taking a glance at the GCVS (Kholopov et al. 1985). Table 1 lists the number of variables as a percentage of the total number of stars in the sky for different amplitude ranges and for a given range of maximum brightness.

For the 0th to 2nd magnitude this is statistics of small numbers. But fainter than third magnitude it is easily seen that, for a given amplitude range, the percentage of known variables is more or less constant with magnitude until, from a certain magnitude, it rapidly decreases. Obviously, this threshold magnitude indicates that below this limit many variables are awaiting discovery. Concerning the internal errors of photographic plates (see the above light curves of HZ Her's comparison stars) we are able to detect variables down to an amplitude of $0\overset{m}{.}2$ on the basis of plate archives. Variables with this amplitude seem to be known almost completely down to the 4th or 5th magnitude. Compare this **bright** limit with the limiting magnitude of, e.g., the Sonneberg Plate Archive (14^{m}).

The general shape of Table 1 will be substantially alterated after completion of the HIPPARCOS project. This satellite discovered about 10 000 to 20 000 variables, mainly in the range of small to very small ($0\overset{m}{.}001$) amplitudes.

One may now ask oneself, Why do we need to know more and more variables, if we are not able to study all the known variables sufficiently? There are at least three reasons.

1. For some sub-classes of variables we only know a few objects (sometimes only one). Knowledge of more of these peculiar objects is urgently needed (e.g. double-mode Cepheids).
2. There might be unknown types of variablility (particularly micro-variability and very-long-term variability).
3. There has been no complete investigation into the statistics of variables for 25 years (the last one was done by Richter (Richter 1968); This would certainly give a new insight into the role of variables in our Galaxy and in the life of a star.

In this relation, I want to emphasize that we should keep in mind that today's prominent objects were mostly discovered through systematic investigation (with comparatively small telescopes!) and rarely by accident. This fact seems to be forgotten very often.

As for plate archives, it is obvious that every one of them represents a vault of undiscovered treasures. In the past we were not able to dig for all of those treasures owing to a lack of suitable measuring machines. But now let us dig with scanner and computer!

As a byproduct these activities also promise a comprehensive insight into other astronomical objects such as asteroids, comets, optical bursts and perhaps even MACHOs.

4. Problems/limitations of computerised monitoring with plate archives

The digitisation of photographic-plate archives meets with some difficulties apart from the access to the archives themselves. We mention only a few of them:

- Most of the plates do not contain a calibration wedge. The photographic density-intensity-relation has therefore to be estimated by indirect methods.
- The colour sensitivity, absolute sensitivity, and grain size of the photographic emulsions vary over the years. This could simulate variability. The statistical errors related to these effects yield the limit of detectibility of variables.
- Because of the rather bad state of some plate archives, we fear that log-book data got lost; there might also be typing errors in the log-books, which is very critical for some particular types of variables. For instance, wrong Julian dates cause biased light-curves (or even more confusion).

5. Future projects

We have seen that we must digitise long-lived plate archives in order to do very-long-term monitoring of a large set of objects. This approach has to be seen in connection with modern CCD-based activities in monitoring.

CCDs are advancing to take over the task of sky patrol and sky monitoring since the data are already stored in digital form and CCDs are much more sensitive. But it would be a fatal mistake to stop all contemporary photographic patrols now. Instead, as long as CCDs are rather small in size (compared with big photographic plates), CCD patrols and photographic patrols should run parallel. This is also important for a smooth adaptation and calibration of the different colour sensitivies of CCD and photographic emulsion to assure homogeneity.

Future activities should therefore cover the following fields:

1. Scanning all plate archives of the world (time needed: 10 to 20 years).
2. Tests of large CCD arrays for use in sky patrol (see e.g. Kroll & Lehmann 93).
3. Establishment of a world-wide sky patrol as an international project.

Acknowledgements

I owe a dept of gratitude to Hans-Jürgen Bräuer of Sonneberg Observatory and Hilmar Duerbeck of Münster University for many helpful discussions. In particular, I would like to thank Chris Sterken for organising this meeting, and all its participants for adopting a resolution pro Sonneberg Observatory, which is threatened with closure by the end of 1994.

References

Alcock, C. et al., 1993, *Nature*, 365, 621

Bräuer, H.-J., Fuhrmann, B., 1992, *The Messenger*, 68, 24

Hoffmeister, C., Richter, G.A., Wenzel, W., 1990, *Veränderliche Sterne*, Leipzig

Kholopov, P.N., et al., 1985, *General Catalogue of Variable Stars*, 4th edition, Moskow

Kroll, P., 1993, *Newsletter of the IAU Working Group on Wide-Field Imaging* 3, 46

Kroll, P. & Lehmann, T., 1993, *Newsletter of the IAU Working Group on Wide-Field Imaging* 3, 30

Kroll, P. and Neugebauer, P., 1993, *Astronomy and Astrophysics*, 273, 341

Maisack, M., Kroll, P., Weber, T., 1993, *Mitt. über Veränderliche Sterne*, 12, 147

Richter, G.A., 1968, *Veröff. d. Sternwarte Sonneberg*, 7, 236

Richter, G.A., 1992, in *Viña del Mar Workshop on Cataclysmic Variable Stars*, ed. N. Vogt, 12

Udalski, A. et al., 1993, *Acta Astronomica*, 43, 289

DISCUSSION

Hall: The similar photographic archive at Harvard suffers from the well-known Menzel gap. Does the Sonneberg archive have any similar gaps?

Kroll: No, it does not. These were plates taken even during the war but, of course, less frequently.

Harmanec: You showed us indications of possible slow variations of some objects based on many decades of photographic observations. Could you comment, please, how sure you are about the reality of these changes? One would expect that the "colour system" of the plates was changing over the years, too, as different types of emulsion, processing, etc. were used.

Kroll: That is right. However, changes in emulsion type, etc. should yield jumps in the light curve instead of gradual increase or decrease. Also, these changes should affect many stars. Perhaps, the sample of 6 stars used in my Fig. 2 is too small to show these effects. We have not yet investigated these changes and corrections systematically. However, at least star *c* should be variable.

Le Contel: I would like to mention that every plate coming from the French Schmidt telescope at CERGA is automatically digitized in Paris (contact Dr. Guibert at Paris Observatory).

Young: Having used the Harvard plates, I can say that most of the images are in the edge zones, where the image structure is very complex—affected by astigmatism, field curvature, chromatic aberration, and so on. To compare these images by eye is relatively easy; but to scan these and try to extract magnitudes by some algorithm is not so simple. Fitting a single Gaussian PSF to such images is far from optimal. A great deal of work will be needed to do a good job.

Kroll: The fit we used has 7 parameters which allows also to fit heavily-distorted stars at the plate edge, with lower accuracy, of course. Since we compare the brightness of neighbouring stars (within several millimetres on the plate) with each other, all objects are affected by nearly the same distortion. Using a better PSF should yield better results, which is a main task for the scanning project.

Following this Discussion, the participants of the Workshop have unanimously adopted a resolution in support of the continuation of the scientific activities of Sonneberg Observatory (see p. 373).

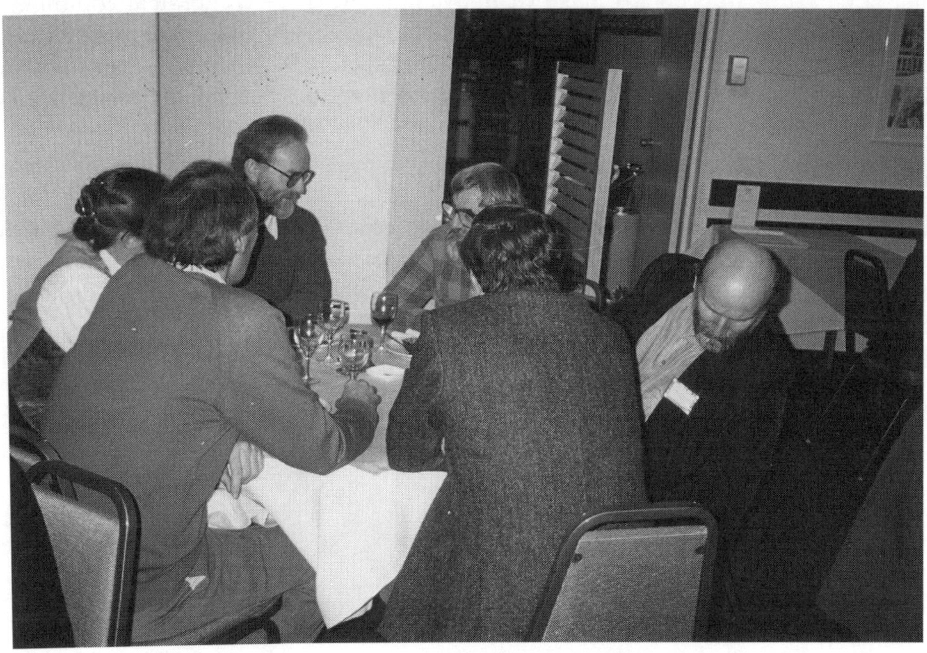

Lloyd, Mrs. Olsen, Olsen, Young
Schwarzenberg-Czerny searching for periodicities

RESOLUTION

The participants of the NATO Advanced Research Workshop on

The Impact of Long Term Monitoring on Variable Star Research,

recognizing the indispensable scientific contributions of photographic sky monitoring in general to our knowledge of stellar variability, structure and evolution, and

acknowledging the outstanding achievements of the Sonneberg Observatory in particular in these respects: its continuation of a photographic sky patrol for more than 60 years, yielding 250 000 photographic plates (now catalogued in machine-readable form) which have led to the discovery of one third of all variable stars known today and, in addition, its compilation of bibliographic catalogues of variable stars,

strongly urge the continuation of such sky patrols allowing for the eventual replacement, if necessary, of the photographic plate by digital detectors,

actively support Sonneberg Observatory's endeavours to continue its work in these fields, especially its initiative to commence the digitization of its own and other sky patrols in order to make them available to the astronomical community as a **"Sky Library"** of the optical history of the sky, and to guarantee international participation in the formation and extension of such a **Sky Library** through the appointment of a suitable committee to prepare a list of detailed specifications, time-table and costs,

and meanwhile

encourage the states of Thüringia and Bavaria, as well as authorities of the Federal Republic of Germany, to explore further the various means at their disposal for ensuring the continuation of Sonneberg Observatory as the natural and appropriate nucleus of the **Sky Library**.

DISCUSSION No 6: PHOTOGRAPHIC PHOTOMETRY

Kroll: It is not our aim to transform one dead archive, the plate archives, which are not easily accessible, into another dead archive of computer-readible data. Therefore, we must do a kind of standard analysis to offer the information in an object data base (or "sky-lib", as Hilmar Duerbeck mentioned). The raw data remain accessible and there might be better reduction procedures in the future, yielding better results. This is the advantage once we have scanned the plates.

Harmanec: I think that the need to digitise old plates can also be supported by the fact that the effects of imperfect fixing and/or washing already manifest themselves as a non-homogeneous yellowish fog on a number of them. On the other hand, I feel that we must find a reasonable balance between the effort to save the old data, and to build new, more accurate and more sensitive detectors and instruments. Observational astronomy is an expensive science and we can succeed only if we shall evaluate realistically the limits of how much we can ask. Repeatedly during this meeting I noted some excitement about the fact that yet longer-term variations were discovered through systematic monitoring. This is an excitement which I also share. However, people who are not in favour of long-term projects may ask: what really new knowledge or insight—in terms of physical interpretation—does that bring? Therefore, the best strategy to defend long-term monitoring should be to present well-defined astrophysical problems and to show very precisely why they cannot be solved without systematic long-term studies. Any attempt to create something like a "trade-union organisation" of long-term observers may become counterproductive since it is not the method but the result that counts, I think.

de Groot: From what I have heard so far, I wonder whether we should go to the IAU to organize things on a world-wide basis. The drawback may be that this will probably involve us in long delays. Many projects progress faster when individuals take initiatives.

Griffin: The IAU has already taken this matter on board, at least in principle. For the past two years an IAU Working Group has been promoting the commencement and coordination of archives of spectroscopic data. In the context of the present discussion, one decision that has been taken is to gather all photographic spectra into one central location, thereby (it is hoped) preserving them (better than in some present conditions), making them all readily available for perusal, and in the

C. Sterken and M. de Groot (eds.), The Impact of Long-Term Monitoring on Variable Star Research, 375–378.

charge of an assistant who could prepare digitized records upon request.

Sterken: The IAU has done more. In 1987, at the Strasbourg Conference "Coordination of Observational Projects in Astronomy" (Cambridge University Press 1988, eds. C. Jaschek and C. Sterken) the participants recommended the establishment of an IAU Working Group which would work under the auspices of several IAU Commissions, and would be a clearing-house for rapid exchange of information between schedulers of telescope time, for defining the types of projects to be undertaken, for synchronizing space- and ground-based observing facilities, and for distributing information between projects and the participants. That Working Group has been established at the XXth General Assembly (Baltimore 1988). Unfortunately, the Working Group did not work, and it was eradicated at the Buenos Aires GA.

Breger: I must defend the IAU to some extent. The IAU is made up of individual astronomers and does not have a huge external organization behind it. It keeps out of national (or even ESO) decisions, but will help with resolutions we supply to them. It will support specific projects (actually all projects) but will not interfere with local decisions.

Le Contel: One of the major problems for astronomy everywhere is that 75 to 80% of the budget goes to staff salaries. So, as we develop more and more projects (especially, but not only, big ones), we can understand that our activities ask for the closure of some projects or observatories. We have to decide by ourselves which ones can be stopped, and which ones must continue.

Griffin: So, let us decide here and now to drop COSTAR.

Le Contel: I did not say nor think that!

Sterken: Two comments. First, In the discussion on closing observing facilities of modest size at large observatories, one should remember that the action always amounts to saving a meagre couple of % on a total budget. Of course, one should always discuss total costs—that is, include all costs of infrastructure and operation. Still, the cost of building a road designed for the transportation of huge steel structures and very large mirrors will also be very large compared to the cost of a road leading to a telescope of modest size. Second, when discussing the importance of science done at large telescopes compared to results obtained at small telescopes, one should make the comparison using a parameter that describes the scientific output correctly. For example, if you measure the economic importance of a harbour by the *volume* of goods that are transferred, the port of Rotterdam ranks first in The Netherlands. If measured by the different parameter of *added value per unit of manpower*, Rotterdam falls to the fifth place. Still, small harbours and large harbours must both exist, since they have very complementary functions and tasks. In the same way large and *small* telescopes should be built and maintained, because of their complementary character.

Young: The effectiveness of small versus large telescopes was investigated a few years ago by Helmut Abt, who found that—by any means (papers, citations)—the small telescopes were more productive than the large ones.

Waelkens: Archiving of astronomical data may be the kind of project which fits into the programmes financed by the European Community.

Griffin: Yes, but the usual programme (Human Capital and Mobility, DGXII) has to emphasize the *scientific* results that are to be the *goal* of an investigation. Support for archiving *per se* might rather be found within a different Directorate (e.g. DGXIII for telecommunication) serving a different goal.

Dravins: On *Data collection and analysis*: There has been, and is being accumulated, a quite considerable volume of observations of variable stars. ESO's LTPV programme is one; others are e.g. the TYCHO programme on HIPPARCOS (whose data are due to become public in a few years' time), and the data collected in dense stellar fields for variability searches due to gravitational microlensing (e.g. the MACHO project). When discussing future extensions of such programmes, a likely question to be posed will be *What fraction of these data has actually been analyzed?* and, based on such analyses, *What types of data are still missing for the further understanding of variable-star physics?*

de Groot: The fact of possessing dormant data cannot be held against a scientist. Even theoreticians possess such data. It is in the nature of most research, and certainly of astronomical research, to accumulate data which may prove less interesting than anticipated. However, many become interesting in the future as our knowledge increases.

Dravins: On *Costs for different projects*: Studies have been made of the cost per published research paper based upon work at larger and smaller, ground-based and space instruments. One, perhaps surprising, result is that the cost does not vary much with the size of the installation. It seems that there is a (cost-driven) regulating effect, such that the observing efforts on space observatories are reduced to a few hours, *versus* a few nights on large ground-based telescopes, some weeks on small ones at good sites, or several months at poor sites. One possible conclusion would be that there is no reason for proponents of either small ground-based telescopes, or of large space installations, to be critical of one another.

Le Contel: In my opinion an observatory, a telescope, a project, can be considered in the same way as a human being: it is born, it lives, it dies (the question of the life time is, of course, important). So we have to face this situation and, at the same time, propose alternative solutions: e.g. the French Schmidt telescope may be closed but the group in charge of it proposed a very attractive alternative which consists in putting in operation a mosaic of CCDs in the focal plane. That offers new perspectives, but has a cost, of course. The scientific interest is *now* at a level where it can compete with other instruments.

Le Contel: For what concerns the small telescopes at La Silla (used for LTM) it is not enough to propose to keep them in operation. We should think about the possibility of installing APTs—which means evaluating how the capital cost and the operating budget compare with the present situation. Another possibility is to go to the South Pole, where nights are longer (no daily gaps in the data for periods as long as 8 times 24 hours). Of course, this is not an ESO project.

Sterken: I am willing to support Le Contel's statement in the sense that—if it is really necessary to formulate fascinating projects to release funds—a South-Pole Observatory is one way to go. But I would like to see it operational before more conventional facilities operating at modest cost are closed.

Le Contel: I agree, but we have to show that we are not afraid of the fact that a telescope may be closed. The problem is to maintain the scientific programme at the higher level which may not be totally possible with telescopes or techniques becoming older and older.

Bruch: If we have an organisational problem in our science for which there is no evident solution—as in the case of archiving, closing down or continuing a place like Sonneberg, or concerning the current difficulties of LTPV—we must pursue several ways to overcome the problems. In the case of the LTPV project the proposed resolution [see the resolution mentioned on p. 361 of these Proceedings (Ed.)] is just one way we have discussed; we wish to examine other ways, too. The resolution is not meant as a simple protest against closing the small telescopes. If you look carefully at the draft you will not find a statement like that. It is a plea for continuing the support to the LTPV group without a specification of how this ought to be done. The workshop participants are not in the best position to elaborate a detailed plan on how to overcome the specific problems. This must be done by the LTPV members in consultation with ESO.

Le Contel: Concerning the plates, I think that it is not enough just to digitize them. We must also try to develop methods for compressing the information, as well as for reading them rapidly. E.g., Bijaoui's group in Nice is developing techniques to separate galaxies from stars and other objects very quickly.

Duerbeck: Digitizing the plates and storing the data on "permanent" media is certainly a good way to ensure *widespread accessibility* to them. It is my experience (and Andy Young expressed a similar opinion) that sky patrol plates have a wide range of quality, of depth of exposure, of image definition, etc., sometimes even varying over a given plate (use of astrographs, etc.). Furthermore, calibration spots are usually not available. A straightforward reduction (image segmentation, image clessification, brightness determination) will, therefore, only be possible in rare cases. We need either a very sophisticated method of analyzing the plates automatically, or a set of simple image-processing tools, with whose help a potential user can analyze interactively the object of interest and its surroundings on all available plate scans.

Williams: (to Duerbeck) Another reason for retaining the original data besides the catalogued objects, is that there may be objects not known to the algorithms recognizing stars or galaxies: objects such as bow shocks around hot stars or gravitationally-lensed images.

LESSONS LEARNED FROM 30 YEARS OF LONG-TERM VARIABLE-STAR PHOTOMETRY

Douglas S. HALL
Dyer Observatory
Vanderbilt University
Nashville
Tennessee 37235
U.S.A.

ABSTRACT. This paper recounts my 30-year involvement with long-term photometry of variable stars of different types. I have observed solo, collaboratively with over 200 other photometrists, both amateur and professional, and with two different automatic telescopes, a 10-inch and a 16-inch. The relative advantages and special problems associated with these different modes are discussed critically with respect to phase coverage, amount of data obtained, and resultant precision and accuracy. The last two sections touch on the questions of archiving of data and co-authorship of papers. The first has been a problem and the secong one has not.

Keywords: automatic telescopes – photometric accuracy – photometric precision – photometers – eclipsing binaries – variable stars – starspots – flare stars – data archiving – amateurs – collaboration

1. Introduction

When Chris Sterken invited me to talk at this N.A.T.O. Workshop, the topic of which is long-term photometry of variable stars, it made me realize that precisely this has been the underlying theme of my astronomical research for my entire professional career to date. So, I would like to recount my personal experiences in this arena and lessons which might be drawn from them. This is the same approach I followed when John Percy asked me 8 years ago to talk about my 22 years of experience doing photoelectric photometry of variable stars with small telescopes. The reader is referred to that paper (Hall 1986) for details and particulars which are relevant to the topic of this Workshop.

My work has involved almost entirely photometry (not spectroscopy, photography, astrometry, etc.). The targets have been stars (not galaxies, planets, asteroids, etc.). Those stars have been variables or suspected variables (not constant field stars or populations of a cluster). And almost every project or problem tackled has had a long-term focus (years or decades rather than single nights or single observing seasons).

C. Sterken and M. de Groot (eds.), The Impact of Long-Term Monitoring on Variable Star Research, 379–392.

On this long 30-year trip, however, I have walked down many different roads, each with its own character (Hall 1986). First was the 24-inch at Harvard's Agassiz Station, which I had at my disposal almost every night for one continuous observing season, in the summer of 1963 . There have been the observing runs at national observatories like Kitt Peak, lasting days or weeks. There has been the faithful year-around and year-after-year use of the conveniently accessed "home telescope", in my case the 24-inch at Dyer Observatory. There has been the development of an international network of photometric collaborators and the coordination of that network to work on many different projects, almost every one long-term. And there has been the acquisition of the first automatic telescope ever to be dedicated to a task involving long-term photometry of variable stars, a 10-inch, and the subsequent acquisition of another, a 16-inch.

My work at times has been done solo. At times it has involved students, both graduate and undergraduate, both at my institution and at other institutions. It has involved other professionals, both astronomers and non-astronomers such as physicists. It has involved amateur astronomers ranging from high-school students to octagenarians, from the working to the retired. And at times it has relied on the essential role played by someone who might be termed a technician.

2. The long-term aspect and the scientific pay-off

For most of the variables I have worked on, long-term photometry was an essential ingredient required to understand a particular star or understand some peculiar phenomenon exhibited which made that star important.

One example were the so-called long-period Algol-type eclipsing binaries. If the orbital period of any eclipsing binary is longer than about a week, it generally requires more than one observing season to complete the light curve at all phases, including the critical eclipse phases. Among the Algol-type binaries, i.e., semi-detached systems in which one star is overflowing its Roche lobe and transferring matter onto its companion, those of long-period are especially interesting in several respects. First, these long-period Algols account for the majority of members of a puzzling group of binaries, those containing so-called "undersize subgiants" (Kopal 1959). The existence of such a class was confounding attempts to understand close binaries in the context of the then-new mass-transfer theory, because mass transfer was presumed to require that one star fill its Roche lobe exactly. It turned out that there was no such class at all. The members were shown to be a mixture of RS CVn-type binaries (evolved systems prior to Roche lobe overflow) and truly semi-detached binaries with SB-1 orbital elements rendered unreliable due to distortion by gas-stream effects and/or with inaccurate light curve solutions (Hall 1974, 1975). Second, as a group they are in the interesting advanced stages of mass loss, where the originally more massive star is now the less massive and the current mass ratio is only 0.1 or in some cases even less (Tout & Hall 1991).

Third, as a consequence of dynamical/geometrical factors, they can and usually do harbor a semi-permanent accretion disk surrounding the mass-gaining star (Hall & Garrison 1972). That disk can be either thin or thick, both optically and geometrically, with the thick case giving rise to the additionally interesting W

Serpentis-type binaries (Guinan 1989). Fourth, here one tends to find the interesting "double contact" binaries, in which the one star overflows its Roche lobe and the mass-gaining star is spun up, flattened, and now rotating at its centrifugal limit. Precise, complete light curves and good computer solutions of them were necessary to prove that such systems exist (Wilson & Mukherjee 1988). Fifth, an apparently uniquely different mechanism for stellar variability, still not understood, manifests itself exclusively in long-period Algol-type binaries, examples being RX Cas (P = 516 days) and β Lyrae (P = 275 days). See Kriz et al. (1980) and Guinan (1989).

The photometric variability of the stars or groups of stars characterized as chromospherically active or heavily spotted or having a strong dynamo has been almost entirely within the province of long-term observation. For reviews see Hall (1987, 1990, 1991a, 1992, 1993, 1994), Hall & Busby (1990), Hall & Genet (1988), Hall & Henry (1990, 1994a), and Strassmeier et al. (1988, 1993).

Many different time-scales are involved in a complete description of the photometric behaviour of these stars, most of them "long". First is simply the rotational modulation which produces the quasi-sinusoidal "wave" in the light curve as the star covered with a longitudinally asymmetric distribution of dark spots rotates on its axis. Rotation periods can be quite short, less than 1 day, but they can be as long as months in some binaries, the extreme example being 109 days in HR 7428 = V1817 Cyg (Gessner et al. 1990) or around a year in some single stars, examples being P = 335 days in HR 1362 = EK Eri and P = 385 days in HD 181943 (Hooten & Hall 1990). Second is the fascinating fact that the agent responsible for the photometric variability, the big spot or spotted region, is long-lived, with lifetimes ranging from a few months to several years (Hall & Busby 1990, Hall & Henry 1994a). Adding frustration to the fascination is the fact that we still do not have a complete empirical description nor a good theoretical understanding of the essential behaviour (lifetime, size, drift in latitude and longitude, rules for recurrence) of these large spot regions. Third, there is additional photometric variability on a time-scale longer than the spot lifetime, years or decades. We do not have a complete empirical description nor a good theoretical understanding of this very-long-term behavior either. An important recent contribution correlated a 40-50 year cycle in mean brightness and in $B - V$ with a 40-50 year cycle in orbital period changes in the chromospherically active binary CG Cygni and used this correlation to provide support for a promising new theory of magnetic cycles operating in one of the two stars (Hall 1991b). However, the case for a correlation had to rely partly on differential photographic photometry from an earlier epoch. Fourth, although one thinks of flares as extremely short-term phenomena, the important recent discovery of the most energetic stellar flare ever observed, on V711 Tau = HR 1099, came about only because that star had been monitored photoelectrically in a systematic manner for more than 15 years (Henry & Hall 1991).

3. Not enough data? Too much data?

In my early years, when my data came from the Dyer Observatory 24-inch and the smaller telescopes at Kitt Peak, two 16-inch and one 36 inch, I was able to keep up with the observing, data reduction, analysis, and writing of papers. It was, however, an exhausting task, even with the help of students in all four steps of the process. That is because, as most of us know, when the students are graduate students at the Master's degree level (not Ph.D.) or undergraduates, the time and energy input by the professor very nearly equals the output from the student. Significant amplification can result only when a graduate student at the Ph.D. level or a post-doctoral research associate is involved, neither of which I have ever had the luxury of. Moreover, apart from the fact that most of my observing projects were long-term in nature, often many years elapsed between the first data point and the publication of a paper, sometimes more than a decade, simply because the overall task from observing to publication was so difficult.

This early phase of my work as a variable-star photometrist working solo came to an end for a variety of reasons, most important of which was simple exhaustion. The frequent week-long or month-long observing runs at Kitt Peak, usually in the middle of an academic semester, the half nights or full-nights of observing at Dyer with classes to prepare for and teach the next morning, the data reduction and analysis and paper writing to be accomplished somewhere in between, all became too much.

My involvement with collaborators in long-term photoelectric photometry projects began around 1970 and has expanded to include a grand total of approximately 210 individuals (120 professionals, 90 amateurs) who have sent me usable photoelectric photometry which has resulted in co-authored papers published. They come from a total of 14 different countries. The impetus for this new tack was three-fold. First, a personal factor, no more sleepless nights were required on my part, because the data came in the mail during the civilized daylight hours. Second, a scientific factor, long-term observing projects which could not be accomplished by one telescope at a single observing site but required a network of telescopes at many different sites could now be done, most of them for the very first time by anyone. Third, a more far-reaching factor, the paradigm of an amateur astronomer equipped with professional-quality photometric instrumentation and obtaining professional-quality data and, thus, acting as a bona fide research scientist, was developed. This cadre of amateur scientists represents a new and independent resource which should be potentially valuable in these hard times of limited manpower, limited observing time, limited financial resources, and still too many variable stars in the sky. At this time, however, the potential is largely just that, potential. If one counts the number of owners of equatorially-mounted clock-driven telescopes 8 inches in aperture or larger and figures about $2000 for a commercially available photoelectric photometer, then this cadre has the potential to number in the tens of thousands. In reality, if one counts photometrists active today, this cadre numbers only a few hundreds.

This phase of my work, especially the collaborating with amateur astronomers, has evolved over the years. As the number of collaborators grew, the cost in my time was sure to grow. Not only did data have to be processed but, before and on a continuing basis, it was necessary to "train the raw recruits". To minimize having to accomplish the training on a one-on-one tutorial basis, my co-author Russ Genet and I wrote a book in 1982: "Photoelectric Photometry of Variable Stars - A Practical Guide for the Smaller Observatory". There is now a second, revised, expanded edition (Hall & Genet 1988). A little before that, in 1980, I co-founded (again with Russell Genet) an organization called "International Amateur-Professional Photoelectric Photometry", which publishes the quarterly I.A.P.P.P. Communications, now in its 53rd issue, and organizes regional workshops and symposia in various countries all over the world. An essential aspect of the amateur scientist paradigm is that the amateur works closely, in a mutually beneficial way, with a professional astronomer, although there are a few instances where amateurs have done meaningful scientific research, including publication of results, entirely on their own (Lines et al. 1988). Because it is not possible for any one individual professional, myself included, to coordinate the potential observing capacity of all of the amateur photometrists who are active at any given time, I had hoped that other professionals would do as I have done and benefit similarly from the collaboration. This has happened in a number of cases, but frankly not in a sufficient number of cases to make full use of the observing capacity waiting to utilized. A regular feature of the I.A.P.P.P. Communications is a running list of all "Papers Published by Amateurs" having to do with photoelectric photometry, with the names of the amateur co-authors printed in boldface type. The amount of data I personally have received from amateurs and used to publish papers jointly co-authored with them reached a peak a few years ago, but at the time of that peak it was virtually my sole source of data, with only a minor part resulting from our telescope at Dyer Observatory or observing trips to Kitt Peak. That volume of data from the amateur collaborations was so vast that, frankly, a fraction of it, though reduced, still awaits inclusion into a published paper.

My use of photoelectric photometry from computer-controlled automatic telescopes can be considered the third, and currently ongoing, phase of my work. The best recent overall summary of this work was in a paper entitled "An Intimate Relation with Two Automatic Telescopes for Nearly Nine Years" (Hall & Henry 1994b). Those two telescopes were the so-called "Phoenix 10-inch" and the "Vanderbilt-T.S.U. 16-inch". On the question of volume of data, perhaps I should say that the metaphor "trying to drink from a fire hose" was coined specifically in reference to this "problem" of data output from an automatic telescope. Considerations of time spent moving to, identifying, and centring on a star, the so-called duty cycle of a telescope, show that an automatic telescope and the most effecient manual operation of a telescope of comparable aperture by experienced observers result in data collection at comparable rates (Walker 1985). That, however, is a best-case comparison. Automatic telescopes can and do observe all night long, every clear night, and every clear hour of intermittently cloudy nights. Manually-operated telescopes tend to suffer down time during the second half of each night, when humans prefer to sleep, during especially cold nights, as a result of exams, homework, parties, vacations, etc., and due to midnight-lunch breaks and bathroom breaks. Moreover, the manually-operated telescope which is readily accessible and, hence, most likely to get consistent use tends to be located at an observing site which is not optimal in other respects (in or near a city and its lights, at low elevation, in the British Isles or in the eastern half of the United States or in continental Europe) whereas an automatic telescope is free to be, and almost always is, located at an optimal site, making for more hours of cloud-free sky each year.

It is difficult to count exactly the total number of published papers based on "my photometry" from the 10-inch (Boyd et al. 1990ab) and the photometry from the 16-inch (which belongs exclusively to me). One measure of the output, repeat, just one measure, is the number of new variable stars discovered. As of June 1992, that number was 47, with 16 of them being in the Yale Bright Star Catalague, i.e., of naked-eye brightness (Hall & Henry 1994b). That represents a large amount of scientific output, but let me admit that this body of data still has not been milked of all of its potential scientific content, relatively more for the 10-inch data but relatively less for the 16- inch data. Even though I am currently collaborating with colleagues at the Tennessee State University Center of Excellence (Gregory Henry, Joel Eaton, and Michael Busby) in the utilization of this 16-inch photometry, we are together in our judgement that the data are coming in faster than we can use them, i.e., we are losing the race. By way of reacting to this "problem", we now have, and for some time have had, a policy of welcoming collaboration with an astronomer at another institution, if the collaboration is of the sort that we send him the fully reduced data, he assumes responsibility for the lion's share of the scientific analysis and the writing of the paper, and we all are included as co-authors on that paper. Meanwhile, of course, we continue to engage in collaborations of the more typical sort, for example, where we contribute photometry to a joint spectroscopy-photometry investigation of a variable star and we assume responsibility for analyzing the photometry and writing "our half" of the paper.

4. Precision and accuracy

First, a short speech about the significance of accuracy in photometry. There is no one dividing line between good versus bad or usable versus unusable or accurate versus inaccurate photometry. If one insists on dividing lines, they should be in terms of what accuracy is needed for the scientific problem at hand. Consider a Mira variable with a full range of 5 magnitudes and a smooth light curve with no "bumps" or "humps", a typical eclipsing variable with an eclipse depth of one full magnitude, a typical starspot variable with a full range of $0\overset{m}{.}2$, and the short-period pulsating variable with a full range of only $0\overset{m}{.}01$ discussed by Don Kurtz at this Workshop. Then consider, respectively, the $\pm\ 0\overset{m}{.}1$ accuracy typical of visually estimated magnitudes, the $\pm\ 0\overset{m}{.}01$ accuracy typical of much differential photoelectric photometry, the $\pm\ 0\overset{m}{.}005$ accuracy characteristic of relatively good photoelectric photometry, and Don Kurtz' phenomenal $\pm\ 0\overset{m}{.}001$ accuracy which surely must be close to the achievable limit. The resulting signal-to-noise ratios, respectively, would be $50 \div 1, 100 \div 1, 40 \div 1$, and $10 \div 1$. Here one might conclude that the "best" photometry (the millimag accuracy) is actually the "worst", if the criterion is defining the shape of the photometric variability.

In treating this question, let me make it clear that I will be talking exclusively about differential photometry of variable or suspected variable stars, i.e., brightness measures made with respect to a nearby comparison star presumed to be non-variable itself. The alternative approach, usually termed "all-sky photometry", sometimes is employed in variable-star investigations and is meaningful to discuss, but I myself have seldom done it. Also, I will be talking about single-channel photometry and not about any form of dual- or multi-channel photometry, although I have respect for the potential benefits of this alternative approach.

My emphasis will be on assessing the relative precision and accuracy which can result, or actually has resulted, from the different types of photometry which I myself have been responsible for carrying out over the last 30 years. The first would include the years of photometry with the same telescope and virtually unchanging photometric equipment at my home observatory, in my case the 24-inch at Dyer Observatory. The second would include the various observing runs of limited duration at Kitt Peak, using a variety of different telescopes, similar but different photometers, similar but different photomultiplier tubes, and similar but different broadband filter sets. The third would be my consolidating of photometry contributed by my international network of 210 different observers who used different telescopes of various types and of various apertures and instrumentation packages which were all unique. Additionally, I will comment on the relative accuracy of the amateurs versus the professionals. And the fourth would be my use of the two previously mentioned automatic telescopes.

Considerations of accuracy will be discussed in more detail than precision and, generally, I will take accuracy to be the rms deviation or standard deviation of single differential magnitude measures from a mean of many measures taken over a relatively long interval of time, in the case of a pair of stars both of which are non-variable, or the rms deviation or standard deviation of residuals from an analytical fit to a variable's light curve, when it is believed that those residuals have not been made significantly larger as a result of the analytical fit being an imperfect description of the variability. As a rule, a "single observation" is in reality a mean of three individual intercomparisons between the variable and its comparison taken in sequence within just a few minutes, but I think many of us do this as a common practice.

The Dyer Observatory photometry approximated the ideal of many years of continuous photometry with the same telescope and virtually unchanging photometric equipment at one's home observatory where all factors pertinent to photometric accuracy and precision are under one's own control. The set-up itself could be described as thoroughly conventional: a Cassegrain of modest (24-inch) aperture, a conventional single-channel photometer, a selected 1P21 photomultiplier (in fact, the very same tube in use continuously for three decades), broadband (UBV) filters on a filter slide (again the very same filters in use continuously for the same three decades), a D.C. amplifier plus strip-chart recorder, later changed to a voltage-to-frequency converter, later changed to a pulse-counting device, and an assortment of interchangeable diaphragm slides but again the very same slides from the middle 1950s to the present.

The resultant accuracy was what I have taken to regard as typical, in the vicinity of $\pm\ 0\overset{m}{.}010$, except in cases of relatively faint stars where photon noise necessarily made the accuracy worse. The major reason why we didn't do a bit better, $\pm\ 0\overset{m}{.}007$, for example, was probably our practice of using all nights which were "photometric" with regard to transparency, rather than restricting ourselves to the subset (about half) of the "very best" nights. Over the years we did frankly encounter intermittent problems, mostly of instrumental origin, which ruined or diminished the accuracy of portions of our photometry, but my impression is that these losses were minimal due to the invariance of our photometric system, the time-tested calibrations of its many parts, etc.

The second would include the various observing runs of limited duration at Kitt Peak, using a variety of different telescopes, similar but different photometers, similar but different photomultiplier tubes, and similar but different broadband filter sets. With my Kitt Peak photometry there was not the luxury of invariant, previously calibrated equipment totally under my control. But I regularly did all the things one should to secure accurate photometry. For every filter-telescope-detector combination used, I determined transformation coefficients in each bandpass using a reliable method such as the cluster-of-red-and-blue-standard-stars technique, and determined extinction coefficients in each bandpass. Twice nightly I calibrated the gain steps of whatever amplifier was being used. I explicitly measured, in seconds of arc, the diameter of the diaphragms which had been used. This all cost extra time, away from observing my variable stars, and extra energy, but it paid off. The resultant accuracy proved, in general, to be no worse than that of the Dyer photometry, even when put to the strictest test, namely, comparing photometry obtained at both observatories and finding no significant systematic differences.

Now we come to the international network of collaborators. They used different telescopes of various type (refractors, Newtonians, Cassegrains, Maksutovs, etc.) and various apertures most of which would be considered small or modest (between 4 inches and 48 inches), different photometers of various type (constructed in a university machine shop, home-made in a basement or garage, commercially obtained, etc.), different photodetectors of different type (photomultipliers, photodiodes, etc.), different filters with different characteristics (bandwidth, effective wavelength, thickness, etc.), different observing sites differing in quality (elevation, darkness of sky, uniformity of transparency, etc.), different amplifiers and recording devices (D.C., voltage- to-frequency, pulse counting, strip-chart recorder, milliammeter, paper and pencil, printed tape, floppy disk, etc.), diaphragms of various diameter, and so on.

For the amateurs, my modus operandi was to have them send me semi-raw data, i.e., times in the form of geocentric Julian dates with the heliocentric correction not yet applied and photometry reduced to the point of differential magnitudes not yet corrected for differential atmospheric extinction nor transformed to the standard photometric system. Then they would send me the latitude and longitude of their observatory site, extinction coefficients representative of their site, and transformation coefficients for each of the bandpasses they were using. Armed with this, I would complete the data reduction and arrive at heliocentric Julian dates and differential magnitudes on the standard system. In this process a few expedient short-cut procedures were used. First, I recommended they determine transformation coefficients by observing a red-blue star pair with accurately known standard magnitudes (Hall 1983). If they reported a suspiciously large coefficient, however, indicating a poor match to the standard system, I instructed them to obtain a more suitable filter for that bandpass. Second, I made a practice of using a realistic seasonal mean primary extinction coefficients for each observer's site and an assumed color-dependent extinction coefficient realistic for each bandpass used. Adverse consequences of these various short expediencies were minimized by my having dictated the comparison star which should be used, with that star selected to match the variable as closely as possible in colour, angular separation in the sky, and mean brightness. The last of these selection factors protected against loss of accuracy due to possible scale errors, such as non-linearity of meters, dead time in pulse-counting equipment, inaccurately-calibrated gain steps in amplifiers, and saturation of photomultiplier tubes.

So, what was the resulting accuracy, determined as I explained in the third paragraph of this section? First, let me admit that all of the variables recommended by me for collaborative observing were sufficiently bright, relative to the generally small telescope apertures being used, that photon noise did not adversely affect the accuracy which could be achieved. It was a little surprising but certainly gratifying that it was around $\pm\, 0\overset{\mathrm{m}}{.}010$ in most typical cases. Needless to say, different variables observed during different observing seasons by different mixes of observers did not yield exactly the same quantitative measure of the accuracy. A few times it was significantly better, as good as $\pm\, 0\overset{\mathrm{m}}{.}007$, but seldom was it as bad as $\pm\, 0\overset{\mathrm{m}}{.}015$. I did screen the data, but in accordance with defensible criteria. If the internal error (the standard deviation of the three closely-spaced individual variable-comparison measures which made up "a single observation") exceeded $\pm\, 0\overset{\mathrm{m}}{.}030$, that observation was deleted. If a lone observation failed the "three-sigma test" (Hall et al. 1986), it was deleted. And if the data from some one observer differed systematically from all of the others by a significant amount and the cause could not be identified or if it contained a large fraction of observations which failed the $\pm\, 0\overset{\mathrm{m}}{.}030$ test or the three-sigma test, all of the data made by that one observer on that variable during that observing season were deleted.

Among this large group of observers I have collaborated with, the accuracy of the amateur data has ranged from better than the typical professional astronomer normally obtains to so poor as to be useless. The very best I have ever encountered was the photometry of V396 Per = HR 1063 by Howard Landis and Howard Louth (Landis et al. 1985), both of whom happen to be amateur astronomers; the rms deviation in their light curve was $\pm\, 0\overset{\mathrm{m}}{.}003$. Some became known to me after sending me photometry which was faultless from the beginning; they required no further tutorial attention on my part. Others required a great deal of tutorial attention, by telephone, by correspondence, by reference to the list of 23 "sources of error" laid out in the book by Hall & Genet (1988), by participation in an I.A.P.P.P. Symposium or Workshop, or by a visit to their observatory; once their various specific problems were solved, however, their photometry improved and remained good. A few, but surprisingly few, never were able to obtain usable photometry, for a number of various reasons, only some of which I was able to identify.

Last come the two automatic telescopes I have used. The question of precision and accuracy has been answered in great detail quite recently by Henry & Hall (1994) and Hall & Henry (1994b). Let me summarize briefly.

In the case of both telescopes, the accuracy, defined in the same way as earlier in this paper, began at the same $\pm\, 0\overset{\mathrm{m}}{.}010$ and continued as such for the first year or so. Both telescopes experienced at least one interval of time, lasting a fraction of a year, when the accuracy was degraded to something like ± $0^{m}_{.}015$ or almost ± $0^{m}_{.}020$. With the 10-inch the problem was a malfunctioning power supply to the photomultiplier which was not recognized for a while and took a while to be repaired. With the 16-inch the problem was a wearing of the gears in the drive system which resulted in a degradation of the telescope's centring and, in conjunction with an unrecognized focussing problem, resulted in frequent inaccurate photometric measures. Both problems were cured and the accuracy returned to its previous ± $0^{m}_{.}010$ level. Then a sequence of additional improvements in technique and upgrades in instrumentation on the 16-inch succeeded in improving the accuracy dramatically, up to the point where it is now doing photometry at the theoretical limit, i.e., limited only by scintillation and photon noise. With the 16-inch aperture given and with the bandpasses and integration times being used, this amounts to an accuracy of ± $0^{m}_{.}003$.

Let us look at the question comparatively. First, we see that photometry from existing automatic telescopes has been as good as typical manual photoelectric photometry and, at its best, can be as good as the best manual photoelectric photometry. Second, things can go wrong and the resulting precision and accuracy can suffer, but these things are mostly similar in nature to the things which go wrong with photometric telescopes which are operated manually. Moreover, since an automatic telescope like one of these two is alway dedicated to the sole task of photoelectric photometry and in most cases is under the total control of one principal astronomer or a single team at one institution, we approach the ideal of the nearly invariant system and minimize problems which result when the opposite is the situation. Even when something does go wrong, the fact that the data have been obtained and reduced in a consistently uniform manner, often makes it possible to remedy a problem after the fact by means of some global operation on the existing data set. Third, because the data are obtained automatically, without passing in front of the eyes of a human being, there is the danger that a problem might go unnoticed somewhat longer than would be the case with photometry obtained in manual operation. Moreover, because the flow of data output is so fast, a large amount of flawed data can accumulate before it is spotted as defective. This special problem can be avoided if care is taken to scrutinize the data consistently at frequent intervals, ideally once a day, on a morning-after basis. The Vanderbilt-Tennessee State team now operating the 16-inch does exactly this, because the data from a night's observing are downloaded daily, by modem, from the master computer on Mt. Hopkins to our computer in Nashville and we do discipline ourselves to look at it almost every day.

Let me close this section with a final point, that collective multi-site photometry can result in a bonus. The strictest measure of photometric accuracy comes when one compares data obtained independently with telescopes, instrumentation, reduction techniques, etc. as different from each other as they possibly can be. Any one data set, especially if it is perfectly homogeneous, can contain within it one or maybe more than one systematic error which might escape attention. Let me recount an example. There have been instances where the same variable was being observed simultaneously by one of my two automatic telescopes and by several telescopes in my collaborative network. In a few of those instances all of the photometry was consistent except that from the automatic telescope. In that way, one of the problems with one of the automatic telescopes, "problem C" as discussed by Boyd et al. (1990ab), was brought to light, and it probably never would have been without the comparison.

5. Co-Authorship problems ?

On virtually all of the papers I have written, which have photometric observations as their basis, there were co-authors, most of whom contributed directly to the actual photometry. In the earlier years these tended to be my students, both undergraduates and graduates, observing with the Seyfert 24-inch at Dyer Observatory. In the later years, these tended to be collaborators, both professional and amateur, observing at their own observatories.

Many of my photometric projects focussed on stars which varied by some mechanism which was not strictly periodic or repetitive in a predictable way. If the non-repetitive variability additionally had a rather long time-scale, say, a month, then a campaign of simultaneous photometry from many different observing sites was required to achieve adequately complete phase coverage, not possible from any one site because of intermittent cloud cover, etc. Projects like this necessarily culminated in a papers with rather large numbers of authors. The record for me was the paper on λ Andromedae by Hall et al. (1991), which has a total of 31 co-authors.

In most of my co-authored papers, I frankly was responsible for most of the actual work, including generally most of the photometry, the analysis of the data, and the actual writing of the paper. Due to the fortunate circumstance of timing, I was never harmed professionally by the dilution effect which might result when one paper is credited to so many individuals. In the early years, before I had achieved the security of academic tenure at Vanderbilt University, my co-authored papers were relatively few, the number of co-authors per paper was relatively small, and those co-authors for the most part were my own students, whose inclusion on a paper is a credit to their professor rather than a detraction. In the later years, I was not harmed, for several reasons. First, by about that time, I had received tenure and was, to first order, protected. Second, Vanderbilt University itself never has employed a rigid mathematical approach to quantifying a faculty member's publication list, such as counting the total number of papers and dividing by the total number of co-authors. Instead, the overall strength of a candidate for promotion and the impact of his work in his area of research is assessed by means of letters from the outside written by other astronomers familiar with that area of research. Third, my general area of research (photometry of variable stars, let us say) is, like any specific area of astronomy, an intimate one where for the most part each player knowns the other players. Consequently, I suspect that most astronomers reading one of my multi-authored papers immediately understand the situation accurately, i.e., recognize my name and my probable role in the paper and surmise that the unfamiliar names are "probably Doug Hall's network of amateur astronomers observing with their small telescopes". Moreover, I suspect, most astronomers agree with me that acknowledging the meaningful contribution by amatuers by giving them co- authorship on a par with profesisonal astronomers is a good and appropriate thing to do.

To date no editor of any astronomical journal has ever objected in any way to one of my papers because of the number of co-authors, even though the listing of authors has sometimes extended to the second journal page. Only one of my professional co-authors has ever objected to having his contribution diluted by the inclusion of a large number of amateur co-authors. In that case, that person had a promotion on the horizon and was at an institution which did employ the sort of rigid mathematical approach to quantifying a publication list mentioned in the paragraph above. Once one of my co-authors questioned his own inclusion, protesting modestly that "surely just one point on one light curve isn't enough for co-authorship". We did not, however, remove his name from the paper.

6. Publishing or archiving the data

First let me say something about the I.A.P.P.P. Communications, mentioned above. I was one of its Co-Editors between the years 1980 and 1984, the Editor in Chief between the years 1984 and 1987, and one of the two Assistant Editors from 1987 to the present. This journal emphasizes papers having to do with photoelectric photometry as a technique and with amateur-professional collaboration, but not so much papers presenting specific scientific results. Therefore, not surprisingly, we do not generally publish tabulations of photometric data.

As an investigator in the area of long-term photometry of variable stars, however, I am keenly aware of the critical need to have available, by some mechanism or another, the actual numerical data resulting from previous photometry of any variable star which is of potential scientific interest in a long-term context.

Over the years I have tried consistently to have all of my own photometry made available, by some mechanism or another, to the astronomical community of the present and future generations. The mechanisms have been various, including publication in tabular form (typically in camera-ready format) in the paper itself, in a data archive maintained by the journal which publishes the paper, in the Archive for Unpublished Observations of Variable Stars maintained by Commission 27 of the I.A.U., and on a magnetic tape stored at and accessbile from the Astronomical Data Center (see Boyd et al. 1990ab). In doing this, I have encountered problems. Reluctance on the part of a journal editor to "waste pages" on tabulation of data has forced me to use the data archive approach, and reluctance on the part of the Commission 27 Archive to accept data acquired with one of my automatic telescopes has forced me to use the magnetic tape approach at a different facility.

Now the other side of the coin. In the most earnest way possible, I urge all other astronomers to make their photometric data available, by some mechanism or another, to the astronomical community of the present generation and future generations. There have been very many instances, painful instances, where I have learned of the existence of photometry of some interesting star (as evidenced by a light curve in picture form, a description in words of observations which were made, a short table of parameters derived in some way from it, a table of normal points resulting from years of photometry phased according to some period, etc.)

but found that original photometry (in the form of magnitude versus Julian date) was inaccessible to me, effectively lost forever, especially in cases where the original observer or author of the paper is no longer alive.

References

Bachmann, K. T., Brown, T. M., 1993, *Astrophysical Journal*, 411, L45

Boyd, L.J., Genet, R.M., Hall, D.S., Busby, M.R., & Henry, G.W. 1990a, *I.A.P.P.P. Comm.*, 42, 44

Boyd, L.J., Genet, R.M., Hall, D.S., Busby, M.R., & Henry, G.W. 1990b, *I.A.P.P.P. Comm.*, 42, 54

Gessner, S.E., Hall, D.S., Lines, H.C., & Lines, R.D. 1990, *Astronomical Journal*, 100, 2017

Hall, D.S. 1974, *Acta Astronomica*, 24, 215

Hall, D.S. 1975, *Acta Astronomica*, 25, 95

Hall, D.S. 1983, *I.A.P.P.P. Comm.*, 11, 3

Hall, D.S. 1986, in *The Study of Variable Stars using Small Telescopes*, ed. J.R. Percy (Cambridge: Cambridge University Press), 63

Hall, D.S. 1987, *Bulletin of the Astronomical Institutes of Czechoslovakia*, 70, 77

Hall, D.S. 1990, in *Active Close Binaries*, ed. C. Ibanoglu (Dordrecht: Kluwer), 95

Hall, D.S. 1991a, I.A.U. Colloq. No. 130, 353

Hall, D.S. 1991b, *Astrophysical Journal*, 380, L85

Hall, D.S. 1992, *Astronomical Society of the Pacific Conf. Ser.*, 34, 27

Hall, D.S. 1993, in *Precision Photometry* eds. D. Kilkenny, E. Lastovica, & J.W. Menzies (Capetown: S.A.A.O.), 168

Hall, D.S. 1994, *Memorie della Societa Astronomica Italia*, in press

Hall, D.S. & Busby, M.R. 1990, in *Active Close Binaries*, ed. C. Ibanoglu (Dordrecht: Kluwer), 377

Hall, D.S. & Garrison, L.M. 1972, *Publications of the ASP*, 84, 552

Hall, D.S. & Genet, R.M. 1988, *Photoelectric Photometry of Variable Stars* (Richomond: Willmann-Bell)

Hall, D.S. & Henry, G.W. 1990, in *Active Close Binaries* ed. C. Ibanoglu (Dordrecht: Kluwer), 287

Hall, D.S. & Henry, G.W. 1994a, *I.A.P.P.P. Comm.*, 55, in press

Hall, D.S. & Henry, G.W. 1994b, I.A.U. Colloq. No. 136, in press

Hall, D.S., Kirkpatrick, J.D., & Seufert, E.R. 1986, *I.A.P.P.P. Comm.*, 25, 32

Hall, D.S. et al. [31 authors] 1991, *Journal of Astronomy and Astrophysics*, 12, 281

Guinan, E.F. 1989, I.A.U. Colloq. No. 107, 35

Henry, G.W. & Hall, D.S. 1991, *Astrophysical Journal*, 373, L9

Henry, G.W. & Hall, D.S. 1994, *I.A.P.P.P. Comm.*, 55, in press

Hooten, J.T. & Hall, D.S. 1990, *Astrophysical Journal, Supplement Series*, 74, 225

Kopal, Z. 1959, *Close Binary Systems* (London: Chapman and Hall)

Kriz, S. et al. [9 authors] 1980, *Bulletin of the Astronomical Institutes of Czechoslovakia*, 31, 284

Landis, H.J., Louth, H., & Hall, D.S. 1985, *Information Bulletin on Variable Stars*2662

Lines, H.C., Lines, R.D., & McFaul, T.G. 1988, *Astronomical Journal*, 95, 1505

Strassmeier, K.G., Hall, D.S., Zeilik, M., Nelson, E., Eker, Z., & Fekel, F.C. 1988, *Astronomy and Astrophysics, Supplement Series*, 72, 291

Strassmeier, K.G., Hall, D.S., Fekel, F.C., & Scheck, M. 1993, *Astronomy and Astrophysics, Supplement Series*, 100, 173

Tout, C.A. & Hall, D.S. 1991, *Monthly Notices of the RAS*, 253, 9

Walker, W.S.G. 1985, *I.A.P.P.P. Comm.*, 21, 42

Wilson, R.E. & Mukherjee, J. 1988, *Astronomical Journal*, 96, 747

DISCUSSION

Sterken: My experience with amateurs is that it is very difficult to have them performing in the way you want them to carry out the job: apparently, they have problems in assigning priorities, and I am very well aware that this is normal, since it is, also for me, difficult to assign priorities to activities I do as a passtime. Among the students observing in the LTPV project, I found that I had more difficulties with those students coming from the *organised* amateur world—that is, amateur societies and clubs—than with the others.

Hall: My reply is only to admit that, yes, amateurs generally do not observe as intensively or consistently as professionals do, but this is understandable because their participation is, after all, voluntary and unpaid. However, their combined output as a group is substantial because we have, for example, 50 part-time (amateur) observers compared to only 2 or 3 full-time professional observers.

Sterken: After a decade of working with APTs, where are the data now? If I wish to use these data, where should I look to find them?

Hall: The full set of 4 years of my photometry of 92 variable stars with the "Phoenix 10-inch" APT has been published by Boyd, Genet, Hall, Busby and Henry (1990, I.A.P.P.P. Communications No. 42, p. 44 and 54). Our $\sim$ 6 years of photometry of $\sim$ 150 variables with the "Vanderbilt 16-inch" APT, however, has not been published.

Griffin: Do you interact at all with the AAVSO?

Hall: The AAVSO, which originally was and still is primarily dealing with visual observations, does work a little with photoelectric photometry collected by amateurs. Some amateurs send some photoelectric data to AAVSO headquarters where they are collected and something is done with them, but I am not sure exactly what. I work individually with amateurs on a collaborative basis, letting them be co-authors on the resulting papers, etc. A few of these amateurs may be sending some of their photometry also to the AAVSO, but that means—to answer your question—I myself do not really interact with the AAVSO itself in any substantial way.

Niarchos: How many automated telescopes are now in operation, what are the plans for building new telescopes of this kind, and what is the approximate cost of one such telescope?

Hall: Order of magnitude 10 exist, in operating status, in the world. I believe 7 are located at the Mt. Hopkins site in Southern Arizona, managed by the Automatic Photoelectric Telescope Service, which is essentially Louis J. Boyd. These 7 range from 10 inches to 1 metre. Some are owned by the APTS itself and used by outside astronomers on a rent-a-star basis. Others are owned by single institutions, such as the Vanderbilt 16-inch. Others are owned by consortia of several institutions. A handful of APTs are nearing completion. Examples: Korea, South Africa, Catania, and the 36-inch SARA telescope on Kitt Peak. A 1-metre APT, complete, as constructed by Autoscope, costs a few hundred thousand dollars.

LONG-TERM ASPECTS OF MULTI-SITE CAMPAIGNS

M. BREGER
Institut für Astronomie
University of Vienna
Türkenschanzstr. 17
A-1180 Wien
Austria

ABSTRACT. Organizing worldwide multi-site campaigns utilizing a large number of participating observatories can be difficult. For some types of multi-periodic pulsating variables, concentrated campaigns from only two carefully selected observatories can be just as valuable. This is demonstrated with new data for the δ Scuti star FG Vir, where the nine frequencies of pulsation can be correctly identified from the measurements originating in only two time zones.

Major long-term applications of multi-site campaigns require that the multi-site data be analyzed together with previously existing data. This raises the problem of the proper frequency sampling, which is examined by a numerical experiment.

Although the photometric frequencies of pulsation for δ Scuti stars are found in the range of 4 to 50 cycles per day, we give evidence for the existence of additional frequencies outside this range. These additional frequencies include combinations and harmonics of the main frequencies. In order to include these very low and very high frequencies, the observing and reduction techniques for multi-site campaigns need to be revised.

Keywords: multi-site campaigns – δ Scuti stars – frequency analysis

1. Introduction

Multi-site measurements obtained on several continents spaced around the globe are necessary to examine the variability of multi-periodic stars in order to eliminate the aliasing problem caused by (regular) observing gaps. This aliasing problem could lead to incorrect choices of frequencies and the masking of multi-periodic structure in the power spectra. Numerical experiments as well as more thorough discussions of this point (Breger 1992) can be found in the proceedings of a previous scientific conference held in Belgium.

Successful multi-site campaigns have been carried out for a variety of variable stars, ranging from δ Scuti and roAp stars on or near the main-sequence to white dwarfs. The present paper is biased in the sense that it represents some of our experience with multi-site photometric campaigns covering the δ Scuti variables with main periods of pulsation in the range of 20 minutes to several days. However, many of the comments apply to different period ranges and other types of variable stars as well.

C. Sterken and M. de Groot (eds.), The Impact of Long-Term Monitoring on Variable Star Research, 393–404.

Here we would like to concentrate on three important aspects of multi-site campaigns as well as the long-term application of the data:

1. Is it scientifically useful to organize multi-site campaigns with only two sites rather than a large of number of observatories spaced around the globe?
2. What are some of the problems and pitfalls associated with analyzing data sets spanning a decade or longer?
3. Why *must* we plan and reduce the multi-site data in such a way to include frequency ranges previously thought to be unimportant?

In this paper we will attempt to demonstrate these aspects with multi-site data in the light of recent (mostly unpublished) results.

2. The case for extensive campaigns using only two sites

Most astronomers and many administrators have already been convinced that multi-site campaigns are necessary. This is an improvement over the situation ten years ago. Multi-site campaigns usually involve the participation of three or more observatories on three or more continents. In principle, one could plan a campaign using only three observatories properly spaced around the globe (e.g. Chile, Australia and South Africa) in order to minimize aliasing problems. In practice, additional observatories are added for two reasons: (i) the weather at one observatory might be poor, while another observatory could obtain data, and (ii) it is often difficult to obtain a large block of observing time from a telescope-scheduling committee. The latter reason is scientifically regrettable: twelve nights of data from a constant set of equipment is much more valuable than two blocks of six nights each with different zero-points (see below)!

Since multi-site campaigns often are expensive and difficult to organize, let us turn the question around and ask: Under what circumstances would single-observatory measurements or campaigns involving only two observatories on two continents be sufficient?

Let us examine the question by using real data from a recent multi-site campaign of FG Vir. This star provides a good example, since its multi-periodic pulsation is quite complex. From 1993 March 12 through 1993 April 5, a total of 162 hours of usable photometric data were obtained on four continents. Some photometry is simultaneous using both the three-star and continuous-observing techniques. The preliminary analysis of the data clearly shows the presence of nine frequencies of pulsation. Due to the multi-site nature of the campaign, no aliasing problem was experienced. FG Vir has a dominant frequency of 12.72 cd^{-1} with a relatively large amplitude of $0\overset{m}{.}022$, while the other modes have much smaller (but still easily detected) amplitudes between $0\overset{m}{.}002$ and $0\overset{m}{.}005$ mag. Could all these frequencies be detected with observations from only one or two time-zones?

This question is explored in Fig. 1, which shows the power spectra after prewhitening the strong, dominant 12.72 cd^{-1} peak. The top panels contain the power spectra for the 139 hours of data remaining after removal of the SAAO data. This artificially introduces a serious, regular gap at the Europe/Africa longitudes. Nevertheless, all the frequencies can still be uniquely identified. If only the 75 hours from a single observatory (McDonald Observatory) are used, the correct identifications become impossible. In fact, the aliasing causes even the largest peak to be spurious! Various experiments with prewhitening trial frequencies would lead to a correct identification of about five frequencies (in addition to the 12.72 cd^{-1} frequency previously prewhitened) from the single-observatory data. As expected, the single-observatory measurements fail to reveal the true spectral content of the star.

The analysis shows that the presence of regular eight-hour gaps, as would occur with campaigns with only two participating observatories, does not mask the true spectral content of FG Vir. Since campaigns with two participating observatories are much easier to organize than full worldwide campaigns, we would like to recommend more use of such campaigns. To be successful, such two-observatory campaigns should follow some precautions:

- The two observatories should be spaced at least eight hours (in time) apart.
- *The campaigns should not be minicampaigns*, although only two observatories are used. The lack of other observatories should be made up by additional data from the two sites. Note that in the FG Vir example shown above, the two-site data contained 139 hours of data used. In practice, this translates to more than twenty nights of data and many more actually scheduled! A δ Scuti star campaign should be scheduled for several weeks or even months (considering unsuitable lunar phases). Such a large commitment is possible for small telescopes.
- The chosen sites should have a high probability of photometric weather. With few sites the danger of observing gaps caused by weather is increased dramatically. A campaign involving a small telescope at ESO and another one in Vienna, for example, would probably be doomed to failure.

We recommend two-site campaigns for those stars for which little pulsation information is available. In many (most?) cases, an extensive two-site campaign should not require an additional larger campaign. On the other hand, we strongly caution against *small* two-site (or multi-site) campaigns.

3. Long-term aspects: joining data from different years

The long-term monitoring of variable stars has an astrophysical impact far beyond the obvious benefits such as improving the precision of the periods and checking previous results. The first item is quite important: multi-site campaigns usually only last between one and three weeks and, therefore, are afflicted with poor frequency resolution. This can be improved by adding (single-site) data obtained at other times, as long as the different data are reasonably homogeneous. The important question of the long-term amplitude and period changes can also only be studied with data spanning many years. Furthermore, the larger amounts of data created by merging the multi-site data with other available measurements can reveal additional frequencies due to improved signal/noise ratios. These strong benefits of merging data from different months or years should offset the disadvantage of additional computing-time requirements caused by the long base-line in time. This disadvantage is slowly disappearing due to the availability of modern, more powerful computers.

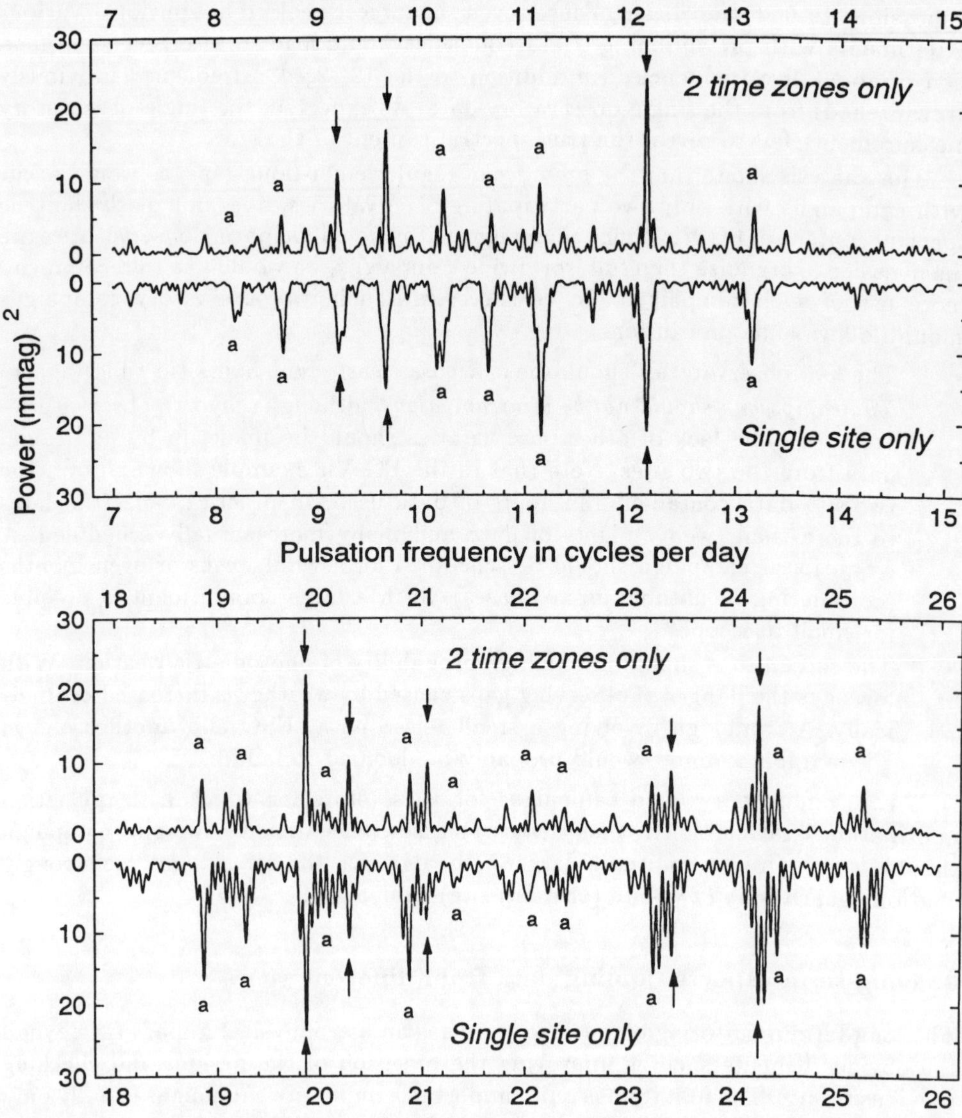

Fig. 1. Power spectra of parts of a multi-site campaign of FG Vir, after prewhitening the dominant frequency of 12.72 cd^{-1}. Arrows indicate the true frequencies, while the letter "a" refers to 1 cd^{-1} aliases. The frequencies can be identified correctly from measurements obtained at two time-zones only. Aliasing prevents the correct identification of the frequencies even for extensive single-site data.

The initial frequency analysis is usually performed by scanning a predetermined range in frequency with a chosen algorithm in order to find the "best" peaks. Fig. 1 is an example of such a power spectrum, which utilized Fourier transforms. Another popular algorithm is the method of reduction of variance, which uses considerable more computing time. All methods have in common that the frequency sampling (i.e. the separation of the trial frequencies or the step size) need to be specified. The value of the frequency sampling should naturally be chosen to be small enough so that the "best" peaks are not missed. To save computing time, this obvious requirement is not always met. Incorrect frequency sampling has been shown to be *responsible for many incorrect results in the literature.*

What is the connection between this problem and the long-term aspects of multi-site campaigns)? The problem surfaces almost only when data sets covering a long time-base are analyzed. What frequency sampling should be adopted? If the observations cover a time ΔT, then the optimum sampling in frequency space, $\Delta\nu$ is related to ΔT by

$$\Delta\nu = 1/(x\ \Delta T)$$

where the value of x needs to be chosen. In his analysis of the application of the CLEAN algorithm, Michel (1993) points out that FFT (Fast Fourier Transform) algorithms automatically use a value of $x = 1$, which can lead to large errors. (This is not the only problem with the application of the CLEAN technique in our field). He emphasizes the need to increase the frequency sampling and calculates that $x = 4$ decreases the errors caused by limited sampling to $\leq 3\%$ in amplitude and $\leq 1/8\ \Delta T$ in frequency. For similar reasons, the period-finding packages used by us permit the selection of $4 \leq x \leq 20$.

The discussion concerning the optimal frequency sampling can be illustrated by a numerical experiment. We have simulated a realistic observing-time distribution with two sets of measurements spaced one month apart with five nights each. The data contain a frequency of 10 cd^{-1} with a variation of 0.01 in amplitude. Fig. 2 shows the results for a Fourier analysis with chosen values of x ranging from 1 to 30. The monthly observing gap, corresponding to the lunar cycle in assigning observing time, should lead to side lobes of 0.03 cd^{-1} in the power spectrum. This expected pattern is shown in the top panel, which oversamples the data. The correct frequency is clearly determined by $x = 10$ or larger, while both $x = 1$ and $x = 2$ miss the correct answers. A value of $x = 4$ still has a reasonable chance of indicating the value of the true frequency, although the possibility of detection of an alias peak cannot be excluded. One can summarize the results by concluding that a choice of x between 4 and 10 represents a good compromise between scientific rigor and computing efficiency.

The minimum value of $x = 4$ places severe constraints on the analysis of data covering extensive periods of time. For data covering 5 years, a scan of the 0 to 100 cd^{-1} range would require calculations at about 10^6 frequencies. Fortunately, time-saving algorithms for Fourier transforms exist and have been proposed by Ponman (1981) and Kurtz (1985), while O'Donoghue and Warner (1982) have addressed the problem of observations obtained in separate blocks separated by very long time-intervals.

Finally, it is probably not advisable to merge δ Scuti star data covering ten years or longer due to the slow long-term period changes present in many of these stars. Small changes of dP/dt between 10^{-9} and 10^{-10}, as found in the star 4 CVn (see Breger 1990), lead to phase shifts of several tenths of a cycle over 15 years. The phase shifts would smear out the variations in a combined solution. The solution to avoid such problems would be to compute amplitudes and phases for data subsets and to compare the results for each frequency and each year of data.

4. Why campaigns should not cover a narrow frequency range only

The observing and reduction techniques used in multi-site observing campaigns are optimized in order to achieve the most accurate information for a chosen range of expected pulsation frequencies. The observational success for the different types of variable stars has, to some extent, been achieved by the specialization. Nevertheless, the results are affected by prior expections of what frequencies should be excited in the star. The problem is caused by the fact that the detection of vastly different periods require different multi-site campaign techniques.

Let us illustrate the problem (and possible solutions!) by examining δ Scuti stars. They pulsate with radial or nonradial p modes. Photometrically the $\ell = 0$ to 2 modes are found. The periods range from about thirty minutes to six hours or frequencies between 4 and 50 cd^{-1}. However, there now exists considerable evidence that δ Scuti stars also show variations with different time-scales, both longer and shorter than the standard frequency range. While the detection and interpretation of these variations are still uncertain, many low and high-frequency variations are linear combinations and harmonics of the periods found in the "normal" frequency range. We believe that the study of these new variations will provide valuable insight in the details of stellar structure.

The existence of peaks in the power spectrum of δ Scuti stars, corresponding to the frequencies of linear combinations and harmonics of the frequencies of known pulsation modes, is not a new discovery. These peaks are observed in Pop. I and Pop. II high-amplitude multi-mode pulsators (see Walraven, Walraven and Balona 1992, Rodriguez et al. 1992, Rolland et al. 1991). They can also be observed in pulsators with amplitudes of a few hundredths of a magnitude (e.g. in X Cae, Mantegazza and Poretti 1992).

4.1. THE CASE FOR HIGH FREQUENCIES, $F > 50\ CD^{-1}$

Let us look at some additional evidence for the existence of relatively *high frequencies* in δ Scuti stars. A recent WET (Whole Earth Telescope) multi-site campaign accidentally discovered a new δ Scuti variable, the star CD -24 7599. The photometric technique used is unusual for a δ Scuti star in the sense that the photometric measurements were made continuously without observing comparison stars with the same photometer channel. This discriminates against the detection of low frequencies, but enables the analysis of the high-frequency domain. Fortunately, the star lies on the ZAMS so that low frequencies are not expected. In Vienna we have analyzed the available 117 hours of multi-site data and find seven frequencies of

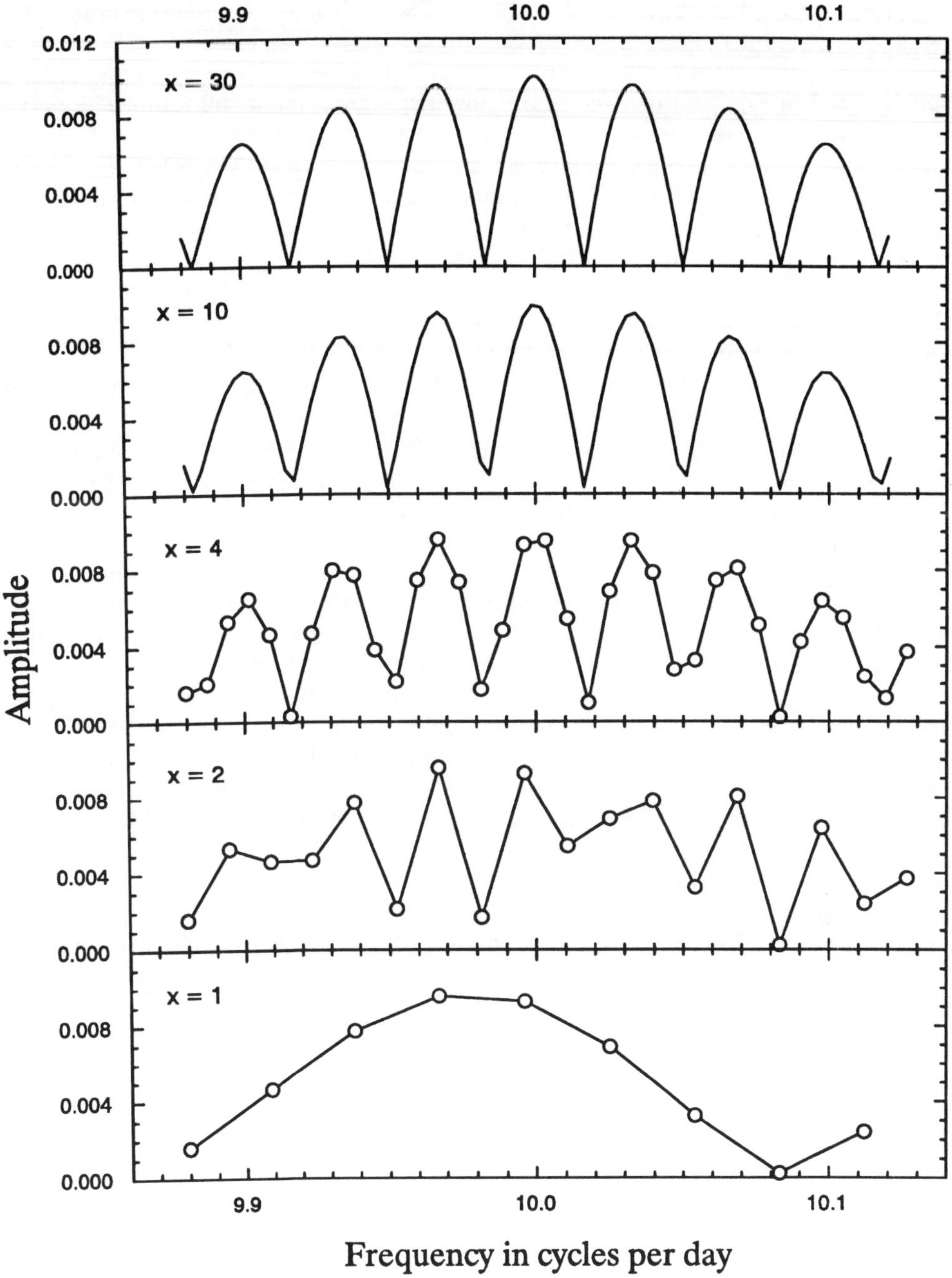

Fig. 2. Effect of different frequency sampling on the correct detection of frequencies. The diagram shows that a standard value of $x = 1$ (see text) misses the true peaks.

pulsation ranging from 27 to 38 cd^{-1}. The results are being prepared for publication (Handler et al. 1994).

However, CD -24 7599 also shows some power at even higher frequencies. This is illustrated in Fig. 3, which shows the power spectrum in the 60 to 80cd^{-1} range. In particular, the existence of $2f_1$ and $(f_1 + f_2)$ appears convincing, where f_1 and f_2 are the two frequencies with the highest associated amplitudes of $0^m.01$. In the top panel we have drawn a line of significance corresponding to a power signal/noise ratio of 12.4 or amplitude signal/noise ratio of 4.0. This very conservative criterion to recognize intrinsic variations reflects the fact that photometric noise does not have a normal distribution and has been experimentally derived (Breger et al. 1993). However, many peaks below the line are probably real as well. The fact that these peaks are maximum at those times when the star has the highest combined light modulation in the 20 – 40 cd^{-1} range may give a clue to the origin of these peaks and argues against a resonance mechanism. The peaks at 74 and 76 cd^{-1}, respectively, with amplitudes of one millimag, correspond to periods less than twenty minutes and *probably would not have been detected with the standard three-star technique.*

4.2. THE CASE FOR LOW FREQUENCIES, $F < 4\ CD^{-1}$

Similar problems at the observational level exist at low frequencies. In many published δ Scuti star papers, the range of 0 to 4 cd^{-1} can be regarded as the garbage heap for observational problems such as incomplete extinction corrections and equipment changes. Many of the detected peaks in the literature are not intrinsic to the star. As will be argued below, the study of this frequency range requires special observing techniques. Nevertheless, some of the detected low frequencies may be real, especially when large data sets from one observatory are available.

Consider the extensive measurements of 4 CVn obtained during 27 nights in 1974 by Fitch, Wisniewski and Bell, which were briefly discussed by Fitch (1980) and analyzed by Breger (1990). We are grateful to Walter Fitch for making these data available prior to publication. The residuals from the $7f$ solution for 4 CVn show a huge peak at 1.40 cd^{-1} with a V amplitude of about $0^m.012$. This peak (or one of its 1 cd^{-1} aliases) has a good probability of representing a variation intrinsic to the star, since it is present in all subgroups of the 1974 data. The peak was not seen in the 1983/4 data, but the star also had much smaller overall amplitudes of pulsation at that time. Due to the aliasing problem, interesting interpretations in terms of linear combinations (differences) of other frequencies or even the presence of g-modes might be overly optimistic at this stage.

For asteroseismological applications it is necessary to separate the real from the spurious frequencies in this range. This affects the planning of multi-site campaigns.

4.3. TECHNIQUES FOR A LARGER FREQUENCY RANGE

The previous discussion has shown that a wide frequency range should be examined, e.g. for δ Scuti stars the range from about 100 cd^{-1} down to less than 1 cd^{-1} needs to be covered. Millimag photometry is still mainly obtained with standard photoelectric photometers containing photomultiplier tubes. At frequencies of 24 cd^{-1} or less, the lack of stability of the atmosphere as well as of the equipment requires that measurements of the variable star are alternated with those of comparison stars. A

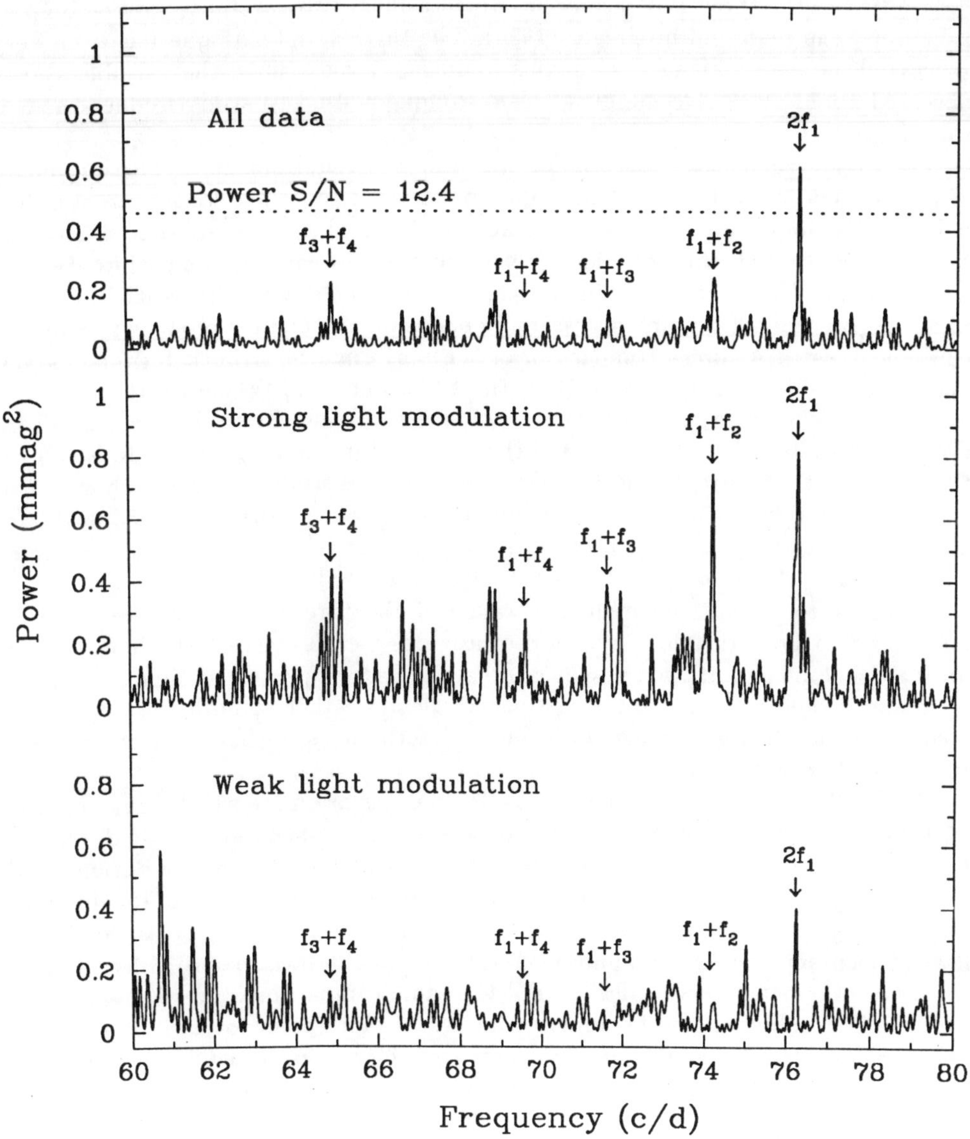

Fig. 3. Evidence for the existence of frequency harmonics and combinations in the high-frequency range for the δ Scuti star CD-24 7599. The effects are most pronounced at those times at which the star has strong light modulation.

description of the three-star technique, which utilizes two carefully chosen comparison stars, can be found in Breger (1993). The three-star technique has been very successful on the millimag level in this frequency range. Since the variable star is observed about every five minutes, this technique is used to study frequencies less than 50 cd^{-1}.

For high frequencies it is possible to omit the requirement of measurements of comparison stars with the same photometer channel, since long-term stability becomes less important. This means that essentially continuous measurements can be obtained. Indeed, this technique is used to study roAp stars and white dwarfs. The frequency limit at which measurements of comparison stars with the same photometric channel become necessary, depends on the quality of the site and the equipment used. A direct comparison of the performance of both techniques has become available: During March, 1993, the δ Scuti and the Whole Earth Telescope networks combined in order to study both the short and intermediate periods of the pre-white dwarf PG 1159-035 and the δ Scuti star FG Vir. Some details can be found in a paper by Breger and Handler (1993). In this study it was also shown that a hybrid technique, utilizing observations of comparison stars only once an hour, causes spurious frequencies. At excellent observing sites, the continuous technique can be reliable down to around 24 cd^{-1} (period of one hour).

In order to extend the frequency range of the three-star technique, we recommend that both the three-star and continuous-measurement technique be used during the campaign. If two telescopes are available for a particular time-zone, both techniques could be utilized during times of clear weather at both telescopes. We recommend duplicate coverage with different techniques, rather than observing a larger number of objects.

The problem and the solution at low frequencies are more subtle, but possibly even more important. Where does the observational problem at $f < 3\ cd^{-1}$ come from? We already mentioned above that problems with extinction corrections lead to spurious power in the low-frequency domain. However, the linking of the different obervatories during multi-site campaigns is another source of error. At the millimag level of accuracy, the different photometer/filter combinations lead to different zero-points at different observatories, even if the same comparison stars are used. This problem is usually handled by applying an average zero-point for each observatory before merging the data. Merging problems are minor when the equipment is stable at every observatory and a large number of nights are available for each observatory. In practice, this is not always the case. *The problem of merging the data from different sites often destroys the information at $f < 3\ cd^{-1}$.*

The problem can be illustrated by a numerical experiment: Consider the extreme case of a one-day period ($f = 1\ cd^{-1}$) and continuous coverage at three observatories with no overlap in coverage (eight-hour data strings). The equipment is stable and a large number of nights are available. Linking the obervatories with their individual zero-points reduces the amplitude to only 32% of the true value! Even worse, irregular coverage leads to a complete falsification of the value of the true frequency.

If one also considers the sad reality that errors in the extinction corrections lead to spurious frequencies in the same range and that, sometimes, individual nightly zero-point adjustments are indicated, it is not surprising that multi-site campaigns are particularly poor performers in the $f < 3\ cd^{-1}$ range. Yet it is possible to obtain reliable frequencies in this range:

1. the equipment and reduction techniques used at the different sites should be as homogeneous as possible. Here we would like to reference an important paper on this subject by Sterken (1993);
2. campaigns should concentrate on obtaining five *or more* nights of coverage from each participating observatory so that the average zero-point becomes more accurate;
3. overlap in the time coverage should be attempted to provide checks that the adopted zero-points are indeed reliable;
4. the temptation to shift the zero-points of individual nights of data should be resisted. If the data are really poor in quality, they should be omitted.

Acknowledgements

This work was partially supported by the Austrian Fonds zur Förderung der wissenschaftlichen Forschung under grant P8543-GEO.

References

Breger, M. 1990, *Astronomy and Astrophysics*, 240, 308
Breger, M. 1992, in *Variable Star Research: An International Perspective*, ed. Percy, J. R., Mattei, J. A., Sterken, C., Cambridge Univ. Press, p. 171
Breger, M. 1993, Proc. IAU Coll. 136, 106
Breger, M., Stich, J., Garrido, R. et al. 1993 *Astronomy and Astrophysics*, 271, 482
Breger, M., Handler, G. 1993, *Baltic Astronomical Journal*, in press
Fitch, W. S. 1980, *Lecture Notes in Physics* 125, 7
Handler, G., Breger, M., Sullivan, D. J. et al. 199, *Astronomy and Astrophysics*, submitted
Kurtz, D. W. 1985, *Monthly Notices of the RAS*, 213, 773
Mantegazza, L., Poretti, E. 1992, *Astronomy and Astrophysics*, 255, 153
Michel, E. 1993, *Delta Scuti Newsletter* (Vienna) 6, 19
O'Donoghue, D. E., Warner, B. 1982, *Monthly Notices of the RAS*, 200, 583
Ponman, T. 1981, *Monthly Notices of the RAS*, 196, 583
Rodriguez, E., Rolland. A., Lopez de Coca, et al. 1992, *Astronomy and Astrophysics, Supplement Series*, 93, 189
Rolland, A., Rodriguez, E., Lopez de Coca, P., Garcia-Pelayo, J. M. 1991, *Astronomy and Astrophysics, Supplement Series*, 91, 347
Sterken, C. 1993, Proc. IAU Coll. 136, 92
Walraven, Th., Walraven, J., Balona, L. A. 1992, *Monthly Notices of the RAS*, 254, 59

DISCUSSION

Harmanec: I very much appreciated your mentioning the problem of a proper choice of the frequency step in period analyses. Actually, it is useful to realize that the frequency step Δf is equal to *the phase difference of two consecutive trial periods* over the interval covered by the data. Therefore, to set $\Delta f = 1/\Delta T$, where ΔT is the time interval covered by the data, means that the two consecutive periods will differ for just one full cycle over the time span covered by the data. In my view, it is therefore necessary to use a frequency step not larger than $0.1/\Delta T$ if you know (or expect) that the variation is roughly sinusoidal. It is even more important to realize that, when searching for periodic variation with a double-wave phase curve, one has to use $\Delta f \leq 0.05/\Delta T$, and so on!

Breger: These are important considerations. We use $x = 10$ for the frequency step of sinusoidal variations. A finer grid might very well be necessary for more complex variations.

Harmanec: Regarding your discussion of the analysis of data from several stations and covering several seasons: I think it is important to say here explicitly that *no instrumental system is secularly stable in colour* (it changes especially quickly after a new aluminization of telescope mirrors). Therefore, if you study phenomena for which there is a colour-dependent phase shift, you cannot combine data from several seasons and/or instruments without properly reducing the information into a standard system. This, in turn, requires at least 2-colour observations. To be fair, however, I must add that the importance of this treatment is the most acute for low-amplitude changes and/or large colour-dependent phase changes.

Young: I was glad to see you emphasize the zero-point and transformation problem. These can, in principle, be eliminated by obtaining data in several overlapping colours, so that the spectrum is fully sampled.

Breger: You are absolutely correct. In the case of short-period variable-star campaigns, the wavelength spectrum is not fully sampled. And it cannot be fully sampled since we concentrate on a specific wavelength region. Thus the problem of the zero-points remain.

Griffin: Are the collaborators and their (small) telescopes already participating as dedicated or "key" programmes?

Breger: Our network is a collaboration between cooperating scientists rather than institutes. This has been working so well that (so far) we have not yet undertaken political steps such as applying for key-programme status. The sole departure from concentrating on the science only, is our attempt to put up a dedicated APT.

Griffin: The success of your multi-site campaigns is the fundamental first step in demonstrating to "politicians" the immense scientific value to be gained by retaining small telescopes as dedicated participants on such programmes.

Poretti: In δ Scuti star studies it is necessary to have an adequate frequency resolution since excited frequencies can have very close values. For this reason, multisite campaigns should span 20–30 years at least.

Breger: I fully agree with you. Multisite campaigns often have terrible frequency resolution since they are relatively short in duration. A solution might be for one of the participating observatories to continue measuring the star during some nights covering several months.

GENEVA PHOTOMETRY AND ITS HOMOGENEITY

N. CRAMER
Observatoire de Genève
CH-1290 Sauverny
Switzerland

ABSTRACT. Long-term monitoring of astronomical objects also requires the long-term conservation of the measuring technique applied to that end. Geneva photometry, in particular, has been defined and developed with the aim of being as conservative and homogeneous as possible. Here, we give a historical overview of the system and discuss several important implications resulting from its homogeneity.

Keywords: photometric techniques – astronomical data bases – catalogs – stars: general

1. Introduction

Astronomical observational researchers are increasingly encouraged to make use of the large and powerful instrumental armoury provided by the collective funding of multinational research organisations. Such observational means allow the fast acquisition of huge amounts of sophisticated data. Well-conceived observing programmes are thus able to furnish, within short notice, the necessary means to enable the rapid production of original publications. This is a highly desirable outcome regarding the current criteria of evaluation of the performance of research groups. It is also of fundamental importance for the progress of our understanding of the Universe. However, it is hardly conceivable that such expensive high-performance instrumentation should be used for time-consuming monitoring purposes. Yet, the momentum acquired by this current trend has the unfortunate tendency to belittle (or, in some cases, completely disqualify), in the eyes of those who hold the power of decision regarding the orientation of astronomical research, the long-term efforts undertaken with the aid of more modest instrumental equipment. As the presentations at this workshop show, there is nevertheless still great demand for the long-term high-precision monitoring of physical processes going on in stars, and it is obvious that exclusive attitudes will not serve the progress of our knowledge, and that some form of consensus will have to be found in the near future.

The main concern of long-term monitoring is that observations, acquired over long periods of time that may amount to several decades, be directly comparable with each other; in other words, that they be homogeneous. In the particular case of photometric data, this means that magnitudes and colour indices which, by

C. Sterken and M. de Groot (eds.), The Impact of Long-Term Monitoring on Variable Star Research, 405–420.

necessity, must reproduce those of a given set of standards, also bear the same long-term photometric significance for objects which may have energy distributions that are very different from those of the standard stars used for the reductions. *In fact, this implies that the transmission functions of the given photometric system be invariant with time.* In other words, appropriate transformations may well at any time reduce the natural measured values of the standard stars to those of the standard system and, likewise, account reasonably well for the colours and magnitudes of the "normal" stars in the batch of reductions. But objects with unusual spectra presenting peculiar lines or other spectral "accidents" that may be evolving with time, as is quite often the case with objects which require long-term monitoring, will always be fraught with a certain amount of uncertainty that is itself a function of the deviation from the standard transmission functions as well as of the reduction procedure used. Seen from such a viewpoint, the conservation and duplication of a given photometric system becomes very difficult to achieve in practice.

There are essentially two main causes of departure from an ideal set of transmission functions. The first is instrumental (transmission functions of the filters, telescope optics and response of the detector), and may be qualified as "avoidable" up to a large extent if great care is taken regarding the stability of the instrumentation which, incidentally, also implies the dedication of a telescope to a given photometer. The other is due to local atmospheric properties, and is unavoidable. Thus, the more widely a given system is used, the more difficult it becomes to ensure its homogeneity. The difficulties inherent to transformation theory in the case of high-precision photometry are treated in an extensive manner by Young (1992), and a good overview of the problems concerning the conformity and homogeneity of photometric systems is given by Sterken & Manfroid (1992).

Considerations of this kind led F. Rufener (1985) to use the terms of "open" and "closed" photometric systems. Open systems are to be understood as being widely reproduced systems such as the standard UBV or the Strömgren $uvby\beta$, systems. Closed systems are, on the other hand, those that rely on a small number of photometers held by a restricted group of observers using a unique reduction process, such as, for example, the Walraven and Geneva systems. The meaning of these terms has sometimes been misinterpreted, and have led to unjustified criticism in a few cases. The Geneva system is the prime example of a closed system, and this is true regarding its duplication and the data-acquisition techniques and reductions. The data, however, are catalogued and made freely available to the astronomical community, and the Geneva photometric data base is, in fact, very "open" indeed.

2. The "closed" Geneva system: Historical background

The definition of the Geneva seven-colour photoelectric photometric system by M. Golay dates back to the late fifties. His aim was to take advantage of the availability of photomultiplier tubes and their high photometric accuracy. The passbands (Fig. 1, Rufener & Nicolet 1988) were chosen with the object of reproducing the general properties of Johnson's UBV system with an almost identical V band but with slightly different U and B bands presenting less overlap at the wavelength

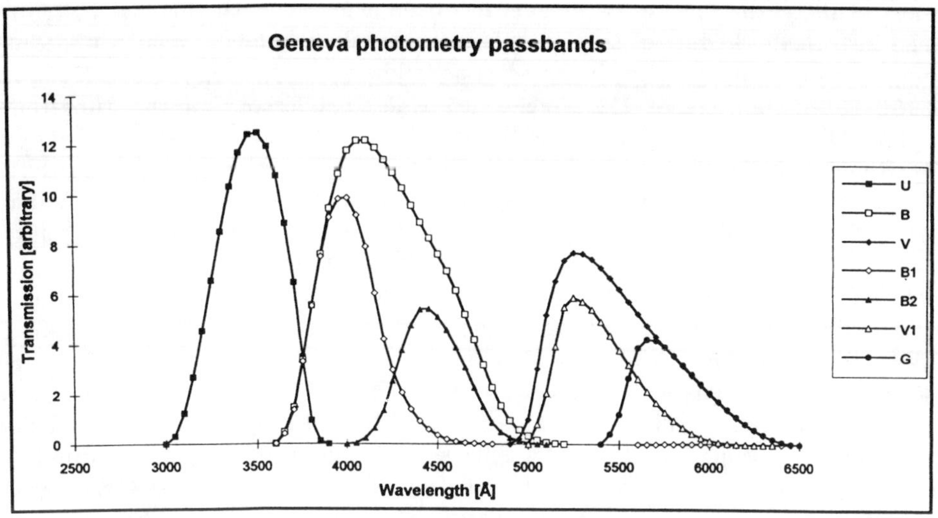

Fig. 1. The passbands of the Geneva $UBVB_1B_2V_1G$ system

of the Balmer jump. Four more bands of intermediate width, B_1, B_2, V_1 and G, were added in view of approximating the stellar classification properties of the Barbier–Chalonge–Divan low-resolution photographic spectrophotometry which was based on a measurement of the Balmer jump, an estimate of the hydrogen-line strengths and of gradients over the Balmer and Paschen continua.

The new photometric system was put into practice by M. Golay and systematic observations were initiated by Golay and Rufener in February 1960 at the Sphinx Observatory (3600m) of the Jungfraujoch High Altitude Research Station with the aid of a first photometer named $P2$ (a prototype made by Golay and called $P1$ had formerly been tested in Geneva). Together with that instrumentation, Rufener also introduced a new measurement and reduction technique based on a twin Bouguer method that makes use of two contra-varying extinction stars, and called by him the "M and D" technique (see Rufener 1964, 1985). That technique allows the determination of "instantaneous" extinction lines at given moments (usually 4 to 6 times) during the night, and allows an accurate determination of atmospheric extinction and of its evolution during the observations.

Five other photometers have been constructed since then. The first four, $P3$ to $P6$ were of an improved design, but were still one-channel instruments that measured through the 7 filters consecutively. Only two of these were equipped with Geneva filters ($P5$ was sold to a French group in 1969, and $P6$ has been kept as a reserve of spare parts). A double-channel instrument called $P7$ (see Burnet 1976), that ensures the quasi-simultaneous measurement through the 7 filters of star and star plus sky background by means of a single photomultiplier, revolving filter wheel and beam-selecting chopper, became operational in 1975. Since 1977 and up to now, $P7$ has been attached to the Swiss telescope at the ESO La Silla observatory and most of the Geneva photometry carried out in the southern hemisphere, indeed most of the entire Geneva catalogue, has been acquired with this last instrument. The historical schedule of the four photometers that "made" Geneva photometry may be summarised, up to the present, as follows:

1960-1968: Photometer *P*2 attached to a 40-cm telescope at the Jungfraujoch Sphinx Observatory (3580m) in the Bernese Alps. Definition of the standard system by Rufener. In 1968, installation of the present 76-cm telescope at Jungfraujoch. Transfer of the 40-cm telescope and *P*2 to Gornergrat (3130m) in the Valoisian Alps. Between 1966 and 1968, photometer *P*3 mounted on the 1m Swiss telescope at the Observatoire de Haute Provence (OHP) in France.

1969-1977: *P*2 for the first two years at Gornergrat, then *P*3 and *P*4 operating alternatively at Gornergrat, Jungfraujoch and OHP (in 1974, *P*2 was acquired by Mons University and sent to their Izaña station on the Canary island of Tenerife). Two short campaigns in 1971 and 1974 at La Silla with P5, temporarily equipped with a set of Geneva filters and attached to ESO telescopes. In 1975, transfer of the 40-cm telescope and *P*4 to La Silla and beginning of an uninterrupted measurement campaign in November. Installation of a 1-m telescope at Gornergrat by the Observatoire de Lyon, France. Sharing of the telescope with the Lyon group; Geneva photometry being done with *P*3.

1977-1993: Installation of the *P*7 photometer at La Silla. In 1980, replacement of the 40-cm telescope by a 70-cm instrument. In 1985, removal of the 1-m telescope from Gornergrat. All subsequent northern photometry has been done at Jungfraujoch with *P*3 and *P*4. In 1979, one of these photometers was used for 4 months on the Spanish telescope at Calar Alto. A new photometer P8, using the rapid-sampling technique of *P*7 but presenting several improvements, is under construction.

3. Homogeneity of the system

The conservation of the system has always been the prime concern for Geneva photometry, and the small number of photometers and observing sites, as mentioned above, have created the conditions that are necessary to attain that goal. The instrumental stability is achieved for *P*2 to *P*4 by using filters coming from the same batch of Schott glass (multilayer interference filters were tested initially, but the presence of extensive wings, secondary bands and surface defects made their application impractical). The detectors are Lallemand 20-stage photomultipliers, used in current-measuring mode and equipped with S-11 photocathodes. The detectors are thermostatically cooled and magnetically shielded. The temperature of the filters is also thermostatically controlled to limit drifts of their colour transmissions. Fabry lenses are mounted to minimise the effects of de-centring in the field limited by the diaphragm, and the photometers are sealed to prevent altering the instrumental response by dust and the condensation of water vapour. *P*7 is equipped with the same filters and shares the same general features of the earlier photometers, but is equipped with an EMI 9502B photomultiplier which is better suited for pulse counting. The accuracy of the measurement is also significantly enhanced by the rapid-sampling technique (see Burnet 1976; Bartholdi et al. 1984). In all cases, stability was increased by the long-term dedication of the telescopes to each photometer.

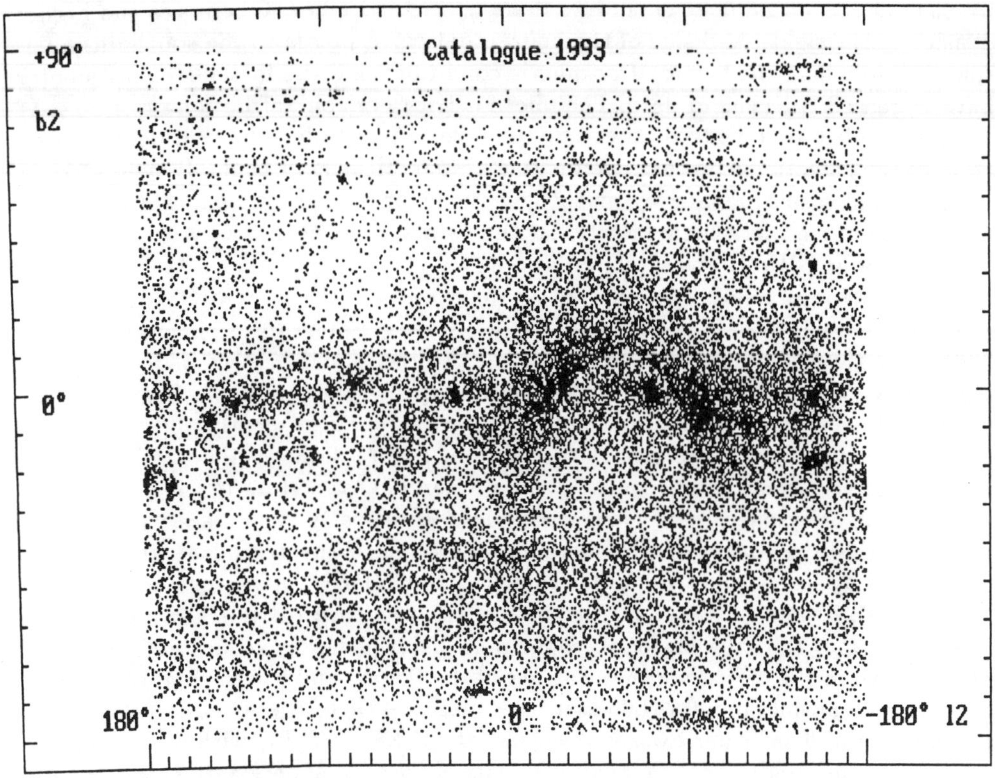

Fig. 2. Whole-sky distribution, in galactic coordinates, of stars measured in the Geneva system

The instrumental stability is supplemented by the constancy of the reduction techniques (Rufener 1964, 1985). The whole data bank is also periodically recalculated, and mean values of colours and magnitudes incorporating the new measurements are computed (process described by Cramer 1989). At the same time, a variety of stability tests based on the inter-comparison of batches of old and new measurements are carried out to ascertain the conservation of the system (see Rufener 1981, 1988). These tests have consistently shown that the homogeneity of the colours exceeds the 10^{-3} magnitude level over the whole extent of the catalogue.

4. Current status of the data

The Geneva photometric data bank currently contains almost 300 000 individual measurements with the corresponding averaged colours and magnitudes of about 40 000 stars. The whole-sky distribution of these data is shown in galactic coordinates in Fig. 2. The figure shows a certain number of selection effects that reflect the past and present observing programmes. The local concentrations correspond to clusters or to a number of surveys done in selected areas. One notes the higher density at the right-hand side of the figure, that corresponds to the southern hemisphere, and where the better atmospheric conditions and the faster $P7$ photometer have increased the yield of data. The photometric "horizon" of La Silla can be seen as an extended arch centred on the South Pole. The smaller but well-defined arch close to the galactic plane corresponds to a systematic survey of the B-type stars of the Michigan catalogue. Surveys in the galactic polar regions are apparent as well as a sampling of southern high-galactic-latitude stars, done with a view to exploiting the forthcoming Hipparcos satellite data.

In Figs. 3–6 we have chosen to display the totality of the current Geneva data in some colour diagrams. The classical U, B, V (Geneva) diagram is shown in Fig. 3. The other three diagrams are based on the 3-dimensional X, Y, Z reddening-free representation which optimises the analysis of the B-type stars (see Cramer 1993, for definition of the parameters and references), but also clearly illustrates the general classification properties of the system. Figure 4 shows the X, Y diagram with the almost rectilinear sequence of the O and B type stars ($Y \sim 0$), and the strong luminosity effect which displaces the evolved early-type stars towards the upper part of the diagram. The distinct separation of the K and M giants from the dwarfs is also visible. By projecting perpendicularly to the former diagram, onto the X, Z plane, we see in Fig. 5 that the sequence of the B stars remains rectilinear, and that the late types appear to be even better separated. The Y, Z plane shown in Fig. 6 is along the sequence of the B stars and has the highly unusual property of causing about 14 000 stars, i.e. almost 40% of the whole catalogue, to be concentrated in a narrow zone centred at $(Y, Z) = (0, 0)$, within a radius of only a few hundredths of a magnitude. The early-type supergiants stand out to the right and the magnetic Ap stars detach themselves perpendicularly below. The separation of the late-type giants from the dwarfs is again very well marked. It is, however, clear that a reddening-free environment does not always reflect in an optimal manner the resolving power of the photometry. It is true here for the early-type stars but, for the later types, the effects of metallicity often vary jointly with the reddening vector (see Grenon 1978, 1981) and the difficulty then lies in the choice of an independent assessment of interstellar extinction.

Some general statistics concerning these data are given in Figures 7 to 10. The evolution of the contents of the catalogue with time is shown in Fig. 7, where the importance of the installation of the $P7$ photometer at La Silla is clearly visible. The distributions of the r.m.s. deviations of colours and V magnitudes are shown in Fig. 8 for all stars with at least two measurements. The second peak at r.m.s. values > 0.01 is due to the ten times wider bins that were used to construct the histogram thereafter. The distribution of visual magnitudes is seen in Fig. 9. When compared with the 1988 catalogue (Rufener 1988), one notices a substantial increase of stars with magnitudes 7 to 9. This is due to measurements related to the Hipparcos mission and to selections based on the contents of the Michigan catalogue. The mean r.m.s. values of colours and magnitudes are related to the visual magnitude in Fig. 10. Here, stars with two or more measurements were considered, and probable variables with r.m.s. values > 0.03 (see Fig. 8) were excluded if $m_v \leq 10$. The larger increase of the r.m.s. values for the colours at visual magnitudes larger than 10 is

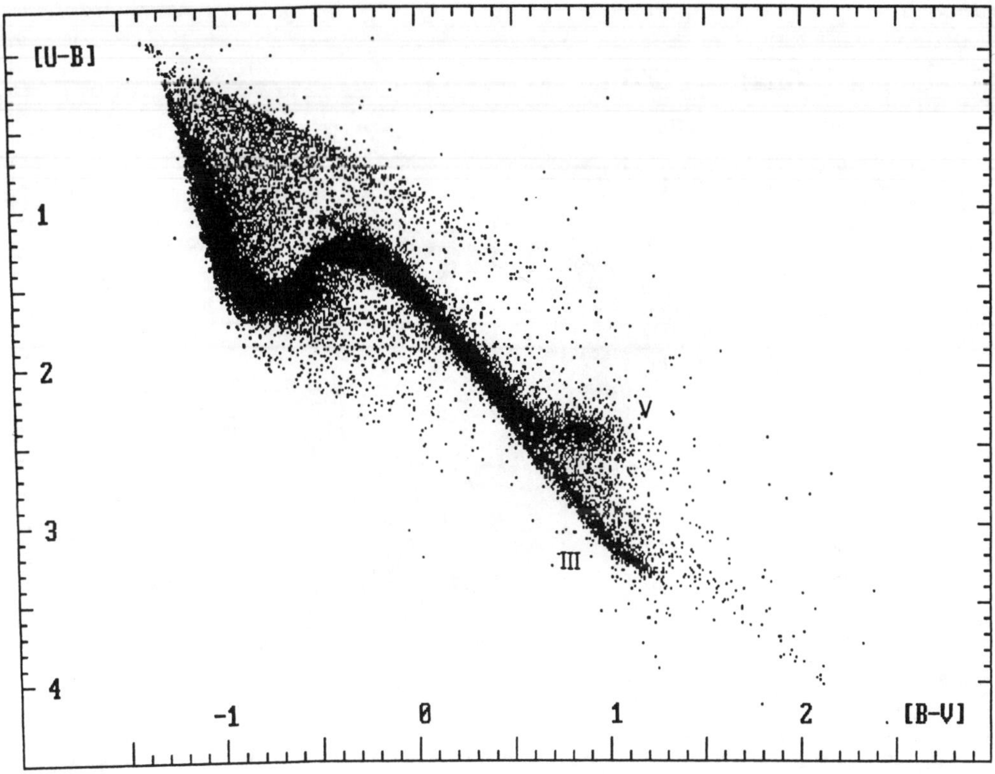

Fig. 3. The Geneva photometric databank, displayed in the $[B-V]$, $[U-B]$ diagram

largely due to an undersampling of the measurement of U for the K-M type stars by the $P7$ photometer, which distributes the integration time evenly between the whole set of filters (see Rufener 1988).

5. Applications that require homogeneity

We mention here some typical cases where homogeneity of the photometry is important.

5.1. LONG-TERM MONITORING AND MICROVARIABILITY

As mentioned above, if long-term monitoring of astronomical objects is to be undertaken with an accuracy level better than one percent for the magnitudes or colours, then this implies necessarily the long-term stability of the general properties of a photometric system. In the case of the Geneva system, the stability is well illustrated by the study of SN 1987A (see G. Burki et al. 1991), where the accuracy of the V-magnitude determination was shown to be better than $0\overset{m}{.}004$ over more than a year of monitoring. Another very good example of the merits of homogeneity

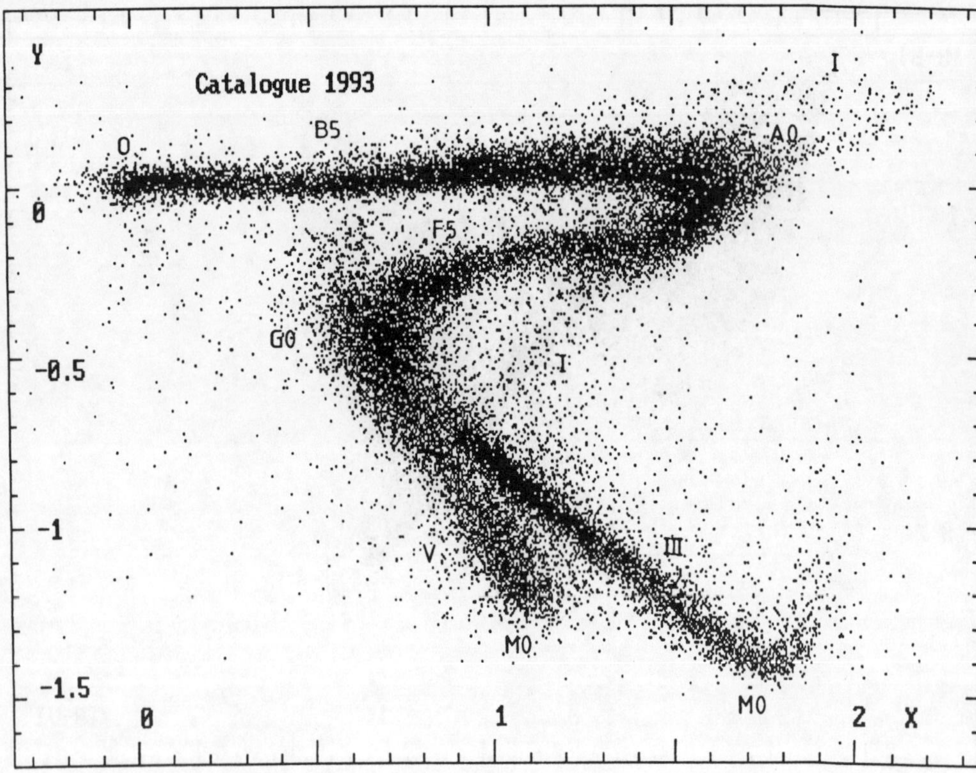

Fig. 4. The Geneva photometric databank, displayed in the X, Y diagram

is given by Waelkens (1991), where data collected between 1976 and 1989 in the Geneva system led to the discovery of three non-radial modes of HD 74560 with $P = 1^{\rm d}.551, 1^{\rm d}.645$ and $1^{\rm d}.737$, with visual amplitudes of $0^{\rm m}.027, 0^{\rm m}.013$ and $0^{\rm m}.008$, respectively. A good general discussion of the detection of microvariability at the millimagnitude level over the whole range of the HR-diagram, enabled by the homogeneity of the Geneva photometry, is given by Grenon (1993) in the context of the Hipparcos mission photometry.

5.2. PHOTOMETRIC BOXES

The technique of photometric boxes is specific to the Geneva system; it was introduced by Golay et al. (1969), and subsequently refined by Nicolet (1981a, 1981b, 1993, 1994). The idea is to extract the maximum amount of information from the photometry by simultaneously using the values measured through the whole set of filters. This is done by considering neighbourhoods in the six-, or less, dimensional space of the colour indices, or else in a three-dimensional parameter space (d, Δ, g or X, Y, Z) if reddening is too important, as is most often the case with early-type stars. The neighbourhood can be centred either on the colours of a given star, or

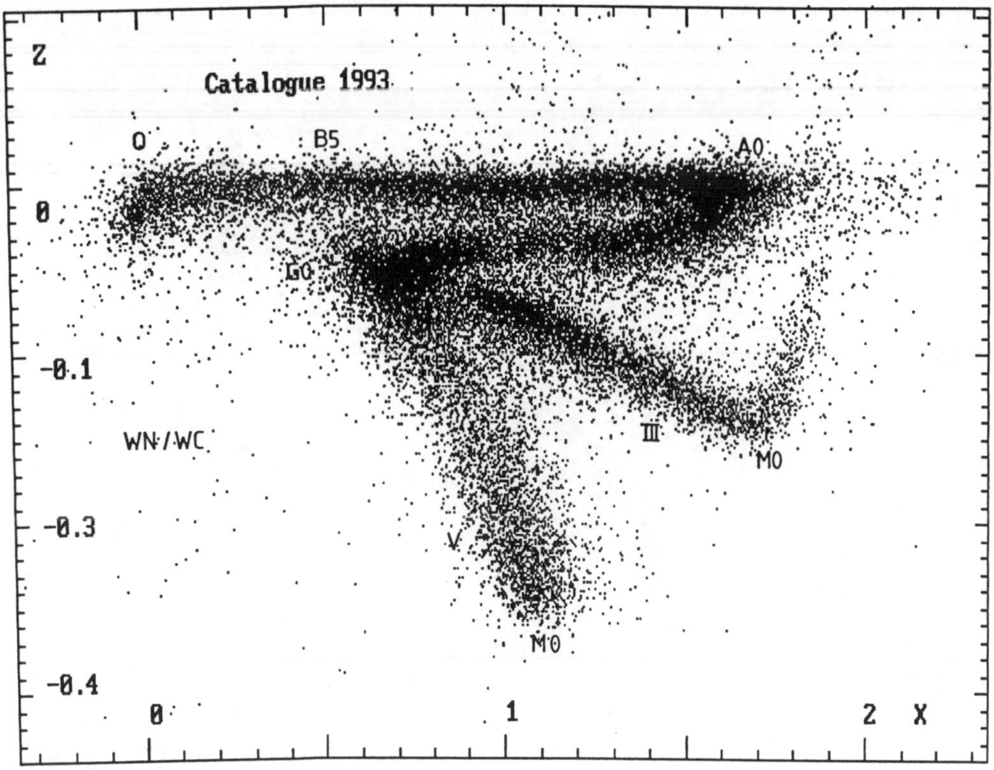

Fig. 5. The Geneva photometric databank, displayed in the X, Z diagram

on a location corresponding to the synthetic colours of a stellar-atmosphere model or else that of a specific calibration point, for example. The distance to the border of the neighbourhood is defined by an appropriate metric which can be adjusted to take into account the different accuracies of the individual colour indices or parameters. The assumption made in this *topological* approach to multicolour photometry is that the properties of any number of stars, or models, or calibration points, etc. lying within the given neighbourhood can be equated with each other, provided the radius of the "photometric box" is small enough. Typically, the radius of a box is of the order of 1 to 2 hundredths of a magnitude in most applications. A box centred on a given star will also contain other photometrically-similar stars coming from various locations within the entire data base, which have been measured at different epochs and positions in the sky. It is therefore obvious that this technique requires a very high degree of homogeneity of the photometry. The box technique has been successfully used to inter-relate the distances of open clusters (see Nicolet 1981b), and the forthcoming Hipparcos satellite parallax data are expected to improve dramatically the distance-scale out to very remote clusters possessing stars which can be photometrically equated with Hipparcos stars.

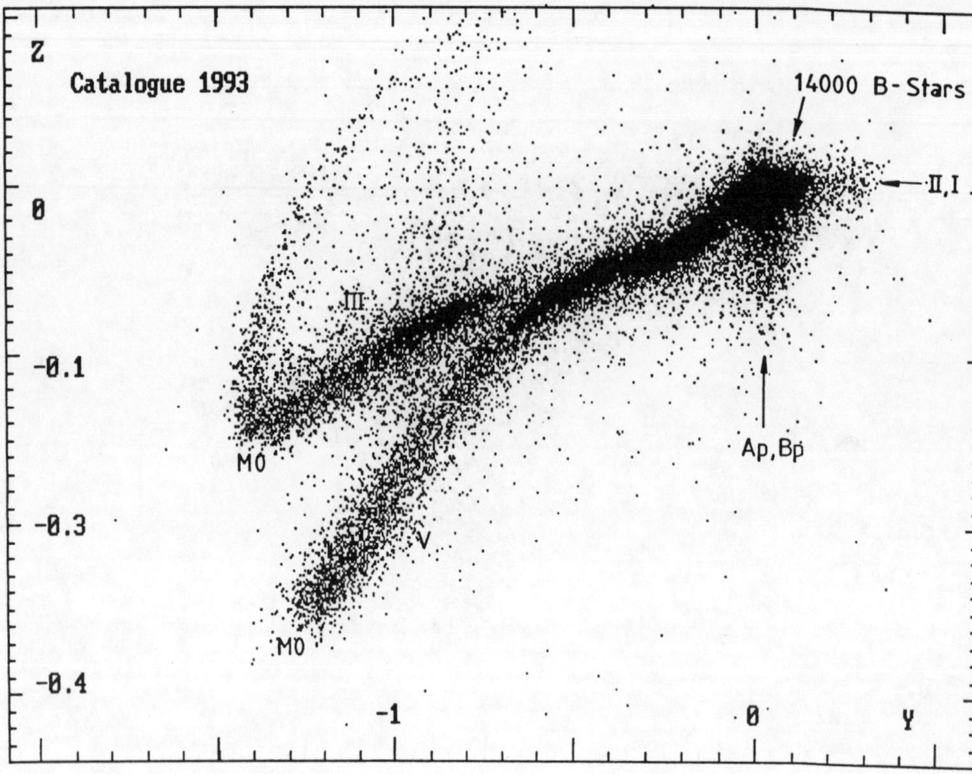

Fig. 6. The Geneva photometric databank, displayed in the Y, Z diagram

5.3. CALIBRATIONS IN GENERAL

The relations which link photometric data with physical parameters such as absolute magnitude, effective temperature, gravity, abundances, etc. depend on the local "sensitivity" of the system. A sensitive photometric indicator, is one which varies with a large amplitude for different values of a given physical quantity. Such a relationship is, however, often restricted to a certain range of temperatures and gravities. Outside that domain it may not only lose its sensitivity regarding the estimate of the parameter, but it most often also loses its "purity" by responding to other physical parameters as well, thus providing ambiguous results. Even within the zone where the indicator is sensitive, the slope of its correlation with astrophysical quantities is quite generally well in excess of unity and depends on the colours of the star, thus being subject to errors in the photometry. In both cases, i.e. be it in zones of purity or of ambiguity, homogeneity is necessary either to ensure that a unique calibration scheme will guarantee the same overall accuracy of the estimate, or to allow the application of a standard procedure to resolve the ambiguities mentioned above for the entire mass of data of the photometric system.

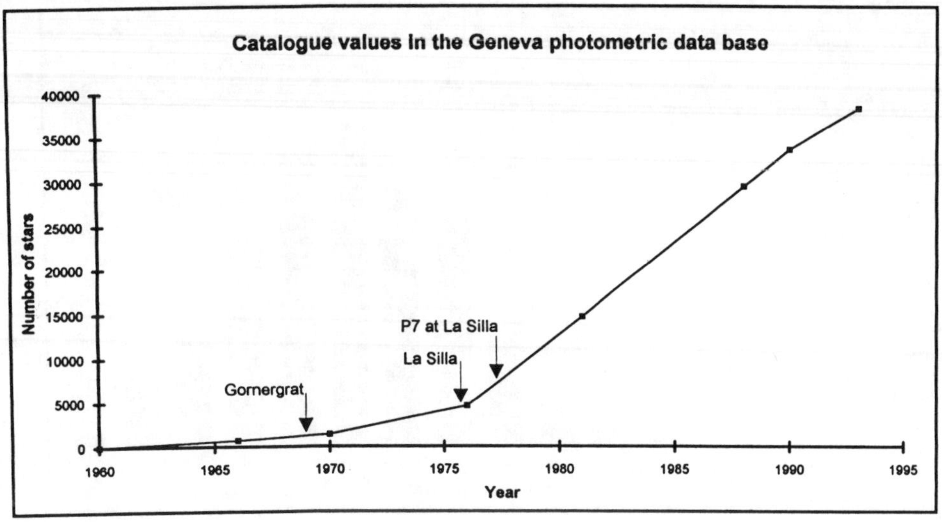

Fig. 7. The evolution of the contents of the Geneva data catalogue with time

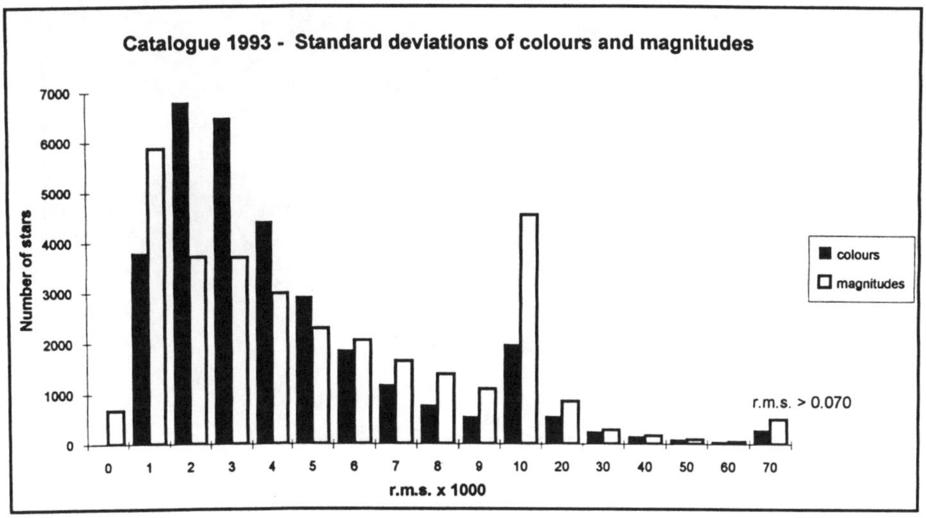

Fig. 8. Distribution of *r.m.s.* deviations of colours and V magnitudes for all stars with more than two measurements

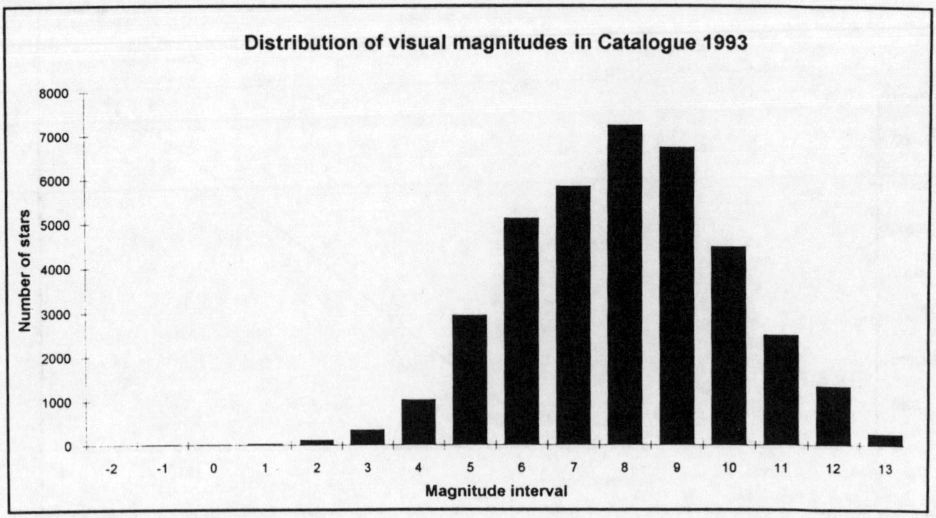

Fig. 9. Distribution of visual magnitudes in the 1993 Catalogue

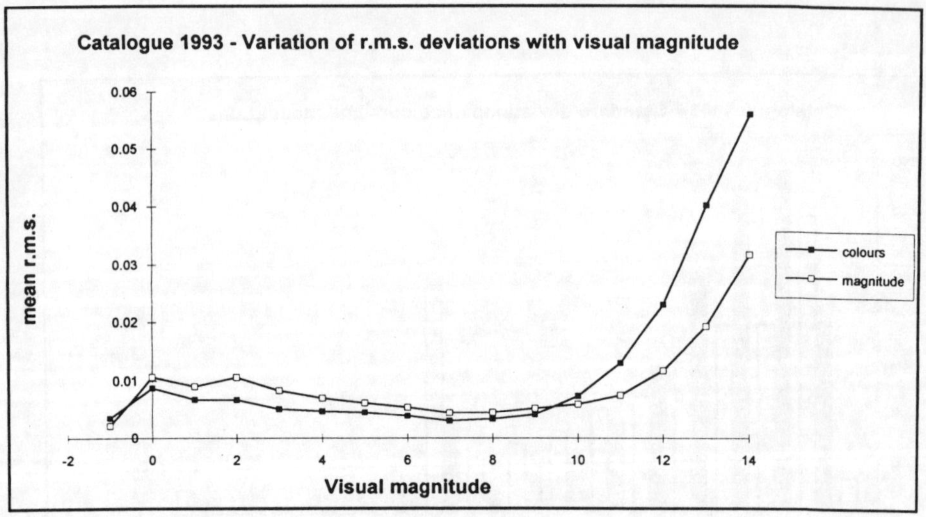

Fig. 10. Mean *r.m.s.* values of colours and magnitudes as a function of visual magnitude

6. Geneva photometry as a "homogenizing" tool

It is obvious that, if sufficiently narrow correlations can be established between the colours or parameters of Geneva photometry and those of a given set of high-quality photometric data coming from another system, then the Geneva photometry can serve to "predict" indices as they would have been measured by that system. It is also obvious that such a process can be used to detect systematic deviations between various sets of measurements, of different origin, but made in clones of the other photometry, and can serve to "homogenize" them relatively to each other. This can be particularly interesting when these sets of data do not overlap, but do overlap with Geneva photometry. We may also mention here that the homogenization is not necessarily restricted to another *photometric* system. Any set of quantities that can be well correlated with Geneva photometry can be linked together in a homogeneous manner. The following three examples illustrate some possibilities.

6.1. CORRELATIONS WITH REDDENING-FREE INDICES: THE β INDEX

The reddening-free X and Y parameters of the Geneva photometry have been shown to be very well correlated with the Strömgren index by means of a cubic polynomial expression in the case of the O, B and early A-type stars (Cramer 1984). The calibration of the resulting (X, Y) estimator was based on 950 values taken from the homogeneous catalogue of β indices published by Grönbech and Olsen (1977). Apart from offering a powerful means of detecting emission in the hydrogen lines via the $\beta - \beta(X, Y)$ difference, this composite "index" also allows sensitive tests of consistency to be made regarding theoretical models seen through both systems (Cramer 1984b), or regarding the conformity of β indices taken from different sources. Indeed, during the work leading to the calibration mentioned above, various sets of data (mainly in clusters) coming from different authors and compiled by Mermilliod and Mermilliod (1980) were seen to show systematic differences with values of the $\beta - \beta(X, Y)$ index reaching $0^{m}.03$ in some cases. Since the range of the β index is only about 0.35 for the early-type stars, such systematic deviations of nearly 10% would affect the astrophysical interpretations based on standard calibrations in a non-negligible manner. This shows that compilations of that kind certainly require some external criteria for checking their conformity and helping to carry out the homogenization.

6.2. THE CASE OF THE CORRELATION WITH UBV PHOTOMETRY

Because of the differences of the two systems mentioned above, Johnson and Geneva U,B,V indices are not directly comparable with one another. An empirical study of their relations (Cramer 1984a) made use of a zone of photometric "purity" in the reddening-free X, Y representation. It was restricted to the intrinsic colours of the B-type stars of classes V to III, and showed that the correlations of the $B - V$ and $U - B$ indices are locally quite linear. As is to be expected, for the relations for both indices to extend over the whole range of the early-type stars requires higher-order terms. A new study done by M. Golay (in preparation) narrowly correlates the *observed* U, B, V indices with the set of observed Geneva indices for the whole range of stellar types. The relations given by Golay effectively allow the prediction of Johnson indices by Geneva photometry at the 1% level when tested with the best samples of UBV data, and allow the comparison of synthetic colours derived in both systems. The comparison of the two sets of synthetic indices corresponding to the 1993 Kurucz models shows a non-linear correlation with some residual gravity effects, thus suggesting that the fault resides in systematic errors affecting the definitions of the pass-bands. Such empirical relations provide a powerful new tool for comparing and homogenizing UBV photometry taken from various sources.

6.3. USE OF PHOTOMETRIC BOXES

Boxes are very restrictive concerning the properties of stars sharing the same photometric neighbourhood. As mentioned, they have successfully been used by Nicolet to homogenize cluster moduli. But they can also be applied to the homogenization of *any* quantity that can be correlated with Geneva photometry. For example, Golay (1994) has shown that boxes centred on the mean *photometric* locations of given spectral types contain fewer and fewer stars classified differently, as the sources of spectral classification considered tend towards higher quality. In other words, if for the most homogeneous spectral classifications, boxes centred on a star of a given type according to that classification contain other stars classified by the same authors, then the types are almost always identical. This kind of convergence implies that the photometry can indeed be useful for testing the homogeneity of spectral types and for detecting systematic shifts between different sources of spectral classification.

7. Conclusion

The Geneva photometry has maintained its homogeneity since its definition in 1960. This is essentially due to the great care taken during the duplication of the system, the engineering of the photometers and the standardization of the observing and reduction procedures. The high levels of accuracy of the magnitudes and colour indices allow the study of light and colour variations of very small amplitude over long periods of time. The overall homogeneity of the data enables metrologically-demanding techniques, such as photometric boxes, to be carried out efficiently. The Geneva colours can, in many cases, be narrowly correlated with indices of other photometric systems and serve to predict values in those systems, thus providing a means of homogenization of photometric data from different sources. Any kind of data, be they photometric, spectroscopic, synthetic, or other, can thus be tied together provided they all correlate well with Geneva photometry. A photometric box may contain a variety of objects, some of which have well-established physical properties; those properties can then be extended to the other members of the box with a high degree of confidence. The ability to standardize data in that manner is already well provided for by the 40 000 stars currently measured, and will get significantly better as their number increases.

References

Bartholdi, P., Burnet, M., Rufener, F., 1984, *Astronomy and Astrophysics*, 134, 290
Burki, G., Cramer, N., Nicolet, B., 1991, *Astronomy and Astrophysics, Supplement Series*, 87, 163
Burnet, M., 1976, *Etude et rialisation d'un photomètre differentiel commandé par ordinateur* Thèse N 232, Ecole Polytechnique Fédérale de Lausanne
Cramer, N., 1984a, *Astronomy and Astrophysics*, 141, 215
Cramer, N., 1984b, *Astronomy and Astrophysics*, 132, 283
Cramer, N., 1989, in *Precision Photometry: Astrophysics of the Galaxy*, (A.G. Davis Philip et al., Eds.) L. Davis Press, Schenectady, 173
Cramer, N., 1993, *Astronomy and Astrophysics*, 269, 457
Grenon, M., 1978, *Propriètés photométriques des étoiles, G,K,M en relation avec la structure et l'évolution galactique* (Thesis), Publ. Ob. Genève, Série B, Fascicule 5
Golay, M., Peytremann, E., Mader, A., 1969, Publ. Obs. Genève, Sér A, No 76, 44
Golay, M., 1994, in *The MK Process at 50 Years*, Vatican Observatory Proceedings. To be published.
Grenon, M., 1981, In Astrophysical Parameters for Globular Clusters, IAU Coll. 68
Grenon, M., 1993, In Inside the Stars, IAU Coll. 137, ASP Conf. Ser., Vol. 40, W.W. Weiss and A. Baglin (eds.)
Mermilliod, J.-C., Mermilliod, M., 1980, Bull. Inform. CDS 19, 65
Nicolet, B., 1981a, *Astronomy and Astrophysics*, 97, 85
Nicolet, B., 1981b, *Astronomy and Astrophysics*, 104, 185
Nicolet, B., 1993, in *Poster Papers on Stellar Photometry*, IAU Coll. 136, Dublin Institute for Advanced Studies. To be published.
Nicolet, B., 1994, in *The MK Process at 50 Years*, Vatican Observatory Proceedings. To be published.
Rufener, F., 1964, Publ. Obs. Genève A, 66, 413
Rufener, F., 1981, *Astronomy and Astrophysics, Supplement Series*, 45, 207
Rufener, F., 1985, in *Calibration of fundamental Stellar quantities*, IAU Coll. 111, (D.S. Hayes et al., Eds.) Reidel Publ. Co., Dordrecht, 253
Rufener, F., 1988, Catalogue of Stars measured in the Geneva Observatory Photometric system (fourth edition), Observatoire de Genève, Sauverny
Rufener, F., Nicolet, B., 1988, *Astronomy and Astrophysics*, 206, 357
Sterken, C., Manfroid, J., 1992, in *Astronomical Photometry - a Guide*, Kluwer Acad. Publ., Dordrecht, 167
Waelkens, C., 1991, *Astronomy and Astrophysics*, 246, 453
Young, A. T., 1992, *Astronomy and Astrophysics*, 257, 366

DISCUSSION

Young: Failure of a Fabry lens to flatten response over the focal plane is usually due to an opaque photocathode, inclined to the optical axis. The Fabry lens converts position in the focal plane to angle at the photocathode. The effect is very pronounced with 1P21 photomultipliers.

Cramer: Actually, we have this condition with our photometers using Lallemand tubes, but we have not encountered this problem.

Van Winckel: Can you obtain the same accuracy with the CCD mounted six months ago? On La Silla you probably have the best possibility to evaluate the advantages and disadvantages of the two (CCD *versus* photomultiplier).

Cramer: This question is still being evaluated. First results show that the transformation to the standard system is only about half as accurate. Gilbert Burki can inform you better on that subject.

Burki: The accuracy of the seven-colour measurements in the Geneva system with a CCD is very good in the natural—that is, instrumental—sytem. The problem we are evaluating now is the transformation to the standard system; we will certainly have to apply colour equations to some of the indices.

Kurtz: Is your catalogue available electronically and how many stars are in it now?

Cramer: The data of the 1988 catalogue are available from the Strasbourg CDS and are on the NASA compact disk of photometric systems. The subsequent data have not beem thoroughly corrected and are in our data base, but can be obtained on request.

Burki: The catalogue contains the averaged measurements; individual measurements, however, are available from Geneva.

Le Contel: I would like to mention how impressed I am with the quality of the Geneva system. In my opinion, this high quality is due in large part to the very high quality of Lallemand's tubes and to the very precise knowledge of their physical properties. Unfortunately, they cannot be used in photon-counting photometers.

Waelkens commands attention – Van Winckel, Cramer, Burki, Jorissen, Schimpke, Bruch

PHOTOMETRIC MONITORING: ACCURACY, HOMOGENEITY AND PRECISION IN EXISTING LONG-TERM PROJECTS

Andrew T. YOUNG
Astronomy Department
San Diego State University
San Diego
CA 92182-0540
U.S.A.

ABSTRACT. Long-term monitoring requires all observations to be reduced to a common basis. The long-term stability of instrumentation is inferior to the precision available during a single observing run. Because the transformation from instrumental to standard system changes with time, standard stars must be measured in every observing run. Thus, one must in fact do "all-sky" photometry, even if local comparison and check stars are used to monitor extinction and zero-point changes. The balance between extinction and transformation errors is discussed, with recommendations for scheduling telescope time most efficiently.

The accuracy, and thus the homogeneity, that can be reached depends on how well the standard passbands can be matched. Manufacturing tolerances on filters and detectors are so large that good matching is possible only numerically, after the observations have been made. Then, to synthesize good approximations to the required passbands, we need instrumental passbands that overlap.

Keywords: photometry – accuracy – precision – homogeneity – transformations – extinction

1. Basics

By "precision" is meant the repeatability of a measurement, usually under fixed conditions. Photometrists often test precision by duplicating a measurement to see how well it repeats. On the other hand, "accuracy" means the absence of error, as measured against some external standard, such as a set of standard stars. Crudely, we can regard precision as the concern of the differential photometrist, and accuracy as that of the all-sky photometrist. However, these distinctions are at the ends of a continuum: as comparison stars become farther removed in position, time, and spectral type from programme stars, there is no sharp demarcation at which we suddenly switch from precision to accuracy; rather, our attention gradually changes to include a broader set of problems.

Unfortunately, "homogeneity" is a notion that, while obvious enough in principle, seems to elude us in practice. Sterken (1993) gives numerous examples of the kinds of difficulties that arise in real programmes. There are many types of varying

C. Sterken and M. de Groot (eds.), The Impact of Long-Term Monitoring on Variable Star Research, 421–436.

systematic errors that masquerade as real astrophysical effects; some are obviously defects in our data, but others leave us in doubt. The most vexing problem is that we can obtain apparently good transformations of the calibration stars to a standard system, only to find disturbingly large discrepancies among the programme objects.

Errors of several hundredths of a magnitude, or even several tenths of a magnitude, are often found; yet the precision of the best photometry is better than a millimagnitude. What is the cause of these large errors, and how can we avoid them?

2. Introduction

The central feature of long-term monitoring programmes is clear from the name: they are *long-term.* That means that instrumental stability is a crucial problem, so that accuracy may be a bigger problem than precision.

Just as one never can step into the same river twice, one cannot use the same instrument on two different observing runs. Filters and detectors age; optics get dirty, and are cleaned, or even re-coated; reduction programmes evolve; the Earth's atmosphere changes with season and from year to year; observers may quit, retire, or even die (as may instruments); and even some "standard" stars may change.

However, some lucky observers can use *nearly* the same instrument over and over again. When an instrument is left on the same telescope for years at a time, the most drastic changes can be avoided. On the other hand, some programmes necessarily must combine observations from many different instruments.

One can trace out a range of programmes from (a) tightly-controlled programmes that involve fixed equipment, used by the same well-trained observers for many years to observe a rather homogeneous group of stars (Lockwood and Skiff 1988, Lockwood et al. 1993); to (b) programmes involving a small number of different but relatively stable instruments, each run by experienced observers, and still restricted to objects with rather similar spectra (Harris 1989, Zappalá et al. 1989); to (c) projects employing heterogeneous and changing equipment, trained and untrained observers, and programme objects with peculiar spectra that differ substantially from the "normal" stars used as photometric standards (Young and Irvine 1967, Young 1974a, Manfroid *et al.* 1991); to (d) attempts to combine observations of a supernova, made by observers whose interests lie outside photometry, using whatever equipment is available at diverse observatories, perhaps supplemented with observations of standard stars and perhaps not (de Vaucouleurs 1993). Unfortunately, most long-term projects fall into category (c).

Many of the problems in differential photometry were reviewed recently (Young et al. 1992), and the reader is referred to that paper for further details. That paper emphasized the need to measure extinction carefully in broadband photometry, because higher-order terms (which are quite appreciable) do not cancel out in "differential" photometry. Even in intermediate-band work, colour terms in the extinction are significant at short wavelengths (Tautvaišienė and Straižys 1985). For example, one should remember that the u band of the $uvby$ system is defined by a glass filter, so that the passband is not much narrower than the U of UBV. In addition to these problems, even narrow-band systems suffer from the instabilities and long-term changes discussed below.

3. Effects of Ageing

The passage of time affects every component in the instrumental system. Consequently, it is not possible to treat observations made at widely-separated times as if they were made by the same instrument. The only way to compare observations that are widely separated in time is to convert them all to a common basis. Usually, this is done by transforming all the data to a standard system.

Let us review some of the problems involved:

3.1. FILTERS

Two types of long-term changes in filters are important: temperature effects, which are reversible; and permanent changes. Filter-temperature effects are discussed in detail by Young (1967), and reviewed on pp. 105–108 of Young (1974b).

In coloured glasses and other absorbing filters, such as dyed gelatin, solutions, and crystals, there are two main mechanisms at work. The sharp-cut-off glasses, and in general, similar steep absorption edges in other materials, display an Urbach tail whose phenomenology is well understood. The position of the cut-off shifts in energy by some fraction of kT for each degree of temperature change; typical values are on the order of 0.1 nm per degree in the visible. The cut-off moves to the red as temperature increases.

The bandpass glasses, and other absorbing media, mostly exhibit absorption bands in which the absorption coefficient has a Gaussian profile. These bands broaden with temperature by a mechanism known in the solid-state literature as "phonon broadening." The broadening is often accompanied by a shift in band centre, usually toward longer wavelengths with increasing temperature. The combination of the two mechanisms usually shifts passbands redward by about

$$\Delta\lambda = 0.5nm \cdot \lambda^2$$

if λ is measured in microns.

These passband shifts change the effective wavelengths of the instrumental system. For example, the typical 0.1 nm shift per degree corresponds to about 0.1% of the $B - V$ colour baseline; so, a 20-degree seasonal shift typically changes the measured magnitudes of blue stars by about $0\overset{m}{.}02$ per magnitude of colour difference, relative to red ones. The shift has the same sign, and similar magnitude, in B and V, so it may go unnoticed in transforming colours. However, it contributes to changing conformity errors, which will be discussed below.

Diurnal temperature effects are likely to be large for stars observed, e.g. at certain USA sites, in late Fall, when the days are warm, and the nights are long and cold. Considerable temperature change can occur during the night at this time of year. Large temperature changes often occur at observatories that are on low hills in dry climates; the Boyden Observatory is noted for this, because the nocturnal inversion reaches the summit of the 'kopje' about midnight. Isolated high peaks, on the other hand, tend to remain above the inversion all night, and have much smaller temperature variations.

Interference filters also show thermal shifts, but they are typically 5 times smaller than for absorbing filters. However, conventional interference filters suffer serious long-term effects caused by absorption of water vapour, because the coatings have a porous microstructure. Adsorbed water alters both the thickness and refractive index of the coating layers, starting from the outermost layer. The result is a detuning of the resonant cavity, with changes both position and shape of the passband. In severe cases, the coatings actually peel off from the substrate. The deterioration can be slowed down by sealing the edges, but no sealant is impervious to water molecules. Until recently, this was an unavoidable problem with interference filters, and manufacturers refused to guarantee that a filter would stay within the original specifications for more than a year or two.

However, a new type of filter construction is now becoming commercially available. The new filters have oxide (rather than cryolite and zinc sulfide) coatings, which can be made very compact by ion-assisted deposition. These coatings are essentially unaffected by water vapour. The catch is that the oxides provide a smaller range of refractive indices, so that many more layers are required. Consequently, the new filters are about twice as expensive to manufacture as the old ones. If astronomers are willing to pay the cost, the new filters offer an escape from the serious instability that has plagued interference filters for decades.

However, no matter how stable the filters themselves are, they may become contaminated. It is a common experience to open up a photometer that has been used for several years, and discover that the filters and other optical elements are covered with a greasy film. Sometimes this is lubricant that has migrated from moving parts; sometimes it comes from outgassing of motors, solenoids, or electronic components, or even paints. The number of photometers that have been put out of action by the accumulation of insects on optical surfaces is also considerable.

One must always keep an eye on instrumental zero-points, which are the most sensitive indicators of optical contamination. One usually expects organic contamination to produce significant ultraviolet absorption, so the zero-point of the shortest-wavelength colour index is the first thing to check. Popper (1982) has emphasized the importance of zero-point values in assessing the match between instrumental and standard systems.

3.2. DETECTORS

Like filters, detectors show both reversible thermal shifts, and permanent changes. Detector-temperature effects are reviewed in Young (1974c, pp. 46–49). Although that review was limited to photomultipliers, the physics is similar for other solid-state detectors, such as CCDs. In all cases, the main effect is the steepening of the Urbach tail at low temperature. As the material becomes more transparent in the threshold region, fewer long-wavelength photons are absorbed in the device's active region, and the red response drops.

The effect is of the same sign as in filters, though of somewhat smaller magnitude. The combined effect may be more serious in variable-star photometry than was previously thought. Recently, Park (Park and Chen 1989, Kim and Park 1993) has observed somewhat larger temperature effects in practice than I calculated from laboratory measurements on filters and detectors. It is noteworthy that these large effects were discovered in "differential" photometry. This problem seems to deserve more detailed investigation.

There are also secondary effects, due to the exponential change in phonon population with temperature. Consequently, the carrier mean-free-path decreases as temperature increases, changing the effective active volume of the detector. Other properties of the semiconductors used for detection are also temperature-dependent. For example, the effective band gap changes slightly with temperature in most materials; lead sulfide is notable for having a very large band-gap temperature dependence. Normally, these effects are small compared with the large changes in the Urbach tail.

Irreversible changes in detector spectral response with time are a more serious problem for long-term projects. Johnson (1962) found substantial changes in his 1P21 on cooling with dry ice; it recovered only partially afterward. Such changes were foreseen by Schmidt-Kaler (1961), who had pointed out that even with Johnson's original instrument, age effects might change the instrumental spectral response. These effects in photomultipliers are probably due to differential pumping of cesium vapour (Young 1974c).

Other detectors also show long-term drifts. CCD workers have begun to see drifts in device behaviour with time, probably due either to diffusion of impurity atoms, or to mechanical creation and motion of lattice defects by thermal stresses. The active region in silicon is very shallow at short wavelengths, because of the high absorption coefficient, so surface contamination and other small effects can appreciably influence the response.

Finally, one must remember that many "detector" problems are actually power-supply or other electronic problems. One tends to forget that power supplies age along with everything else. We recently had a high-voltage supply that had not been checked for many years; when measurements became erratic, it was found that the transconductance of the old amplifier tubes in the regulator circuit had fallen so much that the output was no longer regulated. The effective dead-time parameter in pulse-counting depends on both tube gain and discriminator level; as electronic components age, the intrinsic non-linearity of pulse-counting changes. If the programme and comparison stars differ in brightness, varying nonlinearity and the resulting wrong corrections for it produce apparent magnitude shifts.

3.3. OPTICS

Telescope optics get dirty from time to time, and must be cleaned. Mirrors eventually need to be re-aluminized. Changes of 30% have been reported on re-aluminizing (Lockwood et al. 1993). Dirty mirrors typically lose more response in the ultraviolet than in the visible, partly because some of the "dirt" is actually an organic film—especially at observatories surrounded by pine trees. However, some of the loss in UV reflectance with time is due to increased scattering as the effective surface roughness increases.

3.4. THE EARTH'S ATMOSPHERE

In addition to the well-known seasonal changes in atmospheric transparency, there are year-to-year and long-term changes. In recent years, the eruptions of El Chichon and Mt. Pinatubo have made stratospheric aerosols whose optical effects are obvious to even the casual observer. Sterken and Manfroid (1992a) show some recent examples. There is a large literature on this subject; see (Manfroid 1993) for references.

In addition, the appreciable anthropogenic changes in the atmosphere, such as a steady increase in carbon dioxide, and a decrease in ozone, are producing long-term changes in spectral transmittance. The ozone is particularly important in the visible spectrum. These effects are very serious for the U-B colour index, which by definition contains the atmospheric transmission as part of the UBV system.

Another place where we may expect trouble from decreased ozone is the colour term in the V extinction. In the region from 500 to 600 nm, the Chappuis bands of ozone nearly compensate for the decrease in Rayleigh scattering with increasing wavelength. Consequently, it has been common practice to ignore any colour term in the V extinction. However, as the ozone decreases further, observers may need to include the colour term routinely.

3.5. CHANGES IN OBSERVERS

Sterken (1992a, 1993) has pointed out the decreasing skill of the typical observer, which is a serious problem in the ESO LTPV project. The observer's skill is important in several ways. An unskilled or careless observer will produce increased centring errors; may cause difficulties in subtracting sky; and nearly always neglects the calibration measurements needed to remove the effects of the atmosphere and the instrumental signature from the data.

Centring errors are often made much worse than they need be by bad design choices in instruments. They are large if photomultipliers with inclined photocathodes are used. Uncorrected flat-fielding errors, which can easily be several per cent, cause similar problems in CCD photometry. Centring errors were discussed at some length recently (Young et al. 1992) and need not be treated further here.

Sky measurements were also discussed by Young et al. (1992). However, I have subsequently found that the sky measurements made by typical observers amount to a correction of less than a millimagnitude for bright stars. Therefore, the distance from the programme star at which sky is measured is probably not as serious a problem as I had feared. A good rule of thumb is to measure the sky two diameters of the measuring aperture from the star, or a minute of arc away, whichever is larger. Because of fluctuations in the background, and unseen faint stars, one should be aware that using a fixed sky position for each star will give better precision, but probably worse accuracy, unless the sky area is carefully chosen in advance.

For faint stars, a more serious problem may be to measure the sky frequently enough, if a single-channel instrument is used. The airglow component of the night sky varies with time and position, so one should move back and forth quickly between sky and star positions. A practical limitation to this is the time required to re-acquire and re-centre a star. For further discussion, see Young et al. (1992).

A closely-related problem is the use of different measuring apertures for different stars. While the dependence of the instrumental system on the aperture size is becoming well known to CCD users, one still sees novices using different sizes for different stars. I once saw a visitor at ESO trying to use large apertures for bright stars and small ones for faint stars, in extracting magnitudes from a CCD frame. I suppose he had somehow mixed up "aperture photometry" with iris-diaphragm photometry on photographic plates. As another example, Chris Sterken has shown me observations made by a student who took no sky measurements for an hour and a half, and then took 20 in a row! The things a beginner will do are just incredible. The lesson is that an ignorant user can make a good instrument worthless, or reduce even good data to nonsense.

One of the advantages of automated telescopes is that centring and sky measurements can be made more reproducibly. However, automation does not automatically guarantee improved precision. Sterken and Manfroid (1992b) found that automation initially made precision worse at one telescope, because the automatic centring system did not allow the telescope to settle before taking data. A one-second delay cured the problem (Sterken 1991).

The choice of calibration measurements is often left to the individual observer; but some observers never get enough calibration data, and most do not distribute them optimally. The proper distribution of extinction and standard stars is discussed at length in (Young 1974d, Young 1993a), and will be discussed further below. Fortunately, the photometric planning program distributed with MIDAS embodies these principles, and should allow observers to make better use of the time devoted to calibrations.

However, the only real solution is to put more effort into photometric education, and the training of observers. I learned this during the Harvard-NASA planetary photometry project, where we got much better data from the French station than from the South African one, simply because we had less turnover of observers (Young 1974a).

3.6. CHANGES IN EQUIPMENT

Obviously, replacing filters or detectors makes major changes to the instrumental system. At large observatories like ESO or NOAO, equipment changes frequently. It's quite common to see components like filters and cold-boxes "borrowed" from one instrument for use with another. With frequent changes, instruments are more likely to get dropped or broken, and subsequently replaced. Such an environment is very unfavourable for long-term stability.

Unfortunately, at a large observatory, not everyone is aware of all the changes. Neither the astronomer nor the night assistant may realize that equipment has changed since an earlier observing run. Sometimes, "improvements" are made to instruments—even those that normally are used in a fixed configuration. Changes in components that are inside the photometer (like filters and detectors) are not obvious to the user. The only safe procedure is to measure a full set of standard stars during every observing run, and to re-determine transformation coefficients every time.

4. Transformation Problems

4.1. BASICS

The essence of the transformation problem is now understood (Young 1974d, Manfroid and Sterken 1992, Young 1993b): different instruments measure different things (different parts of the spectrum), so that astrophysically important features in stellar spectra make exact transformations between systems impossible. However, often the situation is not quite like comparing apples and oranges; it's more like comparing oranges and tangerines. So, stars with spectra like those of the standard stars can be transformed fairly well, even using the conventional empirical formulae.

However, programme stars often occupy a much larger volume of the parameter space than do standard stars (Graham & Slettebak 1973, Manfroid & Sterken 1992, Manfroid et al. 1992). The responsibility for this problem rests, in part, with the people who set up photometric systems, beginning with the Johnson-Morgan UBV system, whose original standards were bright, single, unreddened, normal stars used as classification standards in the MK system. Photometric standards should include reddened stars, composite spectra, and chemically-peculiar stars (especially the metal-poor stars of Population II). Unfortunately, many variable stars—Cepheids, eclipsing binaries with dissimilar components, supergiants, emission-line objects like novae, or interacting binaries, and so on—have spectra quite unlike those of any possible standard stars; instead of tangerines, they act more like lemons.

Here is where the real trouble appears. The residuals found in transforming standard stars can be reassuringly small, while the actual errors in transforming programme stars can be shockingly large (Schmidt-Kaler 1961, Sterken & Manfroid 1987): in observations of Mars made with 5 different instruments (Young 1974a), there were zero-point errors of about $0\overset{m}{.}1$; Manfroid et al. (1992) find errors "up to 0.2 mag or more"; Hamuy and Suntzeff (1990) found that "systematic differences between ... different sites can be as large as 0.4 mag."; and, in the most spectacular examples I know of, Tempesti (1972) found discrepancies as large as one whole magnitude—and Chochol et al. (1993) even more. On theoretical grounds, such errors are expected to be worse in systems with narrower passbands (Young 1974d, Young 1992a).

This seems to me to be the major weakness of the ESO LTPV project: the narrow-band *uvby* system was selected, despite both theoretical warnings of transformation problems (Young 1974d) and evidence of them in practice (Graham & Slettebak 1973). In the latter paper, the authors wrote that "systematic differences of up to several hundredths of a magnitude were found between the photometric indices measured with the two filter sets. ... The cause can be traced to the fact that no two interference filters are exactly identical in their transmission characteristics. Serious discrepancies appear when the spectral energy distributions of the programme stars differ markedly from those of the standard stars used to define the system." No clearer statement of the problem could be desired.

The large transformation errors for heavily-reddened stars long known in the UBV system (Schmidt-Kaler 1961) are now known to exist in *uvby* (Manfroid and Sterken 1987). Transformation problems associated with metallicity have been known for a long time; they are particularly severe because of a shortage of metal-poor standards in most systems. Such errors have been usefully described by Manfroid and Sterken (1992) as "conformity" errors: they are due to differences between the instrumental and standard passbands, and cannot be transformed away.

Beckert and Newberry (1989) nicely illustrate the problem: even for noiseless synthetic photometry, there are residuals of individual stars from any mean transformation between slightly different systems. The explanation seems to be that some stars have stronger lines than others of the same spectral class, due to line-broadening mechanisms such as magnetic fields and microturbulence. Such differences between stars show up with full effect in photometry, but are not detected in routine spectral classification, which depends on line *ratios* rather than absolute equivalent widths. Thus, there are star-to-star differences in the conformity errors at any given spectral type or colour index.

These errors cannot be removed in conventional photometric systems, because these systems fail to measure enough astrophysical information to characterize stellar spectra fully. One way to understand that information is missing is simply to notice that existing sets of passbands grossly undersample the spectrum (Young 1974d, Young 1992b, Wildey 1992). An alternative point of view is that different instruments' passband vectors span slightly different manifolds in Hilbert space (Young 1974d, Young 1993b, Young 1993c). From either point of view, the only practical solution to the problem of conformity errors is to add enough overlapping passbands between existing ones that most of the missing information is actually measured.

4.2. EXTINCTION AND TRANSFORMATION

The need to obtain consistent measurements of standard stars means that we must remove the extinction accurately, even if the system is a narrow-band one where "differential" photometry is relatively free of colour terms. Extinction errors can propagate directly into transformation coefficients, because there is usually some correlation between stellar colours and zenith distance. In particular, the concentration of blue stars to the Milky Way introduces a strong correlation at those times when the Galactic plane is near the horizon.

In general, standard stars can do double duty as extinction stars. However, one must bear in mind what is possible and practical. One cannot reduce the conformity errors for any *individual* programme star by observing a very large number of standards; once there are enough standard stars to make the reduction errors smaller than the conformity errors, further observing time should go elsewhere—for example, into measuring time-dependent extinction.

Thus, while the first requirement for a good calibration is to get enough observations of *enough* calibration stars, it is actually possible to measure *too many* individual standards, with too few observations of each. For example, suppose one has a 5-night run at a good photometric site; probably 3 of the nights will be clear enough to measure extinction and standard stars. One usually wants 2 or 3 extinction measurements per hour; so an 8-hour night will yield about 20 observations of standard/extinction stars at large airmasses, and a similar number near the meridian. That's a total of 40 calibration observations per night, or 120 in 3 good nights.

Now, suppose Observer A picks a set of 30 standard stars, and gets 4 observations of each; and Observer B uses 60 standard stars, getting only 2 observations of each star – perhaps observing a different set of 20 standards each night. One observation of each star is used up in determining a magnitude (or colour) in the instrumental system. Thus, A's 4 observations of each standard leave essentially 3 degrees of freedom per star for the determination of extinction; B's 2 per star leave only one. Thus, A determines the extinction with 3/2 times the weight of B's data. On the other hand, B determines the transformation to a standard system with twice the weight of A—or would, if B's transformations are not degraded by poorly-determined extinction.

At first glance, this might look like a classic trade-off between precision and accuracy, because transformation errors are certainly systematic, while extinction errors look more random (neglecting the correlation between airmass and colour discussed above). But in practice, one can never average data together and beat down the errors without limit. There always comes a point where small systematic errors exceed the averaged-down random ones; further increases in precision are useless, if accuracy is the goal.

This point arrives much sooner than most people realize. In a study of many classical investigations, such as Newcomb's and Michelson's measurements of the speed of light, and Cavendish's determination of the density of the Earth, Stigler (1977) found that undetected bias in the data was typically half as large as the standard deviation of one measurement. He concluded, "One of the lessons of this study is that even the greatest scientists, exercising every ounce of their ingenuity, are unable to eliminate all bias. ... a systematic error of about half the standard deviation of a single measurement remains." That means that the systematic errors begin to swamp the random ones when there are more than about 4 observations per parameter. As there are usually only 2 to 4 parameters to be determined in a transformation equation, one cannot expect much useful gain beyond 10 or 20 standard stars.

As synthetic photometry (Cousins and Jones 1976, Beckert and Newbury 1989) shows, typical conformity errors are on the order of 0.01 mag per star for "normal" stars. If we observe 10 standard stars per parameter in the transformation equation, we beat down the reduction errors to about 0.003 mag. But the conformity error is still typically 0.01 mag per programme star, even if the programme stars are similar to the standards (and more if they are not). At this point, it seems pointless to decrease the transformation reduction errors further by observing more standards—the more so, as conformity errors for variable programme stars can be tenths of a magnitude.

So, to return to our example, it might have been better yet to have chosen 20 standard stars, getting 6 observations of each (two per night on each of the 3 good nights). That would leave 5 observations to determine the instrumental magnitude of each standard star, which Stigler's rule suggests are useful. This would determine the extinction slightly better than Observer B did. The reduction errors will certainly be below the conformity errors if a linear transformation is adequate, and will probably still be below the conformity errors even if a cubic transformation is used. In this latter case, 4 parameters must be determined, but Stigler's rule would suggest that as few as 16 standards might be the break-even point.

Finally, these examples emphasize the importance of reducing together multiple nights of homogeneous observations. Multi-night reductions greatly strengthen the determination of extinction, by utilizing the internal precision of data on a stable instrumental system, and by considerably reducing the number of parameters to be estimated. The great benefits of reducing together several nights' observations have been stressed repeatedly (Young and Irvine 1967, Evans 1969, Young 1974d, Popper 1982, Manfroid and Heck 1984, Young 1993a), but still seem to be poorly appreciated by observers as well as by scheduling committees. Evans (1969) has emphasized the importance of site quality for getting enough data in a short enough time to be useful.

4.3. OPERATIONAL CONSIDERATIONS

There is a moral here for observatory directors and scheduling committees. Not only are operating expenses kept down by avoiding frequent instrument changes; if two observers will cooperate on adjacent runs with the same equipment, they can both benefit from higher data quality by reducing their observations together. Scheduling committees should encourage photometric observers who need the same equipment to cooperate in such runs.

Sometimes, the same goal can be achieved by two observers whose objects are well separated in Right Ascension. I have shared nights with another observer whose object was in the morning sky while mine was in the evening; at midnight, I handed the telescope over to him. We got much better coverage of standard stars between the two of us. This type of observing is common on small automated telescopes, but it should also be practised, when possible, by observers at larger photometric telescopes.

The only catch is that both observers must agree to measure through the same focal-plane aperture, to assure a fixed instrumental system. This can sometimes be a problem when one observer has bright stars, and is willing to use a large field stop to ensure better accuracy, while the other is interested in faint stars, and needs a small aperture to ensure good precision. (This is one of the classic trade-offs between precision and accuracy). The same problem arises in CCD data-extraction, of course; the change in instrumental system with measuring-aperture is due to the extended wings of the star image, and thus is an optical problem, independent of the detector used.

4.4. SOLVING THE CONFORMITY PROBLEM

Manfroid and Sterken (1992) have argued against reduced values that "would not be directly based on an instrumental system in the true sense of the term: they would only result from analytical transformations applied to original data obtained in non-standard systems." The result of such transformations, they protest, "would not represent any physically possible instrumental system. Mathematically the integral equations describing the photometric indices would not accept any solution for the instrumental functions."

However, *all* conventional transformations, which involve matrix manipulations of two or more instrumental magnitudes, have this failing. In fact, they do not correspond to any possible instrumental system, but are simply an arbitrary transformation of nonlinear functions of the actual intensity measurements. Only if stellar spectra were featureless continua would these interpolation schemes approximate what would have been measured by a standard passband with a different effective wavelength than the instrumental ones.

However, there *is* a way to transform data to the values that would have been measured by a physically possible system (Young 1993b, Young 1993c). That is to form linear combinations of the intensities measured with the instrumental passbands. This is not a useful procedure if the passbands are disjoint, as in the existing systems. But a system with overlapping passbands can be used to synthesize the standard ones with satisfactory accuracy.

The discrepancies of over one magnitude among different measurements of novae in the nebular phase show how inadequate present photometric systems are. The passbands are not well defined; nor can they be well defined indirectly through the use of standard stars alone. What is needed is a thorough reform of existing procedures on a sound scientific basis, as recommended by Sterken (1992b).

As our control of passbands is limited, the only way to assure success in the long term is to adopt a set of overlapping passbands that allow synthesis of the proper passbands, to a high degree of accuracy, after the observations are made. With such a system, the conformity errors can be made very small. Then it would make sense to observe more standards and make the reduction errors smaller; however, we would still want 5 or 6 observations per standard. The point is that a balance must be struck between different error sources.

In this case, one would want to reduce all the random errors by using a larger telescope, thereby getting more photons, and less scintillation noise. The introduction of transformable photometric systems will give photometrists a need for time on larger telescopes. In a sense, it is only the large conformity errors in existing undersampled systems that have justified the use of small telescopes for photometry.

5. Conclusions

Long-term projects require continual monitoring of standard stars, and hence, of extinction; they cannot be just "differential" photometry. Even so, conformity errors between different instruments, or even nominally the same instrument at different times, mean that observations cannot be completely homogenized in the existing undersampled systems.

To improve long-term homogeneity requires much closer attention to passbands than has been given in the past. The only practical way to reduce conformity errors is to synthesize the standard passbands from a densely-sampled set of instrumental ones.

References

Beckert, D.C., Newberry, M.V.: 1989, *Publications of the ASP*, 101, 849

Chochol, D., Hric, L., Urban, Z., Komžik, R., Grygar, J., and Papoušek, J.: 1993, *Astronomy and Astrophysics*, 277, 103

Cousins, A.W.J., Jones, D.H.P.: 1976, *Monthly Notices of the RAS*, 81, 1

Evans, D.S.: 1969, *Astronomy and Astrophysics*, 3, 247

Graham, J.A., Slettebak, A.: 1973, *Astronomical Journal*, 78, 295

Hamuy, M., Suntzeff, N.B.: 1990, *Astronomical Journal*, 99, 1146

Harris, A.W., et al.: 1989, *Icarus*, 81, 365

Johnson, H.L.: 1962, *Astrophysical Journal*, 135, 975

Kim, H.-S., Park, H.-S.: 1993, *Astronomical Journal*, 106, 1679

Lockwood, G.W., Skiff, B.A.: 1988, AFGL-TR-88-0221, Air Force Geophysics Laboratory, Bedford, Mass.

Lockwood, G.W., Skiff, B.A., Thompson, D.T.: 1993, in *Stellar Photometry—Current Techniques and Future Developments*, C.J. Butler and I. Elliott, eds. (Cambridge Univ. Press), 99

Manfroid, J., Heck, A.: 1984, *Astronomy and Astrophysics*, 132, 110

Manfroid, J., Sterken, C.: 1987, *Astronomy and Astrophysics, Supplement Series*, 71, 539

Manfroid, J., Sterken, C., Bruch, A., Burger, M., de Groot, M., Duerbeck, H.W., Duemmler, R., Figer, A., Hagemann, T., Hensberge, H., Jorissen, A., Madjesky, R., Mandel, H., Ott, H.-A., Reitermann, A., Schulte-Ladbeck, R.E., Stahl, O., Steenman, H., vander Linden, D., Zickgraf, F.-J.: 1991, *Astronomy and Astrophysics, Supplement Series*, 87, 481

Manfroid, J., Sterken, C., Bruch, A., Burger, M., de Groot, M., Duerbeck, H.W., Duemmler, R., Figer, A., Hagemann, T., Hensberge, H., Jorissen, A., Madjesky, R., Mandel, H., Ott, H.-A., Reitermann, A., Schulte-Ladbeck, R.E., Stahl, O., Steenman, H., vander Linden, D., Zickgraf, F.-J.: 1991, "First Catalogue of Stars Measured in the Long-Term Photometry of Variables Project (1982 – 1986)", ESO Scientific Report No. 8, (European Southern Observatory, Garching)

Manfroid, J., Sterken, C.: 1992, *Astronomy and Astrophysics*, 258, 600

Manfroid, J., Sterken, C., Gosset, E.: 1992, *Astronomy and Astrophysics*, 264, 345

Park, H.S., Chen, K.Y.: 1989, *Astronomical Journal*, 97, 1807

Popper, D.M.: 1982, *Publications of the ASP*, 94, 204

Schmidt-Kaler, Th.: 1961, *The Observatory*, 81, 246

Sterken, C., Manfroid, J.: 1987, in *Observational Astrophysics with High Precision Data*, Proc. 27th Liège International Astrophysical Colloquium (Liège, Belgium), 55

Sterken, C.: 1991, *The Messenger*, 65, 65

Sterken, C.: 1992a, *Vistas in Astronomy*, 35, 139

Sterken, C.: 1992b, *Vistas in Astronomy*, 35, 263

Sterken, C., Manfroid, J.: 1992a, *Astronomy and Astrophysics*, 266, 619

Sterken, C., Manfroid, J.: 1992b, in *Automatic Telescopes for Photometry and Imaging*, A.S.P. Conference series 28, edited by S.J. Adelman, R.J. Dukes, and C.J. Adelman, 57

Sterken, C.: 1993, in *Precision Photometry*, edited by D. Kilkenny, E. Lastovica, and J. W. Menzies (South African Astronomical Observatory, Cape Town), 57

Stigler, S.M.: 1977, *Ann. Stat.* 5, 1055

Tautvaišienė, G., Straižys, V.: 1985, *Soviet Astronomy*, 29, 234

Tempesti P.: 1972, *Astronomy and Astrophysics*, 20, 63

de Vaucouleurs, G.: 1993 (personal communication)

Wildey, R.L.: 1992, *Publications of the ASP*, 104, 290

Young, A.T.: 1967, *Monthly Notices of the RAS*, 135, 175

Young, A.T., Irvine, W.M.: 1967, *Astronomical Journal*, 72, 945

Young, A.T.: 1974a, in *Exploration of the Planetary System*, edited by A. Woszczyk and C. Iwaniszewska (D. Reidel, Dordrecht), 253

Young, A.T.: 1974b, in *Methods of Experimental Physics*, Vol. 12, Astrophysics; Part A: Optical and Infrared, edited by N. Carleton (Academic Press, New York, 1974), 95

Young, A.T.: 1974c, in *Methods of Experimental Physics*, Vol. 12, Astrophysics; Part A: Optical and Infrared, edited by N. Carleton (Academic Press, New York, 1974), 1

Young, A.T.: 1974d, in *Methods of Experimental Physics*, Vol. 12, Astrophysics; Part A: Optical

and Infrared, edited by N. Carleton (Academic Press, New York, 1974), 123
Young, A.T., Genet, R.M., Boyd, L.J., Borucki, W.J., Lockwood, G.W., Henry, G.W., Hall, D.S., Smith, D.P., Baliunas, S.L., Donahue, R., and Epand, D.H.: 1992, *Pub. Astron. Soc. Pacific* 103, 221
Young, A.T.: 1992a, *Astronomy and Astrophysics*, 257, 366
Young, A.T.: 1992b, in *Automated Telescopes for Photometry and Imaging*, (A. S. P. Conference Series, vol. 28) edited by S. J. Adelman, R. J. Dukes, and C. J. Adelman (Astronomical Soc. of the Pacific, San Francisco), 73
Young, A.T.: 1993a, in *MIDAS Users Guide* (ESO, Garching)
Young, A.T.: 1993b, in *Stellar Photometry—Current Techniques and Future Developments* (IAU Colloquium 136), edited by C. J. Butler and I. Elliott, (Cambridge University Press, 80
Young, A.T.: 1993c, in *Precision Photometry*, edited by D. Kilkenny, E. Lastovica, and J. W. Menzies (South African Astronomical Observatory, Cape Town), 25
Zappalà, V., Di Martino, M., Cellino, A., Farinella, P., De Sanctis, G., and Ferreri, F.: 1989, *Icarus* 82, 354

DISCUSSION

Sterken: There are two very widespread misconceptions about transformation and standardization. It is not always understood that

- just as it is impossible to transform the sound produced by a second-class violin to a Stradivarius standard, it is impossible to transform data collected in a poorly-defined photometric system to an outstanding standard—even if very sophisticated numerical algorithms are applied;
- the parameter space of standard stars is not identical with the parameter space of the programme stars. This should not be so surprising since this happens also in other fields of science. In medicine, for example, drugs and medicines are tested on animals; yet they are applied to humans, though it is very well known that the parameter spaces are far from identical.

Hall: Incredibly, I have learned of some photometrists, both amateurs and professionals, who convert count rates or deflection length into differential magnitudes with an incorrect formula. They use $\Delta m = -2.512 \times \log_{10}(n_1/n_2)$ which is wrong of course.

Breger: Is there an attempt for the new CCD photometric systems to avoid the errors made in the case of previous photoelectric filter systems?

Young: Not yet. I believe this will be necessary if large systematic errors are to be avoided.

Kurtz: You are calling for narrow filters to give the same slope across the band as wider filters such as *UBV* filters. Did you mean to suggest that the wider filters can be *synthesized* from the narrower filters so that it is not necessary to observe through the wider filters at all?

Young: The slope can be found with filters of the same width, or narrower. In either case, the desired passbands can be synthesized.

Harmanec: I think it is fair to say that your insistence on a multicolour system with overlapping passbands is certainly justified for objects with very complicated energy distributions like novae. At the same time, I am convinced that for the majority of stars with reasonably smooth energy distributions, you can get good results with relatively simple, although non-linear, transformation formulae for systems like UBV or *uvby*. The important thing is to convince the observers of the need to evaluate the probable impact of some simplification on the precision of their results. Just a very simple example: if the coefficient of the colour dependence α in the transformation $B_{st} - B_{ins} = \alpha(B - V) + \ldots$ changes by 0.05 from season to season, then the magnitude difference of two stars which differ by, say, $0\overset{m}{.}1$, will *systematically* change by $0\overset{m}{.}005$ from one year to the other in an, otherwise stable, instrumental system. If one studies long-term variations of objects that vary by 1 mag or more, one can forget about it and a consistent use of the instrumental system may give satisfying results. In a search for small-amplitude long-term changes, however, you cannot neglect such an effect.

Young: You are quite right. I might add that, *if* the star's energy distribution is smooth, the overlapping bands can be used to add *precision*. If the star turns out to be peculiar, they allow *accuracy*. Thus, the use of multiple, overlapping bands allows the observer to balance random and systematic errors.

The Surgeon-general warns you that spotted stars can damage your health

Two generations of dutch astronomers: oldest and youngest participant – Walraven, van Loon

STRÖMGREN PHOTOMETRY, GALACTIC RESEARCH, AND VARIABLE STARS

E.H. OLSEN
Copenhagen University Observatory
Brorfeldevej 23
DK-4340 Tølløse
Denmark

ABSTRACT. Strömgren developed the four-colour *uvby* photometry and the Hβ photometry to create a tool of great potential in studies of galactic structure, composition, dynamics, and evolution. Using this tool, many large-scale surveys have been carried out, and a few of them are described here. The long-term monitoring inherent in these surveys has resulted in the discovery and investigation of a number of interesting and important variable stars.

Keywords: Strömgren photometry – *uvby*β – long-term monitoring

1. Introduction

The four-colour *uvby* photometric system was designed and tested by Strömgren in 1959 at Lick Observatory, and the first use of the system took place at Mt. Palomar in 1960, resulting in the still unpublished Strömgren-Perry (1962) catalogue. At the same time, Crawford developed the Hβ index at Kitt Peak (cf. Strömgren 1983).

This photometric system did not come out of thin air. During the previous decade, Strömgren had been working with interference filters, and in 1952 he defined the narrow-band c, l photometry (cf. Strömgren 1952). This photometric system was designed to measure the hydrogen features in stellar spectra and was therefore very suitable for B, A, and F stars. The index c measured the Balmer discontinuity and is the forerunner of today's c_1 index. The index l measured the strength of Hβ and is the forerunner of the present β index. Together, they made possible the derivation of the two main parameters of stellar atmospheres, viz. $T_{\rm eff}$ and M_V.

During the 1950s, the importance of metallicity became more and more clear, and Strömgren (1957, 1958) introduced the m index in 1957. This index measured the depression in the violet part of the spectrum, relative to the blue-yellow gradient. It is the parent of today's m_1 index. It is not particularly tied to B, A, and F stars, but is useful also for G and K stars, with their increasing number of densely-packed metallic lines in the violet part of the spectrum. And, of course, the blue-yellow gradient, today's $b - y$ index, is useful for all types of stars.

From the very beginning, Strömgren's aim was to develop a tool, which would

C. Sterken and M. de Groot (eds.), The Impact of Long-Term Monitoring on Variable Star Research, 437–444.

be useful in galactic research. The goal was to determine the three fundamental stellar atmospheric parameters for large samples of normal stars in order to study galactic structure, composition, dynamics, and evolution. This requires large-scale surveys of stars, and these have a tendency to become long-term observing projects, and therefore they connect nicely with the subject of this workshop. One might say that we have been monitoring at least the standard stars for more than 30 years, and, not surprisingly, we have found a few variables among the standards.

The 1960s and 1970s were characterized by the solid ground-work done by Crawford and his collaborators in establishing a net of standard stars and calibrations in terms of intrinsic colour, absolute magnitude, and metallicity (cf. Crawford 1960, 1975, 1978, 1979; Crawford & Mander 1966; Crawford & Barnes 1970). In parallel with Crawford's calibration work, numerous problems in galactic structure and evolution were attacked with the new tool of $uvby\beta$ photometry by Strömgren, Crawford, Perry, Barnes, Bond, Knude, Twarog, and many others. Problems studied were past and present distributions of dust and gas; properties of open clusters; correlations between age, metallicity, and kinematics; the helium content of Population I stars; age determinations for B, A, and F stars; and birth places of B stars. Sadly enough, in the 1990s this type of galactic research, envisaged by Strömgren, is becoming more and more difficult, as small telescopes are rapidly becoming an endangered species.

2. Large-scale surveys

I would like to describe a few long-term observing programmes which have led to the discovery of several interesting variable stars. The goals of these programmes are studies of the history of our Galaxy, based on large, complete, all-sky samples of stars.

2.1. THE F-STAR CATALOGUE

The first of these programmes is designed to study the correlations between age, metallicity, and kinematics of F stars. The first inkling of such an observing programme is found in one of Strömgren's (1958) papers at the Vatican conference on stellar populations. Strömgren asks, whether or not the observations are compatible with the assumption of a unique relation between initial chemical composition and age of a star. In retrospect, this question is rather prescient, considering that such a unique relation has, until recently, often been taken for granted by people modelling the evolution of our Galaxy. Strömgren concludes tentatively that there is a detectable scatter in initial chemical composition among stars of the same age and notes that further investigations along these lines may lead to more information regarding the correlation between epoch of formation and chemical composition at formation for stars of our Galaxy.

At the conference on interstellar matter held at Princeton in 1961, Strömgren (1962) had the first results from the new Strömgren-Perry (1962) *uvby* catalogue and concluded that "stars of nearly the same chemical composition as the Hyades stars have been formed out of the interstellar medium through a period ... presumably longer than 4 billion years". The evidence for metallicity scatter among stars of the same age was still only tentative. Strömgren (1962) also explored the correlation between age, metallicity, and kinematics, and he concludes his presenta- tion by mentioning a planned *uvby* observing programme of F8 and G0 stars down to 8th magnitude, "so that [a] rather ... detailed analysis of the correlations between age, chemical composition, and space velocity would become possible". The same programme is discussed in his George Darwin Lecture (Strömgren 1963). The observations were initiated by Perry (1969), who published *uvby* data for more than 800 F stars.

After Strömgren's return to Denmark in 1967, a four-channel *uvby* spectrometer was designed, built, and used on the Danish 50-cm reflector (now renamed the Strömgren Automatic Telescope) on Cerro La Silla, ESO, Chile. Some of you present here have had opportunity to use that instrument. The first results were a survey of the brighter O-G0 stars in the southern sky (cf. Grönbech & Olsen 1976).

In January 1974, Strömgren gave a talk in Brorfelde, where he described the potential of $uvby\beta$ photometry in galactic research. I was fascinated by the prospect of learning details about the early evolution of our Galaxy, and when one of my colleagues, right after Strömgren's talk, told me about the new, very efficient cross-correlation methods for determination of radial velocities, I soon after went to Strömgren and told him that I would like to observe the 10 000 brightest F stars with the specific purpose of finding a large, kinematically unbiased sample of intermediate Population II stars within 100 parsecs.

To select the stars, we used the all-sky HD catalogue (Cannon & Pickering 1918-1924) and selected all stars brighter than $m_v = 8.3$ and with spectral types from A5 to G0, inclusive. The early-type limit was chosen to ensure that even very metal-poor F stars were included. Also, the stars earlier than F5 could be used for investigations of the metallicity range among the younger Population I stars and studies of the distribution of interstellar matter near the Sun.

The observations were made between 1976 and 1981 from La Silla and Kitt Peak and published ten years ago (Olsen 1983). A number of spectroscopic investigations have confirmed that about 95% of the photometrically selected metal-weak F stars are indeed metal-weak (cf. Abt 1986, Gray 1989).

Radial-velocity measurements for about 10 000 F stars are nearly copleted now, mainly observed by Mayor, Andersen, Nordström, and their colleagues in Geneva. For a majority of the stars, new proper motions have been determined by the CAMC on La Palma. Thus, this large material is ready for a detailed discussion.

For a carefully selected subset of the F-star catalogue, high-resolution, high S/N spectra have been analysed with model atmospheres by Edvardsson et al. (1993). They have determined the abundances of 13 elements, and correlated them with ages and galactic orbits. To me, their investigation of the chemical evolution of the galactic disk is a quantum jump in galactic research and will probably be the most lasting result to come out of the F-star catalogue. It is fulfilling all the goals set by Strömgren around 1960 in his planning of this programme.

2.2. THE G-STAR CATALOGUE

In 1978 I became interested in the so-called "G-dwarf problem", i.e. the discrepancy between the observed metallicity distribution of long-lived stars and the distribution predicted by simple, theoretical models (cf. Schmidt 1963). The observed distribution is based on about 150 stars from various catalogues of variable quality and with some selection biases. It seemed to me that a larger and more homogeneous material was needed. Based on the correlations with MK classifications, which were evident among the primary *uvby* standards, I soon convinced myself that *uvby* photometry would be an excellent tool, both to separate late-type dwarfs from giants and to derive the metallicity of the dwarfs. It therefore seemed entirely possible to isolate all G dwarfs down to some limiting magnitude. From the HD catalogue all G5-type stars down to $m_v = 8.6$ were selected, while numerous smaller catalogues of stars with analysed spectra and subdwarfs provided calibration objects. I should mention that the HD catalogue only defines the types G0, G5, and K0 in that range.

The observations were made from 1984 to 1987, and the catalogue has just appeared in print (Olsen 1993). A slight extension of the range of *uvby* photometry to stars that are cooler, more luminous, and more metal-poor than the original Crawford-Barnes (1970) *uvby* standards has been necessary.

Based on the catalogue, it will be possible to give a nearly complete, magnitude-limited sample of perhaps 1500 "photometric dwarfs" in the range G2-G8 with photometrically-determined metallicities. The next step will be careful MK classifications from slit spectra, especially for the stars in the metal-poor tail of the metallicity distribution, to weed out contaminants like composites and barium giants. Also, radial velocities and new proper motions are being determined. A thorough discussion of the sample will require information on the kinematics of the stars, and also on the detection of binaries.

2.3. A VOLUME-LIMITED SAMPLE OF LONG-LIVED DWARFS

The magnitude-limited G-star catalogue, just described, is not expected to be complete with respect to spectral types, say, in the interval from G2V to G8V. This is due to the fuzziness and large discontinuities of the HD types, and, as long as we only had that catalogue, it was difficult to define a handy observing list, which would result in a well-defined dwarf sample to, say, K2. Fortunately, we now have at least part of the Michigan Spectral Catalogue. Using the first three volumes (Houk & Cowley 1975; Houk 1978, 1982), I have recently observed all potential G0 to K2 dwarfs south of $\delta = -26°$ and within 40 *pc*, i.e. with a magnitude limit dependent on spectral type. Among the G0 to K2 stars, I selected not only all dwarfs, but also all stars with an uncertain luminosity class. Of 3900 stars observed in *uvby*, about 1200 seem to be "photometric dwarfs" (Olsen 1994).

3. Variable stars found in the surveys

I will finish by mentioning a few of the most interesting variable stars found in these surveys.

HR 1861: This is a primary *uvbyβ* standard. The spectral type is B1IV. In comparisons between several successive instrumental systems of the Danish 50-cm spectrometer, this star was often the most troublesome with the largest residuals. I made a statistical test on the residuals for this star and for two other standard stars observed shortly before or after HR 1861 on the same nights. The probability of HR 1861 being constant was estimated at only 0.001, while for the two other stars the result was 0.34 and 0.39 (cf. Olsen 1977b). The y-range of HR 1861 is slightly less than $0^m_.01$, the type of variability is not known. It may belong to the low-amplitude B-type variables discussed by Waelkens at this workshop.

AY Ceti: I was reminded of this star, when I met Douglas Hall for the first time, a few days ago. He immediately remembered that I was one of his numerous co-authors on a paper concerning HR 373. Grönbech and I had used this primary four-colour standard during the first two years at the Danish 50 cm, but then discarded it due to variability (cf. Olsen 1974). Our observations were later used to derive a photometric period, which seems to differ drastically from the orbital period (cf. Eaton et al. 1983). The star is now known as AY Ceti; it is a binary consisting of a G- type giant and a white dwarf.

Four-colour standards: Several other standards, both primary and secondary, have been found to be variable (certain, probable, or possible). They are HR 2707, 3084, 4133, 7152, and 8799. Also HD 17925, 20630A, 36395A, 81997B (= BD -2°2902), 152391, 174080, and 184711 (Olsen 1974, 1983, 1993, 1994).

HD 17576AB: This visual double star showed an enormous ultraviolet excess when it was first observed for the F-star catalogue, and it was subsequently observed regularly. Some speculations concerning the visual companion were published (Olsen 1979) and nearly simultaneously Darius and Whitelock (1978) gave a spectral classification ($G0V$: + hot subdwarf). They had detected the object in the ultraviolet sky survey with the TD-1 satellite. Gray (1989) has classified the object $G2 : V$: with "washed out" lines. The *uvby* observations show that the primary is variable. During nine days in March 1979 it gradually brightened by $0^m_.06$ in all colours. It was at the bright level in both August 1976 and September 1979 (cf. Olsen 1980). To me, the visual companion, at $1''_.8$, looks bluer than the primary, and is probably the source of the ultraviolet radiation. The visual magnitude difference $3^m_.2$ leads to $M_V(B) \sim 8$, assuming $M_V(A) \sim 4.8$.

BR Crucis: The very large variations of this cool Ap star were found accidentally, because faint companions to bright stars were routinely observed by Grönbech and Olsen (1976). Follow-up monitoring revealed the photometric ranges 0.08, 0.25, 0.05, 0.12, 0.12, 0.16, 0.45, and 0.02, in, respectively, u, v, b, y, $b-y$, m_1, c_1, and $u-b$ (cf. Olsen 1977c). Light curves are shown by Renson (1978), who derived a period of $2^d_.873$.

TZ Fornacis: HD 20301 was found to show double lines on a plate taken at the ESO 1.5m coudé spectrograph for radial-velocity determination (cf. Andersen & Nordström 1977). Regular monitoring started soon after at the Danish 50 cm, and—by sheer luck!—the star became $0^m_.12$ fainter from the first to the second night. Since both photometry and spectra indicated the system to be composed of two giants, its potential significance was immediately obvious. More than 100 *uvby* observations during 1975-1977 revealed three eclipses leading to a period of either 38 or 76 days (Olsen 1977a). After a massive observational and theoretical effort, a definitive analysis of this evolved, but detached, system has now been made by Andersen et al. (1991). This represents the first highly-precise determination of absolute dimensions for a normal giant star. The masses and radii of the two stars, F7III and G8III, have been derived with uncertainties of about 2% or better.

References

Abt H.A., 1986, *Astrophysical Journal*, 309, 260

Andersen J., Clausen J.V., Nordström B., Tomkin J., Mayor M., 1991, *Astronomy and Astrophysics*, 246, 99

Andersen J., Nordström B., 1977, *Astronomy and Astrophysics, Supplement Series*, 29, 309

Cannon A.J., Pickering E.C., 1918-1924, Harvard Ann. 91-99

Crawford D.L., 1960, *Astrophysical Journal*, 132, 66

Crawford D.L., 1975, *Astronomical Journal*, 80, 955

Crawford D.L., 1978, *Astronomical Journal*, 83, 48

Crawford D.L., 1979, *Astronomical Journal*, 84, 1858

Crawford D.L., Barnes J.V., 1970, *Astronomical Journal*, 75, 978

Crawford D.L., Mander J., 1966, *Astronomical Journal*, 71, 114

Darius J., Whitelock P.A., 1978, *Nature*, 275, 428

Eaton J.A., Hall D.S., Henry G.W., Hopkins J.L., Krisciunas K., Landis H.J., Louth H., Olsen E.H., Renner T.R., Stelzer H.J., 1983, *Astrophysics and Space Science*, 93, 271

Edvardsson B., Andersen J., Gustafsson B., Lambert D.L., Nissen P.E., Tomkin J., 1993, *Astronomy and Astrophysics*, 275, 101

Gray R.O., 1989, *Astronomical Journal*, 98, 1049

Grönbech B., Olsen E.H., 1976, *Astronomy and Astrophysics, Supplement Series*, 25, 213

Houk N., 1978, Michigan Catalogue of Two-Dimensional Spectral Types for the HD Stars, Vol. 2, Univ. of Michigan, Ann Arbor

Houk N., 1982, Michigan Catalogue of Two-Dimensional Spectral Types for the HD Stars, Vol. 3, Univ. of Michigan, Ann Arbor

Houk N., Cowley A.P., 1975, University of Michigan Catalogue of Two-Dimensional Spectral Types for the HD Stars, Vol. Spec 1, Univ. of Michigan, Ann Arbor

Olsen E.H., 1974, *Information Bulletin on Variable Stars*, No. 925

Olsen E.H., 1977a, *Information Bulletin on Variable Stars*, No. 1317

Olsen E.H., 1977b, *Information Bulletin on Variable Stars*, No. 1332

Olsen E.H., 1977c, *Information Bulletin on Variable Stars*, No. 1367

Olsen E.H., 1979, *Astronomy and Astrophysics, Supplement Series*, 37, 367

Olsen E.H., 1980, *Information Bulletin on Variable Stars*, No. 1770

Olsen E.H., 1983, *Astronomy and Astrophysics, Supplement Series*, 54, 55

Olsen E.H., 1993, *Astronomy and Astrophysics, Supplement Series*, 102, 89

Olsen E.H., 1994, *Astronomy and Astrophysics, Supplement Series*, in press

Perry C.L., 1969, *Astronomical Journal*, 74, 705

Renson P., 1978, *Astronomy and Astrophysics*, 69, 351

Schmidt M., 1963, *Astrophysical Journal*, 137, 758

Strömgren B., 1952, *Astronomical Journal*, 57, 196

Strömgren B., 1957, *Astronomical Journal*, 62, 319

Strömgren B., 1958, in: O'Connell D.J.K. (ed.) Semaine d'étude sur le Problème des Populations Stellaires, *Ricerche Astronomiche, Specola Vaticana* 5, 245

Strömgren B., 1962, in: Woltjer L. (ed.) The Distribution and Motion of Interstellar Matter in Galaxies, W.A. Benjamin Inc., New York, p. 274

Strömgren B., 1963, *Quarterly Journal of the RAS*, 4, 8

Strömgren B., 1983, *Annual Review of Astronomy and Astrophysics*, 21, 1

Strömgren B., Perry C., 1962, Photoelectric *uvby* Photometry for 1217 Stars Brighter than $V = 6.5$, mostly of spectral classes A, F and G, Institute for Advanced Study, Princeton, New Jersey, preprint

DISCUSSION

Delgado: Do you know of any prospects to enlarge the sample of available standards, including more reddened B-type stars?

Olsen: Well, there are only a few slightly reddened B stars among the primary *uvby* standards. If one tries to extend the set of standard stars to reddened B stars, one would be trying to re-create the original, but non-existing, instrumental *uvby* system for these stars by some kind of mathematical extrapolation, and that is a very dangerous business, as we just heard from Andy Young. It would be better to create a new system for these stars, perhaps using an existing set of *uvby* filters. I have recently reduced *uvby* photometry of supergiants, for which there are primary standards, but it was certainly not easy to match the system for these, often reddened, stars on the fringe of the original system (cf. Gary & Olsen, 1991, *Astronomy and Astrophysics*, 87, 541).

Burki: The last star you mentioned (TZ For) is a very interesting case: SB2, eclipsing binary, known rotational velocities. Thus we know all the physical parameters of the stars together with the orbital ones. And this system allows to describe the process of circularization and synchronization.

Olsen: Yes, the G8III star has probably been to the top of the giant branch and is now in the core helium-burning phase, and its rotation is synchronized with the orbital motion.

Breger: The "official" calibrations of the B, A and F stars are more than ten years old. What is done now to update these?

Olsen: There is a number of newer calibrations. For F and early G stars Edvardsson et al. 1993 (*Astronomy and Astrophysics*, 275, 101) discuss newer calibrations. There is also a number of papers by Nissen, Schuster, and their collaborators dealing with the same spectral range (cf. Schuster et al. 1993, *Astronomy and Astrophysics*, 97, 951, and references therein). Jakobsen has made calibrations for B and early A stars, but unfortunately they are not published. For the difficult range A0-A3, Grosbøl 1978 (*Astronomy and Astrophysics, Supplement Series*, 32, 409) has made calibrations.

Hall silenced by Mrs. Sterken

GENERAL DISCUSSION

Sterken: The workshop's topic was on the impact of LTM on variable-star research. We have seen this impact in many fields of variable-star research, each in its own specific region of the H-D diagram. We should resume the discussion on what the impact is of *reducing* LTM activities.

Griffin: The prospect of yet another battle that needs to be fought for the sake of small to moderate-size telescopes is rather disheartening. As a community of astronomers we need to get our act together as a matter of some priority, in order to address these issues democratically and avoid the repeated need to fight rear-guard actions. We have a forum: the IAU, which was the first international scientific union and is still the envy of some other scientific disciplines; but we do not use it to anything near its potential in such situations. As a community we should try to decide our own priorities (regarding national and international collaborations/programmes) on *scientific* grounds and present them in a positive frame to the "politicians", instead of waiting for the latter to pronounce cuts and leaving us the harder job of getting political decisions reversed.

Waelkens: A resolution like "small telescopes are important" sounds a bit too modest. What one should claim is that stellar astrophysics is important.

Breger: A compromise between the two might be useful. This compromise might stress the importance of specific LTM programmes which could be done with small telescopes.

Young: Anything that can be done well on a small telescope can be done better on a larger one. So I think you should defend telescope *time*, rather than *small* telescopes.

Sterken: I agree, in principle, but not completely: not everything that can be done well on a small telescope can be done better on a larger one, see for example the MEKASPEK experience (Mantel, this Workshop) where the telescope is too large for observing even a $6\overset{m}{.}5$ star. In addition, replacing several small telescopes by a large one *always* is a *reduction* in telescope time!

Williams: The issue is not so much small telescopes or stellar astronomy; the issue is that long-term observing programmes are necessary for certain types of problems which cannot be tackled in "snapshot" mode; e.g. luminous objects which vary

C. Sterken and M. de Groot (eds.), The Impact of Long-Term Monitoring on Variable Star Research, 445–447.

over decades or variable stars with many modes of variation. Such problems are not limited to stellar astronomers and can be undertaken with large telescopes with appropriate observing modes.

Harmanec: I think that, if we want to get support of the stellar community for long-term programmes, we should not deny the importance of large and expensive projects for significant progress in astronomy. Rather, we should explain that long-term projects also represent an inevitable ingredient of stellar research.

Sterken: Right. LTM at small telescopes complements the work done at larger facilities. In addition, some of the problems to be studied with large telescopes would not even be known to exist if it were not of preceding LTM work at smaller telescopes.

Cramer: The general impression that one gets from these discussions concerning the further continuation of LTM with small telescopes is that there is a lack of communication between us and people who have the power of decision. Therefore, we should take seriously your suggestion of widening the discussion to include other participants.

Griffin: What does this group/community do/plan to do with its observational data? Is there an effective archiving strategy already in operation, is the situation somewhat disjointed, or what?

Sterken: As I see it, the aperture-photometry (scalar) data are reasonably well-archived. 2-D data (spectroscopy and imaging) are the real problem. It looks as if several observatories (HST, ESO, etc.) make a large effort to archive, but these archives are not yet homogeneous. It may take years to reach a versatile level of archiving worldwide.

Hall: You said that in general there is not a problem with archiving of photometry from LTPV. But I do not agree. I myself have experienced major problems, or at least frustrations, in trying to get my LTPV data published and/or archived. Details are given in the printed version of my talk, so I will not repeat them here in this comment.

Sterken: My own experience is that I would reduce and analyse data and publish the results. The data itself, though accurate and clean, reside on a tape or disk, but are not ready for archiving, since archiving implies an accurate description of the data. Such a description takes time; therefore, few people do it. I see the way out of the dilemma by preparing a short two- or three-page description of the data, which should be published before one can submit the data to the archive. Such a description is crucial, but does not guarantee correct use of the data. For example, few people realize that the Geneva catalogue does *not* list magnitudes in the 7 bands, but only 6 colour indices (which are all relative to B, i.e. $U - B, B_1 - B$, etc.). Though this is explicitly described by Rufener in the introduction to the Catalogues, there are cases where this fact is "discovered" during the analysis, or even at the stage of interpretation, of the data.

Cramer: The recent Geneva papers all say explicitly what the parameters are.

Williams: Data from the JCMT are being archived as from this year; the catalogue to this archive is accessible on Internet. This is very similar to the La Palma telescopes archive, also accessible on line. We hope to set up a similar archive for UKIRT data in the near future.

Griffin: The raw data need to be kept in a data-bank, but it is the reduced/transformed data which are the formats that can be taken up immediately by other astronomers, and which are therefore the correct constituent of an archive. When made suitably available, archival data form the bases of wider scientific endeavours than could otherwise be possible. However, if the astronomical community, recognizing this potential, wants its realization, there has to be a united commitment of resources (chiefly manpower plus hard- and software). The concept is part of the international attitude concerning centres like Sonneberg, Münster, and the WFPA at Sofia: if the community recognizes the value of these activities it will have to ensure their support.

Hall: We are hearing discussions about the question of archiving, raw *versus* reduced or instrumental *versus* transformed data and the problem of necessary detailed documentation accompanying the archived data. But I want to stress that these are second-order concerns. The real damage—fatal damage—is done when data are not published or archived at all. Such data are lost totally and forever; it is the same as if the observations had never been made.

Harmanec: The disadvantage of "extensively organized" data archiving is that it may not be easy to keep the data description and/or reduction quality under control. What if we were to proceed in two steps: i) observatory archives, and ii) central archives? Just to inform you: we do have a computer list of Ondřejov 2 m photographic and Reticon spectra, we do archive all spectra recorded with our 5-channel microdensitometer, and we do have all-sky and differential archives of about 40 000 *UBV* observations obtained at Hvar.

Le Contel: I would like to mention that a group in Nice, Besançon and Grasse recently decided to enter into a database all the measurements obtained in the world on about 20 000 visual binaries and collected by P. Couteau. This archive will be accessible through the CDS in Strasbourg.

Sterken: But there are institutes, like the Rechenzentrum in Heidelberg, where archiving (of *Astronomy and Astrophysics Abstracts*) is carried out on a strictly commercial basis. We pay for this service by buying the publications. Maybe we need more such enterprises?

RESOLUTION

The participants of the NATO Advanced Research Workshop

The Impact of Long-Term Monitoring on Variable-Star Research

(November 15–18, 1993. Ghent, Belgium),

recognizing i) the fundamental significance of long-term observations for solving many important problems in astrophysics, as impressively demonstrated in the course of the present workshop; and ii) the unequaled value of extended observing runs with small telescopes for the education of young astronomers in order to secure observational data of consistently high quality;

acknowledging the great wealth of information collected by many working groups involved in long-term observational campaigns covering a wide variety of investigations which require long time bases such as (among many others)

the Sky Survey at Sonneberg Observatory

the Long-Term Photometry of Variables project (LTPV) at ESO

the Whole Earth Telescope (WET);

realizing the potential of long-term observations achievable with small instruments for the identification of scientific problems and targets to be pursued through different approaches with the upcoming new generation of very large telescopes;

urge all political and administrative bodies and scientific peers involved in decisions about measures influencing the future of long-term projects with instruments of all sizes and using different techniques to promote to the best of their ability the advancement of this basic and important work for the benefit of the entire astronomical community.

Ghent, 1993, November 18

C. Sterken and M. de Groot (eds.), The Impact of Long-Term Monitoring on Variable Star Research, 449.

Index

Object index

WR146 = HM19-3 – 77ff

WR147 = AS431 – 77ff